Blackie

Captain Harold F. Blackburn, A Pioneering, Twentieth Century Pilot In Peace and War

by

William F. Cass

Branden Books,
Boston

Library of Congress Cataloging-in-Publication Data

Names: Cass, William F., 1942-
Title: "Blackie" Captain Harold F. Blackburn, A Pioneering, Twentieth Century Pilot in Peace And War / by William F. Cass.
Description: Boston : Branden Books, 2016. | Includes bibliographical references and index.
Identifiers: LCCN 2016006297 (print) | LCCN 2016008002 (E-book) | ISBN 9780828326223 (paperback : alkaline paper) | ISBN 9780828326223 (E-book) | ISBN 9780828326230(E-Book)
Subjects: LCSH: Blackburn, Harold F., 1901-1989. | Air pilots--United States--Biography. | Air pilots, Military--United States--Biography. | Aeronautics--United States--History--20th century. | Jet transports--Piloting--History--20th century. | World War, 1939-1945--Aerial operations, American.
Classification: LCC TL540.B555 C37 2016 (print) | LCC TL540.B555 (E-book) | DDC 629.13092--dc23
LC record available at http://lccn.loc.gov/2016006297

Paperback ISBN 9780828326223
E-Book ISBN 9780828326230

Branden Books
PO Box 812094
Wellesley MA 02482

www.brandenbooks.com

Photo Credits, Front Cover:
Captain Harold F. Blackburn: Seventh Photographic Reconnaissance Group.
Aircraft Insets (P-12, B-24D, Photo Lightning, and Lockheed Constellation, respectively): United States Air Force, National Archives, Seventh Photographic Reconnaissance Group, and Trans World Airlines.
Boeing 707: TWA Photo/Jon Proctor Collection.

Front Cover Design: Pam Dooley Graphics, Gulph Mills, Pennsylvania.

Table of Contents

Introduction

The first seed for this book was planted in 1951 when the author's mother enrolled him in a Cub Scout pack in small town, southern Pennsylvania. That put Billy Cass on the trail to Eagle Scout and also eventually resulted in several college summers working at Philmont Scout Ranch, the Boy Scouts of America's fabled High Adventure property in northern New Mexico's Sangre de Cristo Mountains. One of Philmont's lesser mountains is the 10,242 foot high Trail Peak, site of the 1942 crash of a B-24 Liberator bomber in which all hands had been instantly killed. The author climbed the mountain twice in 1962. The wreckage made a lasting impression particularly because the author's father had ferried many B-24s from the wartime Willow Run plant in Michigan before becoming a B-26 Marauder pilot and going overseas in 1944. Blackie had been in charge of the four-engine transition school in Albuquerque where the ill-fated bomber was based and narrowly missed being on the doomed flight himself.

After climbing the mountain again in 1991, the author prepared a two-part article on the crash for the regional magazine, *High Country*. The article raised as many questions as it answered and evolved into a book project in the mid-1990s. The core research for that book, *The Last Flight of Liberator 41-1133,* necessarily involved contact with TARPA, the TWA Active and Retired Pilots Association. Through TARPA, the author was able to interview several men who were both well acquainted with Blackie and experts in TWA history.

Following that book's publication, the author mentioned to the Blackburn family in 2001 that he would like to do a biography of Captain Blackburn someday. Unfortunately, three other biography projects intervened. Research for *The Last Flight* had suggested that Blackie would make for an interesting biography which turned out to be an understatement. The author was initially unaware of the many personal and professional struggles Blackie went through along with his disappointments and triumphs. All of that drama was amply revealed through personal interviews and the wealth of information within the Blackburn family and the repository of Blackie's aviation memorabilia, the American Heritage Center at the University of Wyoming.

During the writing of the *Last Flight of Liberator 41-1133*, it became apparent that Blackie was much more than just a B-24 instructor pilot and airline pilot. His life truly mirrors most of twentieth century aviation, both military and civil from the World War I era through international jet transport age. He was also a pioneering pilot as his being a recipient of both the prestigious Mackay and Collier Trophies confirms.

While the author would be the last writer to minimize the contributions made by fighter and bomber pilots in World War II, the fact remains that those two categories have almost been over-published. Blackie's wartime journeys highlight the contributions made by two largely ignored types of pilots: those flying transport and photo-reconnaissance aircraft. Without the support of the latter two types of aviators, the former pilots would have been hard-pressed to achieve as much as they did.

As with any good drama, Blackie is the protagonist facing many antagonists: a broken family, conflict over what he wants to do with his life, the Army Air Corps which initially rejected his flight training application, the Great Depression into which an unemployed Blackie was thrust, and military bureaucracy which thwarted his attempts to go on active duty during World War II. Other forces shaping Blackie's life, beyond his intense flying experiences in World War II, include Lady Luck who fatally abandoned so many of his pilot friends, his employer Trans World Airlines which was its own worst enemy, and what is today's Federal Aviation Administration with its rules on airline pilot retirement age. Trans World Airlines forms a major, frequently self-destructive character in this book and is the source of deep frustration for Blackie, who is grateful for being hired by the company during the Great Depression and simultaneously repelled by its corporate waste and board room intrigue.

It is inevitable that the authors of many biographies frequently become so familiar with their subjects that the lead character almost becomes a relative. With *Blackie*, Captain Blackburn became something between a young grandfather and an older uncle. The research and writing never turned into drudgery. Quite to the contrary. The process (of watching an achievement-oriented young Hal Blackburn evolve from frontier athlete and scholar to college student and musician) became addictive. The author's having been an entrepreneur and commercial pilot made Blackie's orchestra leader days and aviation career that much more meaningful. Captain Blackburn's wartime service as a transport and combat photo-reconnaissance pilot opened new avenues for an author raised on fighter and bomber pilot narratives. Blackie's retiring to a Pennsylvania farm brought back many memories for the author whose earliest years were spent on a far northern Pennsylvania dairy farm. Several of the author's clients over the years marketed products and services to the world's airlines, with a few exceptions. One of the most notable exceptions was Trans World Airlines which made writing about that airline as a major influence on Blackie a fascinating process.

How easy it has been to imagine myself as Blackie's co-pilot on a Lockheed Constellation flight over the North Atlantic in the 1950s, flying off his wing on a photo-mission to Germany in 1944, or accompanying him on a training exercise in an open cockpit Curtiss B-2 Condor bomber in the early 1930s. Such daydreams will, of course, never happen, but some semblance of Blackie's passage through those times and his contributions to twentieth century aviation can be approached through pictures and words. Hopefully, what follows in *Blackie* will do justice to his story.

William F. Cass
West Chester, Pennsylvania
December 7, 2015

Acknowledgements

First and foremost, we have Blackie himself to thank. Considering the wealth of information he left behind, as kept in the caring hands of his family, one can easily conclude that very few biographers enter into a project with so many resources at hand. Blackie's son Bob, daughters Betty, Bonnie, and Be, grandchildren, (particularly grandsons Chris Blackburn and Karl Beard), and step-son Stan Delahoyde, all cooperated to the fullest extent with the author's innumerable requests for insights, recollections, anecdotes, and photographs.

Several of Blackie's fellow pilots, now all deceased, added considerable dimension to the project by enlightening the author on Blackie's flying, wartime service at TWA's Eagle Nest Flight Center and its Intercontinental Division, and the airline's postwar expansion. This biography would have fallen well short of its mark without the expert input from retired TWA pilots Joe Carr, Ed Betts, Howard Kincheloe, Hutch Thurston, and Walt Gunn.

Pamela Shoop Sweeney, whose father was Colonel Clarence Shoop, one of Blackie's closest friends, graciously shared her father's unpublished memoir which is the foundation of the chapter on Blackie's flying with the Seventh Photographic Reconnaissance Group over Nazi-occupied territory in World War II. John Blyth, who was a young pilot with the Seventh Photographic Reconnaissance Group in 1944 (and is most likely that unit's last standing pilot), was most generous with his time in answering the author's many question about the group's operations, aircraft, and personalities.

The author was welcomed at the TWA Museum in Kansas City and given a private tour by its director, Pam Blaschum. The Museum's archivist, John Mays, supported the research wholeheartedly and added many crucial documents to the museum's online archive which was of tremendous value to the project.

Ginny Kilander at the University of Wyoming's American Heritage Center gathered the archive boxes containing Blackburn memorabilia and managed the efficient scanning and copying of many documents. Researchers George Cully and Norm Richards at the Air Force Historical Agency and the National Personnel Records Center, respectively, quickly obtained and copied documents relating to Blackie's military service during the 1930s and 1940s. Tracey Howerton of the Missouri State Historical Society, which is the keeper of remaining TWA corporate records, facilitated copying key records, including one of the project Rosetta stones: Blackie's statistical record of TWA's Intercontinental Division wartime activities. Ann Hussey (senior historian at the Air Force Air Education and Training Command) scanned several important files relating to Blackie's training at Kelly Field and thus added more depth to Blackie's year as an aviation cadet.

While websites have become an indispensable and well recognized implement in any biographer's tool kit, the author was thunderstruck at the depth of the major websites relating to Blackie's life trajectory. Among the most valuable was the State Historical Society of Missouri's collection of decades of the TWA employee newspaper, *Skyliner*. *TARPA TOPICS*, the magazine archive of the TWA Active and Retired Pilots Association, contained considerable anecdotal information about Blackie as well as providing good history on the Intercontinental Division and TWA's introduction of the Lockheed Constellation. The remarkable lives of Jack and Helen Frye as president and first lady of TWA are thoroughly documented in the "Sedona Legend" as hosted by Randall Reynolds. Knowing what was going on in the Army Air Corps of the early 1930s was a critical research objective and that goal was reached thanks to the Air Force Historical Support Division's website archive of the *Air Corps Newsletter.* Tom Kracker's Luftwaffe Aircrew Archives contained capsule biographies of many of the Lufthansa pilots with whom Blackie worked with in the late 1950s. Among the many other webmasters to whom the author is grateful include those hosting the following standouts: Aviation Safety Net (crash analyses), Jack McKillop's Army Air Force Combat Chronology of World War II, the several websites about abandoned airfields in America and Europe, and Time Table Images (TWA schedules from the very beginning). The profiles of the remarkable captains Blackie flew with in the mid-1930s could not have been recreated without Wes Bunker's recollections found at the TWA Seniors Club website.

The late Chope Phillips recalled details of guiding Blackie to the bomber crash site on Philmont's Trail Peak, near Cimarron, New Mexico, where Blackie's close comrade, Robert Redding, was killed. And Russell Seidel, a Berks County, Pennsylvania, farmer shared his recollections of the progressive farmer-neighbor who also happened to be a prominent airline pilot.

Because the author's flying days are now well into the past, it became important to have the technical passages in the manuscript critiqued by a current airline pilot. Fellow former Philmont ranger Captain Roy Paget, who flies today's descendants of the Boeing 707 on the same transatlantic routes that Blackie did, graciously pointed out the author's shortcomings in conveying those cockpit routines critical to Blackie's flying the Connie and Boeing 707.

The author and Blackburn family are very grateful to have *Blackie* joining the distinguished list of titles at Branden Books. Now under the direction of Editor Adolfo Caso, Branden Books is into its second century of bringing new and award-winning reading to a wide variety of audiences.

Chapter One
Born to Fly

The journalist was on his way out of Manhattan very early on the morning on 18 November 1961, a day with temperature in the mid-40s made more chilling by the wind which was blowing at nearly twenty miles per hour. Although the visibility was good at almost fifteen miles, a high cloud cover blocked any sunny rays from brightening his journey westward toward Allentown in Pennsylvania. The fields along the way had recently been harvested, leaving only a tan stubble which was not much lighter than the dull-colored woods whose pristine autumn colors had already receded. It would be a two-plus hour drive to his destination: a secluded farm in central Berks County, nearly 60 miles north-northwest of Philadelphia. Offsetting the dreary time on the road was the journalist's knowledge that he would enjoy a hearty lunch beside a warming fireplace when he reached his destination.

His assignment was to interview a man his editors thought would make a very interesting subject for one of the magazine's in-depth personality profiles. Because he had precise instructions, the journalist had no problem in finding the farm although there was no name on the mailbox – just the rural route from Mertztown and box number (Rte.1, Box 436). Beneath the numbers, in small script, the words "Hidden Brook Farm" appeared in the lower half of the galvanized metal mailbox. "Hidden" was an understatement. Dense woods formed the southern and eastern boundaries while views of the property from the north and west were obscured by wooded hills and tree lines along boundary fences.

A passing motorist may have thought that the farm owner was Pennsylvania Dutch because so many farmers in the county were – maybe his name was Hank Schwartzenbrenner or something like that. Many of his neighbors had names like Huffnagel, Angstadt, Eckenroder, Detweiler, Kuhlengel, or Pfautz which indicated origins in Germany. Unlike his neighbors, this "Hank Schwartzenbrenner" was a transplant from far away. His real name was hardly German; it was actually Harold Fredric Blackburn and he was proud of his Scots heritage. He was even prouder of his family, beginning with his wife Martha who was anything but Pennsylvania Dutch. She was doubly different: Texan and French. All but one of their children had flown the coop, been educated, and gone on to rewarding careers and families of their own. They all enjoyed coming back to what Blackburn lovingly called "The Pore Farm." It was a poor farm once, but Blackburn had turned it into a showplace.

The Pore Farm was average-sized for Berks County, and maybe a bit smaller – just under 100 acres. It had originally been a dairy farm, but no longer was. Yet it was a going concern, and the recently harvested crop fields provided grain for a fine heard of Black Angus beef cattle. Even a cursory glance would tell that this farm was very well-managed. The fences were freshly white-washed, there were no run-out vehicles rusting on the lee side of the sturdy-looking stone barn, and the now nearly dormant fruit trees, gardens, and flower beds showed signs of great care. The only sign that might have given rise to pause was the one on the barn. Pennsylvania Dutch hex signs on barns are hardly out of place in Berks County, and their geometric, intricate designs with a half-dozen or more vibrant colors were missing on the large circle that appeared on the western end of the barn. The white star with a red center on a blue disk seemed curiously simple in a land where many hex signs reminded viewers of kaleidoscopic floral patterns. Tourists, if they could have seen the hex sign, would have easily recognized it only if they were well acquainted with the national insignia on the wings and fuselages of Army Air Corps planes in the 20s and 30s.

After turning onto the lane, the journalist drove through the woods on the farm's south side and emerged beside a cornfield which had been harvested only several weeks before. Blackburn grew corn, of course, and oats, and wheat in addition to alfalfa and clover. Blackburn had seen fields full of crops and trees his neighboring farmers had only read about – olive trees from horizon to horizon in Spain, cork trees in Portugal, cotton in Egypt, grapes in vineyards covering the hills in Italy, coffee in Ethiopia, and hops in both English and German fields.

When working in his fields in late autumn, Blackburn was dressed in a warm, leather jacket, but the rest of his outerwear had been purchased at an Army-Navy store, including the very non-descript, fleece-lined hat. When plowing, he frequently glanced backward to make sure that he was plowing a straight line, parallel to the last row. He did this in spite of the fact that his sense of direction was extraordinarily well-developed. Unlike most of his farmer neighbors Blackburn occasionally stopped at row's end after turning, raised his head, shielded his eyes against the sun, and looked skyward for jet contrails. Not many farmers in those parts ever did that. Nor could many neighboring farmers read a sky as well as he could. For them seeing the high mares' tails of a cirrus formation said only that change was coming in a few days. Farming neighbors were used to a prevailing, gentle breeze from the northwest. They might see that the weather vanes on their barns pointed to the east and knew therefore that they would not have to pray for rain because it was already, surely on the way. Or they might deduce as much if they had a falling barometer, but more likely than not, they had only one outside instrument: a thermometer. Blackburn's tractor, like those of his neighbors, had only a few instruments – just the basics: temperature, oil quantity, ammeter, and possibly a speedometer with the top speed of thirty-five mph. He was used to constantly inhaling the readings from scores of round gauges on his preferred instrument panel which was laid out horizontally instead of the tractor's stark, vertical configuration.

He too welcomed thunderstorms because they brought rain for his crops, but unlike his neighbors, Blackburn had lost friends who had been killed, not by lightning, but by the storms' powerful winds. His neighbors reacted ambivalently to the wind so long as it did not blow down their crops. The man in the leather jacket harnessed wind to his advantage whenever he could. He knew many of the world's great winds too. He had sought shelter from the blowing sands of the *Khamsin* in Egypt. He was wearing boots made where the *Pampero* blows – over the Pampas of Argentina. He had been pushed around by the *Galerna*'s squalls between England and Gibraltar. The south-blowing *Mistral* became his friend and helped carry him past the Alps unlike the *Zonda* which challenged his passage over the Andes to Chile. Of all of the world's great winds, he was alternately blessed and vexed by the Polar Front Jetstream depending on which way he was traversing the North Atlantic Ocean. A very few of Blackburn's non-farming neighbors might visit the New Jersey shore in the summer as throngs of people from the mid-Atlantic do. He had crossed the width of that great ocean hundreds of times – by both the North and South Atlantic. He had looked down upon many of the world's great seas and bays: the Mediterranean, the North Sea, the Bay of Biscay, the Red Sea, the Arabian Sea, the Adriatic, and the Tyrrhenian too. His familiarity with so much geography and weather was hardly surprising considering his having bored holes in the skies above five of the earth's seven continents: only Australia and Antarctica had eluded him over the decades.

Although the nation's television weathermen were only just starting to talk about "the Jetstream" in their broadcasts, Blackburn had first experienced its winds over twenty years earlier. Dense fogs rarely worried members of the local Grange Hall, but they were usually a menace to this farmer when he was away from the farm.

A century before, runaway slaves on their clandestine journeys to freedom had passed through the rolling hills of Berks County. They knew how to navigate using the North Star with which this farmer's neighbors and children might have some familiarity. Unlike them, the man pausing at a turn row and looking skyward was intimately familiar with much more than Polaris. Antares, Deneb, Orion, Regulus, Sirius, Vega, and many others were as familiar to him as old friends.

Blackburn's neighbors had been amazed when first invited into his home when he and his family first moved in. The transformation had been nothing short of remarkable. The two-story, traditional Pennsylvania Dutch fieldstone farm house, falling down when he bought it, dated back to the mid-1700s, but the Blackburns had fast-forwarded the building by two centuries and expanded it considerably. The carpeting and wall decorations told a story which revealed the owner had certainly not grown up or spent that many years in Berks County, Pennsylvania. Yet his life's story had certain parallels. True, he was not from eastern Pennsylvania, but hailed from a farming community in western Nebraska, where corn, beets, and northern beans were the crops that brought in the cash. His education had been at Colorado A&M – with the "A" meaning "Agricultural. Two of his life's closest friends were big-time ranchers in Kansas.

The carpeting told part of the story. His distant travels had taken him throughout North Africa and the Near East. Closer to home, Navajo rugs indicated his affinity for the American Southwest. Over in a corner, the pairs of boots on the edge of the carpet came from South America. The art on the walls was eclectic to say the least: African, Egyptian, Latin, Mediterranean, and more. The pictures and portraits told the final story. One photo showed two pilots, one of whom was Blackburn, rejoicing beside a Lockheed Lightning, obviously glad to be alive after returning from a harrowing mission over Nazi-occupied Holland during the war. Another showed a much younger Lieutenant Blackburn about to take off in one of America's most advanced bombers of the day in the early 1930s. One of the most recent portraits showed Captain Blackburn of Trans World Airlines, a senior check pilot flying one of TWA's big intercontinental jets.

He had flown the famous and the near famous alike, but his favorite passengers were friends and family. He had flown the very top American military brass during World War II, but he always said that his most important passengers, by far, were the first wounded soldiers from D-Day that he flew back to hospitals in the United States. Few pilots had flown more different types than he had. The great bombers of World War II were in the pages of his logbook: the B-24 Liberator and the B-17 Flying Fortress. Technically speaking, he was not a fighter pilot by training, but he had flown the Lightning and the legendary British Spitfire along with the Air Corps' great bi-plane fighters of the 1930s. The groundbreaking Boeing 307 Stratoliner was in his logbook, and many pages recorded flights to five continents in the Douglas C-54 Skymaster. His reputation was made on the Lockheed Constellation and Boeing 707, but he paid his dues on the venerable Douglas DC-2 and DC-3 on which he became a pilot and check pilot, respectively, for TWA. The first throttle and stick his hands manipulated were those of a Curtiss Jenny. He was terrified of heights and his hands would literally shake while repairing a roof or gutter, yet his steely determination carried him through countless wartime crossings of the Atlantic in spite of horrific weather, icing, mechanical problems, and threat of *Luftwaffe* interception from Iceland to North Africa.

He was very careful with machinery. In nearly 27,000 hours of flying covering over six million air miles, he never so much as put a dent in an airplane or injured a passenger. Blackburn, better known as "Blackie" by his long time pilot friends, was a man of remarkable paradoxes. He was five feet ten inches tall, but his military bearing made him seem well over six feet. He was soft spoken, yet the many pilots who served under him saw him as an instantly commanding figure. His humor was usually dry and self-deprecating, but he could be satirically biting as well. His employer, the mighty Trans World Airlines, had tried to convince him on two occasions to become a vice president. They were turned down. Managers did not fly, and flying was at the core of Blackie's being. His career was literally that of twentieth century aviation itself. He, Harold Fredric

Blackburn, was almost exactly two years old when the Wright brothers had made their first flight so long ago in 1903.

Lunch on that chilly, eighteenth of November was warming, but not Pennsylvania Dutch in character. Far from it – it was a Tex-Mex meal. Blackburn's wife, Martha, had grown up in San Antonio, and Blackburn was fond of southwestern fare. After passing the early afternoon in conversation, the journalist and pilot-farmer got into Blackburn's car and started their journey to what is today's John F. Kennedy International Airport in New York. In one sense, it was an almost routine assignment for the writer because there had been a prequel of sorts five years before when he had written an article about the same pilot flying to the same destinations for the same magazine.

The journalist's name was John Bainbridge, and he was a staff writer for *New Yorker*, the magazine for audiences who were opinion leaders, well-traveled, interested in the arts, and sufficiently upscale not to be very concerned about the cost of a first class ticket to any destination in Europe. The new wrinkle was technology-driven; the first article was written when the glamorous, propeller-driven Lockheed Constellation was the queen of the commercial skies. The new article would feature the latest technology – the Boeing 707.

Yet there were some differences. Blackburn was now five years older than when he first met the journalist, and within a few weeks he would be at the end of his game by federal law. He, like all pilots in their late fifties, was struggling with the fact that the day he turned sixty years of age he would be forcibly retired. What made matters worse was his passing all three of his periodic physical exams earlier that year with flying colors – he even passed the vision tests without the use of glasses. When the journalist was writing the first article, the matter of retirement was never mentioned. Bainbridge had the popular misconception that airline pilots usually retired to places along ocean shores where they could enjoy blue water sailing, to towns high in the Rockies where they could ski, or to the South of France or some other exotic location where their aerial journeys had taken them. But no, Blackburn considered raising Black Angus steers, sheep, and prizewinning vegetables as the next best thing to harnessing the jet stream south of Greenland and getting what amounted to a partially free ride down to Northern Ireland.

Of the entire crew nearing a dozen people, Blackburn would spend the majority of his time interfacing with his first officer and the flight engineer. The latter, Bernard Hanson by name, was from northern Wisconsin where his father had settled after growing up in Sweden. Hanson was in his sixteenth year of TWA employment and was a rare hybrid: both a flight engineer and a pilot, although in recent years he was flying strictly as a flight engineer. Actually he was a check flight engineer just as his pilot was a check pilot who evaluated other pilots. Like many of his pilot contemporaries, Hanson had first flown early in World War II in which he served as a flight instructor for Army aviation cadets. He joined TWA in 1945 and had been an instructor at the airline's training center during the early days of the company's transition to jets. Only recently he had moved from TWA's domestic Convair 880s to the ocean-hopping Boeing 707-331 Intercontinentals.

The first officer, was almost twenty-five years younger than Blackburn, although he would eventually go on to fly 20,000 hours with TWA and retire after flying the big 747s. His name was Henry Gastrich, but everybody called him Hank. He had joined TWA eight years before, immediately after leaving the Navy where he had flown propeller-driven attack aircraft during the Korean War.

As the pilot and journalist neared Idlewild Airport, intensive preparation for the flight was already well underway. In TWA's massive Hangar Twelve both on the main floor where the Boeing 707-331 Intercontinental jet was parked and in the weather and dispatch offices located in the same building, dozens of TWA employees were working in what amounted to a well-choreographed process to prepare the big jet for its upcoming eight-hour flight. The journalist was familiar with the checklists used by the flight deck crew, but had never seen the maintenance checklist used by crews which had already been at work on the Boeing 707 for several hours. Just to make sure there were no misunderstandings about when the plane would be ready to fly, the maintenance supervisor immobilized the control yoke with wire and placarded it as well. That red sign and restraining cable would not be removed until he was completely satisfied that the big bird was ready to fly. The 707's exterior and interior had been scrutinized using a 150-line checklist before the cleaning crews had come inside to vacuum and shampoo the carpeting. After cleaning, the five lavatories and two galleys could easily pass a white gloved inspection. The 707 had recently emerged from its "base overhaul" in Kansas City where it had all but been disassembled and put together again and repainted. Many of its parts had been X-rayed to reveal potential problems, and technicians had even crawled around inside each of the 21,260-gallon wing tanks for inspection and cleaning. The only semi-routine inspection undertaken while the pilots were receiving their briefings was a G-14 check which was performed after every four roundtrip flights to Europe. In that inspection, the plane's landing gears were coated with a fluorescent dye; any existing cracks in the gears would be revealed when the dye was wiped off the shiny metal surfaces.

While all of the inspections and provisioning were winding down, Captain Blackburn led his flight deck crew through meetings with the company's meteorologists and dispatchers. The two meteorologists on duty that afternoon were Bill Walther and Jan Terebelski of TWA's Wind Analysis Unit. Terebelski spoke with a noticeable accent as well he might have. A member of the Polish Army, he escaped after the collapse of Poland in 1939 and had been employed by TWA in 1953. Walther had joined TWA immediately following World War II in which he served as an Army Air Force meteorologist. Terebelski's deliberations that afternoon would be radioed to TWA flight offices in Europe for use by west-bound crews. Like Terebelski, Walther's expertise was forecasting the behavior of the jet stream – then an inexact science considering he lacked weather-modeling super computers and earth-circling weather satellites. As best he could, Walther charted the flow of the jet stream over the North Atlantic as the first step in generating that which the captain needed – the minimum time track. That track, or the most optimal route between New York in Paris would provide the captain, and his navigator, with a series of waypoints designed to harness the jet stream's great power as a tailwind. When that was not possible, Walther's deliberations should provide the captain with at least the means of dodging the worst of the predicted headwinds. That afternoon, the jet stream resembled an "S" that had fallen onto its left side with one curve moving from New England up to northeastern Greenland where it changed its mind and sharply descended over Iceland and down toward Ireland and the continent. In its curves, the jet first cradled a low pressure system with southwesterly winds off Nova Scotia's coast and out to sea while a high pressure system on the eastern side Greenland promised northwesterly winds. At altitude, the winds were really gales that would deliver blessed tailwinds, thus creating great efficiencies.

Walther's weather computations were taken up by the next key player – Bill Hussey, the dispatcher who was an old friend of the captain's. Hussey had been based at Payne Field in Cairo during the war when Blackburn had landed after an epic photo-reconnaissance flight over Nazi-held Crete. Blackburn had flown into Cairo as a Lockheed Constellation captain many times since the war and occasionally had Hussey on his flights to Europe as an observer. That was company policy and ensured that dispatchers stayed in touch with what happened on the line. Hussey was, to use a contemporary expression, the "go to man" for the flight because he was responsible for collecting all of the flight's variables (weather, weight of passengers, fuel, cargo, etc.) and pulling them together in a detailed flight plan upon which he and pilots had to concur before any flight could be initiated.

Soon enough the paperwork was completed and the crew of co-pilots, flight engineer, navigator, purser, and hostesses were briefed on emergency procedures and the flight's details. The navigator was completing his all-important work: converting the minimum time track into waypoints he would mark on his map and convey to his captain along with any mid-course corrections. He would remain silent for nearly three hours because Blackburn would be using conventional radio beacons for navigation until passing over Gander, Newfoundland. The journalist too would remain in the background until well after the takeoff, thus appearing almost as a Japanese wife walking silently behind her husband. The code-like radio exchanges once in the cockpit were a blur to the journalist – ground control, TWA, clearance delivery, tower, departure, and sector controllers. The journalist had a rudimentary understanding of aviation terms. He knew that a "VOR" was a very high frequency, omni-range directional radio facility used for navigation and that the American landscape was so littered with them and their little white cone-like towers that they became, in effect, small towns on an aerial map filled with high and low altitude highways. Bainbridge would have been hard-pressed to copy down and flawlessly repeat the clearance heard in his earphones.

"TWA Eight Zero Zero cleared to Paris Orly via Idlewild VOR one-two-zero radial and the Hampton VOR two-five-two degree radial, then direct Hampton, proceed outbound on the zero-niner-six radial until intercepting Nantucket VOR two-six-seven degree radial, then direct Natucket, control corridor eleven-forty-three. After takeoff, climbing left turn heading two-niner-zero degrees. At 800 feet, turn left one-six-zero degrees to maintain 5,000 feet. Expect radar vectors to the Idlewild one-two-zero degree radial. Contact tower on one-nineteen-one for takeoff clearance."

The journalist's strength was in the verbal realm. Blackburn was very comfortable in that milieu, but decades before had also easily grasped the essences of two disciplines that were foreign to the writer, i.e., those of solid and spherical geometry which are the crux of instrument flying overlaid with the dimensions of time and speed. Although there was understandably little discourse between the author and aviator while overflying Long Island, once the pilot turned northeast for Nova Scotia after passing Nantucket the two men began talking about the flight and the aspects of it that both thought might be interesting to those reading *The New Yorker*. Blackburn was widely considered a virtuoso stick and rudder man, but nowadays about the only way to demonstrate that was a feather-light landing. As the essayist spoke, he did so in an arena bathed by the red glow coming from the faces of scores of instruments of which some, such as the Bendix Doppler radar facilitated decision-making. Others such as the Kollsman Integrated Flight Instrument System and the autopilots were reducing the pilots' role to that of monitors, thus creating an atmosphere where there was more time for conversation.

Although Blackburn was a gifted conversationalist, he became more retrospective on this flight. That was only natural on this crossing because his career as one of TWA's pioneering pilots would end in just several weeks upon his sixtieth birthday. Thus stretches of the conversation were not based on the finer points of a travelogue between New York and Paris, but rather where the pilot, Captain Harold Fredric Blackburn, had journeyed and adventured since his childhood on Nebraska's western prairie when it was still a raw frontier during the twentieth century's dawning.

Like many British surnames, Blackburn derives from a place name, based on the old English words "blaec" (black) and "bourne" (brook or creek). Thus the original meaning was a dark-colored creek. The earliest record of Blackburn as a place name goes back to 1206 in Lancashire where today's small city of Blackburn is located roughly twenty miles north-northwest of Manchester in the

English Midlands. "Blackie," as he was called by friends, had an interest in genealogy and really identified with the other Blackburn, a small town roughly between Glasgow and Edinburgh in West Lothian, Scotland. People named Blackburn appear on tax rolls in both England and Scotland in the early 1200s; some were prosperous landowners, others were educators, and at least one was fined for failure to pay taxes. The first Blackburns, Lowland Scots, arrived in Virginia in the mid-1600s, initiating a migration of Blackburns that continued well into the nineteenth century. Blackie's direct ancestors found themselves fighting on the losing side in the English Civil War in 1651 in which the forces of Oliver Cromwell defeated the army of Charles II. Those Blackburns had to leave Scotland in a hurry, hastened to Ireland in an open sailboat, and briefly lived in County Armagh before leaving for America where they arrived in Virginia 1700. Some of the newly arrived Blackburns were involved in shipping, but Blackie's direct ancestor became a farmer, settling not far from Mount Vernon. Blackie's branch gravitated to the Midwest and had reached Indiana via Kentucky where John T. Blackburn (Blackie's great grandfather) was born in 1824. After establishing himself as a dry goods retailer in Crawfordsville, Indiana, he married Mary Ann Powell who gave birth to two sons and two daughters, of which the first son, Hiram, was Blackie's grandfather. The Powells were well educated people who made their marks in medicine and education. Hiram graduated from what is now the University of Illinois and also graduated from the University of Michigan law school. Blackie's father followed Hiram Blackburn, a prominent lawyer and state representative, by graduating from the University of Illinois, but did not go into law; his doctoral degree was in dentistry.

Blackie's parents, Dr. Charles H. Blackburn and Marion Foland Adkins, were married in Danville, Illinois, on 12 May 1897 but shortly moved to Waukegan, Illinois, where their first son, Don, was born in 1898. Blackie arrived on Christmas Day 1901 in Urbana, Illinois, where his father had set up a dental practice in 1898. In September 1902, Dr. Blackburn picked up stakes and moved the family to Mitchell, Nebraska, which was essentially like stepping back in time compared to Urbana. Indeed, Mitchell was still in what could be described as the immediate post-frontier period of western American history. The town's population stood at roughly 400, and few of its citizens were wealthy or educated. The reasons for Dr. Blackburn's heading for the frontier are mixed; he saw a better financial future in a small town setting and had a pioneering spirit. Time would tell that he also was not opposed to solving problems by running away from them.

Mitchell was convenient because his brother Will Blackburn operated a general store there. Although Mitchell had one doctor, there was need for a dentist, hence the move out West. Blackie was too young to remember the details of moving from Urbana to Mitchell, but it was a memorable journey for his parents – by horse and spring wagon (essentially a large buckboard wagon improved with springs for a smoother ride). Or so goes the legend in the Blackburn family.

The more likely scenario is arriving at Will Blackburn's home on a spring wagon after having taken the train from Chicago to Scottsbluff which is located only a few miles to the east of Mitchell. Unfortunately, Will Blackburn's store burned to the ground in March 1903. For the first few months, Dr. Blackburn and his family stayed at a rented home in Mitchell.

In early 1904, Dr. Blackburn filed for and was granted a 640-acre homestead on Dutch Flats, a large tract of land northwest of Mitchell. The land there was actually first settled by German, not Dutch, immigrants. Much later in life, Blackie would again settle on land in "Dutch" country that had been settled by Germans, but that second homestead was far away in the East and decades later. While most people associate the Homestead Act with Abraham Lincoln and the post-Civil War scramble for land in the West, the Act and its several amendments were still very much alive in the early 1900s. The Kincaid Amendment, under which Dr. Blackburn acquired his land, was strictly limited to homesteads in western Nebraska and differed from the original Homestead Act in two ways. There was no residency requirement and the individual homestead acreage was 640, not 160. The amendment was passed for several reasons of which not the least were to bring more land into local tax rolls and to minimize overuse of public lands by large cattle ranchers. While the elder Blackburn had plans for a real home on the property, he started with a truly modest dwelling which was, in fact, a sod house, or "soddie" made from buffalo grass sod (which turned green in summer). That was followed a year later by an improvement which was called a "shanty," essentially wood and tar paper home, although the Blackburns continued to live in town until the "little house on the prairie" was finished. Considering the scale of the dwelling, it is clear that Dr. Blackburn never really considered the temporary shanty homestead as little more than an investment property.

Dr. Blackburn's wanderlust was still active, and the family moved to San Antonio in September 1904 and Dallas in January 1905 before returning to Mitchell in September 1905. Among their new friends in San Antonio were the Bondurants, a family that would eventually become an important part of Blackie's life. Several times over the next few years, the Blackburns would join the Bondurants for summer vacations on the Texas Gulf Coast. That Blackie was still in elementary school during the last of those vacations meant that he ignored the youngest Bondurant child, a little toddler named Martha. There would come a time in the future when Blackie certainly would not ignore the vivacious Martha. After returning to Mitchell, the family took up residence in a very modest and as yet uncompleted house on the homestead, but they stayed there only until the onset of cold weather when they moved back into town. Blackie's winter clothing was typical for boys on the windswept prairie: high-laced boots, sheepskins, and canvas jackets lined with fleece. Little girls wore long underwear just like the boys, but also wore black stocking and high button shoes. What few affluent men there were in town wore heavy black coats of dog or bearskin, typically with large wooden buttons.

Baseball was the principal form of amusement for Mitchell adults during warm weather. Every Sunday afternoon, the Mitchell town team played another western Nebraska community at home or traveled by spring wagons or train to other communities. Mitchell was served by the Chicago, Burlington & Quincy Railroad, but there was only one train a day. One of Blackie's childhood heroes was the best baseball player in town, Lew McCalmott, the local blacksmith. McCalmott could throw and bat a baseball much farther than anybody in town. He was also illiterate, and when he received a letter which he thought might contain sensitive information, he would take it to the only man in town he truly trusted: Dr. Charles Blackburn.

Blackie grew up to speak "American English," which linguists say is the form of English spoken by Americans with no regional accent. Blackie's diction was essentially without accent, but his pronunciation, basically that of his parents, differed from his rural friends who said "crick" for "creek" and "git" for "get." "Genuine" was pronounced "Geniew-Wine." Blackie always said "Indians" instead of "In-juns" of which there were still many in the area, including the Minatare, a sub-branch of the Sioux.

Food put on the Blackburn table was typical for the era and locale: basic meat and potatoes. The local buckberry, which Blackie never found anywhere but western Nebraska, produced a jam that he always favored. Many families produced jerky, which was also popular during hunting trips as was parched corn which is dried field corn (not sweet corn), that has been roasted with butter and salt. The parched corn swelled up, but never completely burst like popcorn. A sound set of teeth was essentially for snacking on parched corn which had a nut-like flavor. "Krogan" was a meal made from antelope shot out of season. Hunting and fishing were important, and Blackie was adept at each. Catfish were the usual quarry in what Blackie and his friends called the "bar pits" which were ponds always located close to the railroad tracks. Years later when in college and studying surveying, Blackie finally learned the origin of the term bar bits. They actually were "borrow pits," so named because they were where railroad construction crews went to "borrow" dirt to improve the railroad track beds. The same bar pits offered good duck hunting. It was not unusual for Blackie to join a couple buddies and come home with thirty ducks after a few hours lurking in the reeds next to the water.

When spring arrived in 1906, the family moved back to Dutch Flats where brother Charles was born on 11 November 1906. Again, when winter approached, the family left the homestead and rented a home in Mitchell. Within several years, Dr. Blackburn was becoming well established in Mitchell and was able to buy a new home in May of 1909. The new house was on Broadway just a few blocks west of the center of town. On 22 November of that year, the last of the Blackburn children was born – a brother named Joseph.

Dr. Blackburn, an active Mason, was proving to be an asset to his community which he served as mayor, school board chairman, and, temporarily, postmaster. He finally sold the homestead in the spring of 1913 and used the proceeds to buy a new model, six-cylinder Studebaker car. Dr. Blackburn's dental practice was thriving to the extent that Blackie's mother went to work in his office in the spring of 1914. In June 1915, Blackie's brother Don graduated from high school and within a few months had an appointment to the United States Naval Academy where he became a midshipman in early summer 1916. Unfortunately, life as a middie did not work out for Don who enlisted in the regular Navy in August 1917.

For the summer of 1918, Blackie was employed by the United States Reclamation Service (USRS) as a laborer which was arranged by an uncle, Claude Fulton, who had moved to Scottsbluff to work for the USRS regional office located there. Now known as the Bureau of Reclamation, the USRS then had several major water management projects ongoing in its Great Plains Region which included western Nebraska. Although Blackie was really a ditch digger, the time he spent around engines and construction equipment would serve him very well in the years ahead in ways that he could not imagine at age sixteen. During that summer, Blackie earned $60 a month building the dams and canals that would extend irrigation into new areas around Mitchell. Blackie was a day laborer who lived at home; most of the workers lived in a USRS labor camp known as Camp Kiowa, located about five miles northwest of Mitchell. Therefore Blackie was able to avoid one of the unpleasant realities of many unwashed men living in close proximity to each other: lice or what were more frequently referred to as "pants-rabbits."

When he was not digging with a shovel, one of Blackie's favorite pastimes was a game called "shinny" which was a variation on field hockey. For sticks, boys used old hickory ribs from derelict covered wagons. Then Blackie found a set of his father's long-abandoned golf clubs which he briefly put to use. Unfortunately, he eventually knocked the heads off nearly all of the clubs and had to go back to hickory ribs. Although many of Blackie's friends later in life were golfers, he never took it up. Still, in just a few years, Blackie would find sets of golf clubs most useful for purposes other than that for which they were intended.

America had been at war for over a year as Blackie returned for his senior year in high school in September 1918. Like most American schoolboys, he was reading about American doughboys fighting in France and the Allied pilots above them in the clouds dogfighting against the Kaiser's best fighter pilots. Blackie wanted to get into the fight and therefore enlisted in the Army on 9 November 1918 hoping to become a truck driver. He never got much farther than taking the oath because two days later, the war was over. So Blackie continued in his senior year at Mitchell High School where he was president of the senior class and consistently on the honor roll.

He also was a very enterprising young man. Blackie was a chauffeur for the local physician, Doctor Faulk, whom he would drive on house calls after school. Blackie became intrigued by tennis, a sport that would become one of his lifelong interests. Unfortunately, there were no courts at school, and Mitchell's municipal budget precluded building any courts. So Blackie and his best friend Dave Hilyard built one of their own, including grading and surfacing the court, building the net posts, and improvising their own net.

Blackie was active in the 4-H and won a contest with his design of a unique hutch which had an automated feeding system for its rabbits. In addition to playing baseball, he was a standout musician which is not surprising because music was important in the Blackburn family. Indeed, all of the boys played a musical instrument, and there was a piano in the Blackburn home too. Blackie was popular with the young ladies, although most were disappointed because Blackie had a steady girlfriend, Helen Jones, who was a sophomore. She was quite attractive with a lovely smile, perfect teeth, and sparkling blue eyes. She was the only one to whom he ever brought a box of candy. Helen's father Archie, a veteran of the Spanish American and First World Wars, was the postmaster of Mitchell and founder of the local American Legion post, but also was a multi-tasking entrepreneur. He was a part-time accountant, owned the Crystal Theater in Mitchell, and played the trombone in his own band for which his wife, Myrtle, was the pianist. In those days, the movies were silent and Myrtle played background music on the theater's piano until the "talkies" arrived. Blackie had a reputation as a good piano player, but years later he would discover that Helen had just as much if not more talent with the ivories than he did. Blackie occasionally played in the Jones band, but was more visible, as was Helen, in local theatricals and musicals which played at the theater and the high school. Blackie enjoyed the movies, arrived at the Crystal early, then waited with the rest of the audience until Helen and her sister Laura arrived, and then took their seats in the front row. In matters of the heart, odds favored the boys because in the student body at Mitchell High School in 1919, there were almost two girls for every boy in the class.

1919 was a pivotal year for Blackie because late that autumn he was taken aloft for an introductory flight by a barnstormer flying war surplus Curtiss JN-4 "Jenny" trainer. The pilot had flown in France during the war and would become an aviation legend in the decades ahead. His name was Eyir Sloniger but he was better known just as "Slonnie." Sloniger realized that "Eyir" was a bit unusual, so he signed his name as "Ira" on the little certificates he gave his passengers after landing. He was a fellow Nebraskan and was earning a living as a barnstormer, performing at fairs and carnivals while not landing outside communities and selling rides for a few dollars per passenger. Eventually, Slonnie would become one of Blackie's competitors by joining American Airlines and becoming its number one seniority pilot. Like Blackie, Slonnie would fly C-54s for the Air Transport Command during World War II. Slonnie hopped out of the Jenny after landing on that memorable Fourth of July, eager to head to a nearby cafe for lunch. Blackie asked if he could just sit in the plane until Slonnie got back. So Blackie sat there for the better part of an hour, completely transfixed by the experience. It was as though Blackie had just been branded, and so it was. That twenty-minute flight took place in a broad field below the bluffs that gave Scottsbluff, Nebraska, its name. Today that site is a golf course. Starting at an early age, Blackie usually had a camera with him, and the record of that flight still survives. They show downtown Scottsbluff and its grid of streets, the North Platte River, the old North Platte River Bridge, and the railroad tracks – all from an altitude of 800 feet.

The flight with Slonnie was not the first time Blackie had been right up close to an airplane and pilots. Several months before the introductory flight with Sloniger, Blackie and his buddy Dave Hilyard had gone to Alliance, a small town about sixty miles northeast of Mitchell, to watch some barnstormers who made up a flying circus which the government sponsored as means of whipping up some patriotic enthusiasm for its final Liberty Loan bond drive. The pilots included both Royal Flying Corps and the French *Armée de L'Air* aviators in addition to Americans. The pilots, who wore leather jackets, flying helmets, goggles, and flowing silk scarves, instantly became Blackie's heroes. They were rolling, looping, and holding the spectators spellbound. No rides were offered, but even so Blackie decided that day that he had to become a pilot.

Blackie graduated from Mitchell High in June 1919 but not long after suffered a serious setback. It was rheumatic fever for which treatment in those days was bed rest. Blackie was extremely lucky because he suffered no long term complications of which heart issues are the most serious problem. In fact, decades of frequent, comprehensive flight exams would show that Blackie had the strongest of hearts. There was one complication of his long term bed rest; he would not be headed for college in the autumn of 1919. Then, a dose of insult was added to injury with the removal of his tonsils. Following his recuperation, Blackie went on what amounted to a graduation trip – to Florida with Dr. Faulk. At about the same time, Blackie's dad became a partner in the local drug store which provided Blackie with more part-time employment. And in his spare time, he assembled a small dance band that was definitely becoming very popular – "Blackie's Bunch."

There were other diversions beyond his part-time jobs to keep him busy until he could start college in 1920. Like most young men in Mitchell, Blackie was an outdoorsman. Hiking and camping were favorite activities, and there were plenty of bluffs to climb in the immediate area. The North Platte River was nearby as were numerous ponds for fishing. The same venues were productive for more duck hunting. Blackie was not a clairvoyant, but the many early mornings and late afternoons shooting ducks were developing another skill that would come in handy years later – leading other airborne targets with a machine gun instead of a twelve gauge shotgun. Blackie was not one to shrink when a buddy said, "I dare yuh." One of Blackie's more interesting dares took place on a hunting trip in which another hunter dared him to swim across a creek. It was duck hunting season and the dare was to swim to the other side naked. Blackie collected on the bet and was photographed in a demure pose on the other side. Blackie shrugged it off, but kept the picture, put it in his scrapbook, and captioned it, with typical Blackburn dry humor, as "natural history." Frogging around in cold water was an endeavor with which Blackie was familiar; in the winter he ran a trap line for muskrats. The trapping and part-time jobs revealed several facets of Blackie's character – he was highly motivated, goal driven, and while not a tightwad, he appreciated the value of a dollar and was becoming a real go-getter.

Blackie's Bunch was gaining momentum, and not just as pick-up combo for the occasional dance party. By early 1920, the group was playing all around Scottsbluff, including one venue which Blackie would come to regard almost as a second home: the National Guard armory in Mitchell. On 30 July, 1920, Blackie enlisted in the Nebraska National Guard and was assigned to the howitzer company of the 134th Infantry Regiment. He would serve in that unit for a full, three-year enlistment period, and much of that time was spent at the armory. The unit had a distinguished record and could trace its origins back to the Nebraska frontier of the 1850s when residents formed volunteer militias to deal with Native Americans who resisted pioneers who were establishing farms and ranches in western Nebraska. The regiment fought in the Civil War, primarily in western campaigns, including being under the command of General Ulysses S. Grant at Fort Donaldson and Pittsburgh Landing. The unit was called into national service again where it saw action in the Philippines during the War of 1898. It was pressed into service just eighteen years later when it was under the command of General Pershing in the Mexican Border Expeditions of 1916-17. In World War I, after training in the United States, the regiment landed in France for its final training, but before it could move up to the front, its ranks were reduced below combat strength when many of its soldiers were sent as replacements to other divisions. Although Blackie could not have known it at the time, the 134th Infantry Regiment became part of the Thirty-fifth Infantry Division during World War II when it landed at Omaha Beach a month following the invasion. It did not miss any fighting after that point and marched its way to the front, into Alsace-Lorraine, then up to the Ardennes, and finally into the Rhineland.

Blackie got what military indoctrination he could that summer, but in September there was no time for drilling at the armory in Mitchell. His buddy Dave Hilyard graduated from Mitchell High with the class of 1920, and both were off to Colorado A&M in Fort Collins. There was the usual freshman hazing complete with the new arrivals wearing little beanies, but with Blackie being such a good-natured person, the brief lack of civility on the part of upperclassmen was quickly brushed off. Blackie continued his military training in Fort Collins by joining the Army ROTC where the curriculum was based on giving cadets a brief introduction to all combat arms, although the emphasis at Colorado A&M was overwhelmingly on artillery.

Blackie pursued a degree in civil engineering, and for all practical purposes, his major out of the classroom was in music. He played clarinet in the college band and the ROTC band as well. The college band, most visible at athletic events, occasionally became an orchestra with the addition of a piano and a string section. The orchestra was of the caliber that it frequently played in Denver and also went on an annual three-week tour in Colorado and neighboring states. Blackie became triple-rated; he could play the saxophone in addition to the clarinet and piano, but in college he branched out into the drums. The ROTC band was dominated by brass and woodwinds as might be expected, but they also had a strong percussion section to which Blackie occasionally loaned one of his drums.

ROTC summer camp was enjoyable if for no other reason than Blackie could fire weapons much more impressive than his Winchester model 97 twelve gauge shotgun. The thirty caliber Springfield rifle was not much more imposing than the 30:30 rifle Blackie used for hunting antelope or deer, but the Browning water-cooled machine gun certainly was. For the first time Blackie holstered sidearm, a Colt .45 semi-automatic pistol. His uniform was standard for the period: Smokey Bear type field hat, leggings, high-collared tunic, and Jodhpur style pants. The major part of ROTC back on the campus and at summer camp was very consistent with Blackie's assignment in the Nebraska National Guard: field artillery. Thus, he became very familiar with the American-manufactured version of the venerable French 75 mm howitzer.

In the summer of 1921, Blackie returned to Mitchell and worked at the drug store when not leading Blackie's Bunch. Returning to Colorado A&M in September, he was initiated into Sigma Phi Epsilon fraternity which opened yet another musical avenue; the fraternity had a small dance band that played not only at its own events, but frequently played for other Greeks, campus organizations, and many other school activities. The five-man combo billed itself as the "Sigma Phi Syncopatin' Phiends." Blackie was the drummer while others played piano, banjo, trumpet, and saxophone. The group quickly became popular beyond the campus and played in Fort Collins at a dance hall, the Downtown Dancing Nook, and at other parties and dances in the Fort Collins area. The musicians billed themselves as masters of the "jazzfest" but naturally played a lot of waltz and foxtrot dance music. More importantly, several members of the group, including Blackie, were evolving into something beyond a small group of talented college musicians; they were becoming a professional band known as "The Silver State Serenaders." Six musicians formed the group which included an accordion and French horn. The band's leader, Keith Boltz, was employed by the Lyric Theater in Fort Collins, where the Serenaders frequently played. Keith Boltz married Miss Jessie Reece in the summer of 1921 and honeymooned in Estes Park, a location that would play prominently in Blackie's future.

Blackburn Family Collection

Blackie during his days with the Silver State Serenaders.

As a young musician, Blackie was playing in what could easily be described as one of the most significant periods in American music – an era when the Jazz Age was rapidly gathering momentum in mainstream America. It is beyond the scope of this book to define the Jazz Age or to define jazz itself for that matter because musicologists still disagree among themselves. Suffice it to say that Blackie was a strong proponent of the new music that was frequently met with disapproval, especially by young musicians' elders because it was easily associated with "alarming" trends, i.e., availability of alcohol at speakeasies, "flappers" dancing the Charleston, loose morals, etc. Blackie was playing in a geographic part of the country that was still far from universal electrification, thus the availability of phonograph players in many rural homes was several years away. Moving pictures were without sound and what music was played was live, frequently as an adjunct to the film or as a part of the evening's entertainment. Radio was still in its infancy and had a long way to go before its reach matured. In short, given those conditions, live music flourished and dancing was a much more common form of entertainment than it is today. The scene was set for Blackie to ride the new wave in American music.

The 1921-22 school year was another good one as far as Blackie was concerned. His grades were excellent and he had a very busy schedule considering his involvement with music, fraternity, and ROTC. Although he could be considered athletic, Blackie had no time for the school's teams although he attended nearly every major sporting event – as a member of the band. He was on his fraternity's intramural baseball team and played in that exciting season of 1922, first thrashing Sigma Chi in a double-header with scores of 28 to 8 and 17 to 7 before going on to the fraternity championship by beating the Phi Delta team 7 to 1 (for the second year in a row).

Blackie's ROTC weekly activities were dominated by playing in the band, marching, and drilling during the school year which was capped off by attending summer camp where firing the howitzers was the principal activity. Many of the Army's other arms got little more than textbook treatment, including the Signal Corps, Chemical Warfare, Corps of Engineers, Medical Corps, and Armored Corps among others. However, on 26 May 1922, the Air Service did get some local recognition at Colorado A&M when six Jenny biplanes landed in a small field adjacent to the school. The Jennies had flown in from Fort Sill in Oklahoma and were on what amounted to a tour of schools with ROTC departments. This was of great interest to Blackie who witnessed the aftermath of the flight's arrivals. The field was short, narrow, and there was such a strong crosswind blowing that two of the Jennies cracked up. The damage was so substantial that maintenance crews later arrived from Oklahoma, removed the engines for rail shipment back to Fort Sill, and made a big bonfire out of the wood and canvas remains of the two broken biplanes.

Blackie did not return to Mitchell for the summer vacation until July because in June he was in Estes Park, Colorado, as the drummer with the Silver State Serenaders. The major attraction in the area was Rocky Mountain National Park which had opened in 1915. The popularity of the park ushered in a building boom in the area in which many hotels, inns, road houses, and restaurants opened, not only in Estes Park, but also along the roads leading to Estes Park from Fort Collins and Denver. The Serenaders had been booked by the Halfway Place, a popular road house in the Big Thompson Canyon, for the month of June. Many of the establishments in the area were popular with college students and young people for a reason beyond just some good, live music: alcohol. While Prohibition had become the law of the land in 1920, Colorado had gone dry in 1915, but that hardly stopped young people from drinking and dancing in locales where the long arm of the law was spread thin.

While Blackie was enjoying a lively and profitable summer, life back in the Blackburn household in Mitchell was not so rosy. The Blackburn marriage was falling apart. Dr. Blackburn had asked for a divorce, but Marion absolutely refused. On top of the marital discord, Dr. Blackburn's investments were so disastrous that he was unable to meet ordinary financial obligations. His solution was to simply walk out, abandon his family, and start elsewhere which is exactly what happened in the summer of 1922 when he fled to Denver and later Chicago. His departure had little impact on Blackie's brother Don who was away in the Navy, but for Blackie, his mother, and two younger brothers Charles and Joe, aged sixteen and thirteen respectively, life changed dramatically. Returning to Colorado A&M for his junior year was out of the question. Blackie and Don would now have to support the family. Beginning on 19 August, Don's Navy paycheck would not go as far as it once had in supporting the family. He had returned briefly to Mitchell to marry Lee Hemby. Fortunately, Blackie's Uncle Claude Fulton was able to get Blackie a job as an assistant mechanic at the Reclamation Service's maintenance machine shop in Scottsbluff. Charles was able to earn some money at Blackie's old drug store job. Blackie's Bunch, which included brother Charles, brought in a little money playing for dances at the Mitchell Armory and other local events.

Blackie played center on the howitzer company's basketball team which was one of many Guard unit and municipal teams in western Nebraska. The team won three-quarters of the games it played that year, but what made some games memorable was the extraordinarily low level of scoring which suggests that the Mitchell Guard team and several of its opponents fielded defenses of the highest caliber. One game in particular was remarkable. A member of Blackie's team committed four fouls and was forced out of the game with the town team from Melbeta (about seven miles southeast of Scottsbluff). Unfortunately, only five young men from the Mitchell Guard team had traveled to Melbeta, so there was nobody on the bench who could go into the game. The Melbeta coach was sporting and offered to withdraw one of his players, but Blackie's coach refused the generous gesture. Nonetheless, the Mitchell Guards went on to win the contest 21 to 14.

In very early 1923, Blackie parlayed his skills as a USRS mechanic into a job closer to home – in Mitchell as a mechanic with B&C Motors, the local Ford dealership owned by Hank Baumgartner and M.C. "Shorty" Cannon. B&C had the lion's share of the community's car sales and regular service business for one very simple reason: The Ford Motor Company sold more cars than all other manufacturers combined and the Model T was in its heyday. Up until 1923 there was some truth in the old characterization about options on new Fords, "You can have any color so long as it is black." New innovations for the Model T were slow in coming to the market in previous years, but dealers in 1923 had more to offer customers. There was the new "Fordor," Ford's update of the Model T with four doors. New coupes were offered with front-hinged doors and stylish, slanted windshields. They also came with a turtle-backed rear section complete with a trunk. For $65, an optional electric starter was available; otherwise, it was get out and crank'er up to get started. Another innovation appealed to Mitchell residents whose roads were anything but smooth: steel firewalls replaced the old wooden firewalls in 1923. Most of the cars Blackie worked on were Ford's best-sellers: open touring cars and runabouts. Both sold for slightly under $300.

His pay was meager but he spent only several months in the garage because the owners quickly realized what a winning personality Blackie had and that his technical knowledge was useful in more locations than just under a car. They moved him from the backroom to the showroom as a salesman at $100 a month. Blackie was leading Blackie's Bunch at dances held at the Mitchell Armory every Friday night – and beginning to build a reputation as a talented musician among competitors. He turned down another band leader who was looking for a polished piano player; the money was not there, not at five bucks a night.

The local newspaper ran what amounted to an "advertorial" based on Blackie's commentary on contemporary music:

The rumor that jazz is dead is four years old now. Jazz is not dead or even dying, but some changes have been made. When melody united jazz with harmony, the result was syncopation. That's what the valley's best orchestra, "Blackie's Bunch," plays at the armory every Friday night.

Some people called him "Hal," but the most common nickname was of course "Blackie." In other quarters he was picking up another moniker – "Jazzy" Blackburn. Not surprisingly, Blackie's handbill advertising played up the jazz connection with promises of, "Easy Goin' Jazz, Captivating Syncopation, and Entrancing Harmony."

Working fulltime and playing for dances had one negative impact. Blackie's enlistment period with the Nebraska National Guard expired in July 1923, and he saw no way of being able to keep up his commitments. He was given an honorable discharge on 29 July 1923. That was, however, not to be the concluding chapter in Blackie's service to the National Guard of the United States.

Now with a little money in his pocket even after helping with the household finances, Blackie again heard the sky calling to him in the person of another barnstormer who landed a Standard J-1 biplane in a field just outside of Mitchell that autumn. The "Standard" was very much in the Curtiss Jenny's shadow because it simply did not perform as well. The Standard J-1 had been ordered by the Army as a back-up to the Jenny because the Army assumed it would need more trainers than the Curtiss factory could turn out. The lack of performance also explained why there were so many J-1s available at low prices on the surplus market. The pilot, Lowell Yerex, was from Wellington, New Zealand, had a degree from Indiana's Valparaiso University, and had flown with the Royal Flying Corps, shooting down several Germans before he himself was shot down and captured. He returned to the United States and barnstormed for only a few years before heading south of the border to become an airline entrepreneur. Before he was forced out of aviation in the late 1940s, Yerex had launched and operated airlines in Central America, the Caribbean, and Brazil. Blackie invested $150 with Yerex and got ten flying lessons for his money, although he did not progress to the point of going solo. Blackie's best buddy, Dave Hilyard, also had been bitten by the flying bug and was just as anxious to learn to fly as Blackie was.

It was from Lowell Yerex that Blackie learned that war surplus Jennies, in their original crates, were on sale at Love Field in Dallas for only $500. In fact, "surplus" hardly describes the glut of Jennies that were available. During the war, the military at first assumed that only about 3,500 Jennies would be needed, primarily for training. However, the Army and Navy wound up buying well over 6,000 of them which explained why hundreds of them were being sold to barnstormers and aviation enthusiasts across the country. The $500 also included ten hours of flight instruction, or what Blackie figured would be enough to transform him into a real barnstormer.

Blackie now had a plan; he and Dave would drive to Dallas, Blackie would buy a Jenny and quickly get checked out on it, and Dave would drive the car back to Mitchell while Blackie flew home. A bitter disappointment awaited the pair when they arrived in Dallas with $500 in cold cash in Blackie's pocket. Yes, there were hundreds of pristine Jennies to pick from for 500 bucks, but the engines had to be purchased separately. Blackie did not have a dime over $500 in currency with him and Dave couldn't cover the extra expense of the Curtiss OX-5 engine which cost another $250. Dejected Blackie and Dave drove back to Mitchell where Blackie tried to borrow the $250 needed to transform him into an aviator. Relatives and friends at the Ford dealership and drug store did not share Blackie's enthusiasm, seeing it as a bad investment.

Over the next few years, Blackie was constantly dreaming about his flights with Yerex and Sloniger; those memories were kept alive by Blackie's reading material. There were plenty of pulp novels about the great flying aces of the war and one of Blackie's favorite magazines, *Popular Mechanics*, always had a few articles on aviation. Indeed, American and British publishers had launched magazines such as *Flight* and *Aviation* only a few years previously. In the mid-1920s, Blackie tried to get back into flying, but realized that taking a few lessons here and there as time and money permitted was not the way to make flying his life's work. So, Blackie tabled his flying ambitions for the time being.

In late May 1924, Keith Boltz called Blackie and begged him to come back to the Serenaders. Boltz had spent his honeymoon the previous summer at the Baldpate Inn near Estes Park had made many contacts with other hotels, dance halls, and restaurants in the area. He had signed enough contracts for live music in and around Estes Park, including the upscale Stanley Hotel, to keep the band booked solidly throughout the summer. By late August, the band made enough contacts with out-of-state visitors to go on tour in Kansas and Missouri at the close of the season. Although Blackie was now on the road, the major portion of his income went back home for the support of his mother and two younger brothers.

While in Kansas City, Blackie joined the local chapter of the American Federation of Musicians (AFM). Unions were something he had never encountered as a college student or while leading Blackie's Bunch around Mitchell, but now that he was playing in bigger leagues union membership was expected nearly everywhere he played. His membership would eventually be extended to local unions in Colorado, Wyoming, California, Nebraska, Montana, and Oregon. Blackie's attitude toward unions was essentially ambivalent – to be able to play music meant that he had to join the AFM. While he was an employee playing the drums, the matter was entirely clear cut. It became much less so in the future when Blackie became a band manager and owner, although having one foot in each camp while he was in the music business never created any problems for him. That would not be the case in another twenty years when he was employed in a very different industry.

Blackie's tour in Missouri and Kansas ended with the holiday season in December 1924. He had been noticed by a larger band led by Ted Rose whose ten-man Royale Gardens Orchestra, had just started playing in the mountain West. Earlier that year, Rose had stopped on his way to California to play in Billings, Montana, and saw some potential in the upper Rockies. Although Rose's territory defined his orchestra as a regional organization, the band had played in Chicago and he personally had experience playing on both the West and East Coasts. Rose was no lightweight and had been a member of Harry Yerkes' Happy Six Band. That band actually had nine members, but there was no confusion about its track record: it had recorded with both Columbia and Victor back in New York. Rose had also played with the Art Hickman Band in San Francisco. Hickman was quite likely the best known band leader on the West Coast although he occasionally went back East for recordings and a limited number of engagements in the New York City area. Many of the best known band leaders of the mid-920s, such as Duke Ellington and Guy Lombardo were known for their unique "sound" or variation of jazz as was Ted Rose, a man whose original training was in classical music. Rose's particular sound, in his own words, "was a blend of the essential rhythms of jazz with idioms and harmonic shadings of the symphony." The result was something between jazz and the classics to which dancing was very natural, i.e., it still preserved dance rhythm in the music that was original in melody and sweet in harmony.

Blackie signed on as the new drummer when the Rose band opened at the Arkeon Dance Academy in Casper, Wyoming, in January 1925. With Casper as the band's base, Blackie was able to occasionally drive the roughly 140 miles back to Mitchell where his mother still lived with Blackie's brother Joe who was then a junior in high school. In the mid-20s, Casper's population of was approximately 15,000 which ran it a very close second to Wyoming's largest city, Cheyenne.

Ted Rose had built expectations and cleverly positioned himself as "The Paul Whiteman of the West" at a time when the real Paul Whiteman, a Denver native, was topping the national charts with such hits as "Rose Marie," "What'll I Do?" "There's Yes, Yes in Your Eyes," and "Someone Loves You After All." Within a few short years, Whiteman's organization numbered nearly thirty musicians, but in 1924 it was essentially the same size as the Ted Rose Band. Like Whiteman, Ted Rose billed himself as the king of jazz, although there was a long line of other band leaders who would take issue with either band leader about that claim. Unlike Blackie, Ted Rose would never play in the Mitchell National Guard Armory, but would, and did, play in more upscale venues such as the Crystal Ballroom high atop the Hotel Gladstone in downtown Casper.

The Arkeon Dance Academy was located in downtown Casper too and had a brick façade and wide marquee with hanging pennants. Four broad doors opened to reveal a dance hall much larger than one would have expected; the place could easily accommodate several hundred dancers. T.J. McKeon, owner of the Arkeon, was a born salesman who marketed the Arkeon through a seemingly inexhaustible series of promotions. There were special dance nights for teachers, visitors, silk stocking dances, drawings for a new Hupmobile, Christmas dinner dances, Halloween dances, novelty and confetti parties, and one that appealed to the ladies: Charleston bag dances. The Ted Rose band was just one of several which had long term contracts with the Arkeon, and when another band was playing the Arkeon, Rose took his band on the road for other engagements. Blackie was already familiar with many of the locations Ted Rose was playing, i.e., the bigger hotels in cities such as Denver. However, there were occasional down periods between engagements which Blackie put to good use by becoming a salesman at Medniek's clothing store in Casper starting in May 1925. That was typical of Blackie; he could not understand how some people could be content to be idle. The "go-getter" in Blackie also had another motivation; he was still a major source of support for his mother and younger brother.

While Blackie was playing at the Arkeon, a new start-up airline was formed strictly for carrying the mail between Salt Lake City and Los Angeles. It was Western Air Express (WAE) and within several months the company would also get into the passenger business. This was all lost on Blackie who, at the time, could not have realized that several of those Western pilots would eventually become his closest teachers and friends.

In September 1925, Ted Rose took the band on a two-month tour of Montana to play in Bozeman, Billings, and Butte. During this period, Blackie's entrepreneurial juices were flowing because a tremendous opportunity suddenly appeared at the end of the tour. The Ted Rose band went broke while in Butte.

Thus, a new band started playing at the Arkeon in November: Hal Blackburn and his Orchestra. In addition to the suddenly unemployed members of the Ted Rose band, Blackie knew nearly all of the talented musicians in the area and still had his Silver State Serenaders contacts, thus he was quickly able to recruit six of them into the new band. Like Blackie, two members of the new band were double-rated. Jack Bowring was the piano and accordion player while Newton Nelson played an unlikely couple: violin and trombone. The banjo player was Dewey Powell. Lloyd Derby, Lloyd Webb, and Van Pickett played the clarinet, saxophone, and trombone respectively. Unlike many of the big bands of the 30s and 40s which had singers on their payrolls, Blackie and many other regional band leaders in the 20s hired local singers with whom they or the dance halls were familiar.

As a businessman, Blackie did not follow the Rose approach of concentrating on the more upscale venues. Blackie would play those as well and, more importantly from a revenue standpoint, the many lesser events such as county fairs, American legion celebrations, dances at Elks halls, i.e., whatever it took to keep the band playing as many jobs as possible. Fortunately, Hal Blackburn and His Orchestra and its repertoire proved to be such a hit that they were monopolized by the Arkeon. While Blackie provided the music, the Arkeon occasionally brought in other entertainment; one popular pair was the DeCecilitos, a dance act from San Francisco whose specialty was the tango and the whirlwind dance. Blackie was kept so busy in Casper that he really did not start touring until May 1926 when he went on a summer tour that was limited to Wyoming. When the summer ended, Blackie took the band farther afield and really started living out of a suitcase, playing as far north as the Canadian border and almost to Mexico City where after crossing the border the band found conditions so primitive that they were content to find a few jobs in Laredo just long enough to cover their expenses back to Wyoming.

Blackie promoted the band as being eager to please in an upscale sort of way. One promotion had the men dressed in tuxedos as waiters with Blackie carrying a big tray of various delectables; the headline ran, "Hal Blackburn and His Orchestra, Serving Music to Your Taste." In another poster, it was, "Steppin' Out with Hal Blackburn and His Orchestra, You're Invited." Blackie positioned the band by saying, "They Go Miles To Make Smiles," creating "Music That Makes You Dance." Although the last time Blackie ever used golf clubs was to play shinny in the fields near Mitchell, he did not hesitate to use several golf bags as upscale props in a promotional photo of the band, their touring car, and instruments in front of the Arkeon. The promotions were working as confirmed by glowing editorial in the newspapers of cities where Blackie played. To borrow a phrase later used by Les Brown, Blackie's organization was definitely becoming a "band of renown." Increasingly throughout the band's days in Wyoming, the smaller city newspaper articles presented the Arkeon in

Casper as by far the leading ballroom in the state if not the entire region. As the principal band at the Arkeon, Blackie was basking in a very nice halo effect.

The band earned its money in a variety of ways. When booked for a private engagement, the fee was negotiated in advance. When playing a dance hall or other venues where tickets were sold, the band got a percentage of the front gate. When booked for jobs that extended over several days, such as Fourth of July celebrations, the band might play one night for free, but get a graduated take of the ticket sales on subsequent days. Tickets to dance halls typically cost an average of fifty cents, sometimes more, sometimes less. Some dance halls charged only men an admission price, but not ladies. Many dance halls hired women to dance with men. Sponsors of dances at public events typically did not require an admission fee for women.

Hal Blackburn and His Orchestra had a diverse repertoire that numbered nearly 150 different pieces of music. Because so many of the engagements were in dance halls, waltzes and foxtrots comprised a significant portion of the band's music. The Charleston was all the rage, and the band not only played the Charleston, but also demonstrated how to do it the right way and the wrong way. As might be expected, there were nearly two dozen seasonal numbers that were played on patriotic holidays and the Christmas and New Year season. The balance of the music was based on what was played by the big, nationally known bands, including Paul Whiteman, Ted Weems, Isham Jones, Fred Waring, Ted Lewis, and Vincent Lopez among many others. Blackie's play list therefore included many of the top hits of 1926 as well as the immediately preceding years. Typical numbers included "Bambalina," "Moonlight in Mandalay," "Remember," "Riverboat Shuffle," "Moonlight and Roses," and "Wonderful One." Ragtime and jazz were important parts of the repertoire which Blackie kept as broad as possible since he wanted to appeal to a wide audience. He was also not above the occasional specialty number and hearing a cowbell, kazoo, or a gunshot (with blanks of course) was not unheard of when the Blackburn band played. There were two other songs which may have been indicative of what was starting to become an obsession: "Beyond the Clouds" and "I'll Fly to Hawaii."

When 1926 came to a close, the band had put many, many miles on its two Chevrolet touring cars which had to negotiate roads of less stature than today's I-25 or I-70. Since the band was covering plenty of rough roads, the cars always carried several spare tires per car. In its 1926 tour the band quite literally crisscrossed the state of Wyoming. The year began in Thermopolis on 13 May at the Plunge Pavilion which was a large facility featuring what was about to become Wyoming's second largest swimming pool. The largest swimming pool in the state was the Lavoyeland "Natatorium" (a word rarely heard anymore – meaning swimming pool) located in Salt Creek, a town with a history made confusing by the fact that it was a "company town" variously known as Midwest and Lavoyeland. The company was the Midwest Oil Company of Casper. With oil flowing from the Salt Creek Oil Field, there was money flowing with it; hence the construction of several entertainment venues including the Salt Creek Dance Hall and the Lavoye Theater.

The tour continued to the Lander Open Air Dance Theater, the Worland Dance pavilion, the Town Hall in Kaycee, and the Masonic Temple in Rawlins. In many instances, Blackie provided the music and shared the performance with others. In Whiting, Blackie provided the music at the Whiting Theater while the audience enjoyed live performances by Vilma Banky and another up and coming actor named Ronald Colman. While Wyoming was the home territory, Blackie occasionally had to make take advantage of other opportunities even though they were geographically awkward. Thus, he was pleased to spend a weekend back home in Mitchell where the band had been hired by the American Legion to play at the Fair Ground Pavilion. Another interruption occurred when Blackie played at the Hotel Semlo in Salt Lake City. Then it was back to Wyoming via Montana where the band played in Butte for a Red Cross benefit dance and Miles City at the Twilight Inn before heading down to play in Buffalo, Riverton, and Basin at the City Hall, Masonic Hall, and Open Air Pavilion respectively. Occasionally, the band would come back to a city and play at a different venue as was the return to Lander where the group played a dance at the National Guard Armory. Guernsey and Greybull were on the circuit – both for dance hall jobs.

Unlike Ted Rose who had set his goals too narrowly, Blackie defined his market broadly and included the corporate and institutional markets as well. Thus Hal Blackburn played for a sugar beet refiners cooperative, trade unions, and car dealerships among others. On 10 October 1926, the tour was over and the band re-opened at the Arkeon back in Casper where it remained until February 1927.

One of the mid-course detours taken by the band during the 1926 tour was to Klamath Falls, Oregon, which was the home of Jack Bowring, the band's piano player. The band had played at the municipal auditorium in Klamath Falls and also the Open Air Pavilion in Medford which was about fifty miles to the southeast of Klamath Falls. Blackie liked what he saw in Klamath Falls in terms of existing business and growth potential. He therefore made the momentous decision to leave Casper when his latest contract expired with the Arkeon and moved the band to Klamath Falls. He was also uncomfortable with his mother living alone in Mitchell, especially since Don and Charles were no longer living there. It was decided that Blackie's mother and brother Joe would move to Klamath Falls, a city that was growing faster than any other in Oregon in the late 1920s. Fueled by the timber industry, agriculture, the railroad, and new highways, Klamath Falls was about the same size as Casper, but by 1930 it would become larger than either Casper or Cheyenne.

The Blackburn home on West Broadway in Mitchell was sold to Mr. and Mrs. J.V. Keimig who moved into the house from their ranch outside of Mitchell. A public sale was held in which the furniture and many other items were sold including Marion Blackburn's sewing machine, the refrigerator, bicycles, a typewriter, a saddle, a phonograph, and what had to cause considerable sadness, the family piano. Some dental equipment was also sold. Then Joe drove his mother to Klamath Falls with a brief stop in Denver. Blackie was waiting for them in Klamath Falls where he had rented a small Dutch colonial style home at 1018 Washington Street, just north of the business district.

Blackburn Family Collection

"Serving Music to Your Taste" as only the Hal Blackburn Orchestra could in 1927.

The band's touring days were almost over, because Blackie had become the business manager of two properties in Klamath Falls: the Winter Garden (a ballroom) and the Altamont Dancing Auditorium. The band played a regular dance schedule at those facilities and also rented them to organizations which in turn hired the band for entertainment. Although Blackie and the band played practically every night of the week, he still had enough time in the morning and afternoon to work another job. As he had done in Casper, Blackie became a clothing salesman, this time at the Oregon Woolen Mills store in Klamath Falls. The men in the band were also able to find day jobs, typically giving private music lessons, working in music or other retail businesses, etc. At this point, Blackie was personally netting nearly $400 per month from the band alone, which was a very good salary in the late 1920s, and with typical Blackburn frugality had saved much of his income. His only extravagance had been the occasional flying lesson which had not gotten him well on the way to a license simply because his lifestyle precluded consistently flying once or twice a week until he could log the required number of hours for a license. Aviation did not become part of the local scene in Klamath Falls until much later, and the nearest airport was in Medford, but that was more than two hours away by car. The dream of becoming a pilot was still there and as strong as ever, but Blackie simply was not in a position to do much about it.

Although touring with the band basically ended when the group moved to Oregon, there were a few exceptions in the summer and autumn of 1927. The Klamath Falls city fathers were anxious to see the completion of a north-south highway which would bring more traffic, trade, and tourists into the Klamath Falls area from northern California. They and the local chamber of commerce had linked up with the chambers in San Francisco and Sacramento to accelerate the construction of what was becoming known as the Klamath Falls-Weed Highway (today's Route 97). This alliance, based on the expectations of increased traffic and business between northern California and southern Oregon, resulted in a mid-July visit in which 150 businessmen from Klamath Falls were hosted by the chambers in Sacramento and San Francisco in what was billed as a "good will tour." As part of the tour, which took the night train from Klamath Falls on 13 July, the delegation brought along the city's finest orchestra to provide entertainment at the lunches and dinners while visiting the California cities. For Blackie, it was a real eye-opener, especially seeing so much aerial activity over the big cities with their military and municipal airports.

After spending two days at the Hotel Sacramento, the group entrained for San Francisco and two days at the Hotel Californian where Blackie played at the lunches and dinners. More rewarding to Blackie was the band's late evening playing at the El Patio Ballroom, one of the upscale dance halls on Market Street – and being broadcast live over radio station KPO in San Francisco. When the tour ended, the band did not put its instruments back in their cases. They returned by rail with the delegation on the Cascade Limited and played on the way back to Klamath Falls.

By the autumn of 1927, Hal Blackburn and his Orchestra, which numbered nine musicians at that point, had established itself as one of the stronger regional bands in the west. They were not quite up there with the national likes of Paul Whiteman or Fred Waring, but they had enough stature to be recognized by W.A. Quincke and Company, a major musical publisher based in Los Angeles. In its trade advertising to the music industry, the company featured sixteen of the leading bands in the west, including Hal Blackburn and His Orchestra. Because the band had so much business in Klamath Falls, Blackie became very selective about out-of-town engagements and limited the band's engagements to northern California where he played at the grand opening of the spectacular Alhambra Theater in Sacramento and the Hotel Redding in Redding.

On the surface, Blackie seemed to have it made. In spite of his day job, he had enough leisure time to qualify for Red Cross certificates in lifesaving and first aid, and he was able to go fishing and deer hunting which is not surprising considering opportunities for sportsmen in that part of Oregon. He enjoyed the satisfaction that came from seeing his mother well cared for, was leading a band that was respected in its territory, made a comfortable living, and, because he had been invited to join the Lions Club, could be said to have been accepted as a member of the local community. Under the surface, there was considerable turmoil. Leading a band was not what he wanted to continue doing in life – pure and simple. Granted, the exhaustive touring days were behind him, but he was coming to hate the lifestyle: work all day and work most of the night for six days a week. He wanted to be a pilot and not just any pilot – he decided that only the best would do and that was to become a pilot in the Army Air Corps. He did have some obligations, however, but New Year's Eve was the only one that he could not avoid. Hal Blackburn and His Orchestras, *all three of them*, played at Altamont, the Winter Garden, and at the Pine Grove Ballroom in the Willard Hotel on Main Street. Blackie was on a fast circuit between all three venues leading the core Blackburn band and the other two quasi-Blackburn bands until 1928 was a fact for all three audiences.

One might easily speculate on how far Blackie could have gone in the music world if he had wanted to. He possessed a charming personality, certainly had the technical ability to lead a larger band, had proven entrepreneurial skills, and for the time being was playing in a favorable economic climate. Could he have taken his orchestra into the league where the national bands played? Apparently, Blackie was asking the same question because he went on a solo tour of California several weeks after the very successful Sacramento and San Francisco chamber of commerce tour. He went to the major hotel ballrooms and upscale dance halls in San Francisco, Oakland, Los Angeles, and even down to San Diego. It was impossible on that jaunt not to notice how Crissy Field, March Field, and Rockwell Field in those cities respectively were putting so many bombers and pursuit planes aloft into the sky. That trip may very well have been the death knell of Blackie's professional music career.

On New Year's Day 1928, a Sunday, Blackie sold the band to his clarinet player for $200, including sheet music, stands, and instruments. He had just turned twenty-six years of age and, for the time being, kept his daytime job as a clothing salesman. Then Blackie, his mother, and Joe started planning a move to Los Angeles where Blackie would join the Air Corps as an aviation cadet. Many years later, Blackie described himself as being "green as grass" when it came to his aspirations of becoming an Army flyer, admitting that he then thought all that he had to do was express an interest in becoming a cadet and pass a physical exam. The years as a band leader were not wasted because in less than fifteen years he would be able to apply the many skills he had acquired as an entrepreneur in much, much larger organizations. He had become an adept negotiator, a shrewd judge of talent, a personnel manager able to inspire confidence in employees under occasionally difficult circumstances, and a businessman with bottom line responsibility. Probably the greatest skill he polished as a band leader was that of managing interpersonal relationships within a diverse group of people. Because there were occasionally as many as three Blackburn orchestras, Blackie had to deal with the full range of human shortcomings, including petty jealousies, overgrown egos, artistic tempers, inconsistent or inadequate job performance, and the many other personnel and financial problems which vex managers of both small and large organizations. Time would tell that Blackie had learned his lessons well in the Mountain West.

When he moved the family to Los Angeles in the spring, he rented an apartment at 1201 Shenandoah Street from Mrs. Alice Smith, a seventy-two year old widow, and her divorced daughter, Sibyl Riley, a secretary. Blackie's family never addressed Alice as Mrs. Smith who was a transplant from Missouri – it was always an endearing term, "Grandma Smith." Blackie picked up odd jobs while waiting to take the Air Corps entrance exams and then hearing back from the Army on whether he had been accepted as an aviation cadet. He knew that he had passed the rigorous physical exams taken at Fort MacArthur because those results were instantly obvious, but was much later shaken to learn that his math score on the written tests had disqualified him. Still, the rejection held promise of acceptance if he could get a satisfactory score when the test was next offered. At first, Blackie probably could not have anticipated how the odds were stacked against him and any other aviation enthusiast wanting to become a cadet. At the time Blackie was applying, 9,000 other young men were too, and only about 600 would be accepted. Becoming the one of the seven percent who were accepted into the program was no guarantee of success. The success rate gradually improved in the late 1930s and early 1940s when the Air Corps desperately needed more pilots, but that was years away. Blackie may not have originally been aware that when he was eventually accepted into the program at March Field in 1930, four out of five members of his primary/basic training class would fall by the wayside.

Blackie decided to hit the books, take enough algebra, trigonometry, and geometry to ace the math section on the cadet exam and possibly finish his bachelor's degree. Fortunately, he had saved enough money from his band days to afford the luxury of becoming a student again. He therefore enrolled in the summer term at the University of Southern California although he looked upon USC only as a temporary expedient because later in the summer he was accepted at the University of California at Los Angeles which gave him standing as a junior based on his record at Colorado A&M and USC. Because he was basically living on savings, Blackie joined the California National Guard's 160th Infantry Regiment to bring in a few dollars. That unit traced its history back to the War of 1898. It was better known as the parent regiment of the "Lost Battalion" and its fame from World War I; in World War II the regiment fought in the South West Pacific Theater and at war's end was fighting in the Philippines.

Not too long after Blackie arrived in Los Angeles, there was an item in the *Los Angeles Times* on 16 May 1928. It was not front page news, but appeared in the business section and was about the formation of a major, new corporation in which Charles Lindbergh was the technical advisor. The company, called Transcontinental Air Transport (TAT), had been incorporated the day before with the goal of establishing transcontinental air service supplemented by railroad service over some sections to reduce the time it took for pure train travel across the country – the new service could take a passenger from one coast to the other in just two days, not three as with the railroads. In practice, the planes flew passengers in daylight and the trains carried passengers at night or in very bad weather. It required close coordination between the two railroads, the Santa Fe and The Pennsylvania, and massive investment in new airway facilities and terminal buildings. Significantly, the airline had been incorporated as a *passenger* airline. In the beginning it had no intention of carrying air mail which was the life's blood of most airlines in that era. TAT was a major start-up company in those days and was capitalized at $5,000,000, of which twenty percent went to weather observation and forecasting facilities which provided coverage of the entire air route and 100 miles on each side of it. The resulting communications system utilized radios for contact with aircraft, teletype and telephone between air terminal and weather stations, and also tied into the railroads' own private telephone exchanges.

The new airline started operations with ten Ford Trimotor aircraft and ten spare engines. The pilots were selected very carefully; they were the crème de la crème of airmail flyers, senior pilots from aircraft manufacturers, or experienced military pilots. The co-pilots were all junior officers who had graduated from Kelly Field.

As an older student with definite goals in mind, Blackie did well academically, especially in his math courses where algebra was his most important subject. He was unaware of it at the time, but the other math courses in solid/spherical geometry and trigonometry would be of tremendous value in just a few short years because they helped him visualize spatial relationships in his head, a definite asset when he got to the point of making instrument approaches using the radio range system of the early 1930s. He was also required to take courses outside of his major although he had been allowed to transfer many of the credits previously earned in history, English, and biology. He filled out his schedule with ROTC and courses in economics, Spanish, and geography where he earned Bs in all but Spanish which resulted in a C, which is surprising considering how Blackie's interest in linguistics blossomed little more than ten years later. In the late spring of 1929, he took the aviation cadet entrance exam again and passed it, but that hardly meant that he could report for duty the following Monday morning. The earliest class he could join was that starting at March Field located in Riverside, just west of Los Angeles, in February 1930. Blackie was also at a crossroads financially and could no longer afford to be a full time student. He had no interest in going back to the music business, but decided to combine business and pleasure in a job that would also be advantageous when he did become a cadet. He became a mechanic again – working on airplanes, not automobiles. He had another reason for withdrawing from UCLA; as the head of the family, he also had to financially help his younger brother Joe who entered UCLA that autumn.

While visiting the many airports in and around Los Angeles – and there were many indeed – Blackie had struck up a friendship with Homer Weber, owner of Spartan Airport. Weber was about Blackie's age and had been bitten by the flying bug after briefly serving in the Navy late in World War I. He returned to the family business, a bakery, following the Navy, but quit that to get into flying as a barnstormer and employing his wife Pauline as a wing walker. Weber had been one of the many pilots employed in *Wings*, the immensely popular silent movie about Word War I fighter pilots that had actually been filmed in Texas. Weber continued his association with Hollywood by flying in more of the aviation films which became a regular part of Hollywood's offering in the 1930s. In March, Blackie worked out a deal with Weber in which a part of Blackie's pay would be in the form of flying lessons. As 1929 progressed, with Weber becoming more involved with film production, Blackie became the assistant manager of the airport which was located less than a mile southeast from where Blackie lived on Shenandoah Street. The airport is long gone and is now a residential area with shopping malls nearby. The Westside Neighborhood Park is now located on top of the old runway. Weber eventually went to work for Lockheed Aircraft and went into the Army Air Force as a stateside service pilot flying C-47s during World War II.

Blackie's flight instruction began on 12 April with a half hour flight in a Spartan C-3 biplane. "Spartan" was actually an Oklahoma-based aircraft manufacturer that would become famous for its advanced "Executive," a very high performance single engine monoplane introduced in 1935. The two "Spartan" names, the airport and the manufacturer, were quite unrelated. In the late 20s and early 30s, Spartan's C-3, which was offered with a choice of several different engines, was very popular with flight schools. Blackie was averaging two lessons per week in April and May, then usually one per week in June and July. With almost thirteen hours of instruction, Blackie was sent off solo on the Fourth of July. Just three days later, Transcontinental Air Transport inaugurated its first flight out of Los Angeles although the event was of no major interest to Blackie at the time. That first flight certainly made headlines in a big way. Charles Lindbergh, in California, pressed a button in Los Angeles and when heard in New York, it authorized the departure of the "City of New York" train from Penn Station to Columbus, Ohio, where passengers detrained and emplaned for the West in the company's Ford Trimotor, "The City of Columbus." The airlines, always anxious to use celebrity endorsements in those days, were loaded for bear on that day. Amelia Earhart christened the train in New York and Gloria Swanson and Mary Pickford christened the planes, but not with champagne. Since Prohibition was the law of the land, grape juice was used instead.

Blackie was more interested in getting his private pilot's license; that came on 7 November with Flight Inspector James Hall at Mines Field. Nearly all of his flying time was in one of the C-3s, powered by a Siemens or Walter radial engine of 125 hp, with the only exception being a few flights in a Waco Nine biplane which was powered with a Curtiss OX-5 engine of 90 hp. Blackie's first solo cross country was to San Diego and back, although most of his flying was confined to the greater Los Angeles area, including flights to Mines Field and Clover Airport in particular. Although Blackie was then a rated pilot, he still hung out at airports, hands gripping the fence, watching planes come and go. He was also an enthusiastic spectator at air shows and decades later recalled one in particular at Mines Field which eventually became Los Angeles International, although it was little more than a wheat field in the mid-1920s. One of the more memorable airshows he attended was at Mines field in 1928 when the Navy's flight demonstration team, "The Three Sea Hawks" put on quite a show. Blackie introduced himself to the leader, Lieutenant Daniel W. "Tommy" Tomlinson, a 1917 Naval Academy graduate. Unknown to either man at the time, their paths would cross many times in the future beginning in 1935.

A week after getting his private ticket, there was another development in America's fledgling airline industry that even then, as today, was marked by mergers. On 16 November 1929, Transcontinental Air Transport bought Maddux Airlines, a regional carrier with routes extending south from Los Angeles to San Diego to Phoenix.

Although Blackie was, technically speaking, a very junior pilot at that point, many of his friends and acquaintances were much more experienced, including the best known stunt and race pilots of the day. One of his friends was Margaret Perry, an heiress who had just purchased the Culver City Airport, a field that was a favorite of movie stars and moguls, including Howard Hughes who flew out of Culver City in the early 1930s. Perry was a sportswoman who excelled at polo, golf, and tennis; she could afford the finest things in life since her family had been invested in numerous profitable ventures other than the stock market.

Culver City Airport was opened in 1927 under a different name, but was purchased the following year by the younger William Gibbs McAdoo, Jr., not to be confused with his father of the same name who had been a U.S. senator from California and President Wilson's Secretary of Treasury. There was another Wilson connection: he was the elder McAdoo's father-in-law. McAdoo, the younger, had great designs for Culver City which he planned to use as the western terminus for an airline he hoped to launch in 1930. The plan fell through as did his need for the aerodrome which he sold to Ms. Perry, the Hollywood socialite and pioneering lady flyer, who offered Blackie a job as manager of the airport. Blackie had to turn her down since he knew that he would be going into the aviation cadets in the very near future. Ms. Perry was an early leader in the Ninety-Nines, the women's pilot organization. Thus such leading aerial ladies as Amelia Earhart, Florence Barnes, Louise Thaden, and Jacqueline, "Jackie,"

Cochran were frequently seen at Culver City along with many of Hollywood's brightest lights. Blackie and Jackie became very good friends, and even decades into the future a box of chocolates always arrived at the Blackburn home just before Christmas – sent by Jacqueline Cochran.

From his association with Margaret Perry, Blackie also became acquainted with many of the nation's leading stunt, racing, and test pilots, including Jimmy Doolittle, Roscoe Turner, Paul Mantz, Benny Howard, Tex Rankin, and a man whom he never would have guessed would be his employer someday – Howard Hughes. Many of them and some of Hollywood's stars turned up at the lavish parties the wealthy heiress Florence Barnes, better known as "Pancho," gave at her two mansions in greater Los Angeles. Blackie was an infrequent guest, but only at the San Marino mansion which was located just south of Pasadena; that was reserved almost exclusively for Pancho Barnes' flying friends. Her Hollywood friends tended to be invited to the ocean front mansion down in Laguna Beach where guests frequently partied even if Pancho Barnes was not there. Although Blackie was not a regular guest, he was an excellent asset at a party because he was a good conversationalist, could dance a little soft shoe if the occasion arose, was the best piano player in nearly any crowd, and he could make a bourbon and water last nearly all evening. Although most of America was tolerating Prohibition, Pancho Barnes ignored it and flew in liquor weekly from Mexico.

Blackie's flying time accelerated significantly in the three months between passing his private pilot test and his going into aviation cadets. There were changes at home too. Brother Joe had started as a freshman at UCLA and his mother had a new job as the manager of a small apartment building. Blackie was "field visiting" in the Los Angeles area, getting in more cross country time, including flights to Bakersfield, San Francisco, and San Diego. He also flew frequently as Margaret Perry's personal pilot, but actually logged more hours practicing for what he knew was right around the corner – becoming an aviation cadet at March Field. On 15 February 1930, Blackie learned what was most likely the most important lesson in his flying life and came very close to killing himself in the process. He was flying the return leg of a cross country to San Francisco and like many new pilots was at a very low altitude because that is where the sensation of speed could most be enjoyed. He was also at the most dangerous stage of a pilot's development, i.e., under 400 hours, and enjoyed elementary aerobatics, sudden wing overs, and flying through the fleecy tops of cumulus clouds. Blackie had been on one San Francisco roundtrip before, but was still checking his position on his map, glancing at the instrument panel, and then studying some notes on his map when he just happened to do something which he should have been doing all along – look outside the plane. He had all but flown into rising terrain, and another few seconds of not maintaining situational awareness would have crashed. He had had only one lapse before when on 29 January he ran out of gas, but fortunately was well within gliding distance of

Weber's field to pull off a dead stick landing. Those two incidents matured Blackie very quickly and were the basis of his subsequent reputation for rigid vigilance and discipline in the cockpit.

At about the same time, Western Air Express made an acquisition – Aero Corporation of California. The real goal was not Aero Corporation of California which was the western distributor of Fokker Aircraft. Rather, its subsidiary, Standard Airlines, was the target as was its southwestern air mail route. Coming into the WAE fold were several Standard pilots, including Jack Frye and Paul Richter who had been partners in a flying school. Frye and Richter built the business up to the point where it was flying air mail and later passenger service between Los Angeles and Phoenix and Tucson. The names Frye and Richter meant nothing to Blackie in January 1930, but they certainly would in just a few years.

Chapter Two
Mister Blackburn, Harold F., 6544831, Aviation Cadet (A Dodo)

Before reporting to March Field, Blackie was required to resign from the California National Guard which he had joined as soon as he had moved to Los Angeles. His resignation from Company C, 160th Infantry Regiment was accepted on 11 February on the basis that he was enlisting in the Regular Army as a privte. His enlistment in the Army proper was done on 20 February at Fort MacArthur in Los Angeles. The fort was named for General Arthur MacArthur, the Civil War general who was the father of General Douglas MacArthur. Blackie made a wise choice by joining the Air Corps considering what was awaiting the 160th in little more than a decade into the future; the regiment fought as part of the Seventh Infantry Division in two costly Pacific campaigns: the invasions of New Britain and the Philippines.

Although Blackie's focus was strictly on military aviation in early 1930, much was happening in the civil sector as a result of the deepening depression. The nation did not need as many airlines and to survive the remaining airlines had to merge. Maddux was absorbed by TAT, becoming TAT-Maddux. Western Air Express absorbed Standard Airlines and its management of Jack Frye and Paul Richter. The stage was thus set for the eventual merger of TAT and Western Air Express.

The Army Air Corps normally paid a cadet's travel expenses from his home to his primary training field. Blackie did not bother with the paperwork because his first assignment, March Field, was just west of Riverside, only several miles from where he lived in Los Angeles. If he or any of his new associates thought they would be flying right away and doing a lot of marching at first, they were disappointed: wrong on both counts. It would be over two weeks before they sat in an airplane cockpit, and they ran, not marched, everywhere. The closest Blackie got to an airplane was helping push upperclassmen's planes from the hangar out to the line.

In the spring of 1930, the Los Angeles urban sprawl had decades to go before it crept to the March Field boundary fence. The field was in the form of two side-by-side triangles, forming a square. One triangle comprised the flying field, and the other was for hangars, barracks, and administrative buildings. "Field" was the operative word because there were no paved runways – just a field that produced great clouds of visibility-robbing dust during the dry, summer season. In the early 30s, March Field had ten large hangars with roofs painted in a striking, red and white checkerboard pattern. The field officially opened in March 1918 as a training base on which 103 Curtiss JN-4 "Jenny" aircraft were based and was named for Lt. Peyton March, a young aviator who had been killed in a flying accident in Texas. Lt. March's father was the Army Chief of Staff. In 1923, March Field was closed because a decision was made to concentrate training activities in Texas. With the Army Air Service expansion in 1926, March Field re-opened in 1927 because the facilities at the other primary training base, Brooks Field in Texas, could not accommodate the larger numbers of new cadets. For at least two of the new instructors, March proved to be a good early springboard to bigger and better things. Lts. Hoyt Vandenberg and Nathan Twining both became Air Force Chiefs of Staff. One of their students went on to become chief of the Strategic Air Command, General Curtis LeMay. The commander of the primary flying school was Capt. Joseph McNarney. His path and Blackie's would cross over the North Atlantic in little more than a decade.

Blackie had joined a very exclusive group considering the fact that just the pre-enlistment physical examinations alone could result in more than ninety percent of applicants being rejected. The rejection rate would continue throughout the upcoming year, albeit not quite so substantial as ninety percent.

It was a good time to be a cadet in most respects because the Air Corps Act of 1926 had put new muscle into the training of Army pilots, including a new generation of training aircraft which replaced the aging, World War I era Curtiss Jenny biplanes. Unfortunately, the economy of the early 1930s would dampen military career opportunities for most successful cadets. To better understand Blackie's cadet experience requires a detour back to the pre-World War I era and also to America's most popularized aviation cadet era: World War II.

Seeing storm clouds on the horizon, the Army Air Corps dramatically changed its pilot training program in 1939, realizing that it did not have the facilities and manpower to start training the tens upon tens of thousands of pilots it would need for what loomed on the horizon. Rather than try to conduct primary training itself, cadets were sent to aviation academies in the private sector. The number of these schools grew exponentially in the years leading up to American participation in the war and indeed well into the first years of the war. Some of the country's best known aviation academies, such as Embry-Riddle, the Ryan School of Aeronautics, and Parks Aviation College, became contract schools for the Army as did many of the larger flight schools that were part of the Civilian Pilot Training Program which had been launched in 1938. While the instructors and facilities were civilian during primary, there was a military cadre at each facility; the Army personnel were responsible for military indoctrination and evaluation of cadet progress in the air. The washout rate was high. From 1 July 1939 to 31 August 1945, the Army Air Force graduated 193,440 pilots through its aviation cadet program. During that period, over 124,000 cadets washed out of pilot training, especially during the primary phase, for an overall failure rate of approximately forty percent.

The aviation cadet program evolved during the war, but at the beginning a cadet logged nearly seventy hours in ten weeks of primary training. That phase was designed to weed out those who were really not suited to become pilots and provided the successful candidate with a fundamental capability to fly a simple airplane, including elementary aerobatics. If lucky, the cadet might get a couple of hours of cross country flying in before moving to basic training at an installation where the staff was strictly Army Air Force. In basic, a cadet moved up to a more sophisticated, modern airplane with much more horsepower in which he perfected his flying skills by learning formation flying, simple instrument flight, night flying, and longer cross country navigation. And he added another seventy hours to his logbook over the ten weeks of basic. Advanced training, also taking ten weeks and bringing the logbook total up to roughly 210 hours, took place at another airfield where cadets learned to fly the AT-6 advanced trainer with a constant speed propeller, retractable landing gear, and a 600 hp engine. As the war progressed, pilots identified as likely bomber or transport candidates near the end of basic went to an advanced flying school where they learned to fly simple twin-engine trainers.

Later in the war, once advanced training was completed, the most promising pilots went into transition training which was five weeks for fighter pilots and ten weeks for bomber pilots. Early in the war, the best pilots frequently became instructors while many graduating cadets with average skills became co-pilots in bombers or transport aircraft. The new bomber first pilot was also sent to an operational training unit before joining a bomb group or a replacement crew center. As the war progressed, training was streamlined to the extent that a cadet might fly the old bi-plane PT-17 Stearman trainer in primary, the AT-6 in basic, the B-25 Mitchell in advanced, and come out of the program with fewer hours logged than the cadets of the early war years. Because the "needs of the service" were paramount, an enthusiastic young graduate at a single-engine advanced school, such as Luke Field in Arizona, for example, might be crushed when he expected a posting to some fighter transition school only to learn that his new orders were to Salt Lake City where he would join a newly formed heavy bomb squadron as a co-pilot. Such were the manpower needs of the service at the moment of graduation.

By contrast, the pilot training run-up to Word War I was poorly organized – and with good reason. Nobody really knew the value of military aviation or even how to employ it properly. The Army's first pilots had been trained by America's very first pilots: the Wright brothers and other pioneers, including Glen L. Curtiss. By the outbreak of World War I, Army aviation was still just a section in the Army Signal Corps and years away from having the same stature as combat branches such as infantry or artillery. In those days, a pilot was considered qualified if he had completed a five-minute flight during which a fifteen mph breeze (or better) was blowing, reached an altitude of 2,500 feet, taken a passenger aloft to a height of 500 feet, and landed within 150 feet of a designated point. The successful candidate also had to complete a reconnaissance mission flown at 1,500 feet of altitude to an objective at least twenty miles away. That was all in addition to satisfactorily demonstrating a dead stick landing. During this pre-war period, training bases (usually of short duration) were located in College Park, Maryland, Augusta, Georgia, Houston, and two locations that Blackie would come to know very well: San Antonio and San Diego.

In the summer of 1914, the Aviation Section of the Signal Corps had been formally established, but it was pitifully small with a budget to support only sixty officers and 260 enlisted men. And that was at a time when European governments were spending heavily to develop their air arms. The first practical application of this new capability came in 1916 during the Mexican Punitive Expedition when the 1st Aero Squadron took its Curtiss JN-3 biplanes to Columbus, New Mexico, where they were used for communications and photography, but not bombing or strafing troops of the Mexican rebel, Pancho Villa. At about the same time, training of military pilots was progressing from the use of Wright Brothers-type pusher aircraft to the Curtiss JN-4 "Jenny," an updated version of the JN-3. The training program was evolving into the three stages (primary, basic, and advanced) through which Blackie was about to progress.

The American entry into World War I resulted in crash programs that produced over 11,000 pilots, bombardiers, and observers by the time the war ended in November 1918. Rather than reinvent the wheel, the Army essentially copied the Canadian system that used university facilities for what amounted to two months of ground school which was followed by another two months of flight training in a primary school which put fifty hours in a candidate's logbook. Successful cadets got their commissions and wings after fifty flying hours which was followed by a one-month course specializing in pursuit, observation, or bombardment. Then, with a total of ninety hours, young lieutenants were sent to the front.

All cadets could ask for what specialization they hoped for after graduation, but it was the commanding officer who made final decisions. Thus, the youngest, most aggressive pilots typically went to pursuit squadrons while older pilots noted for attention to detail and possessing superior navigation skills and determination went to bombers. Blackie, fitting the latter to a tee, was destined for the heavy brigades long before he was even asked for his own opinion on future assignments.

Nearly two dozen fields were used to train pilots for WW I, but those basking in the sunbelt's fine flying weather dominated, especially Kelly Field in San Antonio. As it turned out, the facilities for advanced training proved inadequate, resulting in many American pilots receiving their advanced training in French schools. Following the Armistice in November 1918, the Air Service strength plummeted from nearly 140,000 officers and men to just under 10,000 and remained at that level until shortly before Blackie reported to March Field as a cadet. By the mid-1920s, there was growing appreciation for the role bombardment aviation could play in the country's defense as had certainly been demonstrated by General Billy Mitchell's successful bombing of captured German battleships and obsolescent American battleships in 1921 and 1923.

Mitchell's friends in Congress, formed as the Lambert Committee, had very ambitious plans which included the creation of a separate air arm similar to the Royal Air Force, i.e., free of any attachments to the Army. All three branches would be controlled by a department of defense. A separate Congressional committee headed by Representative Dwight D. Morrow recommended an upgrade for the Army Air Service, but it fell far short of the Lambert recommendations. The Morrow committee did recommend a change of name for Army aviation. It became the "Air Corps" which was to benefit from a five-year plan including an inventory of 1,800 aircraft and nearly 17,000 officers and men along with the construction of a comprehensive, new training facility that would become Randolph Field. The grand plan was never realized because the necessary funding was never appropriated. Congressman Morrow had a little explaining to do in several years concerning his more conservative approach to Air Corps growth. His future son-in-law, Charles A. Lindbergh, had recently completed training at Kelly Field as a fighter pilot when the Air Corps Act was enacted on 2 July 1926.

Nevertheless, the Air Corps Act did shape the cadet training Blackie would experience. The Air Corps now had an Air Corps Training Center, located near San Antonio and a syllabus that would still be largely in effect when Blackie reported to March Field. Primary and Basic were given at March Field and Brooks Field in Texas. In the four-month long primary phase, a cadet received enough instruction in a simple biplane to make him the master of that aircraft, i.e., elementary aerobatics along with takeoffs and landings in crosswinds. Armed with the fundamentals, the cadet moved on to four months of basic flying in a heavier plane where cross country, night flying, formation flying, and instrument flight were emphasized. Classroom instruction in primary focused on key aeronautical subjects such as meteorology, navigation, engines, airframes, and radios in addition to military indoctrination.

At advanced flying school, cadets logged about 175 hours in a curriculum which was much more oriented toward warfare, i.e., squadron duties, aerial photography, bombardment, etc. Prior to Blackie's arrival at advanced flying school, there was considerable debate over just what kind of pilot should come out of advanced training: a specialist whose training was strictly in one of the specialties (attack, bombardment, observation, or pursuit) or a generalist who had some experience in all types of military aircraft. As it turned out, Blackie's experience would be a compromise in which he was introduced to all types but went on to specialize during the bulk of his advanced training. In truth of matter, even after gaining their wings and commissions, young Army flyers were still considered apprentices getting on-the-job training during their first assignments after graduation.

The 128 cadets entering Blackie's March '30 Class came from every walk of life across America. Although two years of college were required for entry in the program, many had graduated from such prestigious schools as Cornell, Stanford, the University of Michigan, and Johns Hopkins. Others arrived from engineering schools with Georgia Tech, Cal Tech, and Rensselaer among them. In addition to Blackie, there were other Aggies, including several from Oklahoma and Texas. Many, like Blackie, listed a social fraternity on their applications. Here and there were members of Scabbard and Blade, the ROTC honor society. Most cadets came from state universities and colleges with the exception of few men who were more than first among equals: West Point graduates who were Regular Army officers. Being a West Pointer, however, brought no advantages to the cockpit; the West Pointers washed out of the program at exactly the same rate, if not slightly higher, as the men from junior colleges, four-year schools, universities, and technical schools.

Blackie brought to his cadet days several personal characteristics which were both advantages and disadvantages. He already knew how to fly, had a private pilot's license, and had 101 hours in his logbook which meant that it was unlikely that he would wash out for lack of inherent flying ability. However, he did not know how to fly *the Army way.* Failure to adapt was occasionally the basis for failure. Previously, Blackie was a "take charge" kind of guy, but now he was just about the lowest form of life in the Air Service, a "dodo" (so called after the flightless, awkward, and extinct bird native to the island of Mauritius in the Indian Ocean). The sudden transition from civilian to military cadet held few surprises for him because he had had two years of ROTC and several years in the Nebraska and California National Guards. Blackie was one of the older if not the oldest man on the course, thus he had the maturity that helped him avoid sophomoric stunts that frequently washed out a youthful pilot. The other side of that age sword was the fact that he would be denied opportunities that went to younger men in the future.

Many of the men in Blackie's cadet class had bachelor's degrees and one or two had actually earned master's degrees. Blackie, not possessing a college degree, was very conscious of the fact that any of his superior officers would always favor the college grad when it came down to a choice between two cadets. For that reason, he drove himself relentlessly in every aspect of cadet life – in the air and everywhere on the ground from the rifle range to the classroom.

Upon arrival at March on 27 February 1930, the 128 men in the class were divided into four flights of thirty-two men each with one flight instructor assigned to every six cadets. Of the 128, nine were already officers – members of the regular Army who had graduated from West Point.

Blackie's two closest friends during his cadet days were Kenneth Hadley Greeson ("Kenny"), a solidly built, enthusiastic young man from Marysville, Missouri, and the lanky Clarence A. Shoop ("Shoopy"), whose home was Clarendon in northwestern Pennsylvania. Greeson had just completed the first semester of his junior year at what is now Northwest Missouri State University where he had been the head cheerleader, a member of the glee club, and a member of Sigma Tau Gamma fraternity. Shoop, educated at Denison University in Ohio, was a skilled outdoorsman who enjoyed hunting and fishing; like Blackie, he also was a talented piano player. Aerial fate, so cruel to some while generous with others, would take the life of one of Blackie's new friends in just a few short years. The other cadet buddy would open the doors to some of Blackie's greatest wartime adventures and later climb very high in both military and aerospace industry circles. As it turned, strictly by chance, Kenny and Blackie were assigned to the same cadet flight.

In the meantime, all three friends were constantly focused on not being eliminated from the program. Cadet upperclassmen constantly hazed the newcomers and instilled fear of "washing out" or doing a "180." That fear was pervasive. One of the class humorists even wrote about it and the check pilot who gave struggling cadets their wash ride.

"That Old Gang of Mine"

A-ridin' the check plane,
That's a pretty certain sign
Those check-out rides are breaking up
That old gang of mine.
There goes Jack, there goes Jim
Down to meet the board
Hoping vainly there's a chance
But they're headed down the road.
I get a funny feeling
When I see that washday line.
Those check-out rides are breaking up
That old gang of mine.

Getting fewer in the mess hall
Getting fewer on the line,
Those check-out rides are breaking up
That old gang of mine.
There goes Pete, there goes Bill
Down to Riverside.
Once away, gone to stay,
They've taken their last ride.
I get a funny feelin'
When I see that wash-day line.
Those check-out rides are breaking up
That old gang of mine.

When arriving at the main gate, Blackie found himself in the company of many other young men all of whose worlds were about to change drastically. Harold Frederic Blackburn had been comfortable with "Hal" or "Blackie," but now he was going to become "Mister Blackburn" and would not be accorded the respect that he was used to as a band leader. In fact, he had to ask permission to ask a question and Heaven help him if he did not precede and follow a question or any other utterance with the word "sir." In short order, Blackie and the other misters were marched to the administration building for a very brief interview with Captain James M. Bevans, Commandant of Cadets who later retired as a Major General after many years of service in training and personnel assignments. The interview was short and sweet; Blackie was told which flight he would join along with his barracks assignment. The first few days were spent desperately trying to satisfy upperclassmen demands and drawing clothing and equipment not to mention drilling while not double-timing it between barracks, classrooms, base supply, and mess hall.

After arrival at March, there were inoculations and more examinations, both physical and mental, through which several cadets were eliminated from the program. One of the more interesting tests involved a machine called the Ruggles Orientator, although a better name for the contrivance would have been the Ruggles Disorienatator. At first glance the device looked as though it might have come from a medieval torture chamber, but in reality was a pilot's seat attached to a gimbaled frame in which it was possible to simulate the many unusual attitudes in which a cadet might find himself as with a spinning or tumbling aircraft. The purpose was to quickly identify those cadets who suffered from disorientation or such motion-induced weaknesses as vertigo. Failure to qualify on the orientator was grounds for washing out. Those who successfully dealt with the machine noted that it was occasionally a paying proposition – that is when cadets forgot to remove loose change from their pockets before being strapped in.

The first six weeks, particularly the first four, were the toughest, but by the end of that period fifty-one cadets had soloed. On the other hand, twenty-two had washed out by 1 April and that was just the beginning. Being an early soloist was no guarantee of getting one's wings either; of the-fifty one who had soloed by 6 April 1930, only twenty-four went on to get their wings at Kelly Field on 27 February 1931. Blackie's day began early – very early at 5:45 A.M. which was soon followed by running to the mess hall for breakfast at a table for ten, including two upperclassmen who made sure that cadets in their charge ate a square meal, i.e., sitting and eating at attention. Ground school ran for most of the afternoon and was followed by one hour of vigorous, outdoor exercise. After supper Blackie studied ground school subjects until the 10 P.M. bed check. Blackie's written exams were always given on a Saturday morning, and the authorities, recognizing that all work and no play made Cadet Jack a dull boy, built some leisure time into the program. Thus once his dodo stage was over, Blackie had some off duty time on the weekends, but always had to be back for the 10 P.M. bed check.

Inspections, of which there were many, began immediately: personal inspections, barracks inspections, shoe inspections, locker inspections, bunk inspections, full dress uniform inspections on weekends, and rifle inspections. Infractions resulted in being "gigged," and a sufficient number of accumulated gigs (demerits) could result in dismissal. Usually a gig resulted in immediate punishment in the form of more running, marching for an hour in a punishment tour, rapid physical exercise, or policing the barracks and grounds. Gigging was invariably administered by an upperclassman then into the second half of his service at March, i.e., a cadet going through the basic flight training stage. Giggable offenses were legion: lateness, stain on clothing, a sloppy salute, shoes and buttons not polished to a reflective luster, messy locker or trunk, dirty rifle, and the list of possibilities went on and on. Failure to properly answer a question about the time of day without looking at a watch would result in a walking tour (one hour marching in the barracks area). Lord help the poor cadet who had a humorous response ready when asked about the first name of Lieutenant March for whom the field was named. "Sir, his first name was 'Forward,' sir!"

A few cadets decided that the Mickey Mouse hazing wasn't worth it; so they dropped out at "at own request," which was unfortunate because the worst of the hazing only lasted about 30 days – "Hell Month." Nor did they realize that building discipline was the real purpose behind the hyper-Freshman days hazing. For others, like Blackie, the desire for silver wings burned so brightly that no upperclassman, no matter how petty, obnoxious, or miserable, would ever get in the way. Still, the occasional upperclassman could carry matters to extremes. In one instance, Blackie and several other cadets were ordered to cut grass in front of their barracks. Unfortunately, no lawnmowers, hedge trimmers, shears, or even scissors could be located quickly enough to suit the upperclassman, who supervised several cadets on hands and knees breaking off little bunches of grass, one blade at a time.

Physical training, one hour of it daily, was part of the program and was augmented by what amounted to intramural athletics, including baseball, basketball, swimming, tennis, and volleyball. Most of the athletic contests were played between the upper and lower cadet classes, but those intramural games also included opponents from the regular squadrons based at March. Blackie did not play on any intramural baseball team as he had for his fraternity at Colorado A&M or football as he did as a National Guardsman, but did resume recreational tennis and swimming again. Much to his surprise, he did not qualify for the cadet tennis or swimming teams primarily because most cadets selected for those teams were very recent college jocks whereas Blackie had been away from competitive sports for well over six years. The cadet teams faced very stiff competition from March Field's resident fighter squadrons whose pilots were young, very aggressive, and extremely fit athletes. On the other hand, athletic talent on the baseball diamond was not necessarily a good indicator of future success as an aviator. Of the eleven men qualifying for the cadet baseball team, only two eventually won their wings, and both were good friends of Blackie's: Elmer "Bud" Gorman and Harvey Berger.

By the same token, being elected as a class officer was not always a good indicator of superior flying talent. The popular Bud Gorman eventually became the cadet captain and another close friend (and swimming star) Albert "Pinky" Stainback became the class historian, but of the other five cadet officers, three washed out before the class moved on to advanced training at Kelly Field. The training officers were more clairvoyant when they appointed the cadet leadership for positions of cadet captain, lieutenants, adjutant, and first sergeant. They only erred in one choice; the cadet captain William Nichols was eliminated during basic flight training and was replaced by Bud Gorman.

Unless foul weather intervened, flying was conducted in the morning; classroom instruction dominated the afternoons. In total, Blackie received 308 hours of classroom instruction, beginning with radio code, theory of flight, and air traffic rules. One subject was held outdoors: armament. That was a subject which would continue at Kelly Field, but at March Blackie's practical experience was with the Lewis machine gun, a weapon that had been used by the Air Service in World War I but would soon be dropped in favor of the Browning machine gun. To make machine gun firing more realistic, the guns were mounted so they could be rotated, creating the illusion of movement. The finer points of leading a moving target were practiced on the shotgun range.

There were two subjects in the primary stage that were not graded: only a pass or fail was issued and the latter case was a ticket to the base gate, suitcase in hand. Thus, physical training at March was an easy pass for Blackie and nearly all other cadets. The other non-numerically graded subject was radio code. For a quick-thinking master musician like Blackie, the dots and dashes of the Morse code made for a nice variation on previous musical themes. During the first few weeks, cadets practiced with the dots and dashes for as much as an hour a day.

Parachute orientation came early in a cadet's training, and it was a morbidly vexing subject for Blackie because he was, indeed, acrophobic, a condition that is actually fairly common among pilots. That seeming paradox is explained that while just flying along on a bluebird day at thousands of feet above earth or even flying aerobatics, the affected pilot fears not because he is in *control of the situation*. Yet the same pilot becomes queasy looking over a restraining rail, at, let us say, the uppermost level of the Empire State Building – or even the edge of a subway platform. In Blackie's case, the first manifestation of acrophobia occurred when he climbed the Mitchell Township water tower at age twelve. Having reached the highest point, his grinding fear prevented his attempting a descent. Darkness approached. Figuring that a shaking return to *terra firma* was better than remaining at altitude in the gathering cold and darkness, Blackie gripped the ladder for dear life and returned, trembling, to Mother Earth. The fear never left him decades and thousands of flying hours later – he maintained throughout his flying days that given a choice of bailing out or dead sticking it in, he would choose the latter every time.

Parachute training at March took place in front of a hangar wall exhibit with a master sergeant instructor explaining all of a parachute's components which were separately displayed: pilot chute and its bridle lines at the top, pack shell, folded canopy, shroud lines, and rip cord in the middle, and at bottom the first aid packet. Cadets were taught packing a chute, how to bail out without hitting the plane's tail, when to pull the ripcord, and how to go into a crouch and roll on landing. That part of parachute instruction was mandatory. What was not required was an actual jump; that was voluntary. Blackie, who was determined to be at the top of his class, volunteered to be among the first jumpers; he grittily overcame his fear for the occasion. To minimize the number of flights required for the voluntary jumps, two cadets were strapped aboard the top wings of a DH-4 outside of the ailerons' span while one cadet was seated in the fuselage aft seat. They were flown to altitude where the cadet behind the pilot went over the side followed at safe intervals by those on the wings who were signaled when to release their bindings. It was Blackie's first, last, and only parachute jump.

After the first month, most cadets used their free time on weekends to unwind in Los Angeles where there were many attractions. Cadets were paid $75 a month, but Blackie was not into living the high life in L.A. with that income; instead he was contributing to the support of his mother (and would for the balance of her life). For those not attracted by Tinseltown or L.A.'s bright lights, there were recreation halls at March Field. As might be expected, some cadets brought musical instruments to supplement those already on the base, and it did not take Blackie long to find out who the other musicians were and organize them into a small combo which played some of the tunes from Blackie's professional band days. The early band had seven members and certainly taxed Blackie as an arranger – Blackie played the sax and had no trouble accommodating trombone, drums, clarinet, and Clarence Shoop on the piano, but the remaining two instruments certainly were an odd couple – a tuba and a guitar. Nevertheless, Blackie's (Latest) Bunch proved popular during cadets' off-duty hours. By the time he left for advanced training in Texas, Blackie had organized a small orchestra that played at dances and other March Field activities.

Flying started by mid-March and was initiated by Blackie's first meeting his flight instructor along with the other five students in his section. It was just a hangar session beside one of the Consolidated PT-3 biplanes on which the cadets would start actual flying the next day. What followed was essentially preflighting the PT-3 "Trusty" (so named quite simply because it was trustworthy, especially for spin recovery). The instructor cautioned Blackie and his mates to do exactly as told in the air via the Gosport Tube which was basically a flexible pipe which ran from the instructor's facemask to the student's earpads. It was a

British invention that was the standard on primary trainers. Blackie was admonished to maintain a certain RPM and to constantly monitor the oil pressure and temperature, which was not that difficult considering how limited the PT-3 instrument panel was. Bolted onto the front of the PT-3 was one of the most popular and reliable engines of the 1930s, the nine-cylinder, 200 hp Wright J-5. The PT-3 and PT-3A Trusties that Blackie started on cruised at 80 mph and had a wingspan of 34 feet and a length of 28 feet. To put that into modern perspective, today's best-selling general aviation airplane, the comparably dimensioned Cessna 172 (a high-winged monoplane which your grandmother could easily fly), cruises at double the speed with only three-quarters of the horsepower while carrying twice the passengers.

Once the morning formation and breakfast were over, cadets were run, flown, or taken in a van to their respective stage houses to commence their first lessons. The stage house was essentially an open-sided, wooden line shack with a wind sock on top; it was the small scale equivalent of a breezy, squadron ready room. To relieve congestion and reduce the possibility of mid-air collisions, much of the flying was done at various satellite fields, each of which served just two groups of cadets with their respective instructors. To maintain some order, each cadet group took one half of the auxiliary field and the space above it strictly as its own. As there was only one instructor and one Trusty for every five or six students at first, the cadets not flying had to wait their turn, read, study, push planes into position, crank props, watch those who were flying, or tell jokes. In one instance, the Chinese Air Force became the butt of a joke based on its maneuvers: "One Wing Lo," "Tail Hi," and "Washee Out." Little did Blackie or any of the other cadets realize just then that their class would later be joined by an officer of the Chinese Air Force and that the Oriental gentleman would play a major role in the lives of several members of Blackie's Kelly Field graduating class.

The auxiliary fields offered one intermittent luxury that was even less available back at March Field: the bounty of knowing that a good landing was imminent by listening to the landing gear hitting flower and weed tops just moments before the actual landing. March and the auxiliary fields were essentially just that – unpaved fields. On 17 August, the few flowers, weeds, and grasses at March were pretty well dried out when an exhaust spark ignited the remaining grass and oil-soaked earth. It threatened to become a real barn burner, or rather hangar burner, since the wind was from across field. The conflagration was well beyond the capability of the March Field fire engines to put it out, and Blackie was able to watch a unique solution to the problem. Some quick thinking instructors saved the day by taxiing PT-3 trainers in front of the fire, turning ninety degrees to the right, and revving up with the brakes on. Very soon their wind forced the fire to retreat and burn itself out.

Soon it was Blackie's turn to climb aboard the PT-3 along with his instructor who knew that Blackie already had a private pilot's license. Under most circumstances, the cadet put his hands and feet very lightly on the controls and just followed along as the instructor took off, climbed out, and went through gentle turns. At the completion of the turns, the instructor released the controls and slapped his hands together over his head which was the signal for the student to get a feel for the controls. The usual mistake was gripping the control stick much too tightly, forcing it about, and pushing too hard on the rudder pedals. Rather than waste time reinventing the wheel for Blackie, the instructor left the practice area after only a few minutes and instructed Blackie to return to the field and shoot three landings. The first was workmanlike because nearly all of Blackie's time was in the Spartan C-3, a heavier biplane with more performance in spite of less horsepower. At the conclusion of his third, satisfactory landing, Blackie was sent aloft solo – an event that was announced in the mess hall at lunch along with the names of several other soloing cadets who had previously earned pilot licenses. His solo was also preceded by the instructor tying a long, white cotton ribbon onto the tail of Blackie's PT-3, warning the observing world that a fledgling dodo was about to take to his wings for the first time. Blackie's instructor, Lieutenant Laur, did not regard his student as a natural born, stick and rudder man. He gave some advice that remained with Blackie forever. "Relax, don't be so tense." After that, Blackie's flying became much more natural, and his handling of the controls became easier, more fluid-like. Some of the tension was released just by going solo and one of its immediate distinctions; after the solo, a cadet could wear his goggles strapped to his flying helmet. Before that goggles had to be worn around the cadet's neck.

Cadets not soloing after ten hours of dual instruction were headed for a check ride with the stage commander. Very few cadets passed that ride, and the court of last resort was a final check ride with the chief flying instructor. Only rarely did a cadet get through that trial successfully. Those getting a final down check appeared before an officers' review board which evaluated the cadet's lack of progress and then invariably gave the crestfallen cadet an honorable discharge and ordered him to pack and leave immediately.

As might be expected, flying accidents attended the training of neophyte pilots, but most of those misadventures were simple, typically involving ground loops and taxiing accidents. But within several weeks, Blackie witnessed his first serious flying accident. Cadet Eugene F. Springer was involved in a landing accident on his solo flight, proving the old maxim that "a thousand takeoffs means a thousand landings – *in one way or another*." Springer had taken off on his solo flight and misjudged his landing to the extent that he completely removed the plane's landing gear. Realizing that one bad, very hard landing was enough, he took the PT-3 around and bailed out after reaching 1,500 feet. The Air Corps had decreed that a bailout was justified only upon one of three grounds: fire, complete loss of control, or structural failure. Thus, Springer was unsurprisingly and subsequently washed out.

On 26 May, there was another flying accident, a fatal one this time. The victim was Lieutenant Harland DeKaye, an upperclassman and West Point graduate. The twenty-six year old from Kenosha, Wisconsin, was six miles east of March attempting a forced landing at an auxiliary field at Dry Lake when his plane stalled out of a steep turn at low speed and spun in. The accidents would continue during advanced training and even onto active duty. Blackie, however, never joined the circle of those who were the subject of accident investigation boards – in his subsequent career of over 26,000 accident-free flying hours, he never so much as put a dent in a wingtip.

DeKaye's death was the first cadet fatality while Blackie was March Field. The second cadet fatality involved Malcolm M. Coleman, a twenty-two year old from Virginia who was killed in an automobile accident on 13 July. Coleman and seven other off-duty cadets were driving toward Victorville in two cars to attend a water carnival near Trona. Coleman, a passenger in Cadet Edgar Danielson's car was thrown clear in a collision with another car whose driver fell asleep at the wheel and crossed into Danielson's Lane. Police exonerated Danielson who, along with the others passengers in his car, Cadets Lewis Hunter and Walter Pharr, suffered only minor bruises and lacerations. Luck continued to ride with Danielson, Pharr, and Hunter – they got their wings at Kelly Field seven months later. DeKaye's and Coleman's deaths were the first of the many deaths that Blackie would eventually become hardened to in his long flying career.

As a result of his previous experience, Blackie logged a higher than average number of solo hours during his first two weeks of flight instruction. In the next stage, he would be logging thirty minutes of dual followed by thirty solo minutes as the instruction moved beyond the simple stalls, eights on pylons, chandelles, spins, slips to a landing, and crosswind landings with which Blackie was already quite familiar. Now he was moving into unfamiliar territory, including loops, barrel rolls, Immelman turns, strap-straining slow rolls, and precision landings to a pre-determined point on the field. The precision landings were power off involving 180 and 360 degree turns from 500 feet; the latter were easier to accomplish, but the former required much more finesse at the finish and resulted in occasional hard landings and ground loops. By the end of four months, Blackie's airwork included snap rolls and many simulated emergency landings.

Although Blackie's primary flight experience was designed to provide him with just the essentials, the primary stage included a hint of things to come. He was checked out in the two aircraft that he would be flying in basic: the DH-4 (a light bomber used in World War I) and the Douglas O-32, an early biplane in what would become the first in a long line of Douglas observation ships. Blackie also was given several hours of dual time in what would be emphasized during advanced training: night and formation flying.

In July, Blackie became an upperclassman with over seventy hours in his logbook after passing a stage check with the flight commander who threw everything he could at Blackie. Those curves included such challenges as cutting the engine at delicate moments, i.e., the middle of a pylon turn at 400 feet. By that point, Blackie was wary enough to constantly keep an eye out for an emergency field and was therefore never caught flatfooted unlike some of his acquaintances. Of the 128 cadets and commissioned officers who joined Blackie in the March '30 class primary flying class, only forty-eight went on to basic flying school which would claim another fifteen cadets but also add another 110 hours to Blackie's flying time.

On 16 July 1930, Blackie was aloft in a PT-3 Trusty, completely oblivious to a major development in the airline business. On the other side of the country, Postmaster General Walter Brown, had been interested in streamlining the airmail business and looked to a potential merger as one way of establishing coast-to-coast airmail service. The merger candidates were Western Air Express, Transcontinental Air Transport, and Pittsburgh Aviation Industries Corporation (PAIC). Stockholders at TAT and WAE each became 47.5 per cent owners and the shareholders of PAIC took the remaining shares. The new entity was incorporated on that day as Transcontinental and Western Air Inc. or T&WA which for the sake of continuity in the balance of this book will simply be abbreviated to "TWA."

To celebrate becoming an upperclassman and to applaud the outgoing upperclassmen graduating to advanced training at Kelly Field, a gala ball was held for both groups complete with formal uniforms, a hired dance orchestra, and imported dates. No longer a dodo, Blackie helped greet the incoming batch of dodos, but was now much more pre-occupied with flying advanced aircraft types and the introduction to formation flying, night flying, and instrument flight. His uniform changed slightly, although the change had great meaning symbolically. He traded in his red name badge for a blue one which marked his new status as an upperclassman.

Armament was a major subject in Primary and consisted of instruction in the Lewis and Browning machine guns and in trap and pistol shooting, and sights and gears. Blackie, the old duck hunter, excelled in this phase of training where cadets learned their tail gunners' jobs by firing at moving targets on the ground from the standard open rear turret then in use on bombing and attack aircraft. To develop a sense of leading a target in flight, hours were spent on the shotgun range using Browning semi-automatic shotguns. In the 1930s, aviation cadets were also instructed in the use of the Army's principal weapons ranging from mortars to field artillery. The most accurate cadets were awarded the Army's marksmanship badges based on their scores. Those awards were in the shape of a Maltese cross with suspended bars representing the type of weapon for which the wearer had qualified.

Blackie requalified again after his National Guard service where he was rated as rifle marksman. This time he qualified higher, as a sharpshooter, with what was then the standard infantry weapon, the 1903 Springfield 30:06 rifle. He also qualified as a sharpshooter with the .30 caliber air-cooled machine gun and the .45 caliber Automatic Colt Pistol, but there was a droll touch in that latter distinction, reflecting a lingering influence of the cavalry. Blackie's qualification was with "pistol, dismounted." One can only wonder, if generations later, the Army considered, if even briefly, reinstating the old "pistol, mounted" category for its helicopter flight crews. Nevertheless, armament was a good subject for Blackie who earned a score of ninety-eight based on forty-eight hours of instruction in the subject.

Blackburn Family Collection

Aviation Cadet Harold Blackburn shortly after arriving at Kelly Field in late 1931.

Basic flight still involved the routine of flying in the morning, ground school in the afternoon, and physical training in the late afternoon. By the time Blackie reached the basic stage, he was required to plan a cross country flight based on hours of ground school navigation exercises, i.e., to derive a compass heading based on variables including the local magnetic variation, the magnetic quirks specific to the particular plane he was flying, and the predicted wind. He was expected to compute his ground speed, and therefore his estimated time over all waypoints and the final destination. These were very thorough flight plans and had not been preceded by any long range dual flight instruction. A typical cross country would involve Blackie's landing at a destination roughly 200 miles away, but the attending observations and paperwork were extensive. For example, his flight from March to Santa Maria, California, involved twenty-two checkpoints for which he had to note the actual time over the waypoint and contrast that with his pre-computed time over each one. Blackie computed his flight to Santa Maria (north of Los Angeles, along the coast) as taking one hour and fifty-five minutes. His actual time was one minute less, and his worst checkpoint time calculation was only two minutes off; most were right on the money. Of Blackie's 111 solo hours in basic, sixteen were devoted to cross country work, nearly all of which was flown in the Douglas O-32.

As his eight months at March progressed, the curriculum became more focused and Blackie's grades, already in the 90s, soared. His score in maps was 99, meteorology 94, military customs 99, and personal equipment of pilot 98. Only in engineering (airplane engines and maintenance) did he falter and score less than a 90 which is puzzling considering the fact that he had been both an automotive and aircraft mechanic. Regardless, his ground school test scores and stage flight checks resulted in his being the top man in the March '30 class.

To celebrate their upcoming graduation from basic flight training, the outgoing upperclassmen had a celebratory dance at the Redlands Country Club on Friday 3 October. Having been something of an impresario in his previous life, Blackie, along with Arthur Strunk, Lewis Hunter, and Pinky Stainback, orchestrated their arrival with dates in a horse-drawn carriage and managed to have their entry to the country club's front door announced by trumpets. The conveyance proved to be so popular that most of the party-goers got a ride in the carriage before the evening was over. Fortunately for the cadets, considering how fine the food and drink were, the regular Saturday A.M. inspection was cancelled.

On 10 October 1930, twenty-eight cadet members of Blackie's original cadet class and the two successful West Pointers completed basic training and received orders to report at Kelly Field, San Antonio, Texas, for advanced training along with thirty-two men arriving from Brooks Field. Thus the washout rate was seventy-seven percent (eighty-four percent among the nine West Pointers in the class). Blackie's Class of March '30, with a graduation rate of twenty-three percent seems to have been somewhat ill-starred because its success rate was by far the lowest among the twelve classes that graduated while March was a primary training field from November 1927 through February 1932. During that period the overall March graduation rate was thirty-nine percent. Randolph Field, which became the sole primary training center in mid-1931, enjoyed a substantial increase in cadet graduation rates, averaging fifty-six percent for its first year of operation. During this period, the Army Air Corps trained a handful of cadets from foreign air forces – particularly those from Latin America, although there were a few from Spain, Germany, Siam, and China. The foreign graduation rate was sixty percent which may have been boosted by the need for diplomacy. Getting as far as advanced eliminated, for the most part, any fears about washing out. Of the sixty-two men joining the cadet detachment at Kelly with Blackie, only two were washed out of the four-month long program.

It was while Blackie was in transit between March and Kelly that TWA completed its inaugural coast-to-coast flight in late October 1930, including a stop at a city that would become very important to the airline and Blackie personally: Kansas City, Missouri.

Upon arrival in San Antonio via train from L.A. along with the entire March contingent, Blackie was assigned to the bomber section and began flying the Keystone LB-5 twin-engine bomber on 29 October after two days of drawing equipment and settling in. The checkout was casual to say the least. The machine had a throw-over yoke in the cockpit, and Blackie's instructor took the bomber around the pattern for a couple of landings, handed the yoke over to Blackie, and told him to do the next three landings. Neither Blackie nor the other five cadets in his group had any problems with the lumbering LB-5 which seemed to drop like a steel safe compared to the feather-light old PT-3 Trusty. There was one similarity that the Keystone shared with the little Trusty, one which could be tolerated comfortably only during summer; both sported open cockpits protected by only a modest windscreen.

Although Blackie would get his wings as a bomber pilot, the next three months would give him some exposure to attack, observation, and pursuit flying. But for the time being, he was flying with an instructor and logged twenty-four flights in the LB-5 during the last three days of the month, primarily in simple airwork and precision landings. Blackie was one of just twelve men assigned to the bomber flight; the observation and attack sections were only slightly larger at thirteen men per group. Twenty-five men went to the pursuit track which is where all of the glory was. All of the cadets had grown up idolizing the gallant pilots of the French-American *Escadrille Lafayette* and Captain Eddie Rickenbacker of the American Air Service with his fearless band of Ninety-fourth Squadron fighter pilots. Those heroes battled the "Bloody Red Baron" who led his Fokker Flying Circus over the western front, but it was only the pursuit trainees who were soon to be living the dream. On the other hand, the bomber students faced much better job prospects with the airlines – someday, that is, when the airlines might be hiring if the Depression ever receded.

A popular poem of the day caught the bomber pilot's fate:

"Flaming Coffins"

Who stands up in the propeller's race,
Who gets oil spray in the face,
Who never gets taken for an ace?
The Bomber.

Who has to stand on his two cold feet,
Who is the Fokker's easy meat,
Who catches hell from the strafing fleet?
The Bomber.

Who, when the flight's dead on the run,
Sees the Huns dive on him, specks in the sun.
Who staves them off with his Lewis gun?
The Bomber.

Who is the man who stops their lead,
Who is the man who comes home dead,
Who is the one never dies in bed?
The Bomber.

The major reason why there were so many more pilots headed for fighter training in Blackie's class was quite simply the Army Air Corps' policy of structuring itself primarily as a defensive force. America had no plans for foreign aggression, hence the emphasis on defending against foreign bombers approaching America's shores or its critical interests elsewhere such as the Panama Canal. As Blackie graduated from the aviation cadet program, the Air Corps was composed of only fourteen groups, typically with three squadrons assigned to a group. One of the groups was composed of dirigible squadrons, all based in Illinois, and three were composite groups of mixed squadrons all of which were based outside of the continental United States (Hawaii, the Philippines, and Panama). The two observation groups were based in Texas and on Long Island, New York. There were only two bombardment groups with one headquartered at March Field and the other at Langley Field in Virginia. The rest, five groups out of the fourteen in the Army, were pursuit organizations.

Although many pilots were jealous of those selected for pursuit training the attack pilots felt less envious because their flying was quite simply the most fun. They flew down in the weeds, just feet above the terrain, and were the fighter-bomber pilots of their day. Following the contours of the ground, skimming over hilltops, and snaking through river beds – all in formation, gave the greatest sensation of speed by far, but it was also the most dangerous form of flying where the potential for accidents was very high indeed.

The next five weeks were spent exclusively with the LB-5 on which Blackie was flying nearly every weekday. Only thirty-five LB-5s and LB-5As were ever built and the performance was modest, hence its relegation to training duties. Like most bombers of its day, it was a bi-plane with a pair of Liberty engines mounted on the lower wing. It could carry only a minimal load: five crewmembers and barely a ton of bombs. Blackie shot many landings on the bomber – 174 in total spread over nearly thirty-six hours in the left seat. As typical of the era, the planes Blackie flew had open cockpits and were very noisy. The leather flying helmets muffled some of the noise, but Blackie and many others fashioned homemade earplugs from wax and used them extensively. There was enough airwork to assure his mastery of the ship, but the last three weeks of November 1930 were the most interesting because he checked out in the observation, attack, and pursuit trainers. The curriculum provided him with twenty-one hours in the Douglas O-2 observation aircraft and six tantalizing hours in the Curtiss P-1 and Boeing PW-9 pursuit ships. The deep-throated P-1 was no longer the standard Air Corps fighter of the period but its 400 plus horsepower engine provided plenty of get up and go compared to a PT-3 Trusty.

For most cadets, first flights in the fighters required considerable caution since they represented a quantum leap forward in performance. Blackie's transition to attack flying was delayed until late January, but as with the Curtiss and Boeing fighters, Blackie's seven hours in the Curtiss A-3 and its 435 hp engine were exhilarating, particularly considering the fact that most of January was spent exclusively back in the O-2s for cross country training and a PT-3 with a hooded second cockpit for instrument training. Blackie's "instrument flying" was at best only an introduction and was designed primarily to help young pilots avoid losing control due to low or zero visibility conditions. The PT-3's air data instruments were minimal and consisted of little more than an airspeed indicator, compass, and altimeter. At the time, much was evolving in navigation systems, and Blackie would soon benefit from new instruments such as the directional gyro, refinements of the basic altimeter, i.e., the vertical speed indicator, and radio navigation which would soon go into widespread use. Not every pilot takes to instrument training naturally, but Blackie did – so much so that he could be related to the proverbial duck taking to water. Little did he know then that his ability to fly effortlessly and solely by reference to the little, circular steam gauges in front of his face would be the basis for major promotions little more than a decade later.

Blackie's days were just as busy at Kelly as they had been at March: up very early, breakfast at six-fifteen, barracks cleaned afterwards, and outside for the morning formation at six-fifty. Flying began at seven and ran until nearly noon. The afternoon was devoted to ground school until three-thirty when one hour of physical training began. Supper was at five-thirty and during certain periods there would be night flying.

The cadet detachment was abuzz at supper on the evening of 17 November over a spectacular accident that many, including Blackie, had witnessed right on the field. Cadet John Magden, another of Blackie's closer friends, was practicing 360 degree spot landings in an LB-5A Keystone bomber when he suddenly lost aileron control. Magden, a U.C.L.A. graduate, later reported to the accident board that he heard a loud snap after which he could get no response from his ailerons although the rudder and elevator worked perfectly well. Inasmuch as he was only at 1,000 feet, Magden bailed out; the LB-5A spun in, crashed, exploded, and was completely burned out. The engineering officer did find evidence that a faulty turnbuckle had frayed the elevator cable to the extent that it snapped, thus freeing Magden from any blame. The accident could have been much, much worse. The LB-5A, still containing about 250 gallons of gasoline in its tanks, augured in just 250 yards from the Kelly Field ammunition dump.

On the very next day, 18 November, Blackie got in eighteen flights, primarily takeoffs and landings in the LB-5, including the precision landings Magden had been practicing the day before, but there was alarming news waiting for him after the last landing. His good friend Kenneth Greeson had been injured in a flying accident that morning. Greeson was in the observation section and had been aloft in a two-place Douglas O-2 with Cadet Elmer McKinney who was the pilot flying when they suffered an engine failure near Camp Stanley, an ammunition depot about twenty-five miles northwest of San Antonio. McKinney selected an emergency landing field and brought the O-2 in downwind and was running out of field when he hit a stump. Fortunately, the plane did not flip or catch fire, McKinney suffered facial lacerations, and Greeson had a badly broken left arm. The accident investigation board exonerated McKinney, but Greeson was washed back one class while his arm mended. Although the O-2 was a write-off, Greeson did get his wings and was sent to an observation squadron at Brooks Field in July 1931.

There was no flying on the 27th – Thanksgiving Day. Army mess halls regularly served what would qualify as one cut above what later become known as "comfort food." Breakfast consisted of unlimited quantities of eggs, bacon, ham, and sausages plus a surprising variety of fresh fruit. There was no "lunch" for Blackie or his mates. It was "dinner" and consisted of steaks, turkey, and chicken along with many vegetables. Cadets with a sweet tooth were in great luck with dessert as the choices were practically unlimited. Blackie's supper was simply a variation on dinner. The chief cook pulled all of this off on an allowance of one American dollar per day per cadet – but then a dollar back in 1930 went much, much farther than today especially considering San Antonio's proximity to so much of Texas' prime cattle country and agricultural lands. Indeed, the dining was so nutritious that many cadets put on a few pounds in spite of physical training.

On Thanksgiving the cooks put on a spectacular dinner with more than turkey and all of the traditional trimmings. Many Americans were feeling the pinch of a deepening economic depression, but for the cadets, Kelly Field's mess hall was a land of plenty. If Marion Blackburn's admonitions to Blackie about eating his Brussels sprouts had fallen on deaf ears, dinner that day also included snowflake potatoes, fresh English peas in crème, buttered asparagus tips, hearts of lettuce with Thousand Island dressing and sliced tomatoes, stuffed celery Roquefort, ripe olives, and mixed pickles. For those few who did not like pumpkin pie, there was minced-meat pie and traditional fruit cake, all washed down with hot coffee and tea. Those still hungry could enjoy roasted pecans, candies, and mixed nuts before moving onto cigars. At the head table were Captain Earl H. DeFord and his assistant, Lieutenant Don Zimmerman, the officers in charge of the cadet detachment.

Blackie enjoyed another, very special meal shortly after Thanksgiving Day, although it began with a curious invitation from one of San Antonio's most prominent families: Dr. and Mrs. William Bondurant. The squadron adjutant informed Blackie that he was to telephone a Dr. Bondurant which set Blackie to thinking why he should call a perfect stranger, although there was something vaguely familiar about the name "Bondurant." Then it slowly came back to Blackie that the Bondurants had been acquaintances of Charles and Marion Blackburn going back to the days when the Blackburn family was still together and spending summer vacations on the Texas Coast near Galveston. William Bondurant was the superintendent of both the Texas Military Institute and the San Antonio Academy. His family was of French origin and had been in America since the early 1700s, moving through the South to arrive in Texas in the late 1800s. While Blackie could claim that he came from Mitchell Nebraskan royalty, the Bondurants came from the real McCoy. The Bondurants could trace their lineage all the way back to the tenth century and included one King of Italy, French royalty, and prosperous Englishmen in their heritage. The first Bondurants to arrive in Virginia were French Huguenots who landed in Jamestown in 1700 before proceeding to what is now Richmond. The Bondurants and Blackburns had become friendly during those Texas coast vacations, but the tie was lost as the Blackburn family had broken up almost ten years before. Blackie recalled that the Bondurants had a son somewhat younger than himself and that there was a younger daughter too, but her name was a mystery.

The vivacious, now twenty-four year old daughter, Martha, was at best lukewarm to inviting an aviation cadet to Sunday dinner, especially because she was engaged to a young surgeon in San Antonio. Martha had been educated at private schools in San Antonio, was a recent graduate of Newcomb College (now part of Tulane University) in New Orleans, and was employed by the Texas Military Institute as a librarian. Her attitude concerning the dinner guest was along the lines of, "Oh Mother; how could you?" Blackie hopped a ride down to the Plaza Hotel where Bill Bondurant had arranged to pick him up and drive to the Bondurant residence. Martha's attitude changed dramatically. Both were smitten. Within several weeks, Martha broke the engagement, but before she could return the ring a rather remarkable event occurred. While the Bondurant family was out on their front lawn talking with neighbors on a Saturday afternoon, burglars entered the home from the back door and made off with clothing and jewelry worth $500. Mrs. Bondurant lost the most, including family heirlooms, jewelry, and eight suits of clothing. The only other item stolen, according to investigating detective Aubrey Hopkins, was a diamond ring belonging to Martha Bondurant.

Years later, Blackie always maintained that it was his uniform, the Sam Browne belt, his peaked cap with the aviation cadet propeller and wings, the boots, and his cavalry-style Jodhpur pants which really turned her head. Indeed, he was a dashing figure in uniform and was quite frequently mistaken for one of the idols of the silver screen, Douglas Fairbanks, Jr., to whom Blackie bore a striking resemblance. Blackie was well-attuned to what comprised a good-looking uniform, and because he was so close to graduation, he and nearly every other cadet on the course headed into San Antonio to order their officers' uniforms. So, it was to Luchessi's Blackie went for his high-topped boots and to Lauterstein's, the tailor, for his blouse and pants. Lauterstein's sewed a job fit for a king complete with silver and gold thread embroidery for the insignia.

His new sartorial elegance and pre-occupation with Martha, however, did not deter Blackie from his quest for excellence at Kelly. His afternoons were taken up with ground school as had been the case at March, but now the curriculum was much more focused. There were sixteen subjects, of which only one was a pass-fail course: Squadron Duties of Junior Officers, which Blackie handily passed. His performance in the other fifteen subjects, amounting to 128 hours, even exceeded the top of class performance at March. Six of these courses were general introductions: attack, bombardment, observation, and pursuit aviation which were in keeping with the philosophy at the time of providing all pilots with the big picture first and then moving into their own specialties. Blackie's average score for the four general courses was 98.25 which was coincidentally his average in the other general courses: Combat Orders, Signal Communication, Military Organization, and Divisional Observation. It was in the bomber specialist courses that Blackie aced with 100s: Bomb Racks, Bombs and Explosives, Bomb Sights, Photo Interpretation, Photography, and Reconnaissance. His final average for advanced ground school was ninety-eight in spite of his lowest grade, an eighty-eight in signal communication. Nevertheless, even with a relapse in his code standard, the overall performance kept him ranked as first in his class.

Formation flying was limited to two ships at first, then three which was the basic building block in bomber flying, and finally groups of six. Radios were not used; the aerial ballet was coordinated by hand signals. One minute, Blackie was just one pilot in a line astern formation, and the next he might be in an echelon. He had to be able to seamlessly transition between a left echelon and a right one or a Vee. He and the other men in his section were also given a chance to lead the entire formation which was observed by one of the instructors who was riding in one of the LB-5s or observing from outside the formation in one of the O-2s.

On one late November day, Blackie was serving as cadet officer of the day when he spotted a vagrant snooping around the barracks. Blackie went out, encountered the man, and hailed an MP to escort the interloper over to the front gate where the gentleman revealed his identity. It turned out to be none other than Brigadier General Benjamin D. Foulois who was making an incognito inspection tour prior to his rapidly impending promotion to Chief of the Air Corps. At first blush, Blackie assumed his cadetship would be instantly terminated for his brusque treatment of a drifter in civilian clothes. Quite the contrary followed; Blackie was complimented for his quick thinking and handling of the matter.

By the second week of December, cadets began flying cross countries. Although slated for bombardment training, Blackie and the other bomber pilots-to-be flew those sorties in the Douglas O-2 instead of the big LB-5s for another reason in addition to economy of single versus twin-engine operation. That was maintenance; the LB-5s had such serious maintenance issues that the entire fleet was grounded not long after Blackie left Kelly. Those cross countries were day flights at first and soon included return legs to Kelly well after the sun had set. Blackie and fellow cadets from March Field had some adjustments to make when starting cross countries in Texas. The California landscape around March was liberally littered with good landmarks – unlike the monotonous, flat prairie around Kelly. In addition to the city of San Antonio, there was one good landmark though – all of construction at newly dedicated Randolph Field and its "Taj Mahal," the very visible Central Administration Building northeast of San Antonio. Randolph would absorb the cadet primary and basic classes from Brooks and March Fields after they enrolled their last incoming classes in July of 1931.

Blackie added to his knowledge of Texas geography by flying to and from Laredo, Midland, Dallas, El Paso, Houston, and Waco. Two cadets were assigned to each plane on some cross countries and alternated legs as pilot in command. Whether on a solo or dual cross country, cadets had to land and obtain a confirming signature at each waypoint airfield before returning to Kelly. Cadets were also getting more instrument time, although that was all conducted in the by then rather unexciting old PT-3 Trusty which had a collapsible hood installed over the aft cockpit to remove any visual cues. With the absence of gyro instruments, about all that was expected of a cadet's instrument proficiency was to be able to safely fly in cloud or especially dark nights with no visible horizon without becoming disoriented and spinning in. That could be accomplished by reference to altimeter, airspeed indicator, and turn and bank indicator alone, but the appearance of artificial horizons and directional gyros in a few years that would significantly facilitate instrument training.

The pilots in the pursuit and attack tracks flew obsolete fighter and attack aircraft on their cross countries, one of which ended badly for Cadet Arthur Kingham on December sixth. The accidents vividly demonstrated the types of trouble inexperienced pilots could get into, particularly when instrument flying. On 6 December Kingham was flying an old Curtiss P1-D on a night flight between Waco and Kelly when he became disoriented in low cloud at 1,100 feet and lost control. Emerging from the cloud base, he realized just how low he was and that he would never be able to get out of the spin in time. So he jumped; the chute blossomed just seconds before Kingham's feet hit the ground.

A week later, it was Cadet John Hodgson who had a brush with death caused by inexperience. Hodgson was in the pursuit section and was practicing aerobatics two miles west of Kelly when he shed his upper wing in a power dive. He too was flying a P-1D which was essentially a stock P-1 Hawk with a more powerful 500 hp engine. Although part of the wing struck him in the jaw, Hodgson was able to bail out at 2,000 feet.

February, Blackie's last full month as an aviation cadet proved to be a disastrous period in terms of flying accidents. On the third, Harvey Regan bailed out of a Douglas O-2 while on a cross country between Dallas and Midland when his control stick became jammed. Unable to budge the stick, Regan checked his chute harness and bailed out. On the very next day, Cadet Joseph Donnelley was on a multi-leg cross country to Fort Bliss, near El Paso, when his engine started cutting out. Trying everything, he still could not coax more than 400 rpm out of the sputtering engine. He considered bailing out, but was reluctant to lose his ship especially since he was over relatively level terrain somewhere southeast of Roswell, New Mexico. Donnelley rode it in and was very lucky to survive when the plane rolled over after hitting the ground, bending the fuselage and removing the wings. Donnelley's problems were not over by a long shot, although search planes were sent out looking for him. Fear of rattlesnakes and scorpions kept him close to the plane where he spent the night in the remains of the cockpit, but after thirty hours he headed west on foot and was eventually found by a rancher who took him to Roswell. Other than a bruise on his face he was in good shape; he had made good use of his Curtiss A-3's emergency kit's rations. Donnelley was found blameless – the subsequent investigation confirmed the fact that the ship's gas tank was contaminated with water.

Remarkably, there was yet another accident the next day, although it was not as serious. Samuel Stephenson removed his Thomas-Morse O-19's landing gear in a spectacular ground loop after a cross country to Midland on 5 February. The engineering officer was of the opinion that the landing gear was weak to start with, but the accident review board discounted that supposition and blamed the mishap on pilot error. Had Stephenson not been a West Pointer, his poor judgment could have been a hanging offense, but he still got his wings three weeks later.

The course was accident free for only a week, but then Cadet Harvey Berger wiped out his landing gear bringing a Boeing PW-9 fighter in for a night landing at El Paso. He got off blame-free because the accident board felt that the six hours he had flown during the day was fatiguing to the extent that Berger's judgment was impaired. More importantly, he was also blinded when landing directly into the glare of the airport floodlights. Two days later, Cadet John Kroeger was involved in a mid-air collision on 10 February when his P-1D came into contact with the instructor's ship while pulling out of a formation dive. The damage was minimal and the instructor's plane, with a dent in the fuselage, was easily repaired as was Koeger's ship which suffered minor damage to a wingtip. Kroeger was lucky not to be washed out, but graduation was very close and he had been rated above average as a cadet, so the accident investigation decided not to throw the book at him. It was not Kroeger's last accident – he had another four years later as a reservist flying a P-12 at Langley Field in Virginia where he removed a wing in an intentional groundloop to avoid hitting a hangar. Again luck was with him since the board did not discipline him for what it acknowledged was poor judgment in continuing a landing that should have been aborted.

All of the accidents over the four months involved nearly fourteen percent of Blackie's cadet class. The accidents were inevitable, considering the fact that the young men were still relatively inexperienced to be flying what were fairly high performance aircraft of the day. Some of those men, like Kroeger, who had already been subjected to an accident investigation review board would prove to be accident prone over the run up to World War II considering the number of written-off ships they walked away from (three more crack-ups for Kroeger in as many years).

Blackie's flying skill kept him safe from pilot-induced accidents, but he did have one lapse of judgment while at Kelly. It occurred during his pursuit orientation when he was enjoying some extreme aerobatics in a Curtiss P-1D. Time slipped away from him as looped and rolled, playing tag with the cloud tops thinking nobody could see what he was doing. He had kept the P-1D out for much longer than the time allotted to him and had been instructed to do nothing beyond elementary airwork, i.e., turns, stalls, chandelles, etc. He was, however, not alone. The chief flying instructor of the pursuit section observed it all, joined up on Blackie, and signaled him to return to Kelly at once. Cadet Blackburn paid for the illicit pleasure by walking off several hours with a Springfield rifle over his shoulder.

The Army was about four years away from adopting flight simulators for use in instrument flight training, although Edwin Link had created the first Link trainer in 1929. Kelly Field, however, was using simulation, although not for instrument flight training. Blackie's experience with a simulator was in a two-story building which contained two very large "simulators." The first was a huge canvas, twenty by thirty foot, relief terrain model which very accurately depicted a practice area at Fort Sam Houston just northeast of San Antonio. The scale was such that it appeared to be 5,000 feet below the viewer. The model, designed to appear as thirty square miles of terrain, included woods, creeks, plowed fields, fields with crops, valleys, farm buildings, and two opposing armies. The good guys were in olive drab and the enemy was in red, and hundreds of miniature lights embedded in the canvas simulated firing artillery. The fundamental purpose was to teach communications between air and ground units over a series of different problems since the forces below could be placed in different situations between exercises.

While the terrain model was of particular use to the observation trainees, the attack pilots and Blackie's bomber group had their own dedicated simulator which involved a seven and one-half by fifty foot moving map which could be viewed from above. This map was marked with important objectives such as bridges, bomb dumps, enemy forts, and railway terminals. Although Blackie was not being trained as a bombardier, he had to be very familiar with that precision activity. The would-be bomb aimers sat in a chair above the map which appeared to move toward the cadet as it would appear during flight. The controller, at a nearby pedestal, could control such variables as aircraft speed, wind, and bomb weight. The bomb release mechanism was wired to the controller's desk from which it could be calculated whether the bomb had hit the target. Beneath the carpet and assigned targets were many small lights. By noting the flash, a cadet could tell if he had scored a hit or a miss.

Blackie and the other bomber students also "bombed" the obscura hut, a small shack located well beyond the Kelly boundary fence. The canvas-topped hut housed a camera obscura (a lens which projected the image of the approaching bomber down onto a plotting board). With the hut's canvas roof removed, Blackie made his final approach and radioed the controller in the hut the moment he released his imaginary bomb. Based on the airplane's image relative to the control board's epicenter, the controller could quickly correlate Blackie's speed, altitude, and actual wind conditions to tell if Blackie had scored a direct hit, a near miss, or had completely missed the target. Blackie would not spend much time flying with live bombs; most of that would come after assignment to an operational squadron. He did become very familiar with all of the bombs in the Air Corps inventory at the time which ranged from simple twenty-five pound fragmentation bombs all the way up to the nearly fifteen-foot long, two-ton demolition bomb. He was able to fly one "mission" to bomb an "enemy aerodrome" at Fort Stanley. Although the base primarily served as a bomb dump, surplus and irreparably damaged aircraft were occasionally deposited at a remote location at the facility to serve as targets for cadets in bomber training.

The course was coming to a close by late February. On the twenty-third, Blackie was inducted into the Institute of the Blind based on his successful completion of the instrument flying course. The fancy certificate proclaimed that he had completed the training "without peeping" out from under hood. Not all of the cadets immediately took to instrument flying, but it was one of Blackie's strongest attributes. His musical and managerial talents were of great value in instrument flying, considering the need for flawless multi-tasking, and being able to control his aircraft in the many dimensions required, i.e., not just pitch, roll, and yaw, but speed, altitude, and time simultaneously. Imagine it as being required to solve a life or death solid and spherical geometry/trigonometry problem with a stopwatch being held to your head while simultaneously looking at some hieroglyphics (the instruments) and being subjected to a lot of noise, turbulence, freezing cold or stifling heat, and a demanding instructor to get a flavor for that higher form of flying.

On Friday, 27 February, Captain Ray Giles, the base operations officer, endorsed Blackie's logbook, certifying Cadet Blackburn as an "Airplane Pilot." Originally, the rating had been "military aviator," but the name changed as the Air Corps evolved, particularly after the Air Corps Act of 1926 was passed. "Pilot" was not the only Air Corps rating Blackie earned. In time, he would be rated as a command pilot signifying years of service and extensive flight experience. His cadet training was the foundation for his several other ratings: observer, expert bombardier, and expert aerial gunner.

Blackie now had 314 hours of Army flying in his service logbook (excluding civilian time), including forty-one hours logged that month on bombardment and observation aircraft. His exposure to attack and pursuit aircraft was enough time to give him a flavor of those missions with roughly seven hours in the former and seventeen hours of the latter. The afternoon before, there had been a party in the base recreation hall in which Blackie discovered that Clarence Shoop was not the only other talented piano player to have become an aviation cadet. Eric Danielson, the cadet who came through the deadly car accident back at March, gave Blackie a real run for his money as the piano virtuoso in what was a celebration looking forward to the following day's event: getting lieutenant's bars and those treasured silver wings.

On Thursday evening, a banquet honoring the graduating cadets was sponsored by the San Antonio chapter of the National Aeronautic Association at the Plaza Hotel in downtown San Antonio. Friday morning dawned, not bright and sunny as hoped for, but windy and raining hard which cancelled one of the celebration activities, a massive aerial review over Kelly Field that was to have started at nine A.M. The cadets were not the only party disappointed by the cancellation; San Antonio's mayor, C.G. Chambers had made arrangements for buses to transport nearly 500 people out to Kelly to witness the aerial armada thundering over the field.

By eleven A.M., all cadets and their guests were assembled in the post theater. There was a bit of a delay, but then the station band started playing as the brass marched down the aisle. That band was newly arrived and was one of only seven bands in the Air Corps, having just been transferred in from the Edgewood Arsenal in Maryland. Among the dignitaries officiating at the wings presentation were General Charles Danforth, chief of Air Corps training, Major

Frederick Martin, commander at Kelly, and the guest speaker, Colonel Gideon H. Williams, commander of the Ninth Infantry Regiment at Fort Sam Houston. Martin was on a fast track to a career that would take him to the rank of major general, but his ascent was cut short on 7 December 1941 – he was then Chief of the Hawaiian Air Command and was one of several high-ranking officers blamed for the debacle of American unpreparedness at Pearl Harbor.

After the invocation by base Chaplain Burling, General Danforth spoke briefly about flying safety and financial responsibility (a rather high percentage of the cadets had just taken out hefty loans for the purchase of new cars) along with paying strict attention to the counsel of older, wiser pilots. Colonel Williams was commander of the glory-covered Ninth Infantry, a unit assigned to another much honored organization, the Second Infantry Division. One of his observations was unique and certainly struck home with his audience – that the airplane was rapidly replacing the second half of the old infantry-artillery team. The cadets were listening to him, but what they were looking at was glittering silver – pilots' wings which were aligned in neat rows atop a table on the stage.

Upon hearing his name read by the adjutant, Captain Hiram Thomas, each cadet strode across the stage, saluted, shook hands with and then received his diploma from Colonel Williams and his wings from Major Martin. As Harold Fredric Blackburn received his wings, it was the proudest moment to date in his life, and it was shared by his two leading ladies: his mother and his fiancé, Martha Bondurant. Blackie was now a lieutenant in the United States Army Air Corps Reserve.

There were a few light moments. Frequently, as a cadet's name was read, a girlfriend would break into loud applause. In a few cases, several adoring young ladies would applaud the same cadet from different locations in the audience which, of course created a few venomous, catty stares between the young women. There were a few awkward moments as well; Cadet Donald Kessler's right hand was bandaged and Cadet Walter Edwards got flustered and tried to shake hands with his left hand. Cadet Laurence Williams strode forward and got his diploma and wings. Then Captain Thomas called Cadet Maurice Williams forward and Colonel Williams quite literally did a double take. He was looking at the second half of identical twins, the Brothers Williams. The Williams twins were both headed for fighters at Mather Field in Sacramento. But there was a tragi-comic moment waiting. The final man to be called, Hugh Worthington, was about to be sent to a bomber squadron at Langley Field in Virginia, but unfortunately the last pair of wings to be awarded had just been given to Cadet John David Williamson – somebody had counted the sets of wings incorrectly and was one pair short. Worthington looked visibly shattered, but was assured by Martin that a shiny new pair of wings would be rustled up within a few minutes.

There was an important transition that applied to most of the new pilots, but not to six others just getting their wings. All save those six were discharged from the Regular Army as enlisted men and with that their old serial numbers were retired. Blackie traded out 6544831 to become O-280829. That passage did not apply to the five men who were already officers – the West Point graduates of whom nearly all became general officers by the end of World War II. Nor did it apply to Mr. Chia Mei Hu who would be returning to the Chinese Air Force. He was one of nearly fifty foreign students to be enrolled in U.S. Army aviation schools in the preceding decade. He would also prove in just a few short years to be a pivotal player in the lives of three of Blackie's fellow members of Kelly class of March '30.

The partying was not over. The new lieutenants sponsored a dinner dance at the El Mirador ballroom atop the Plaza Hotel Saturday evening with 135 people attending. Miniature planes hung from the ceiling and one wall was decorated with many blue lights in the shape of a pair of pilot wings. Each table was decorated with acacia blossoms and more miniature airplanes.

Then the squadron assignments came through. The fighter pilots were pretty well split up between Selfridge Field in Michigan and Mather Field in California. Nearly all of the observation pilots were headed east – to Mitchell Field on Long Island. The attack section was all sent to Fort Crockett and the Third Attack Group, except two pilots who were posted to Scott Field in Illinois.

The bomber boys were divided between assignments to Langley Field in Virginia and where Blackie was headed – to an absolute plumb of an assignment: Rockwell Field, San Diego. New pilots were sent where they were ordered, but were consulted for their preference. The top ranking cadets could count on their wishes being granted, and in Blackie's case, there was little doubt because his name was first on the station preference list. The resident bomber unit at Rockwell was the most famous in the Air Corps – the Eleventh Bombardment Squadron which was flying the heaviest bomber in the service, the Curtiss B-2 Condor. Several of his closest friends were headed to Rockwell, including Bud Gorman, John Magden, and Wally Waldorf. There was another very meaningful benefit from being newly commissioned; Blackie's monthly pay of $75 as a cadet had nearly tripled to $205.

After graduation, those who had been friends at Kelly stayed in touch over the years. Few flew combat in what was the then unimagined, approaching war, but others tuned in through their own jungle telegraph. Many became airline pilots once the Great Depression receded, and quite a few went to TWA including Blackie's close friend Kenny Greeson. Others eventually flying for TWA included John Magden, Elmer Gorman, and the Williams twins among them.

Chapter Three
Flying with Jiggs

Although Blackie had his orders for San Diego, he was not going very far for a while. He was assigned to temporary duty at Dodd Field which was part of Fort Sam Houston on San Antonio's near northeast side. His one-month stay there was essentially a holding pattern because his new station was not ready to accept its influx of junior officers, and Kelly Field had no quarters available for recent graduates. Thus, most of the month of March was one of down time – one in which Blackie only logged two flights totaling twenty minutes in a DH-4. During this period, on 31 March, there was a flying accident in Kansas – one that assumed the proportions of a national tragedy. Football hero and legendary Notre Dame, football coach, Knute Rockne, was killed in the crash of a TWA plane while on his way to California. The plane, which disintegrated in mid-air, had been one of WAE's Fokker F-10s, a high wing monoplane with three engines and a passenger capacity of twelve people. Unlike the steel and aluminum-framed Curtiss Condor Blackie was to start flying in a few weeks, the Fokker F-10 had a wooden skeleton. Inasmuch as dry rot was found in the wing of another Fokker, the entire fleet was grounded and later allowed to fly subject to frequent inspections, but its death knell had been sounded. More importantly in terms of Blackie's future, TWA top management consisting of Richard Robbins and Jack Frye assisted by technical advisors Charles Lindbergh and Tommy Tomlinson drew up requirements for what they saw as the next generation of civil transports: three engines, metal structure, twelve-person passenger capacity, range of 1060 miles, and speed of 150 mph. The result was the Douglas DC-1, a plane that Blackie never flew, but its successor, the DC-2, was another matter.

The Air Corps in which Blackie went on active duty in late February was a very small organization, consisting of fourteen air groups each of which was typically comprised of three squadrons. Blackie joined a force of roughly 5,700 Reserve officers, 1,200 regular officers, and slightly over 12,000 enlisted men. There were also 240 Reserve officers on extended active duty, a path that Blackie would follow in late 1932. The Air Corps of 1932 was still a small part of the Army in terms of manpower and amounted to only twelve percent of the Army's total strength of 125,522 officers and men. Air Corps strength in numbers would not really start growing until later in the 1930s, but what would impact Blackie and the other Reserve officers in his Kelly graduating class was a Depression-related decline in Air Corps spending of seven percent between 1933 and 1934.

When Blackie headed for San Diego, the pursuit/attack community was still the most important force in the minds of the Air Corps top brass. At that time, excluding the mixed offshore groups, there were eighteen pursuit and attack squadrons but only ten bombardment squadrons. Although the Seventh Bombardment Group was headquartered at March Field, its Eleventh Bomb Squadron, to which Blackie was assigned, was based at Rockwell Field in San Diego; the squadron would return to March Field before the year was out. In joining the Seventh Bombardment Group, Blackie made some very important acquaintances. The group commander was Major Carl Spaatz who would go on to command the Eighth Army Air Force during World War II and then become the first United States Air Force chief of staff. Unimagined at the time was the fact that Blackie, in little more than a decade, would have Spaatz's replacement as group commander, Major Joseph McNarney, as a passenger on a transatlantic flight. Blackie's flight commander was West Pointer Lieutenant William Tunner, later to become General Tunner who was in charge of the many transports flying "the Hump" into China during World War II and the same officer who was in charge of the Berlin Airlift.

To say that Blackie was headed for a country club atmosphere in San Diego would be inaccurate or at least reasonably inaccurate. The city offered many advantages including the locations where Rockwell's golf and tennis teams played – at San Diego's leading country clubs. Opponents came from other military organizations in the area and as far away as March Field in Los Angeles and Crissy Field in San Francisco. Rockwell athletes also played teams from private sportsmen's clubs in San Diego. Rockwell Field was located on the same real estate as the North Island Naval Air Station whose athletic teams were arch rivals for Rockwell's athletes. The rivalry extended well beyond athletics – the competition was for the island itself. Both the Army and Navy had been using the island prior to World War I. President Franklin D. Roosevelt (a former Assistant Navy Secretary) eventually signed an executive order favoring the Navy. Thus, by World War II, the Army presence there had become history.

San Diego was a good duty station for the exact same reason that both the Army and Navy had originally established bases there – excellent flying weather. Although the Army Air Corps' budgets essentially dictated how many hours an aviator could fly per year, Blackie did not have to put up with rotten flying weather associated with such other "maritime" bases as Langley Field near the Chesapeake or Mitchell Field on Long Island. Nor did he have to cope with the harsh winters of Selfridge Field in Michigan or the shimmering heat of Brooks and Crockett Fields in Texas. San Diego was already an aviation-minded town when Blackie arrived. In addition to the military aviation presence, San Diego had a new airport, Lindbergh Field, to which Charles Lindbergh had granted his name. Lindbergh's plane, *The Spirit of St. Louis*, had been built in San Diego by Ryan Aeronautical which also operated one of America's premier flying academies. In just a few short years, another manufacturer would establish a major facility in San Diego. In 1935, Consolidated Aircraft left Buffalo and its less than ideal flying weather for sunny San Diego.

Although Blackie now had his wings and lieutenant's bars, he and the other new arrivals were very much regarded as apprentices. According to the Army's definition of "airplane pilot," the holder of that rating was qualified to give instruction to student pilots, observers, and other crew members (active duty or Reserve) in the operation of airplanes, powerplants, instruments, and armaments. And the new pilot's flight duties could include the movement of men, mail, and materials in flights involving the following missions: "cooperation with other government agencies, ferrying aircraft, aerial scouting, reconnaissance, bombing, convoy, patrol, photography, mapping, pigeon training, and tactical maneuvers." As radio technology progressed, the pigeons were phased out.

Lieutenant Tunner's definition, of a newly rated pilot's responsibilities, was much more to the point: function as a co-pilot. Blackie's group of new lieutenants assigned to the Eleventh Bombardment Squadron got off to a slow start and logged only a few flights in DH-4s after arriving in San Diego in late March. Beginning in April, the tempo picked up dramatically as Blackie was flying almost daily. Before becoming a part of the squadron, Blackie had to become a student again and that meant soaking up some of the squadron's history and familiarizing himself with the plane he would be flying, the Curtiss B-2 Condor.

Over the years, Blackie would frequently be asked that of all of the aircraft he had flown, which one was the favorite. His usual response was, "The one I'm flying right now." In private, however, he would always say that he had a soft spot in his heart for the Curtiss B-2 Condor. The heavy, two-engine bomber was not a completely new design as were more famous bombers to follow such as the B-17 Flying Fortress, but was the result of evolution. The Condor owed much of its design to the NBS-1, a joint venture between Martin and Curtiss. The NBS-1 was a bi-plane, two-engine bomber with a pair of Liberty engines built into the lower wing, just above the landing gear. It had seen service with most of the Army's bombardment squadrons, including the Eleventh, and the total built eventually exceeded 100.

Blackburn Family Collection

Blackie in early 1932 with a Curtiss B-2 Condor in the background.

The Condor, ordered by the Army in 1928, was of similar appearance but was much more powerful with its two 620 hp Conqueror engines, use of steel in its fuselage and wing framework, and aluminum ribs. The plane had a few other unique characteristics including one that was unique to the Eleventh Bomb Squadron: its paint scheme. Although the Air Corps was rapidly moving to blue and yellow colors in the early 30s, Blackie's squadron retained the olive drab fuselage. The wings were yellow and the tail contained the usual blue vertical bar on the vertical stabilizer with alternating horizontal white and red stripes on the rudder. The aft fuselage contained a red band with white border, but what set the Eleventh's planes completely apart was the zig-zag red flame pattern with white border on the nose. Dead center on the nose was the individual plane number.

There were several other identifying features of which one would cause an attacking fighter pilot to seriously consider his approach to the bomber. Its defensive armament included what was a fairly standard gunner's position in the nose but also, uniquely, a gunner's station in the after section of each engine nacelle. The nose gunner did double duty; he also functioned as the bombardier whose station was just below the front gunner's. The Condor was relatively clean for a bi-plane bomber of its day with two exceptions. The tail was actually a quad unit with two vertical stabilizers and two horizontal stabilizers. The structures adding considerably to the drag were two great radiators mounted between the engines and the upper wing. All-up weight for the plane with a ninety-foot wingspan was eight tons and top speed was typical for the period: 132 mph. To put that into perspective, the Douglas DC-2 which Blackie would first fly in another four years, had a similar gross weight, engines roughly eighteen per cent more powerful, and wings only five feet shorter, but the Two's top speed was 210 mph. The Two offered many more creature comforts including an enclosed cabin which the B-2 did not have. Unfortunately, not all of Blackie's flying in the B-2's open cockpit would be in the sunny, balmy skies over San Diego. Although the Condor did have some new technology, including an intercom, it was a design from another era and would serve only for another three years after Blackie joined the squadron.

Curtiss did manufacture a civilian version of the Condor for airline use, but it was rapidly outclassed by tri-motors from Ford and Fokker. Very few military Condors were produced, and the Eleventh was the only squadron to operate them.

The Eleventh Bombardment Squadron was arguably the best known bombardment squadron in the Air Service and with good reason. Blackie was somewhat familiar with the squadron because its origins were tied to Kelly Field in 1917. By early 1918, the squadron was in Britain for final training with the Royal Air Force before being sent to the Western Front in August as part of the First Day Bombardment Group. By late September it was ready for combat and flew its first operation on 26 September with its American-built DeHavilland 4s when it successfully bombed Etain with no losses. The Eleventh flew another ten missions in October – all with good results and no losses to enemy action. The squadron's luck turned sour in November when it was severely mauled on a mission to Cheveny le Chateau, running afoul of both German flak and fighters. When the war ended a week later, the Eleventh had been the most active of the four squadrons in the group, flying thirty-two bombing missions. Its gunners had also shot down ten German interceptors, but during its short time in combat, it suffered twenty casualties, including ten killed. When it arrived back in America in late April 1919, it had been recognized for service in three campaigns (Lorraine, Meuse-Argonne, and St. Mihiel) known to every American schoolboy, including a Nebraska high school senior named Harold Blackburn.

Like many squadrons, the Eleventh languished in the 1920s although it was moved from its base at Kelly Field to Langley Field in Virginia to participate in the 1923 bombing of obsolete battleships in the Chesapeake Bay. That exercise was a follow-up to similar bombing in 1921 when airpower advocate General Billy Mitchell first conclusively demonstrated the vulnerability of capital ships to air power. Following its tour at Langley, the squadron moved to March Field in California. After a brief period of inactivation the Eleventh was reconstituted in mid-1928 and stationed at Rockwell Field in San Diego.

The squadron had substantial top of mind recognition for another reason. And that became apparent to anybody who saw its squadron insignia and could also pick up a newspaper and read *Bringing up Father* in the comics section. The central character in that series was Jiggs, a working-class Irish immigrant who had suddenly come into great wealth as the result of winning a lottery. Jiggs was forever reverting to many of his old ways, friends, and haunts (the nearest watering hole) and was constantly harried by his wife, Maggie, who faced an uphill battle trying to raise Jiggs' standards and comportment. The comedian W.C. Fields' appearance was very consistent with the Jiggs cartoon, i.e., round, flushed face, top hat, vest, cane, and spats. The cartoon went into syndication several years before America's involvement in the World War in which Jiggs' creator, cartoonist George McManus, served in the Eleventh. Thus the squadron's insignia became a confidently smiling Mr. Jiggs walking sprightly with a large bomb under his left arm.

Esprit de corps was high in the Eleventh Bombardment Squadron, and one component of that hubris was a dog which served as the unit's mascot. One might logically expect the Eleventh, flying the largest bombers in Air Corps service, to have a large, rugged breed as its mascot, i.e., an Irish wolfhound, an Irish setter, or at the very least a Kerry blue terrier. That was not the case – at all. The Eleventh mascot was a tiny rat terrier named, "Parachute" which had his own little jacket complete with the Jiggs insignia. "Parachute" did bring one advantage to his position; his care and upkeep were very simple compared to mascots of neighboring squadrons which included a mule, bucking bronco, and bear.

While Blackie was at Rockwell, one fighter squadron was based on the field; it was the Ninety-fifth Pursuit whose insignia was the bucking mule. There was a good-natured rivalry between the units and one of Blackie's friends in the fighter squadron was Lt. Irvin "Bert" Woodring who was almost exactly the same age. Although Blackie was occasionally mistaken as the younger brother of Lloyd Blackburn, a major at the time and a prominent attack pilot dating back to the war, there was no mistaking about who Woodring was. Very shortly after Blackie and his Kelly classmates arrived at Rockwell, there was an open house on the base as part of Army Day celebrations on 6 April. During an inspection that day, Major General Malin Craig presented Woodring with the Distinguished Flying Cross for his remarkable cross country flight on 15 October 1930 in which he delivered one of two copies of the London Naval Treaty (the Kellogg Treaty) from Victoria, British Columbia, to Newark, New Jersey, under appalling weather conditions. The other pilot, carrying a second copy of the document, crashed and was killed in Montana early in the flight. There was some urgency to the flight due to the need to get the treaty to The Hague in the Netherlands as quickly as possible.

Woodring held various altitude and speed records and was one of three Army pilots who escorted Charles Lindbergh on his national tour in 1927 after arriving back from France. Woodring was also the sole survivor of an aerobatics group called the "Three Musketeers" which was popular at air shows across the country. He also was one of the pilots in the Army's first successful aerial refueling in 1929. The Musketeers were rivals of a San Diego-based Navy aerobatics group called the Three Sea Hawks, one of whose members, Tommy Tomlinson, Blackie had met several years before in Los Angeles. Tomlinson later became TWA's vice president of engineering. Blackie, however, fraternized with few pilots of the naval persuasion in those days. Tomlinson lived until old age; Bert Woodring did not. Some racing and test pilots could live on the edge for years as the life of Jimmy Doolittle clearly exemplified. Bert Woodring was not a member of that fortunate few, and the experimental Y1A-11 fighter he was testing disintegrated in mid-air near Wright Field on 20 January 1933. Woodring was unable to use his parachute which had saved his life twice before.

Although Mister Jiggs had been flying with the Eleventh for years, heraldry at the group level was only just emerging. It was not until early 1932 that the Eleventh's parent organization, the Seventh Bombardment Group's insignia was approved. The design was blue and gold, featuring a winged skull and an arm holding an aerial bomb with the motto in Latin "Mors ab Alto" (Death from Above). The shield below the skull was divided into four sections representing each of the group's four squadrons and something unique to each squadron. The four Maltese crosses for the Eleventh represented its participation in four campaigns during the war. The Ninth was represented with three crosses for its service, and the Thirty-first's art was in the form of a fleur-de-lis representing its wartime activity as a school squadron. Although the design was for the Seventh Group, the artists submitting the design were all from the Eleventh Squadron: Lieutenants Kiesseg and Bowman along with Corporal Seaburg and Private Moleson. As its resident squadrons changed over the years, the Seventh Group insignia changed also. Today the arms holding bombs, blue and gold shield, and motto endure, but it is only the three Maltese crosses of the Ninth Squadron which remain on the crest (the Ninth having served longest with the group). On the other hand, Mister Jiggs still flies with the Eleventh Squadron's B-52s as of this writing.

Beginning in 1925, the Air Corps conducted annual maneuvers, commencing with a small effort at the group level. By the time Blackie joined the Eleventh Squadron, the maneuvers had progressed from the group and wing level to a complete air division. Unfortunately, Blackie and his fellow brand new Kelly graduates had to be content with their role as observers because the maneuvers were also competitions which meant squadron commanders were only going to the war games with their most experienced pilots. In 1930, the annual maneuvers were held at Sacramento's Mather Field and involved four groups including the Seventh. A total of 114 planes were involved, including sixty-four fighters. Among the key objectives were improving the use of radio to control the battle and the use of cargo planes for support.

The plans for 1931 were much more aggressive. Public relations was to be an important part of the maneuvers which were designed to acquaint the American public with Air Corps capabilities. The War Plans Division envisioned over 2,000 aircraft participating in the maneuvers, but as it turned out not quite 700 aircraft actually flew. They came from every squadron based in the United States except one. The air division formed at Dayton, Ohio, on 16 May and after being delayed by bad weather took off for air shows in Chicago, Boston, New York, Atlantic City, and Washington. Blackie's squadron led the airshows with its aerial parade of B-2 Condors, but the airshows were only part of the maneuvers. The tactical assignment was to neutralize an "enemy" naval fleet headed for the New England coast, but that was not known until the air division had reached New York where it was informed that both the American Atlantic and Pacific Fleets had been knocked out and that the Panama Canal had been destroyed. Fortunately, the enemy fleet retired into a fog bank, leaving Air Corps planners in the dark as to where the enemy would strike next. Atlantic City surfaced as the next enemy target, but the air division repulsed the attack on 30 May. Groundwork for the victory had been prepared by the then Major Henry Arnold who had established supply bases from Ohio to Maryland to support the aerial armada. The "victory" parade was a massive fly-by held over Washington.

Army publicists certainly got plenty of ink in the nation's newspapers because they had enlisted the support of two widely known, former Army pilots: Jimmy Doolittle and Charles Lindbergh. The former performed a support role in his position as a technical representative of his employer, the Shell Oil Company. Lindbergh was in the air with his Missouri Air National Guard unit. One of the most striking photos appearing in national newspapers was of the Eleventh Bombardment Squadron's B-2 Condors, over the boardwalk of Atlantic City, flying out to sea to repel the invading enemy. Not all of the publicity was favorable; with the Great Depression on, many people felt that the extravagant spending on an aerial spectacle was ill-advised. Not surprisingly, the Air Corps plans for maneuvers in 1932 were scaled back considerably and involved aircraft at the wing level only. While 1931 had been a disappointment for Blackie and the other new pilots, they would be in the thick of it for 1932.

There was little opportunity for Blackie to continue his checkout on the B-2 in June because nearly all of the Condors were in hangars undergoing maintenance. Although he flew as co-pilot on a flight up to March to ferry a PT-3 back to Rockwell, Blackie's only B-2 flying consisted of logging an hour and half practicing takeoffs and landings at the end of the month. Nearly all of his thirty hours that month were in the DH-4 on cross countries and a tantalizing hour flying aerobatics in a P-12D fighter.

Rockwell did have one activity that would have appealed to Blackie had it occurred twelve years earlier: the sale of 682 Liberty engines and fifty Wright engines – all in Class A or B condition and priced to sell. Had the sale been in 1919, Blackie could have had a good engine for the engineless Curtis Jennie he almost bought. A number of parties watched for upcoming sales of Liberty engines, and among them were well-heeled sportsmen who retrofitted the powerful engines into their racing speedboats. Another group watching for Liberty engine sales was, in turn, watched by the Federal Bureau of Investigation and the Coast Guard. Rum-runners and bootleggers crossing from Caribbean and Canadian shores could easily leave Coast Guard patrol boats far behind in their Liberty-powered boats' wakes. That type of illegal activity was the farthest thing from Blackie's mind as Prohibition would end in just a couple of years. Nor did it ever occur to Blackie that within another six years he would be playing an undercover role for none other than the FBI's director, J. Edgar Hoover.

Blackie's first fitness report was completed by his squadron commander, First Lieutenant Charles Howard on 30 June 1931 and was endorsed by the group commander, Major Carl Spaatz. Howard noted that he had had little time to get to know Blackie particularly as the new pilots were nearly sidelined by the squadron's participation in the eastern maneuvers, but did note that Blackburn "gives good promise of being good material, intelligent and willing." Promotion was slow in the peacetime Air Corps; Howard had been a pilot in the Great War, but was still a lieutenant. Advancement would come however, and within three years he was promoted, assigned to the rapidly growing Maxwell Field in Alabama, and would become one of the thirty-five officers at Maxwell who founded the Order of Daedalians, the military aviation service society, in 1934. He also was a member of the groundbreaking survey flight of B-10 bombers from Washington to Fairbanks, Alaska, and back in the same year. The flight, led by Lt. Col. Henry Arnold, resulted in the Mackay Trophy being awarded to all participants. It would not be Howard's last Mackay Trophy. Sadly, he and his passenger were killed in a flying accident in Texas just two years later.

Qualifying the new pilots on the Condor really did not get under way until later in the summer when Blackie was flying at least every other day and extending his reach to cross country flights outside of California to Tucson and Phoenix. Throughout the autumn, he and the other six pilots from Kelly were flying the B-2s in bombing practice several times a week, more cross countries up to Los Angeles, San Francisco, and Bakersfield. When not in the B-2s, they were back in PT-3s under the hood to maintain instrument proficiency. At the end of the month, Blackie's logbook was checked and signed by the squadron operations officer who certified that the entries were accurate. By this point Blackie has mastered the B-2 and was logging nearly fifty hours a month with the majority as "pilot," as opposed to "observer."

Before entering the cadet program, Blackie had flown as a private pilot right up to the day before he was sworn in, but there was no time for recreational flying while at March or Kelly. However, as an officer at Rockwell, he had enough off-duty time to rent a plane for some flying on his own which usually meant pleasure flights around Los Angeles with friends, or "field hopping" to use a term from those days. One passenger was more than a friend – it was his fiancé, Martha Bondurant. When he had a few days leave, he also flew to Texas where Martha was again a passenger on aerial tours of Kelly and Randolph Fields.

By the end of October, Lt. Howard had gotten to know Blackie much better and was impressed with the way Blackie quickly assumed more responsibility in the squadron. By then Blackie's primary duty was as an airplane pilot, but he had taken on a substantial workload as the squadron parachute officer, statistical officer, supply officer, and assistant operations officer. Howard concluded that Blackie was well-qualified to perform duties in the next highest grade and praised Blackie for his judgment, common sense, and pleasant personality. That fitness report was the last Howard wrote on Blackie in San Diego because on 29 October, 1931, the squadron moved to the Seventh Group's home base, March Field. Rockwell still had a future as an Air Corps base because arriving to take the Eleventh Squadron's place was the Nineteenth Bombardment Group with its Thirtieth and Thirty-second Squadrons.

Blackburn Family Collection

The lieutenant and his lady in a Curtiss B-2 Condor cockpit.

There would be a pleasant surprise later in the year following Blackie's move back to March Field; Major Joseph McNarney who had been in charge of the primary flying school there, would become the new Seventh Group commander.

Before the move up to March, Blackie had become the squadron range officer; it was not an enjoyable assignment and that was reflected in his efficiency report in which his performance as the range officer was rated as only "satisfactory." That was at a time when Blackie's performance on other assignments was frequently rated as "excellent." Part of the problem was the bombing range at Rockwell which was simply an undeveloped corner of North Island. Because it was shared with other squadrons along with the Navy, the range was overutilized to the extent that there were serious air traffic problems. The San Diego city fathers were vociferous in their complaints and considered the range as a gross public menace. To alleviate some of the heat, the squadron frequently bombed targets on the ocean's surface which instantly set off the local commercial and sport fishing captains who complained that the Air Corps was ruining their livelihoods and sport by killing and scaring the fish.

The targets at Rockwell were occasionally "real" – aircraft on the ground, that is they were DH-4s that were no longer considered airworthy. In most instances the targets were simply concentric circles marked in chalk. At sea, the targets were wooden structures. Shortly after Blackie left Rockwell, the Air Corps established new ranges inland, and one of these facilities was on a dry lake northeast of Los Angeles. In the 1930s, Rogers Lake was exactly that, just a big dry lake with a small encampment for less than a dozen enlisted men who operated the bombing range. The adjacent and much larger Muroc Dry Lake eventually became a satellite field for March and would ultimately grow into a larger facility as World War II approached. During the war Muroc became an operational training base for fighter pilots and heavy bomber crews soon to be sent overseas. By then it was known as Muroc Army Air Field, and beginning in 1942 it was also being used as a test base for the Air Force's latest fighters since Wright Field in Ohio was overextended and, especially, because Muroc was so remote that it offered a certain level of secrecy.

Blackie would soon have a connection with the remote Rogers facility, other than bombing it, because one of his old acquaintances from the Los Angeles aviation community of the late 1920s would start a small business adjacent to it. The connection was none other than Pancho Barnes who would launch another business enterprise at Rogers Dry Lake – the Barnes Airport which also became a primary flying contract school for training aviation cadets early in World War II. Blackie's connection with the Rogers and Muroc facilities was minimal compared to that of one of his closest friends, Clarence Shoop, but that would take another fifteen years to materialize.

Blackie's intense training in formation flying and bombing during the late autumn certainly paid off during the Eleventh Squadron annual bombing competition of late 1931 in which he placed seventh out of the seventeen pilots with a score of 1900 which compared to the average score of 1870. What is most significant in that competition is that a fellow reservist, Second Lieutenant Russell W. Munson was the winner, earning 1947 points and beating several first lieutenants and regular Army officers. One can also conclude that squadron leader Charles Howard was doing a good job training his men; Howard placed fourteenth out of the seventeen contestants. Those pilots were used to thinking of some distant enemy forces, factories, or fortifications in some foreign land as their natural target. What they did not know as the year 1931 closed was that Mother Nature would become an enemy and that their targets would be a lot closer to March than some distant shore.

Nowadays when there is a natural disaster and a major humanitarian effort is required, we routinely see the great cargo planes from National Guard units and the Air Force's Military Airlift Command flying mercy missions to all corners of the earth. Those efforts have their origins in flights made by Blackie and the Eleventh Bombardment Squadron going back to the winter of 1931-32 when several southwestern states were crippled by severe blizzards in which mining communities and Navajo, Hopi, and Zuni Indian villages were completely cut off. The Eleventh "bombed" those locations with food and emergency supplies, setting a precedent that would be followed by other Air Corps bomb squadrons throughout the 1930s, including some rather novel missions flown by the Seventy-second Bombardment Squadron in Hawaii. During a major eruption of the Muana Loa in 1935, the squadron bombed the lava flow in an attempt to alter its path in order to minimize loss of life and property in the city of Hilo. The mission was not an unqualified success, but the technique was used once again during an eruption later in the 1930s.

Langley Field-based bomber squadrons in Virginia flew relief missions during the very rough winter of 1935-36. The Fortieth Squadron dropped supplies to small communities on islands in the Chesapeake Bay. The Ninety-sixth Squadron dropped supplies to ice-marooned lighthouses in the bay as did several Army dirigibles. The lighthouse keepers were eventually rescued by dog sleds – in many cases their lighthouses were damaged by surging ice when the spring break-up occurred.

On 16 January Lieutenant Howard led a flight of six Curtiss Condors, including one flown by Blackie to Winslow, Arizona, where Howard was to report to the Supervisor of Indian Agencies. Over the next five days, the squadron dropped 40,000 pounds of supplies to over 20,000 Native Americans living in Arizona, New Mexico, Utah, and Colorado. Considering the snow-covered landscape below, Native American guides were assigned to the planes as observers. One guide in particular, Mr. Clyde Lizner, impressed the pilots. Lizner, a wealthy and well-educated tribal elder, attended the briefing sessions and observed the "bomb" loading very carefully, all while scarcely uttering a word. The next morning, he arrived at the plane on time, fully dressed in winter flying gear, and unerringly guided his pilot to the objective.

The food supplies were packed in 100-pound bags and dropped from the planes' bomb racks and gunners' stations in the engine nacelles. The aircrews later learned that at the sight of the circling bombers, the Indians called out, "*Chity nah tiagie*!" (airplane, chug-chug). As soon as the villagers heard the planes, they went out into open fields to await the drop. Blackie and his fellow pilots flew an estimated 15,000 miles during the period from 17 to 22 January and reached over 22,000 Indians. The weather was bitterly cold to start with – so cold that the squadron's mechanics had to start the engines hourly every night to make sure a dawn takeoff was possible. Through this zero degree weather, Blackie was flying an open cockpit airplane with only a windscreen and a heavy winter flying suit for protection.

John Hunter, Superintendent of the Southern Navajo Agency, later wrote to Lt. Col. Henry "Hap" Arnold, March Field commander, commending the Air Corps for its rescue missions. In Hunter's words, "The Navajos never tire of telling their emotions when they saw the big Army planes flying over their country, and particularly when they saw that food was being dropped." He also noted that the Navajos were also impressed by the fact that "the father in Washington" had not forgotten them by sending "automobiles in the air." Hunter's commendation was echoed by C.L. Walker, Superintendent of the Western Navajo Agency, who particularly wanted his appreciation expressed not only to the pilots, but to everybody including the mechanics who toiled in the open to keep the planes flying.

The weather from mid-November through late January had been extraordinarily severe with one blizzard rolling in after another. Even with the Agency's snow-plow equipped Caterpillar tractors, the snows completely blocked all roads to Native American villages and their local traders. That was only part of the problem – the Navajos, for the most part, did not live in villages. Being semi-nomadic, individual hogans could be miles from the next one, thus complicating the food delivery strategy.

The squadron leader, Charles Howard, brought up many more details and difficulties in his official report. He noted the heaviest snow line as starting thirty miles north of Winslow and increasing dramatically as the elevation increased. Exposed hogans (semi-circular mud and stick dwellings) on windswept mesas were easier to see because the wind created a low pressure area beside each hogan's leeward side which gave an almost broom-swept appearance. Hogans and villages in canyons and valleys were almost impossible to see. The drop zones differed in terms of population density so Howard assigned two aircraft to the less-populated southern area and the remaining four covered the northern area. Both Hunter and Walker participated in the flights, and it was their knowledge of the territory that made success possible according to Lieutenant Howard. Frequently that which was marked as a "village" on a map might actually have been just one or two hogans and a sheep corral. Because it was impractical to drop at every single hogan, the superintendents pointed the way to community centers and the chiefs' hogans.

Because "food bombing" had never been attempted before, the squadron was involved with some trial and error activity before going on the actual drops. Because he was the assistant operations officer, was the squadron bombing range officer and had been the parachute officer, Blackie was part of the team that investigated the best way to "bomb" the villages. Just putting the supplies in a burlap bag and dropping them from 100 feet was found not to work – the bags ripped open. But when the food was placed in a bag that was bound by rope and then placed in another bag tied with rope, the bags did not rip open when dropped. Each bag contained forty-eight pounds of flour, ten pounds of sugar, four pounds of salt pork, ten pounds of coffee, two pounds of baking powder, and ten pounds of dried fruit. The deep snow was a blessing – the bags would have all broken if they had hit frozen ground. Each bag weighed almost 100 pounds; but ten bags were too bulky for the bomb bay, thus the bags were hung on wing racks. Of the hundreds of bags dropped, only four ripped open.

The Indian's reactions varied at first. Some were terrified and quickly sought shelter in their hogans. Others knew what was about to happen and waved arms and blankets at the planes. Most realized the danger from flying 100-pound objects, and came out from shelter only when they realized that the drop was over. Where there were few hogans, the pilots tried to find the one to which many footprints led. Hitting some of the targets required the utmost precision – such as the one overlooking the Little Colorado River, that is to say perched right on the edge of bluff overlooking the river 3,000 feet below.

Food bombing the Hopi tribe was much simpler because that tribe lived in villages which included a tribal center, a store, a warehouse, and frequently an Indian agency office. Thus, reaching the Hopi required only two flights of bombers. The mercy flights also extended into New Mexico—a sight of some of the heaviest snow, to reach the Zuni tribe country south of Gallup. That area had nearly become a disaster earlier when Navajos had passed through the area in their annual pine nut harvest and gotten caught in heavy snows. The Navajos were sheltered by local Zunis until state road crews were able to plow the roads.

During the mercy operation, Blackie flew seven flights – four as pilot and three as a co-pilot with Lieutenant Lyon for a total of nearly twenty-two flying hours. With typical Blackburnian understatement, his logbook description of each flight was simply "cooperation" (implying cooperation with ground-based authority). The National Aeronautic Association and the Army Air Corps thought differently. Blackie became, along with the other squadron members participating on the operation, a co-winner of the prestigious Mackay Trophy for the year 1932. The flights were well photographed, and since many of them were over northern Arizona, the resulting photography, some of it over the near-by Grand Canyon, was of the highest order and very reminiscent of some of the stunning photographs taken over Atlantic City in 1931. Yet as great as the Atlantic City and Grand Canyon photography was, there was even more stunning photography to follow in just a few months.

There was more good news awaiting Blackie on his return from the mercy missions. He was notified by the department of Commerce, Aeronautics Branch, that he had been granted a transport license for Class I and II airplanes, i.e., up to aircraft weighing 7,000 pounds gross weight. The Class III certificate followed in six months.

Blackie's skill kept him out of trouble, but the same cannot be said for seventeen of his fellow Kelly graduates during their first year or so of active duty. Two of Blackie's close friends were involved in taxiing accidents at March Field – John Magden hit a parked basic trainer with his right wing while following his squadron commander to the end of the runway and Elmer Gorman taxied into a parked gasoline truck. In both cases, Blackie's friends were blamed for the accidents because they failed to constantly look for potential hazards. The seventeen pilots were involved in twenty-one accidents, with Arthur Kingham accounting for four accidents alone. The accidents ran the full gamut of causes ranging from lack of situational awareness, loss of control, mechanical failure, and weather problems. Probably the most spectacular accident was that of Richard Moorehouse who was flying an O-1 from a facility that Blackie would come to know well in just a few years: Richards Field in Kansas City. Morehouse was flying low over Lake Lotawana east of Kansas City and struck some power lines that were suspended no more than forty feet above the placid waters. He immediately tried to pull up, but stalled, and went for a chilly swim. It was his misfortune to be observed from ashore by none other than R.S. "Swede" Knowlson, aviation editor of the Kansas City Star whose subsequent summary was attached to the official accident report which concluded that the accident was caused by pilot error.

Thus, out of a population of roughly sixty graduates in Blackie's Kelly class, well over twenty-five percent were involved in at least one flying accident during their active duty assignments. In reality, the overall flying accident rates were coming down while Blackie was at Rockwell Field. Ten years before, in 1921, there was one accident for every 1,718 hours flown in the Air Corps. By 1930, one accident occurred for every 8,700 hours flown. And the improvement continued dramatically into the first half of 1931 when the rate had improved to one incident in nearly 17,000 flight hours. It is generally conceded that the most dangerous period in a pilot's life is the first four hundred hours after which milestone most pilots have built up enough experience to keep them out of trouble. Nearly all of the accidents among Blackie's classmates occurred during those first several hundred hours, and their accident rate fell off sharply after the mid-30s with one exception which resulted in the only peacetime death among Blackie's fellow Kelly graduates.

The intervening months before their next great mission with on-board photographers were spent in training. Blackie was aloft three or four times a week, alternating between pilot and co-pilot, on practice bombing, formation flights, and cross countries. There was a method in all of this and it was dictated by a plan in which all of the training was based on a schedule beginning with a small flight of bombers and then training schedules for the entire squadron. Next in the progression were flights involving the entire group and its squadrons. Then once a year the training moved to the wing level, and infrequently the division level, which was usually in the form of an annual exercise similar to the one Blackie had flown to the East Coast in 1931. He was averaging about fifty hours a month as pilot and about half that amount as co-pilot. Many pilots in the Eleventh Squadron were a bit jealous of the other two bomb squadrons in the group, the Ninth and Thirty-first, which were trading in their B-3 Keystone bombers for a slightly improved version, the B-4. Blackie was not jealous about the new B-4s because their performance left a lot to be desired; the Curtiss B-2 Condor's top speed, cruising speed, and power were all superior. What was turning the Eleventh Squadron pilots a light shade of green was the other squadrons having to pick up their new planes at the Keystone plant near Philadelphia, i.e., logging some real cross country hours. Within just a couple of years, both the B-2 and B-4 would be turned into dinosaurs by the new monoplane bombers with retractable landing gears.

In February 1932, Blackie logged barely eighteen hours of flying, and with good reason.

On Saturday, 27 February, Blackie and Martha were married at the First Presbyterian Church in San Antonio. Although the Bondurants were prominent in San Antonio, the wedding was a small and intimate one officiated by the Reverend P.B. Hines. The maid-of-honor and the bride's only attendant was Miss Janet Files, and the best man was Blackie's close friend from cadets, Lt. Kenneth Greeson. Both Blackie and Greeson were in Air Corps uniforms. Martha wore a suit of French blue wool and carried a corsage of gardenias and valley lilies. Although the pews were not completely filled, the church itself was fully decorated for the wedding with palms and greenery which included several bridal wreaths at the altar. Before the service started at nine, Miss Martha Ragsdale played "O Promise Me" on the organ in addition to the wedding marches.

Years later Blackie, in jest, would tell friends that it was a shotgun wedding which listeners could easily deduce was not true considering daughter Betty's subsequent birth date. Martha always sharply corrected her husband by observing one fact that Blackie omitted – that it was a military style wedding in which Blackie and Martha descended the church steps under the crossed swords held by Blackie's former aviation cadet friends. Blackie and Martha returned to Riverside immediately after the wedding and moved into their new apartment at 3075 Lemon Street at 2:30 A.M. on 1 March 1932, having driven straight through from San Antonio. They remained in Riverside only briefly and were then off on their honeymoon which was a road trip back to Mitchell and environs.

When Lt. and Mrs. Harold Blackburn returned from their honeymoon, Martha's daily life was very different. One of the changes was a more active social life because she was now an officer's wife which meant numerous activities at March Field, particularly at the officers' club. They had just gotten back from their honeymoon when the officers of the Eleventh sponsored an event that only they could, considering their connection with Mister Jiggs' Irish heritage. It was the annual "Corned Beef and Cabbage" dinner dance at the Officers' Club to celebrate St. Patrick's Day. The club was decorated as "Dinty Moore's Place" in keeping with cartoon character Mister Jiggs' fondness for the tavern owned by his friend James "Dinty" Moore. That comic strip tavern became the basis for many other real taverns named Dinty Moore's and was eventually adopted by a Minneapolis butcher as a name for his meat products. That line eventually enjoyed national recognition when the meat packing giant, Hormel, bought out the butcher and took the Dinty Moore brand into every grocery store in America.

Just several days after Blackie's marriage to Martha, Lt. Howard completed Blackie's fitness report for the preceding month. That document, in which Blackie was rated as "excellent" for piloting skill, attention to duty, initiative, intelligence, and judgment, contained one comment that would cause Blackie much anguish for the rest of his days as a reservist. Howard wrote of Blackie, "General all around well qualified officer. It is regretted that age prevents his being regularly commissioned."

By 31 March, the squadron had completed its portion of that year's wing level training exercises which involved a daylight roundtrip to Sacramento and return. Use of radios was becoming more widespread, and in the second phase of training the squadrons were sent to three other airports statewide and then alerted to an approaching enemy. Once aloft, the squadrons were controlled by radio toward the goal of bombing an enemy army approaching March Field. The Seventeenth Pursuit Squadron was involved in the exercise, acting as the enemy's interceptors. In the final phase of wing training, the bombardment squadrons were directed by radio to bomb an imaginary enemy naval force approaching the coast between Oceanside and San Diego.

Lieutenant Howard put an interesting assignment on the operations board for 20 May 1932 – one that would result in some of the most striking aerial images of the interwar period. At first glance, there seemed nothing unusual about a nine-plane formation flight up to Merced where they would remain overnight. There was a pleasant diversion planned: an automobile tour of nearby Yosemite National Park following their arrival. The bombers were to return to March on 22 May. Blackie and the other pilots sensed that the flight was going to be more than just another training mission when several of the gunners were bumped off the flight and replaced with Signal Corps photographers. Seeing the group off was the group commander, Major Carl Spaatz who had just landed the day before following an eighteen-hour flight from Washington in an O-38 after attending conferences at Wright Field in Ohio and the Curtiss plant in Buffalo.

The route to Yosemite was not direct as might be expected but involved three legs which took them over Death Valley and Mount Whitney before landing at the municipal airport in Merced. The photographers were busy capturing what was a unique northerly flight – flying over the lowest point in the United States and the highest point in California on the same mission. Before returning home on 22 May, Blackie had enjoyed an automobile tour of Yosemite before the flight flew over all of Yosemite's great landmarks: Half Dome, Mount Lyell, El Capitan, and Yosemite Falls which, in mid-May, was at its peak flow. In their day, those photos were every bit as dramatic as contemporary images of the Air Force Thunderbirds flying over such backdrops as Mount Rushmore, Pikes Peak, or Mount Rainier.

Blackburn Family Collection

In May of 1932, Blackie flew a Curtiss Condor as part of the Eleventh Bombardment Squadron's photo mission over Death Valley, Mount Whitney, and Yosemite National Park.

Some of Blackie's flying while with the Eleventh was anything but as exciting as flights over major American landmarks, but it was the type of flying that ultimately would prove just as marketable as his experience in heavy bombers. That flying was in cargo planes. In the early 1930s, there was no Military Airlift Command to move men and supplies. Squadrons and bases had access to the cargo planes of the day, and some of these aircraft were militarized versions of the leading airliners just as the Curtiss Condor bomber also had its civilian version. Most of Blackie's hours in cargo planes were logged on the Douglas C-1C which was the Air Corps' workhorse. It carried everything from supplies to litter cases and parachutists, the latter being the most frequently flown mission profile for Blackie. At the other end of the spectrum, he logged only a few hours in a rather rare bird, the Sikorsky C-6 which was an amphibian used primarily for coastal patrol. All of his flying on the C-6 was for routine test flights after maintenance. The two models that looked best on his resume were both tri-motors, the Ford C-4 and the Fokker C7-A. Both of the latter planes were in only limited use in the Air Corps, but they were used by many airlines, both in America and abroad. As with the C-6, Blackie's few hours on the tri-motors were logged for test flights or dropping parachutists.

Blackie was making his mark in the squadron and continued to impress his C.O., Charles Howard, who described Blackie as "one of the best qualified Reserve Officers assigned to this organization. He has mature judgment, an intelligent interest, and pleasant personality. He cooperates to his fullest extent in all activities." Howard, in what was Blackie's final fitness report while on active duty, also noted that in the event of war, he would assign Blackie as a flight leader, not just as a regular pilot.

Earlier in the spring of 1932, Blackie's squadron duties changed, although he was still the squadron range officer and supply officer. He became the squadron mess officer as well. He did not know it at the time, but the involvement with procuring supplies and the management of food items would be a very valuable experience considering his occupational situation not that far into the future. The squadron was changing too – the Keystone B-6 bomber was replacing the venerable Curtiss B-2, although the performance difference between them was minimal. The B-6 was on its way to becoming the standard bomber in the Air Corps until it was replaced by the Martin B-10, a two-engine low wing monoplane with retractable landing gear.

For the first six months of 1932, Blackie was averaging nearly seventy-five hours a month with nearly every hour as pilot, instead of co-pilot or observer. The squadron had not completely converted to the B-6, so many hours were in the B-2. His mission profiles were dominated by formation practice and cross countries, typically up to San Francisco's Crissy Field and down to Rockwell in San Diego. Occasionally there were flights over to Long Beach where he became acquainted with many of the pilots in the Air Corps Reserve Detachment based at the municipal airport located there. The formation practice intensified in mid-autumn as the squadron prepared itself for the aerial display over Los Angeles on 11 November: Armistice Day. Occasionally, there would be a variation such as mapping flights to Death Valley and other locations in California or taking up a P-12 fighter for practice landings on strange fields (any field not on ordinary cross country routes). Increasingly, Blackie was acting as a check pilot working with newly assigned pilots, particularly on instrument flying.

Although Blackie was a reservist on full-time active duty for the first half of 1932, he was instructing in the civilian sector and logging an average of a dozen flights a month in his off-duty hours. Other flights were for pleasure and included taking friends aloft for airport hopping. Those airports were the same that he had flown in and out of three years before as a student pilot himself – with Culver, Wilson, and Spartan among them. The contacts he had made in general aviation were valuable – he certainly needed them because on 1 July 1932, his period of active duty in the Air Corps came to an end. His applications for acceptance into the Regular Army had been denied due to his age, and the Army could not approve any more applications for extensions to his original tour of active duty. Because the Great Depression had eroded Air Corps funding, none of Blackie's fellow, former cadets got regular commissions; only the West Pointers remained on active duty. Blackie was headed for "Civvie Street" at a time when there was absolutely no light at the end of the Great Depression tunnel.

After winding up his active duty service, Blackie had an extra gas tank installed in his Model A Ford and set out with Martha on an extended road trip of which the first half was for pleasure. After visiting the hot springs at Thermopolis, Wyoming, and Yellowstone, they drove down to Mitchell where Martha met Blackie's brothers, Don and Charles and their wives. Martha also met Blackie's old flame, Helen Jones, who had married Art Delahoyde. Don was on leave from the Navy and Charles was an agent for the Van Norman Machine Tool Company of Kansas City. Charles was visiting in Mitchell preparatory to setting up his own auto supply business which he did in Scottsbluff the following year. Charles' wife was formerly Laura Jones – Helen's sister. Blackie joined his brother Charles and Laura and the Delahoydes on a road trip to the Black Hills of South Dakota (before Mount Rushmore was completed).

The second half of the road trip was through Middle America seeking employment somewhere in aviation. Blackie and Martha put hundreds of miles on that Ford driving to Kansas City, St. Louis, and Dallas before winding up the trip at San Antonio. The Bondurant residence became his headquarters for the first of many direct mail campaigns seeking employment as a pilot. There were no takers, not even a nibble. There was nary a glimmer of hope – all expressed regrets that they had nothing available with no prospects in sight. Pacific Air Transport in Cheyenne hardly expected any changes in the foreseeable future. Jack Frye, CEO at TWA, advised that he was keeping Blackie's application on file, and the Aeronautics Branch of the Department of Commerce sent similar regrets. Pacific International Airways of Alaska and Alaskan Airways offered no encouragement. The Treasury Department's Bureau of Customs planned to hire from within, and American Airways hoped to be more encouraging in the future but they, too, were not hiring copilots. Neither was Eastern Air Transport in New York. Shell Oil, where Jimmy Doolittle headed up the Aviation Department, rounded out the rejection list of which the preceding companies were merely the tip of the iceberg.

Dejected, Blackie and Martha soon headed back to Santa Monica, California, and a small apartment over a garage at 917 Eleventh Street. Out of that base, he spent days driving from Seattle to San Diego trying to find a job – along with millions of other Americans. Finally, Blackie decided that he would be tripping over his own gray beard before he got an offer to work for somebody else. Throughout most of the 1920s, Blackie had been self-employed. On the plus side, he was a self-starter who made friends easily, and he had connections within the local aviation community. Instructing students and flying charters would earn some money and his Air Corps Reserve status would bring in a few more dollars. That would sustain Blackie and Martha until he could put together a plan that would take them beyond a hand-to-mouth existence.

Thus Blackie became a "weekend warrior" at the Air Corps Reserve Detachment based at Long Beach which was a composite unit with its resident observation and pursuit squadrons. At first, his Reserves flying was limited to aircraft he had trained on at Kelly: primary trainers and the O-19. He logged from one to four flights every other weekend and occasionally flew the Douglas BT-2B whose 450 hp Pratt and Whitney engine could pull the plane along at 134 mph. The detachment was composed of many other second lieutenants who had graduated from Kelly in recent years; several had been in Blackie's March '31 class, so joining the squadron was almost like having a small reunion. In addition to Blackie, Johnnie Magden, Albert Reed, Charles McKinnie, and, for a brief period, the Williams twins were all in the detachment. Blackie also became friends with another Kelly grad, Joe Mountain, whose civilian employment involved flying aerial surveys for the state of California. There were some professional pilots in the group too, including Alexis Klotz, another Kelly grad, and several others from TWA. Those TWA acquaintances would become increasingly important to Blackburn, Mountain, Magden, and the Williams brothers.

Blackie was busy in the early autumn practicing as one of the formation leaders for the Armistice Day aerial parade. The Long Beach unit was the home for three squadrons: 514th Observation, 478th Pursuit, and 479th Pursuit although the total number of aircraft on the field could barely make up one full squadron. Nonetheless, Blackie, as assigned to the 479th Pursuit Squadron, was able to fly the Boeing P-12 fighter occasionally in what had to be joy rides – landing at distant, auxiliary fields, poking holes in the clouds, and practicing precision landings. On rare occasions, he could fulfill his Reserves obligations at March flying the B-2s again, typically on cross country and instrument training flights.

By the autumn of 1932 Blackie was regularly instructing students at several airports in greater Los Angeles. Many of those old fields are gone now. Where the daisies used to dance in the propwash of classic old biplanes taking off from Culver City Airport, countless thousands of motorists nowadays drive right over the old field; that long gone airport was at what is now the intersection of Marina Freeway and I-495. In addition to Culver City, Blackie flew charters and instructed students out of Clover Field which is best known today as Santa Monica Municipal Airport. Another airport where he occasionally took students on orientation flights was Mines Field; today they call it “LAX” – Los Angeles International Airport.

Blackie also took students into the tiny Wilson Airport which was located on the west side of Burbank Airport. He did so with great caution; the *Civil Aeronautics Administration Airport Directory* warned pilots to be alert for Hollywood stunt pilots who worked out of the little field. Like many other smaller airports, Wilson was gobbled up by a neighbor; in that case it was the extension of Burbank Airport’s east/west runway.

Because he was a married man beset with financial woes, Blackie’s social life was not the same as it had been in his pre-cadet days in Los Angeles. He and Martha rarely attended the old gatherings at Pancho Barnes’ mansion in San Marino, but Blackie was there one night when Pancho telephoned from Indio, just inside California’s southern border. Pancho’s friends used her mansion as a party venue whether she was there or not, although at this point Pancho had squandered so much of her inheritance that the gatherings were pretty much BYOB. Blackie answered the phone that evening and was thunderstruck to learn that Pancho was flying back from one of her adventures in Mexico and wanted to know what the weather was like and, in particular, if and when any fog might be rolling in (she had more confidence in other pilots than meteorologists). Blackie said that if she hurried, she would beat most of the fog. The fog rolled in faster than Blackie could have imagined, and Pancho was flying on top of it. Fortunately, she saw a glowing patch on top of the fog bank and correctly assumed that it was the beacon at none other than March Field where she was lucky enough to pull off an essentially blind, and very hard, landing with only minimal damage.

Largely as a result of the movie, *The Dawn Patrol*, which debuted in 1930, many people assumed Blackie was part of the Hollywood glitterati. On theater marquees where *Dawn Patrol* was playing, the lights lit up one of the movies’ stars, Douglas Fairbanks, Jr. For all practical purposes, Blackie and Fairbanks could have passed for identical twins right down to their moustache style – the “pencil” as popularized by two other silver screen heartthrobs: Errol Flynn and Clark Gable. Adding to the confusion was Fairbanks’ having starred in two other aviation movies of that era, *The Air Mail* and *L’Aviateur*. The confusion with Fairbanks wasn’t exclusive to California. Much to Martha’s consternation, many of San Antonio’s young ladies often had a dreamy look in their eyes when Blackie approached. With his easy affability, Blackie took it all in stride and welcomed the new contacts, especially when some of them became students. Aviation was big business in Los Angeles long before Blackie got there and for many years after he left, and not just in the number of airports. Along with Buffalo, Seattle, and San Diego, Los Angeles was a major center for manufacturers such as Lockheed, Douglas, North American, and their many subcontractors. In Cal Tech, L.A.'s aviation industry had a strong technical partner and research affiliate. L.A. was a fertile ground for Blackie's holding pattern – a waiting game that would eventually lead to a career with the airlines.

In January 1933, Blackie entered into a lease/purchase agreement with the owner of a failing Shell station with adjacent convenience store. There would be lease payments to the owner; Blackie was reluctant to approach his father-in-law for a loan to buy the place because in spite of their good relationship, he did not want to be seen as an indigent drummer and flyboy. The store had an attached residence which solved one of their problems. The location was at 7601 Vineland Ave. in Roscoe which is now known as Sun Valley, a Los Angeles suburb, or more accurately part of the San Fernando Valley. Blackie leased the property for $50 a month, and would eventually purchase the station and store with attached residence for $6,000. In the early 1930s, Roscoe was a truck and poultry farming center with a substantial immigrant population, particularly from Japan, Mexico, and Italy. The hours were long at first, but business improved steadily, particularly because Blackie had been able to significantly expand the facility. Within several months he was able to make a profit of over $300 a month.

During this business start-up period, Blackie's flying hours plummeted. He flew less than five hours a month as a civilian flight instructor in the spring of 1933 and logged no hours to speak of with the Long Beach Reserve unit until late July. On 1 July he was aloft with a student in a Kinner Fleet for instrument instruction in a flight that lasted an hour. Not so far away in Santa Monica, there was a short, very abbreviated flight of a brand new airplane that would revolutionize civil aviation. It was the maiden flight of the Douglas DC-1, the final result of TWA's contract with Douglas to produce the next generation airliner, and it was not quite what Robbins, Frye, Lindbergh, and Tomlinson originally expected in terms of power. The new plane only had two engines, not three. Still, they had approved the design based on Douglas assurances that the new DC-1 would fly perfectly well on just one engine. The first test flight was not encouraging. One engine failed shortly after takeoff, although lowering the nose brought the power back. Then test pilot Carl Clover raised the nose again and both engines coughed and sputtered; Clover lowered the nose and quickly landed. The problem was identified as easily fixed carburetors. The plane was subjected to further testing and then was turned over to TWA six months later for more evaluation. All of the DC-1's flight testing was taking place less than twenty miles south of Blackie's Reserves field in Long Beach. Had Blackie ever been flying a P-12, the hottest fighter in standard use in the Air Corps, and challenged the DC-1 to a race, he would have been soundly beaten. The DC-1 was faster than any pursuit ship in the military until the Boeing P-26 went into service at the end of 1934.

Only one DC-1 was ever built, but based on its performance and safety, TWA ordered twenty-five DC-2s, derived from the DC-1, at a cost of $65,000 each and put the first one into revenue service on its Newark-Pittsburgh-Columbus route on 18 May 1934. TWA's only modification to create a DC-2 from the DC-1 was to stretch the plane to accommodate fourteen passengers instead of twelve. The new plane offered exceptional performance: 190 mph cruise, ceiling of 23,000, and a 1,000 mile range thanks to its design and two Wright Cyclone engines of 790 hp. each.

The DC-1's main competitor was the Boeing Model 247 which first flew on 8 February 1933 and went into service with TWA's arch rival, United Airlines. Because United and Boeing were an integrated company in those days, United had nearly exclusive rights to all 247s produced. That was, however, a dubious distinction because the 247's 550 hp engines could not produce a performance matching the DC-1. The 247 never sold in the numbers that the DC-1 did for another reason; there were only ten passenger seats in the plane which was uncomfortable because passengers in the forward section had to step over the awkward main spar which ran right through the cabin. Blackie would have been elated to fly either one of the new planes, but his applications to the airlines still came back in the negative.

By mid-1934, Blackie was able to hire some help for his retail businesses, especially because he was in greater demand as a flight instructor at several nearby airports, particularly at Culver City Airport. Although he was again flying every other weekend with the Air Corps Reserve, he had not completely given up on the airlines and by his own admission kept on "pestering" United and TWA. In that lobbying, he had one friend on the inside at United. One of his old Kelly buddies, Edgar "Pinky" Stainback, had managed to get a co-pilot job at United which was quite a distinction considering his active duty tour had been spent in Sacramento as a fighter pilot at Mather Field. Although the airline was no longer hiring, Pinky Stainback urged Blackie to keep on knocking at United's door.

Blackie's convenience store became Blackburn's Red & White. There are still a few Red & Whites around, typically in small rural communities located far from a Walmart or other big chains. The Red & Whites operate on the same basis as IGA Stores, i.e., a co-op in which a central warehouse does all of the buying for the group. Culturally, it was an interesting and pivotal experience for Blackie because it was his first real contact with people from other lands. His experience in Texas as a cadet had slightly acquainted him with Hispanic culture and food, but his new neighbors and customers were largely Japanese-Americans along with their parents who had emigrated from Japan. It was a cultural awakening for Blackie who in the years ahead would eagerly grasp every opportunity to learn all he could about foreign cultures. He would later say that of all of his customers, including Italians and Hispanics, he found that those of Japanese descent were by far the most honest and trustworthy.

Blackie and Martha's household culture was about to change in a big way. Daughter Betty Louise Blackburn arrived on Monday, 13 March 1933. Martha was not taken to the hospital in Santa Monica in Blackie's aging Model A Ford. One of Blackie's friends, Harry Mayekawa, insisted that Martha be driven in the Mayekawa family Buick because it would provide a much more comfortable ride.

On 9 February 1934, Blackie flew with a student, Ray Sullivan, on an introductory cross country to Long Beach and Oxnard in the Kinner Fleet. Blackie was in an expansive mood because on 2 February he had finally gotten an offer of employment as a copilot from TWA. The men who were instrumental in Blackie's receiving the offer were Glendale-based TWA pilots, Jack Walsh and Ivan Olson, both of whom had originally flown with Western Air Express. Walsh was an older man whom Blackie first met in Nebraska when Walsh landed a Jenny on the family farm not far from Mitchell. Blackie was to start with TWA on 26 February by attending the next co-pilot training school scheduled for late February in Kansas City. Two days prior to that date, he was notified of his promotion to first lieutenant in the Reserves.

As Blackie and Sullivan were aloft on their three-leg cross country, there were two back-to-back meetings going on in Washington. Those meetings held the seeds of great disappointment for the Blackburns. Even before the Fleet biplane had taken off, General Benjamin Foulois, the Air Corps Chief, had been summoned to a meeting at the Post Office Building by Harlee Branch, Assistant Postmaster General. At first, Foulois could not imagine what the meeting would be about other than it might be attended by other cabinet departments which had some involvement in aviation. Foulois was aware that the Roosevelt administration had begun an investigation of potential mischief in the awarding of air mail contracts. That was exactly the subject of the meeting in which Foulois was asked how long it would take the Air Corps to prepare itself for carrying the nation's airmail. Foulois' response was ten days. Foulois returned to his office, gathered his key officers for a meeting, and ordered them to quickly create a plan to carry the nation's airmail. On that very afternoon, President Roosevelt announced the cancellation of contracts with the airlines and that the Army Air Corps would now have the responsibility for flying the airmail.

At the end of that period, for better or worse, but mainly worse, the Air Corps became the nation's air mail airline. Winter weather was a formidable problem, especially for Army pilots whose instrument flying experience was limited at best. Within the first week, there were two accidents of which one was fatal. By 10 March, there were another five fatal accidents. The Air Corps encountered so much adversity that its airmail operations were temporarily suspended on 11 March in order to regroup to undergo more training, give some routes to the railroads, and replace outdated planes.

Even with that brief breather, the Air Corps was not up to the task – it sputtered along until early June when new contracts with the airlines were negotiated. The task was monumental; there were eighteen different routes involving nearly 12,000 air miles to be flown by sixty-two day flights and thirty-eight night flights. The Army's planes, most with open cockpits and poor instrumentation, were not up to the task. Compounding that problem was the disproportionately high number of very young Army mail pilots of whom many were still in their year of active duty following graduation from Kelly. The Army's most experienced pilots, instructors at Randolph and Kelly, were not tapped for the new task. Many of the aircraft were not even equipped with radio transceivers. Very few of the new airmail pilots were familiar with flying the radio ranges that were now coming on stream across the country.

With their airmail contracts being lost, even temporarily, the airlines promptly froze hiring. Some pilots were furloughed beyond the few who were usually furloughed during the lower traffic winter months. Blackie's employment offer was promptly rescinded. There was a silver lining of sorts in that dark cloud in that the weakness of the Air Corps became very visible. Seeing the disaster unfolding in the nation's troubled skies, Secretary of War George Dern appointed an investigative committee to determine just how well prepared the Air Corps was for assignments in peace and war. One senator hit the nail on the head by wondering how well the Air Corps could deliver bombs if it could not even deliver the mail. The board was called the Baker Board after its chairman, Newton D. Baker, who had been President Wilson's Secretary of War. Among the committee's members were two noteworthy men. The first was Col. Edgar Gorrell, a major proponent of strategic bombing during the Great War. The second man's name was practically a household word in America, and he had become acquainted with Blackie in 1929. It was Jimmy Doolittle. The board's report was a major influence in the modernization of the Air Corps that slowly began later in the 1930s, but that had little immediate impact on the Blackburn family.

Quite by coincidence in late February, Blackie had tried to cover his losses on TWA's last minute retraction of its co-pilot job offer by applying for a year of active duty with the Air Corps for the specific purpose of flying the mail commencing on 1 March 1934. By the time the Army received the application, there was the growing question of whether it had bitten off more than it could chew in the air mail business; thus Blackie's application went nowhere. That the application was gathering dust certainly was not due to a lack of references; Blackie had pulled out all the plugs and top-loaded the question on references by listing General Foulois, Colonel Arnold, Colonel Spaatz, and Major McNarney as references. For the time being, Blackie was back to square one: flying with the Air Corps Reserve, flight instructing as time permitted, and working the long hours that are the curse of entrepreneurialism. His fortunes as a small businessman were improving dramatically which was most timely because Martha was again with child by early summer 1934. With Blackie being such a shrewd businessman, he was well on his way to completing a buy-out of the Shell/Red and White property. He was very busy in the spring of 1934 because he, as the operations officer of the 479th Pursuit Squadron, was playing a pivotal role in the Los

Angeles Army Reserves area war games of 1934. That annual exercise involved both the Air Corps and their brothers in armor and infantry turning back an "enemy" invasion of Los Angeles. Blackie coordinated the Blue Forces aerial component against the Red Forces, a combination of an imaginary power beyond the Pacific shore in league with a bordering nation in Latin America. As the Blue Forces aerial strategist, Blackie reported to his actual detachment commander, Major Don Poler, who had flown with the 148th Aero Squadron in France during World War I.

Blackie's Blue Forces were faced with a situation in which the Red Forces had landed a division-sized force on the coast at Santa Monica, Ocean Park, and Venice during the night of 18 May and were moving on Fort MacArthur from the rear. If Fort MacArthur fell, the Reds would be able to capture L.A.'s oil fields, industries, and harbor. The Blue Forces, supported by the Reserve air squadrons, had to cut off the Reds. In the meantime, Blue bombers were punishing the Red Fleet which had to retire to the south. The Blue aircraft took photographs of enemy concentrations, attacked them, and tangled with defending Red fighters. The dogfights, witnessed by all of Los Angeles below, included spinning aircraft simulating unfortunate pilots who had been shot down. As with any war games there were referees and judges, and at the end of the day their conclusion was that the Reserves pilots along with Blue infantry, artillery, and armor had saved Los Angeles from the Red peril.

Not surprisingly, Blackie's fitness reports from 1934 show that he was a valued member of the squadron. The 479th Pursuit Squadron commander, Capt. Albert L. Fisher, indicated that Blackie uses "plenty of initiative and tact in the performance of his duties" and that Blackie's services would be especially desirable in wartime, preferably in a higher grade. Blackie had started with the squadron as the assistant operations officer and had quickly moved up to the top operations slot. He filled one other position as well, one that would dramatically widen his list of acquaintances; he began serving as the public affairs officer. That duty increased his visibility because he was frequently quoted in the local newspapers. It also involved some other non-flying duties of which one was recruiting young men for a now long forgotten Roosevelt program: the Citizens Military Training Camps (CMTC). That program took young men age eighteen to twenty-four and paid them a nominal wage to attend a military indoctrination camp for one month in the summer. The program was structured to the extent that if qualified young men completed four summers they became eligible for a commission in the Army, i.e., it was the equivalent of a highly compressed ROTC program. Most attendees in California went to the CMTC camp up the coast near the Presidio of Monterey, although Blackie had no part in the camp's program itself. Other than broadening local awareness of Lt. Harold Blackburn, the weekend warrior and local businessman, the squadron PR job did put even more polish on Blackie's stage presence – something that would prove very useful in the years ahead. Although tens of thousands of young men became CMTC participants, the program never lived up to the goals set for it which is one reason why Fort Benning and other Officer Candidate Schools had to go into hyper overdrive on the run up to World War II.

Blackie's old cadet buddy, Clarence Shoop, had become an inspector at a manufacturing plant after having first been a school teacher following his Air Corps tour. Like many other cadets in Blackie's class, Shoop worked weekdays as a civilian but lived for military flying as a weekend warrior. When Blackie got reacquainted with him, Shoop was spending weekdays with the Wilshire Oil Company as its new transportation superintendent, but his weekends were spent at Griffiths Field in Los Angeles as a pilot with the 115th Observation Squadron of the California Air National Guard. The Williams twins had just found work with TWA in Kansas City – not as pilots, but as ground crew hoping they would get a crack at co-pilot jobs if TWA ever started hiring again. Blackie's friend Elmer Gorman, who was an ordained Methodist minister, had become a civilian again and was the pastor of a Methodist church at Parlier, in California's Central Valley.

A major change in the Blackburn family occurred on 14 January 1935. It increased by twenty-five percent with the arrival of Betty's baby brother, Bob. Considering themselves as young moderns, Martha and Blackie did not opt for "Robert," but decided to name the boy Bob because that is what he would be called for the rest of his life. Similarly, Betty's name was just Betty, not Elizabeth. Blackie got home just in time to take Martha to the hospital in Santa Monica. He had been with a student on a short flight to Long Beach and back. At this point, Blackie's instructing and picking up charter flights here and there kept him in the air four or five times a week. The business was going well enough for him to occasionally take some of his best customers, particularly his Japanese friends, up for pleasure flights which eventually stirred up so much interest that several of his passengers soon became students. Most of that flying was done in Fleet biplanes and Curtiss Wright Juniors. The latter was actually of parasol configuration, i.e, the fuselage was suspended from a high wing. The plane's detractors likened it to a flying bathtub. The plane had a forty-foot wingspan but its little eighty-five hp engine limited its performance, but then high performance was not necessary for instructing primary students. The Fleet biplane was Blackie's most frequent choice for instruction, and it easily accounted for three-quarters of Blackie's instruction time. The Fleet was sturdy enough to cope with the most ham-handed student and was available with a choice of several engines of 100 or more horsepower. Infrequently, Blackie also flew a very rare Kinner-powered Meteor, another high-wing monoplane of which only six were ever built. And like most experienced pilots of the 1930s, Blackie flew several models of those classic Waco biplanes, including the Cabin, Tank, and Taperwing models. When flying charters or giving instrument instruction, Blackie flew the four-place Kinner- powered Waco Cabin C Model.

Almost exactly a year after Blackie's offer from TWA was rescinded, he got a phone call from Kansas City offering him a job as a co-pilot. Blackie's friends on the inside at TWA had renewed their efforts to get Blackie aboard even though at age thirty-three he was several years beyond the preferred maximum age of twenty-eight for new co-pilot hires. A very grateful Blackie accepted TWA's new offer at once and prepared to head for Kansas City and training as a co-pilot. Signing on with TWA was the accomplishment of a steadfast goal: much like finally being accepted into the Air Corps cadet program and winning his wings while so many others fell by the wayside.

Chapter Four
On The Lindbergh Line

Blackie became an employee of Transcontinental and Western Air Inc., a company with less than 1,000 employees, on 8 April 1935. He was greeted in Kansas City by the chief pilot, Harlan Hull, who had interviewed him in Los Angeles, courtesy of Jack Walsh and Ivan Olson the year before. Hull was a widely respected, former Marine fighter pilot who had served in China and had started airline flying with Western Air Express in 1930. Hull introduced the six men in the new co-pilot class to Jack Frye, TWA's CEO, a man who would have a major influence on Blackie's career. Like Blackie, Frye had been consumed by flying ever since seeing his first airplane, a Jenny that made an emergency landing on the Frye ranch in Texas. Unlike Blackie, Frye scraped enough money together working at odd jobs to pay for his pilot's license which he earned at age eighteen. He bought his first airplane, an eight-passenger Fokker, with earnings from his days as a Hollywood stunt flyer, and set up an airline with his partner Paul Richter, flying passengers from Los Angeles to Phoenix. Whereas Richter, a former Colorado farm boy, was of small stature and considered to be very intense although with a big heart, Frye was a tall, burly tousle-haired man with an easy-going manner.

One of Blackie's pleasures when starting training in Kansas City was seeing more of his best friend, Ken Greeson, who was into his second year of flying as a co-pilot in the western region. Since their cadet days, Greeson had gotten married, and his wife was expecting a child in the near future. From Greeson and his other former cadet friends, the Williams twins who were by then co-pilots with TWA, Blackie had already learned what to expect during his brief training period. At that time co-pilot training was a combination of regular flights with experienced captains and minimal classroom work. TWA was then hiring almost exclusively military-trained pilots and assuming that they could absorb what they needed to know about TWA operations and procedures while on the job. Therefore, the need for significant ground school was limited until later in the 1930s when TWA had grown to the extent that there were not enough military pilots completing their active duty assignments to fill the need for new co-pilots.

In just a few years, the program was changed to more classroom work at first after which the new graduate would be assigned to one of the stations along the TWA route from New York to Los Angeles. After only several days of ground school, Blackie was given orders by Lee Flannigan in the Kansas City operations office to fly to Newark on 12 April in a DC-2. The captain was Joe Bartles who had three things in common with Blackie; he was a Kelly graduate, he went into bombardment, and he loved Mexican food. He had been with TWA and its predecessor airline for over six years when Blackie met him. Getting an early first flight was important; that established a pilot's place on the seniority list which in turn heavily influenced where a pilot lived, what he flew, and how much money he made. At the same time, Blackie was advised to join the TWA Pilots Association (TPA). Blackie overcame his ambivalence about the union just as he had with the American Federation of Musicians and signed up. The Airline Pilots Association (ALPA) was in its early stages of organization in mid-1935, but TWA pilots were afraid to join, fearing that an ALPA strike would ultimately cost them their jobs. Over the next few years, the TPA became less effective as a pilots' union, and eventually, in July 1939, TWA did sign a working contract with ALPA which by then had become the union for TWA pilots. Blackie's membership in ALPA would, however, prove not to be a turbulence-free flight over the next couple of decades.

Blackie's pilot seniority number became 115, although there were not that many pilots on the payroll at the time. Many, such as Jack Frye and Paul Richter, had gone into management, some had left to fly with other airlines, and a few had been killed in flying accidents. As a co-pilot, Blackie was close to the top of the pyramid. For every pilot flying the line, there were sixty-eight other TWA people on the ground in mid-1935.

Trans World Airlines

Blackie upon joining TWA in 1935.

Blackie's flight to Newark through Chicago and Pittsburgh with Bartles was a real eye-opener; the weather was dreadful. Blackie never saw the ground while in flight, and there was enough turbulence to convince two passengers to abandon the flight in Chicago and take the train east. Two more New York-bound passengers decided the railroad was a better option than continuing on after the stop in Pittsburgh. Those getting off in Chicago or Pittsburgh could not qualify for the Kansas City-New York roundtrip fare of $131.31. Just one way to New York was $72.95 and a Los Angeles-New York roundtrip came to $288.

The DC-2 Bartles was flying revealed one of its shortcomings: less than ideal sealing, especially around the windscreen joints. There was a standard joke about weather reports when flying a DC-2 – light rain outside and heavy rain inside. Blackie, whose sense of humor frequently ran to satirical descriptions, occasionally called the Douglas DC-2, the "Dizzy Two." Bartles was not kidding when he told Blackie matters could be worse. They frequently were and August's thunderstorms and January's snows and icing were yet to be encountered.

Blackie had a two-day layover after the flight and was staying at the Newark Athletic Club where Bartles, who enjoyed a good meal, asked if Blackie would like to join him on a jaunt over to New York City for some fine dining. Blackie declined the offer and opted for a quick meal at the Club's coffee shop in order to save time. Blackie had brought the aeronautical charts along and thought that he ought to spend more time familiarizing himself with all of the instrument approaches on the return route.

The trip back to Kansas City on the fourteenth went much more smoothly, and Bartles actually complimented Blackie on his performance. Of the other matters, Ken Greeson was right; becoming an airline co-pilot was roughly the equivalent of being an indentured servant – the reward of becoming a captain someday was preceded by several years of abysmal wages (around $200 a month in 1935). A new captain at that time was making four times as much. For that reason, Blackie held onto his Shell station and the Red and White store; he just hired a young man to fill in while he was away. Blackie was very fortunate in that his TWA-assigned domicile was Glendale, which was then the municipal airport for Los Angeles. It was very unusual for a new-hire to be stationed in California because that is where most of the captains and senior co-pilots were based at the time. Granted, fog could be an occasional issue, but the average flying weather certainly beat that over the Alleghenies back East. Being domiciled in Glendale also meant that he could continue with the Air Corps Reserve Detachment at Long Beach, although the days of that relationship were definitely numbered.

Nevertheless, Blackie's ground school was compressed and there was a lot to learn. Because of his friendship with Jack Walsh and his own interest in all things aeronautical, Blackie was generally aware of TWA's history and how it was really the evolutionary result of several mergers. What got Blackie's more immediate attention was a Douglas DC-2 that was propped up on stilts in the maintenance hangar. Entering the cockpit, each student was taught how to perform one of his most important duties as a co-pilot: manually lowering and retracting the landing gear with its infamous hand pump. There was considerable resistance in the pump, and it was a real workout for any co-pilot cycling the contrivance twenty-seven times to raise or lower the gear – on a good day. There were bad days which co-pilots learned to fear, especially in the winter when some of the seventy-two parts in the gear pump froze together.

Another mechanical issue for co-pilots was the heating system which used steam generated by engine heat that was ducted to radiators in the cabin where outside ram air was blown over the them to circulate the heat. Blackie was responsible for operating the heating system which had a mind of its own. If he let in too much ram air, the system would freeze. Conversely, not opening it enough would boil the system over. Either failure was an invitation for the captain's temper to similarly boil over. Pilots wearing long johns as Blackie did were not at all uncommon in the winter months, nor was wearing a woolen overcoat while at the controls. Adding to the ambiance of the passenger cabin was the music made by the radiators: a cacophony of clinks, clanks, and gurgles.

Conversely, Blackie had a partial role controlling the opposite problem – sun-baked airplanes taking off with Saharan cabin temperatures. The captains did what they could to minimize the stifling cabins by bending the company rules on climbs and descents. TWA wanted a manageable 300 foot per minute descent, but captains, especially on southwestern routes where Blackie was first assigned, followed their own procedures. That meant climbing as rapidly as possible after takeoff and letting the bottom fall out on approaches with only minimum concern for the passengers' ears: Because there was no air conditioning or pressurization which would allow flight into colder airspace, pilots and passengers just had to put up with it. And there was not a lot of relief considering the average altitudes flown, i.e., typically between 7,000 and 10,000. Blackie, as a co-pilot, did what he was told – open the air intake in the nose as cooler air was reached and close it on the way down to prevent a flow of increasingly warm air.

Taxiing a DC-2 was an exercise in vexation according to Blackie. Turns were made by cycling a handle next to the instrument panel and simultaneously pushing a rudder pedal. Because there was a delay between action and reaction, the DC-2 occasionally went where it wanted to regardless of the pilot's intentions. And woe betide a pilot who failed to lock the tail wheel before landing; the inevitable consequence was an impressive ground loop. Years later, Blackie looked back on the DC-2's brakes, saying, "The brakes were horrible. They weren't bad after you learned, but until then you had to be a contortionist." He also was critical of the vertical stabilizer which he felt contributed to enough instability that some passengers got airsick and hostesses occasionally spilled coffee. Fortunately, "burp bags" were already part of the air traveler's lexicon. On the other hand, Blackie was impressed with another DC-2 advantage; it was the first auto-pilot equipped aircraft he had ever flown.

Blackie always had fond memories of the DC-2, although until he flew the DC-3 there was little to which the DC-2 could be compared. Blackie also had a special place in his heart for the Curtiss B-2 Condor he flew at Rockwell and March, but the two planes were hardly comparable because they were essentially from two different aeronautical ages. For the co-pilot transitioning to the left seat, ground handling could be an issue until the new captain learned the importance of feet on the rudder pedals, left hand on the yoke, and hands on the throttles *at all times while on the ground.* Blackie also had plenty of praise for the revolutionary plane. He thought it was the best handling transport of its time, including its successor, the DC-3 which could not carry as much ice as a Two. The Two's propellers were not of the variable pitch variety, but cycling between their two, high and low, positions shed ice much better than the DC-3's variable pitch props. Ice would become a major issue for Blackie in just a few years when he began flying in the East on DC-3s. Indeed, going into Chicago on two occasions, he made it just in time before the ice accumulation became too much – had the flights lasted just minutes longer Blackie and his passengers would have been dead on arrival. On the first occasion, he had enough altitude to work with and on the second he was actually over the end of the runway just as the ship was about to depart.

As it turned out, one of Blackie's Kelly classmates who had also flown in the Eleventh Bomb Squadron and the Long Beach Reserves fighter squadron, Johnnie Magden, was employed in 1935 at Cal Tech as the manager of the wind tunnel. Magden, who had graduated from UCLA in 1929 with a degree in aeronautical engineering and then received an M.S. from Cal Tech after active duty with the Eleventh Bombardment Squadron, was very critical of the wing design of the DC-3. That was based on his assumption that the wing loading was too high and that the wing tips and wing roots were not well designed. Magden forcefully expressed his concerns to Douglas Aircraft, but his warnings went unheeded which caused a modest ration of heartburn in 1936 when he joined TWA for a nearly three-decade long career which was initiated as a DC-3 copilot.

Blackie appreciated the Two's speed; it was actually faster than the DC-3. The Three had more powerful engines, but it also was a significantly heavier airplane. Although racing airliners while on regular flights was strictly forbidden, on occasions Blackie found himself on a parallel course with a United Airlines flight. United, in the mid-1930s, operated the Boeing 247 which was slightly smaller, lighter, and considered underpowered with its pair of 550 hp engines compared to the Two's 730 hp engines. Decades later, Blackie looked back on the rivalry with United, noting "When we met up with a United 247, we could go by them on one engine."

More on-the-job training followed on the seventeenth when Blackie was assigned to what would become a very familiar route segment: Kansas City to Albuquerque. His captain was George Brill, a former air race pilot who came to TWA from Western Air Express. Brill was an easy-going captain with a smile about as wide as a DC-1 wingspan. Fate was not kind to him; he would die in a Kansas City hotel in 1948 at age forty-six from a heart attack. More ground school followed for the next few days, including a two-hour instrument evaluation flight in a Wright J-5 powered Stearman C-3 on the twentieth. On the twenty-second Brill and Blackburn were off to Los Angeles. Blackie did not have much of a reunion with Martha but he did have a reunion of sorts the next day when he flew from Los Angeles to Kansas City with his friend Jack Walsh. Blackie was very puzzled by this flight in which Walsh seemingly morphed from old friend into a distant, cold captain. It took a while for Blackie to understand what had happened – Walsh was saying, in effect, "I may have helped you get your new job, but from here on in, you will have to make your mark on your own." Once Blackie had proven himself, the old Jack Walsh reappeared; the two remained close friends thereafter. The flights between Los Angeles, Albuquerque, and Kansas City would become, with few exceptions, Blackie's standard route over the immediate future. One exception occurred early; it was a ferrying a DC-2 to South Bend, Indiana on the twenty-sixth.

Early May 1935 began innocently enough with routine flights between Kansas City and Albuquerque. When Blackie got back to Kansas City on 3 May, he was headed for two weeks of ground school and more dual instrument practice in the C-3 Stearman with Chick Fredericks who was emerging as TWA's instrument guru. The Navy-trained Fredericks was a relatively new hire, arriving earlier in 1935 as an instrument instructor. Shortly after Blackie's class graduated, Fredericks was domiciled in Newark and flew the New York to Chicago route although he would be called back to Kansas City when new co-pilot classes were started. Blackie logged nearly fifteen hours on instruments with Fredericks who was also known as "The Professor" due to his instructing talent. The outcome of his training with Fredericks was Blackie's being granted his Scheduled Airline Transport Rating following a written test and a three and one-half hour flight test on 25 May with the local Bureau of Aeronautics flight inspector, Russell W. DeLaney. That rating, later known as an "ATR" (Airline Transport Rating) required good instrument flying skills, which is why the test was preceded with plenty of practice with Chick Fredericks.

Blackie's time with Fredericks was devoted to navigation in the three-place Stearman which had been modified with the addition of a radio range system for navigation and an intercom system for instructor-student communication. At that point, the "Link trainer" was not in widespread use with the Army or the airlines for that matter, but that would change dramatically with the approach of World War II. The objective was to get Blackie up to speed quickly on being able to flying the airways using the radio range system and flying instrument approaches to all airports in TWA's network. Although Blackie was a consummate Army instrument pilot, his abilities did not automatically translate into airline universe where the low frequency radio range system was still relatively new. He had not gotten that knowledge in the Air Corps because it, at least in the mid-1930s, lagged behind the airlines in installing new navigation technologies. Although the Army was hardly unfamiliar with blind flying, as recently as the year before during the air mail scandal, it had few pilots or planes equipped to "fly on the beam." The "beam" was a "highway in the sky" consisting of four on-course signals sent out from transmitters which were located close to an airport and at intervals and intersections along the nation's airways. Although there were cockpit visual indicators for the system, the airlines preferred the aural version in which a solid, continuous tone (not completely unlike a telephone dial tone) was heard when flying dead center in one of the four legs, i.e., where the two signals overlapped. If a pilot, drifted to one side or another, the solid tone would diminish and be replaced by a Morse signal, either an A or an N, i.e., a di-dah or a dah-dit, thus prompting the pilot to make a course correction to get back on the airway centerline. Actually, it was company policy to favor the right side of an airway simply because that minimized the possibility of a mid-air collision with a plane flying in the opposite direction on the same airway.

Air traffic control was in transition when Blackie joined TWA. Earlier, airline company dispatchers, working in conjunction with airport tower operators, controlled flights in their immediate areas with radio while communication between air terminals was conducted by telephone and teletype. In the beginning, flights were visualized in each center by the simple means of a large map table in which little wooden cubes were moved over the airways as reports came in from dispatchers who had received position reports from en route co-pilots like Blackie. The arrangement was beginning to involve into a system, and in 1935 the first air traffic control center was established in Newark. That was followed by centers in Cleveland and Chicago a year later, but pilots of Blackie's generation did not begin to get any comprehensive en route help or control until radar's introduction into the system in the 1950s. Decades would pass, however, before radar coverage was extended to most of the nation's airways.

The radio range system was not perfect. Thunderstorms played havoc with it and, depending on terrain and weather conditions, the beams could bend in the air. Some pilots whose sense of spatial relationships was not strong could mistake one leg for another, leading to embarrassing if not fatal consequences. With the radio range, Blackie could fly between terminals in zero-zero conditions and use the station as a final approach fix when descending to make a landing. The standard instrument landing procedure was to pass over a station (the "cone of silence" because the tone disappeared briefly) and fly what amounted to a guided course away from the station, and then back to it, and over it on a heading that would lead directly to the airport. Each airport had a published minimum descent altitude, and if the airport was not visually identified within a certain time period after reaching the minimum descent altitude, the approach was declared missed in which case the pilot had to return to the station and hold over it, hoping the weather would clear very soon or go to an alternate airport. Blackie, with a mind that was capable of multi-tasking, instantly visualizing spatial relationships, and having a superb sense of timing found this new variety of instrument flying to be as comfortable as an old shoe. But he still had to demonstrate navigating the old fashioned way by calculating a compass course derived from wind speed, direction, the local magnetic variation, and the magnetic peculiarities of the specific airplane he was flying in.

There were also two means of navigating without having figured the heading or relying on the low frequency transmitters. There were the light lines – a series of coded light beacons spaced at ten-mile intervals than pointed the way between major air terminals which also had coded beacons, although the latter flashed a different color to avoid confusion with the airway beacons. Each airway beacon flashed its letter designation in Morse code, thus providing Blackie with a means of measuring distance in addition to knowing he was on the right course. And in certain areas of the country, specifically on the original cross country air mail route, there were large (seventy foot long), whitewashed arrow-shaped concrete markers on the ground showing the way. Those arrow markers originally showed the route from New York to San Francisco via Cleveland, Chicago, Omaha, Cheyenne, Salt Lake City, and Reno. The few of those old arrows that remain are almost exclusively located in the desert west.

Having been primed with all of the knowledge necessary to make an instrument approach did not mean Blackie was promptly expected to fly one on an actual revenue flight. He was expected to observe them closely. No captain ever let a brand new co-pilot fly an instrument approach or a landing anytime soon for that matter. A progressive captain would give a co-pilot a takeoff or landing once in a while in addition to en route flying, but there were some captains who were stingy and rarely made a move to advance their co-pilots' skills. For one of those solitary captains to give a landing was news and would spread through the co-pilot community accordingly.

It was during Blackie's instrument training sojourn in Kansas City on 6 May that a terrible event occurred near Kirksville, Missouri. Blackie got the news that his closest friend, Ken Greeson, had been killed in the crash of TWA Flight 6 at the conclusion of its flight from Albuquerque. Kansas City weather was so foggy that that landing there was impossible because the field was well below minimums for instrument approaches. The pilot, Harvey Bolton, opted for the emergency field at Kirksville, Missouri, where conditions were no better. He did not have enough fuel on board to reach another alternate; his only option at that point was to try setting it down in any flat field he could find. Unfortunately, he ran out of time – and fuel. The plane crashed into a dike near the town of Atlanta, Missouri, which is well over 100 miles northeast of Kansas City and fifteen miles from Kirksville. Five of the eleven passengers died, and Greeson was killed instantly. Capt. Bolton lingered several hours after uttering his last words, "I ran out of gasoline." There was a lot of blame to go around. The flight had left Albuquerque after a briefing that suggested the weather would be marginal for making an instrument approach into Kansas City. The airplane logbook revealed that the crew had been experiencing problems with the radio equipment from shortly after takeoff which, for all practical purposes, meant that they probably could not receive updated weather warnings en route while there was still time to land somewhere safely. The tragedy was a double-barreled blast of bad publicity for TWA. First, the public was still very much aware that it was on TWA which well-known football Coach Knute Rockne had flown to his death just a few years before. The Bolton crash and another one seven years later would suggest that TWA had just about cornered the market on celebrity deaths. One of the passengers on the flight that took Greeson's life was Senator Bronson Cutting from New Mexico. Because Cutting was so frequently seen in the public's eye, the accident became known as "the Cutting crash." Cutting came from a very wealthy eastern family which provided him with a superb education: prep school at Groton and college at Harvard. During World War I he was an attaché at the American embassy in London, but after the war had chosen to move to the Southwest where he became a prominent publisher in New Mexico. He was elected to the United States Senate in 1928 and was very well regarded there and was expected to go much farther in the years ahead. He had been seated very forward in the plane which is where all of the fatalities occurred. The only means of identifying him was by his wallet and a telephone bill found in a pocket.

Some good come out of the tragic accident. The United States Senate launched an investigation led by the chairman of its commerce committee, Royal Copeland. The full report was given a year later, and it lambasted the government for poor funding and its complacency regarding what the committee considered primitive navigation aids. Change eventually did come via the Civil Aeronautics Board later in the 1930s, but those modernizations were of little consequence to those who perished on 6 May.

Greeson's death was the first major loss in Blackie's life; many more would follow. Years later Blackie reflected on Greeson's loss and those of other friends:

I was at our Kansas City base right after it happened, and the way the older pilots treated it rather shocked me. They listened to what happened and made some non-committal remark, and then went on planning their next flight. Over the years, I've become that way myself. Of course you're affected when a good friend gets killed. Regret it? My goodness, yes! But you don't get maudlin about it. You know the risk that he took, and you know you are going out the next day. It rains on the just and it rains on the unjust. I've always said that if anybody has to get killed in an airplane accident, I hope it's the captain. After all, he gets the most pay, the one who makes the decisions, and is the one who can do the most about it.... This is our business. We accept the money. We accept the working conditions. We accept the fun of it. We should also accept the liabilities – the possibility that something will happen.

It was a philosophy that Blackie would have to rely on for the many friends who folded their wings over the upcoming decades, especially during World War II. Blackie always stayed in touch with Greeson's widow, Karol who never remarried, and the little daughter, Karol Ann, who never got to meet her father.

Classroom work continued and focused on federal aviation regulations, the DC-2 pilot's manual for which Blackie had to become intimately familiar almost overnight, another manual on TWA operating procedures (only thirty-six pages in those days), and the many forms which he would have to complete. The course also included, what was for Blackie, a refresher – filling out flight plans. He brought his own circular flight computer and plotter so he did not have to buy one from the company. But he did have to buy his new uniforms. The TWA route structure was emphasized as was familiarity with all of the terminals used by TWA: all of them from Newark to Los Angeles.

The ground school was not in a fine building dedicated to the instruction of newly hired pilots – that would appear in Kansas City several years later. Blackie's instruction was conducted in a corner of the TWA's main hangar where there was a blackboard, a desk for the instructor, Ted Haueter, and chairs for the students. Of that group, Blackie was the only pilot to fly to his mandatory retirement age of sixty. Blackie even outlasted his instructor with whom he had much in common. Haueter's first name was Oscar, but everybody called him Ted. He was a year older than Blackie and had taken his first flight in a Jenny at age nineteen before becoming an Army pilot. After leaving the Air Corps in 1924, he flew the mail for several years, but then became a test pilot and eventually joined TWA in 1934, staying only four years before joining Continental Airlines where he spent the rest of his career. Haueter did his best to recruit Blackie into Continental, but Blackie was a TWA company man, through and through.

It was Ted Haueter who taught Blackie many of the little tricks of the trade – keeping a schedule tucked inside his hat for ready reference and making up a handy little emergency repair kit packed with safety wire, pliers, screwdriver, etc. Co-pilots were also advised to carry a little bottle of mouthwash in their flight bags and to discretely rinse just before serving meals. Haueter taught the new hires to kneel or squat while talking to passengers or serving meals, thus making sure the co-pilot's rump was not rubbing against the passenger on the other side of the aisle.

Years later, Blackie would fondly look back to his early days in the mid-30s in TWA's western region, realizing that he had flown with TWA legends as captains, of whom some had flown in the Great War. Among the legends who sat in the left seat opposite Blackie were Cliff Abbott – a pipe-smoking Hoosier, Army-trained and formerly with TAT. Eddie Bellande – from Western Air Express was very, very popular with Hollywood starlets, until marrying one: Molly Lamont in March 1937. Bellande's background was varied: Navy flight instructor, barnstormer, and Hollywood stunt pilot, Mel Bowen – even keeled and very frugal. Lawrence Chiappino – handsome Californian whose insistence on details shamed even Blackie's precision; "Chippie" was a University of California graduate, former Army flight instructor, and, in his spare time was an amateur architect. Amos Collins – a still waters run deep type. Johnny Graves with a Distinguished Service Cross and Purple Heart from World War I. Howard Hall – best known as "Sonny Boy," started with TAT. The nickname "Sonny Boy" stuck after leaving a Langley Field post dance and finding there were not enough seats in the cars returning the pilots' dates to their residences. A rather large young lady beckoned Howard to sit on her lap saying, "You just sit right here in my lap, Sonny Boy." Ted Hereford – came from Western Air Express on

Fokker tri-motors, was athletic and a real lady killer. He dropped out of the University of California and sold his car to finance flying lessons. It was rumored that only the aroma of gasoline, oil, and engine exhaust could mask the traces of perfume which lingered on his uniform collars. Dutch Holloway – retained one of his airmail habits, flying with an emergency kit complete with bedroll, rations and parachute. He had flown fighters in France at the end of World War I. Milo Campbell also retained a habit from his air mail days, flying with chamois underwear in winter. George Rice – with a hawk-like face and plenty of experience on Fokkers. And Wes Smith – rough, tough and an outstanding instrument pilot who was never intimidated by thunderstorms. Ted Weaver was a former barnstormer who, like Blackie, had first flown a Jenny in 1919. "Tommy" Tomlinson, the Navy's aerobatic ace, had left the service and joined TWA where he was emerging as the engineering guru. Blackie always said he learned much more from his first captains than the best ground school ever had to offer.

Life as a TWA co-pilot was essentially one of servitude, certainly by today's standards. Jack Frye was reluctant to follow United Airline's lead in hiring "stewardesses," the forerunners of today's flight attendants. Stewardesses in the 1930s were, for the most part, adventurous former registered nurses, and until Frye was forced to employ them, co-pilots, including Blackie, became "bull hostesses," responsible for handing out lunch boxes and serving coffee. Many of the duties flight attendants now perform were the co-pilot's responsibility in the pre-hostess TWA. Thus, Blackie had to make an inspection tour in the cabin at thirty minute intervals – unless his services were needed in the cockpit during radio transmission or bad weather. One responsibility of the cabin tour was to point out any landmarks of particular importance on the ground, although pilots were instructed to avoid flying over or near visible aircraft wrecks on the ground.

Although the "Marketing Era" was decades away, TWA, with one major exception, was in the vanguard of sales promotion and positioning for its day. Chief among that promotion was positioning itself as "The Lindbergh Line." How could a potential passenger resist flying with an airline to which Lucky Lindy had attached his name and reputation? On the first timetable Blackie used in 1935, there were names for all TWA flights, and they all had a memorable ring. There was the "Sun Racer" (the westbound coast to coast flight) and other memorable, catchy names included Sky Master, Sky King, Sky Empress, and one to rival the Santa Fe Railroad, TWA's eastbound Sky Chief, all comprising the "The Criterion of Air Travel." TWA played up its real advantage as being "The fastest and shortest coast-to-coast." The literature promoted scenery along the way, saying passengers could count on "An intelligent description of sights along the way by an experienced co-pilot." Sights there were, including the Grand Canyon, Enchanted Mesa, Boulder Dam, and Lake Mead. Sales literature also encouraged passengers to use air mail at every opportunity as well TWA might have since that is where the money was. TWA was not turning a profit on paying passengers, but relied on the airmail for income. When Blackie joined TWA the company was carrying fully one-quarter of the nation's airmail and flying it for one $1.34 per ton (a figure considerably lower than other airlines). Instead of an attractive young hostess serving coffee, tea, or milk, the July 1935 timetable showed a handsome young co-pilot serving what was then then airline haute cuisine, fried chicken.

The one major marketing exception related to Jack Frye's resistance to hiring stewardesses concerned his perception of passengers preferring United Airlines because of the "pretty girls." So he conducted an airborne survey in which co-pilots gave out questionnaires regarding existing TWA customer service among the major airlines, thus probing the value of having a "host" or "hostess" on each flight. TWA was highly rated by seventy-five percent of the respondents, but Frye also saw the handwriting on the wall and in October 1935 the first young women, all registered nurses, went into training at Kansas City. They went into service three weeks later at $125 a month. Only fifty-five women were accepted out of the 2,000 that applied, and each successful candidate was taken aloft in a DC-2 to make sure she wasn't prone to airsickness. The trial flight was not failsafe, and a few of the young women were later let go as a result of subsequent airsickness. Each woman had to successfully complete a personal interview with Jack Frye before going out to the line, and when they did they were called "hostesses" because Frye detested the word stewardess and its corrupted version, "stews." The ladies made life easier for the co-pilots who no longer had to collect tickets or pour orange juice, soup, tea, and coffee. At first, Blackie had very mixed feelings about the value of hostesses, but his attitude did change – to the extent that one of his most cherished gifts upon his retirement twenty-six years later was a plaque inscribed with "To Hal. A Remembrance from Your International Hostess and Purser Friends."

Blackie's first misadventure as a bull hostess happened four months after he became a co-pilot. It was on TWA Flight 2 from Los Angeles to Kansas City on 17 August when a passenger, Mason Wadsworth, complained of feeling woozy and requested something to drink. Blackie quickly served him a cup of orange juice. That was not enough, and Wadsworth demanded coffee with cream and sugar to which Blackie replied that black coffee was available immediately, but the cream and sugar were part of the lunch box and that would not be served for a couple more hours. That was not satisfactory for Wadsworth who insisted on cream and sugar at once. So, Blackie opened one of the lunch boxes and while pouring coffee, Wadsworth lurched into the aisle. The coffee geysered out of the thermos, onto Wadsworth, and a few drops splashed onto the lady next to Wadsworth. Now becoming desperate, Wadsworth then reached into his coat pocket and lifted out a small flask from which he took several long swigs, saying not to worry about the incident. Ever so diplomatic, Blackie invited Wadsworth to the cockpit to view the Grand Canyon, but before long Wadsworth was back in his seat, sleeping it off. Upon arrival in Kansas City,

Wadsworth apologized and even bought Blackie lunch at Milleman and Gilbert's, the popular restaurant in the terminal. During the ensuing conversation, Wadsworth admitted that he had been partying heavily the night before – and that the party had extended right up to his departure from San Francisco to Los Angeles that morning to make the connection with TWA. Perhaps the day of pre-flight breath-o-lizer tests will be the rule someday, because the problem of intoxicated passengers has only gotten worse over the decades.

Pilots and co-pilots flying with mail in the hold when Blackie joined TWA were required to carry a sidearm. That practice had started when TWA operated a small fleet of mail planes, and subsequently pilots were required to carry an automatic pistol or revolver and to have that firearm inspected and serviced at regular intervals. Pilots were instructed to make sure that their pistols' hammers rested on an empty chamber. Within days of his class starting, Blackie was given a Smith & Wesson .38 caliber revolver by Jim Shaunty in the TWA Stores Department. Mail was very important because it was a sure source of income. Whenever a co-pilot computed an overload situation, freight was offloaded first, and if that was not enough, passengers were the next to go. The U.S. Mail enjoyed priority number one.

Blackie had to be the last person on board and was responsible for seeing that the passenger door was secured and that all passengers were properly seated (after he collected the tickets). And he was the first to the rear of the plane after its wheels were chocked in order to open the door and assist passengers off the plane. Because some unknown passenger complained to top management that he thought TWA aircrews were too casual when addressing each other, company policy, as made clear in ground school, was for all employees, whether in the cockpit or terminal to address each other as "Mister." Blackie was also required to help servicing the plane if the field manager was short on staff. As a co-pilot, he was expected to move customers' luggage from the ground crew handlers onto the plane before takeoff and from the plane to the burly gentlemen after landing. There was some baggage always on board with the DC-2, and it never was removed from the ship, nor was its presence revealed to passengers. That baggage was stored in a compartment just below the passenger luggage hold at the rear of the DC-2 and consisted of a half-ton of sandbags. It was the price of afting an otherwise, contrarily forward center of gravity common to the DC-2 design. As with every co-pilot, Blackie was always responsible for seeing that his DC-2 was fully stocked with all supplies needed during the flight. Punctuality was emphasized. Blackie had to telephone the field manager or dispatcher for a check-in and report in person not later than one hour before departure. Most new co-pilots were on station well before that because they were responsible for the paperwork, including the flight plan which by definition required a thorough weather briefing.

Co-pilots had to read the checklists that were used before starting, before takeoff, and before landing. It did not take long to memorize them, but company policy required that the checklists be read from the card carried in every plane. The "Before Starting Engines Checklist" ran to twenty-two items and the pre-takeoff list added another fourteen. Had any of Blackie's captains left the cockpit after a checklist had been completed, the entire checklist had to be repeated again.

Captains like Joe Bartles might casually participate in the weather briefing and suggest an altitude, but it was the underlings like Blackie that took the winds aloft forecast and crunched the numbers to generate the final flight plan which was approved by the captain. Relying solely on alternatives to a flying a heading based on calculating it from the winds aloft forecast, magnetic variation, and the plane's own magnetic peculiarities was unacceptable. The alternatives, navigating by the radio ranges and light line beacons, were both subject to mechanical failure, hence "doing it the old-fashioned way" was the ultimate back-up system. Blackie and all other co-pilots created the paper trail for every flight with one exception. Co-pilots were responsible for the flight manifest, flight log, and engine log, but they were not responsible for signing the flight clearance. That was strictly the captain's responsibility, and no flight could take off until it had been signed by the captain and the local TWA flight superintendent. The clearances under which Blackie flew were not the terse permissions granted by air terminal clearance delivery or ground controllers. Rather they consisted of the flight plan to which pilot and superintendent agreed, the weather forecast, the terminal weather forecasts for stops the plane would make, and the pilot's initialed weather analysis which included the intended flight's altitude. Carbon copies of the clearance and attachments were kept at the departing terminal. The paper trail did not end at the departing airport, but was supplemented with another report of en route weather upon arriving at the next field. If either the captain or the superintendent disagreed on the safety of the upcoming flight, it was promptly cancelled. As a co-pilot, Blackie handled nearly all of the radio transmissions. Captains of that era rarely wore a headset during takeoffs and landings, relying on co-pilots to converse with towers. Blackie was also responsible for calling in route position reports at regular intervals. When a passenger pushed a button that started a flashing light in the cockpit, Blackie had to ask permission to go aft and see what the passenger needed. That was in addition to the regularly scheduled walks through the cabin every thirty minutes.

Through the first half of July, Blackie was still on the Los Angeles-Kansas City circuit – TWA Flight 2 out and Flight 3 back, usually with the sartorially resplendent, ultra-polite Felix Preeg, a veteran pilot who had flown with Maddux Airlines. Blackie got off the plane back in Glendale on 16 July and was really struggling. He had a medical complaint which is said to be most common among two rather disparate groups of people: professional pilots and farmers. In about twenty years, although completely unknown to Blackie at the time, Blackie would be dividing his time between flying and farming, but for mid-July 1935, he absolutely had to have something done about this painful condition. He saw his family doctor late on the sixteenth and was in the hospital the next day for a hemorrhoidectomy. The surgery was completely successful, although Blackie always said that pumping the DC-2's gut busting landing gear was the problem's source. TWA granted him a medical leave of two weeks which was supposed to be without pay. Regardless of the inconvenience, Blackie did manage enough flying with the Air Corps Reserve to make up what he thought would be a lean month and studiously avoided flying aerobatics or high altitude during that two-week interval for obvious reasons. During this down time, Blackie completed an Army Extension Course (Number 30-9, on staff duties), getting a grade of ninety-nine which was worth thirteen credit hours.

Through some clerical error, Blackie was still paid his full monthly TWA salary. Blackie pointed out the paycheck mistake to Lew Goss, the western region superintendent, another old Western Air Express pioneer, who told him to forget about it, but to step forward on a day off when somebody else was out sick and fly that man's schedule. And that is exactly what Blackie did on many occasions.

Such arrangements were facilitated by TWA's being a much smaller, almost family-like organization in the mid-30s – when Blackie checked into the hospital TWA employed ninety pilots and about seventy co-pilots. New co-pilots in TWA's western region logged forty to fifty hours a month, and on the longer haul routes such as Los Angeles to Kansas City, the pilots frequently only flew one major leg, such a LA to Albuquerque, while the co-pilots flew the entire flight.

It was during this period that a quantum leap forward in aircraft development occurred in Seattle. That was the first flight on 28 July 1935 of what would come to symbolize American airpower in World War II – the Boeing B-17 Flying Fortress. Blackie could only imagine what it would be like to fly one since he could not fathom when such a huge plane would ever become available to reservists. And at that point, Boeing had not started in a direction to modify the new bomber for possible use with the airlines. As it would turn out, Blackie would indeed fly the commercial variant of the Flying Fortress within a few years and eventually the B-17 Flying Fortress itself – along with several other piston-engined aircraft which outperformed both the B-17 and its commercial variant. But those flights were very much unforeseeable to DC-2 co-pilot Blackburn in the mid-summer of 1935.

When Blackie resumed a regular schedule, he was flying roundtrips from Los Angeles to Kansas City with a day or two layover in Kansas City and occasionally two days off once back in Los Angeles. From a scheduling viewpoint, Martha was left with two young children part of the time, but Blackie was also home for a day or two each week which allowed him more time to spend with the business, his Reserve squadron, and his own flying – typically taking Martha sightseeing or taking friends like Harry Fujita, a neighbor, on short cross countries. In September 1935, Blackie's TWA flying was changed solely to the route between Los Angeles and Albuquerque with the only exceptions being an occasional ferry flight to Seattle. At the end of September, the close of Blackie's first full quarter with TWA, the company reported a profit of $212,288. There would be more good fiscal quarters over the next couple of years ahead, but they would not last.

What to do in Albuquerque on those occasions when he had to remain overnight? Blackie enjoyed Mexican food of which there many options in Albuquerque, but it did not take long to see all of the local attractions. One activity was popular with TWA pilots and that was taking one of the company's old Boeing Model 95 airmail planes out for joy rides along the Rio Grande, out to the Grand Canyon, over Sandia Peak, or up north for deer hunting. The Boeing was available for any pilots' use, but they had to pay for the gas and oil which is something co-pilots could barely afford to do.

On 17 December 1935, Blackie was flying from Albuquerque back to Los Angeles when another aviation milestone occurred: the first flight of the Douglas DC-3. It was not called the "DC-3" yet because it had been ordered by another airline, American, which wanted a bigger, better DC-2 to replace its now outdated transport version of the Curtiss Condor bomber. American's president, C.R. Smith, had entered into an agreement with Donald Douglas for the development of a sleeper aircraft, hence the development of the Douglas DST (Douglas Sleeper Transport) which Blackie always referred to as the "SDT." The transformation of the DST into the DC-3 transport with seating for twenty-one passengers came quickly – right after the first seven DSTs had been produced. The DC-3 was a larger plane than the Two, with a wingspan ten feet wider and a length five feet longer. It had more powerful engines, Wright Cyclones of 1,100 hp, but the new plane's gross weight reached 25,200 pounds compared to the Two's loaded weight of 18,560 pounds. The new Three could fly from the West to the East Coast in fifteen hours, which was not an improvement over the Two, but it could carry thirty-three percent more cash-paying customers.

Like most pilots, Blackie admired the Three's engines more than the airframe. The stall characteristics were not as forgiving as the Two's and its single engine performance was marginal. Losing one on takeoff was always a cause for concern – and double that concern if the loss happened just after takeoff at a thinner air, higher altitude airport. The Three did have toe brakes which was a considerable improvement over the Two's instrument panel-based handle.

Richard Robbins and Jack Frye were initially not impressed with the DST's test performance because its teething troubles kept the plane from meeting all of TWA's requirements, i.e., able to land safely at all airports served by TWA and to perform acceptably at high altitude fields such as Albuquerque (one mile plus seventy-two feet above mean sea level). Early tests demonstrated the Three's shaky performance on takeoff and landing roll. It was discovered that the oil scavenge system was preventing the Cyclones from developing full horsepower, so the engines were quickly modified to improve oil circulation. Bendix was brought in as the new brake supplier which solved the long landing roll issue. But by the time the fixes were in place, TWA was third in line to take delivery of a Three after American and United.

At about the same time the Three first flew, TWA appointed a new advertising agency, the William Esty Company whose flagship account was Camel cigarettes. TWA abandoned its "The Airline Run By Flyers" approach for the new campaign featuring Hollywood stars and the nation's leading businessmen enjoying the comfort, economy, and efficiency brought to them by TWA pilots and hostesses. Esty was positively clairvoyant about the Hollywood stars, but was destined to lose the account in the near future. Time would tell that as an advertiser, TWA proved to be a "love'em and leave'em" type of account. The account loss hurt – TWA like other airlines was a major national and local market advertiser. Esty soldiered on for another fifty years before disappearing in the merger mania that swept the agency business in the 1970s and '80s. Little did Blackie know at the end of 1935 what the future would hold for him as a pitch man for future TWA print advertisements.

On 26 January 1936, Blackie flew his last several hours with the 479th Pursuit Squadron at Long Beach, leading a wing formation on two flights, totaling two hours, and then an hour of instrument time with a safety pilot. Although his residence was in California, he was in and out of Kansas City often enough to join that city's elite Air Corps Reserve Detachment. He was also able to easily deadhead to Kansas City on TWA flights from either the West Coast or East Coast. There were several advantages of switching Reserve detachments, not the least of which was the availability of more modern aircraft. The Long Beach Reserve Detachment only rarely had the temporary use of a P-12; otherwise the pilots were stuck with PT-3s, the old Trusty from Blackie's primary days, and one O-19. Kansas City had much better equipment and more importantly that city's Reserve unit was comprised of many of TWA's senior pilots and managers, thus it was a good place to see and be seen. And flying out of Kansas City would provide an opportunity to fly to new places and a few old haunts in Nebraska. The Long Beach detachment offered little in the way of new challenges, as was clearly evident in its annual training camp for 1936: a cross country mission to San Diego and return which was followed by classroom sessions on chemical warfare. In fact, the non-flying events may have been the most appealing – attending the San Diego Fair and then a dance back at Long Beach upon returning. As it turned out, the Long Beach detachment did not enjoy much of a future. Its three resident squadrons were never called to active duty. They were de-activated just before World War II and never reconstituted after the war, unlike the Reserve Detachment in Kansas City.

In early April, Blackie found a note in his mailbox at the TWA station at Glendale Airport. Every other co-pilot at Glendale got the same note; it was from the senior pilot, Bill Coyle, requesting the co-pilots to contact him for setting up check rides for advancement to captain. As it was explained to Blackie,

TWA was in talks which would lead to the purchase of Braniff Airways, an acquisition that would dramatically increase the need for DC-2 first pilots. Although headquartered in Oklahoma City, much of Braniff's operations were in Texas as a result of Braniff's acquisition of Bowen Airlines and its mixed fleet of Lockheeds and Vultee V1As. Braniff already fed into the TWA system at Kansas City and Amarillo and had recently acquired new DC-2s for its own routes. Coyle, a former Army flyer who had joined Western Air Express in 1929, explained to Blackie that co-pilots qualifying as captains could also expect transfers to TWA's eastern operations base at Newark, New Jersey.

On 7 April, Blackie was at home in the San Fernando Valley, having gotten back from Albuquerque the day before. On the eighth, he was back in the air with Felix Preeg on a flight to Albuquerque with a stop in Winslow. The newspapers on that day carried headlines about the terrible crash of a TWA DC-2 near Pittsburgh the day before. Flight 1, originating in Newark, had crashed about forty miles southwest of the airport with the loss of twelve of the fourteen people on board. The plane had picked up several passengers in Philadelphia and would have landed in Pittsburgh at about the same time as TWA Flight 21 which was coming non-stop from Newark. The Bureau of Aeronautics blamed the accident on pilot error and the atrocious weather: heavy cloud and icing. TWA management contested the early findings based on how at least one other company pilot had observed the radio range transmitting a faulty beam signal which had him substantially off course (fortunately for him the weather was not zero-zero). Although Blackie did not know the captain, Otto Ferguson, who was killed instantly, he was aware of the airport's reputation because he had flown into Pittsburgh with Joe Bartles during co-pilot training exactly a year before.

In spite of the tragedy, there were two positive outcomes. Firstly, it hastened the Bureau of Aeronautics moving toward a proper air traffic control system. Secondly, the heroic story of the flight hostess was widely publicized. Hostess Nellie Granger was ejected from the plane where she had been seated near the tail. She soon recovered from unconsciousness with lacerations and bruises, looking at the sky. She found two passengers who were still alive, administered first aid, and then noticed telephone poles in the valley below the mountainside into which the plane had flown. She hiked down and followed the telephone lines over five miles to a farm house where she phoned TWA's dispatcher in Pittsburgh and later led rescue personnel back to the wreckage. Over the next few years, Blackie flew with several dozen different hostesses while domiciled in Newark, but Nellie Granger was not one of them although she had returned to TWA's Atlantic Region as a senior hostess. TWA rewarded her heroism with a Caribbean vacation and moved her to the prestigious Sky Chief flight.

Nellie had plenty of the right stuff. Five years later when she was the hostess on a DC-3 flight landing in Harrisburg, the pilots, Bernie Lloyd and Tom Lynch, could not overcome a jammed landing gear. So, they bellied it in, and

Nellie had the passengers evacuated almost immediately after the Three slithered to a stop.

Blackie and Wilton B. "Red" Miller were to be the first co-pilots checked out as captains which involved much more than adding a stripe to one's sleeves and moving over to the left seat. The flight tests were also a make or break event in a pilot's career since failing a captain's check ride was essentially the kiss of death. To become a captain, Blackie and Miller had to negotiate a series of check rides with Bill Coyle, several flights with line check pilots who would test them on instrument flying and knowledge of airports and emergency fields, and finally a flight with one of the company's top pilots, the chief pilot or one of his assistants. The prospective captains would also have to pass a check ride with a flight examiner from the Bureau of Air Commerce in order to obtain their 3B ratings (essentially a DC-2 type certificate).

Much of Blackie's three hours of being tested by Coyle on 16 and17 April was devoted to routine activities such as takeoffs, landings, and flying instrument approaches. The flights with Coyle also showed that Blackie was an expert on the DC-2's various systems and was thoroughly capable of flying the plane in all types of situations, including recovery from abnormal attitudes while under a hood, single engine operations, stalls, and steep turns. So as not to interfere with traffic at Glendale, most of the takeoffs, landings, and airwork were flown out of Mines Field and Palmdale Airport. With that first step successfully negotiated, Blackie could look forward to the 3B flight test and a series of line and instrument checks while flying a DC-2 on regularly scheduled flights between Los Angeles and Kansas City. Unfortunately, Co-pilots Blackburn and Miller never got past their first flights with Bill Coyle. It was not for lack of flying ability; the acquisition talks with Braniff soured and there were several bad TWA accidents very shortly thereafter. Blackie, ever the stoic, took it in stride and soldiered on as a co-pilot, but there was one benefit. The captains, with whom he regularly flew (Preeg, Coyle, Paseo, Morehouse, and Bellande) were starting to treat Blackie more as a peer instead of a lackey. The new status also translated into Blackie actually flying the DC-2 on instruments and doing more of the takeoffs and landings.

After the false start on making captain, Blackie was busy in May, logging ninety hours on flights between Albuquerque and Los Angeles with stops in Boulder, Nevada, and Winslow, Arizona (TWA actually operated two routes between LA and ABQ). Although he had not been advanced to captain, one other aspect of Coyle's visions of the future as discussed with Blackie and Miller did come to pass rather quickly; Mr. and Mrs. Harold F. Blackburn and family were leaving California and moving to New Jersey where Blackie would start flying out of Newark in early June. The Blackburn family was not just moving, they were also about to sell a business – at a tremendous profit. Some of that windfall went into the purchase of a brand new Buick, but much of the profit was saved because Blackie was still a co-pilot with the big money captaincy well into the distant future.

Flying from New York City's Airport, located in Newark, New Jersey, was attended by several new challenges, or more appropriately, opportunities. Blackie's night flying was about to increase significantly as was his experience with thunderstorms and icing. He also would be flying with a new set of TWA legends, nearly all of whom had been Army pilots, air mail flyers, and/or had flown with TWA's predecessor airlines. Among his first captains was World War I veteran and former Western pilot Hal Snead, the tall, he-man type originally from Glendale who was a line check pilot on his way to becoming a chief pilot. There was Jim Eischeid, a big, burly man who came across as a tough professional football linebacker. Eischeid, who was a world class deep sea fisherman, would become chief pilot at LaGuardia in a few years when New York City finally got a major airport that it could call its own. The most important influence on Blackie's starting to fly the New York to Kansas City route was Harry J. "Jack" Zimmerman who had graduated from Kelly in 1929 and went to Transcontinental Air Transport immediately after his active duty obligation was finished. Zimmerman was from Fremont, Ohio, where his family owned a cutlery manufacturing business. He was educated at Culver Military Academy and had completed his sophomore year at Dartmouth when the family business failed which cut Zimmerman's education short and resulted in his becoming an aviation cadet.

Zimmerman was known as the master of the calculated risk. Much of that mindset was due to Zimmerman being a fighter pilot being trapped in a transport pilot's body. As he completed advanced at Kelly in the pursuit track, he was appalled to learn that he would be going to Langley Field as a co-pilot – on bombers! It was the old story again: the Army personnel officer's reliance on the alphabet to make assignments. Because the vacancies in pursuit squadrons had all been filled before the personnel officer got to the Ys and Zs, Zimmerman had to take what was left over which was the posting to a bomber squadron. Blackie was impressed with Zimmerman's talent, but never tried to emulate some of Jack's antics. On one occasion Zimmerman made a seriously unconventional final approach with Blackie on board as co-pilot. Rather than a gradual descent, Zimmerman arrived much too high on the initial approach in a DC-2 and deliberately stalled the ship only to recover on what was a normal glide path to the runway. Zimmerman, who in a few years would be the chief pilot of TWA's Atlantic Division, had already survived one bad crash. That accident was blamed on a mechanical failure, and although Zimmerman quickly returned to flying, he would not always be so fortunate.

Several of Blackie's close friends were in the limelight on June first. Milo Campbell and co-pilot Frank Busch from his Glendale days brought the first commercial flight into Los Angeles' new airport at Burbank. The twins from Kelly Field, the Williams brothers, became an even bigger hit with the TWA publicity operation when it was discovered that the company had also employed identical twin hostesses: Marjorie and Margaret Stuecken. The two sets of twins were frequently relieved of flying duty to go on the TWA trade show and convention circuit, an activity Blackie would never dream of – although it came to pass many years later when he retired.

Blackie's first flight co-piloting out of Newark took him to Chicago with an intermediate stop in Pittsburgh on 3 June. Unlike the ill-fated Flight 1 of 7 April, the weather on Blackie's 3 June flight was completely unremarkable. He remained overnight in Chicago, and on the fourth flew a roundtrip to Kansas City, spending the night in Chicago again. On the fifth, he returned to Newark. That schedule would dominate Blackie's logbook entries over the next two years and vary only when he was on flights that stopped in Columbus and Indianapolis or when he was assigned to a special charter flight.

That first flight out of Newark (five hours to Chicago with a stop in Pittsburgh) was captained by Don Terry, one of TWA's former Navy flyers and a crackerjack with Morse code. Terry would become an ocean-hopping associate of Blackie's during the war and eventually became a chief pilot. But within TWA pilot ranks, he was best known for his superb handling of a very tense, potentially catastrophic situation that played out on 1 May 1939 as he was preparing to land at Pittsburgh after a flight from Chicago. TWA's chief test pilot, Tommy Tomlinson, had been up in the cockpit en route with Terry and the first officer, M. H. Gassing but left the cockpit as the final approach was initiated so as not to distract pilots during the landing. As he returned to his seat in the passenger compartment, one of his pants legs was ripped by something in the baggage hold. The "something" was a young male panda bear that had just broken out of its chains and taken a sharp-clawed swipe at Tomlinson. Rather than chase Tomlinson, the now irate bear then barged into the cockpit. The animal, temporarily nicknamed "Pan," was intended to become the suitor of Mei Mei, a female panda then on loan to the Bronx Zoo. Shepherding the bear was Professor A.E. Best of West China University who was to bring the Bronx Zoo staff up to speed on the mating habits of the *Ailuropoda melanoleuca* (the giant panda bear) in hopes that a baby panda would arrive sometime in the years ahead. Inasmuch as the DC-2 was on final approach, something had to be done very quickly, so Captain Terry grabbed a leather map case for protection and tried to shove the nearly berserk bear back into the baggage hold. Remarkably, he was able to accomplish this, but not before First Officer Gassing was seriously bitten on the left wrist. Professor Best's best advice came a bit late, "He'll calm down quickly if we can find some bamboo." While ursine-aerial disaster had been avoided, the liaison of Pan and Mei Mei did not go according to plan. Being put on a slow boat out of far away Shanghai, tolerating a noisy, bumpy airplane ride across America, and then being thrust into an environment which did not match his chilly, damp mountainous home, Pan did not grow into his intended role. In any case, he was not Mister Right for Mei Mei and died the following year. Mei Mei followed her intended in death two years later in 1942.

Blackie's estranged father had followed his son's career and learned from friends that Blackie was regularly flying into Chicago where the elder Blackburn had moved after briefly residing in Denver following the abandonment of his family. Dr. Blackburn summoned his son for a meeting in Chicago in which the father tried to impress Blackie by giving him a twenty dollar gold coin. Blackie was not having any of it and promptly departed the reunion. At that point, Dr. Blackburn may have felt the need to make amends, possibly knowing that his days on earth were numbered.

The summer of 1936 passed uneventfully, and although flying over the Alleghenies and the Midwest's sunbaked fields with their attending thunderstorms, Blackie was involved in relatively little instrument flying. His night hours, however, were adding up rapidly. On 17 July, Blackie was flying as Jim Eischeid's co-pilot on a DC-2 flight from Chicago to Newark when one of the twentieth century's pivotal events took place. That was the opening of the Spanish Civil War which was, in effect, a dress rehearsal for the larger conflict that would erupt in Europe little more than three years later. Blackie, like nearly all Americans, was not directly affected by the Spanish Civil War, but its prosecution in which Russia, Italy, Germany, and to a much lesser extent, France, participated did have a major impact on the debate concerning America's role in world events. The war and its aerial operations did contribute to an eventual build-up of the Army Air Corps, but that was still several years away.

There were good reasons for Blackie to be content with his work in mid-1936. TWA was profitable and business was good. TWA had more passengers per mile flown than any other airline. For the first half of 1936, passengers, who carried, 39,037 of them, were up 28.9 percent over the first half of 1935 and TWA was investing in new equipment. New, shielded loop antennas were going onto TWA aircraft. That made reception much better and improved navigation accuracy. Given two different radio transmitters, a pilot could determine his exact position by triangulation using new cockpit instrumentation. At the end of the year, TWA would order stronger engines for its aircraft – the newest 1,200 hp Wright Cyclones.

Blackie's first flight with his new Air Corps Reserves unit in Kansas City unit was on 19 September 1936 – an hour and a half flight around Pittsburg, Kansas. There were so many pilots in the detachment that its summer camp was held in two, back-to-back sessions which Blackie just missed. Had he joined several months earlier, he would have accompanied others, not to a location in Kansas or Missouri, but to Camp Ripley, just northwest of St. Cloud, Minnesota. That facility was a major training base for the National Guard, and provided the Kansas City pilots with many opportunities for a very diversified refresher experience. Every pilot got in some firing at a towed sleeve as well as stationary ground targets in addition to dropping sixteen dummy bombs on a 100-foot circle target. Much of the other activity could best be described as army cooperation due to the many exercises working with artillery and infantry, including radio liaison, dropping and picking up messages, and photography. All pilots went on a night cross country, participated in formation flights, and practiced spot landings from 90, 180, and 360-degree approaches.

The detachment had a few older aircraft such at the PT-3, but was also equipped with several P-12 fighters and the BT-9 trainer (a forerunner of the AT-6). The unit was also trading in its Douglas O-35s for brand new, factory fresh Douglas O-46s. Over the next several years, Blackie would log many hours in the O-46 which was quite advanced for the mid-1930s. The ship was a high wing monoplane with fixed landing gear, a three-blade prop, and a 725 hp engine. Its defensive armament consisted of one fixed and one flexible .30 caliber machine gun. In just a few short years, Army Air Corps leaders realized that reconnaissance pilots flying O-46s would become sitting ducks at a top speed of 200 mph, hence the shift to modifying the latest fighters with the addition of cameras. But in late 1936, Blackie had no inkling of what was coming in terms of photo-reconnaissance or what opportunities would come to him in that specialized form of pursuit flying. For the balance of 1936, Blackie logged only four more flights with the Kansas City detachment, but that would change dramatically in early 1937 when he spent the first two weeks on active duty with that unit. Many pilots who were reservists, especially co-pilots, spent their annual vacation period flying in the Reserves for which they were paid. In a co-pilot's case, the Reserve pay was a good deal because in those days TWA vacations were unpaid and the Reserve pay was higher than a co-pilot's salary for half a month. Most of those two weeks in early February were devoted to airwork, short cross countries, leading formation flights, and giving rides to enlisted men.

The detachment was located on Richards Field at Fairfax Airport which was just across the river from TWA's headquarters at Kansas City's municipal airport. The Army was not the only resident military organization. Indeed, there was another squadron, one which enjoyed a friendly rivalry with the Army pilots. That was a Navy Reserve fighter squadron commanded by none other than Tommy Tomlinson who, in addition to being TWA's chief test pilot, was also the company's assistant director of operations. The Navy unit was known in local aviation circles as the "TWA Squadron" because so many of its pilots and enlisted personnel worked for TWA. Although Jack Frye's first military service was as a corporal in the Army, he became an officer in the Navy Reserve as did Paul Richter. The Navy pilots flew Grumman F-3Fs, and although it was a tubby little, retractable gear biplane fighter that was rapidly becoming obsolete, it was still hotter than anything the Army reservists were flying. The Navy pilots flew well and had won the prestigious Davis Trophy, awarded for top operational performance, three years running. Fairfax was also home to a Marine Reserve unit, 10-MR, which flew observation aircraft and was almost exclusively composed of TWA pilots. Conspicuous by their absence were the Williams twins who preferred not to fly observation and training aircraft during their two weeks of active duty each year. They traveled to Barksdale Field in Louisiana to fly more modern fighters.

The Air Corps Reserve dated back to 1920 and the aftermath of World War I in which Reserve units were established in several of America's largest cities, such as New York, St. Louis, Philadelphia, and Boston, along with a few existing air bases including March and Crissy in California. By the mid-20s, Reserve facilities had been built at other municipal airports, including Kansas City and Columbus. Activity at most of these facilities never reached the level that Army planners envisaged due primarily to a lack of funding, shortages of Regular Army officers for command and training, and the inability of many reservists to make time for flying training and two weeks of summer camp. The original plan called for a pilot to log forty-eight hours a year, although seventy-two hours was set as the first goal, i.e., just enough for many of them who were not airline pilots to be dangerous to themselves and others. In practice during the 1920s, few pilots logged anywhere near forty-eight hours and that was at a time when reservists were only permitted to fly cross countries to destinations less than thirty miles from their home stations. Fortunately, by the time Blackie was attached to the Kansas City Reserve Detachment, the situation was starting to turn around. For example, reservists could fly cross countries to points 250 miles from the starting point.

Blackie was fortunate to find himself in one of the country's most active Reserve detachments and one commanded by a superior officer: Captain H.C. Wisehart, a West Pointer and seasoned aerobatic pilot. Wisehart took advantage of Blackie's skills and enthusiasm and promptly made him the unit's engineering officer. In his first fitness report, Blackie earned an "excellent" in each and every category (over fifteen criteria). Wisehart's personal assessment of Blackie was, "This officer is especially willing, loyal, dependable, and efficient. He has a pleasing personality and is an excellent officer." Wisehart had such confidence in Blackie, that he made him the commander of one of the two sessions of the 1937 summer camp which, again, was held at Camp Ripley in Minnesota.

The extension of cross country distances meant little to Blackie and the other TWA pilots who were in the detachment because they were able to fly Reserve aircraft at detachments in other cities as well. In Blackie's case that was a tremendous benefit because it allowed him to become even more familiar with emergency landing fields on TWA routes. That knowledge would become an absolutely essential part of his checkout as a captain – someday.

After settling into his new schedule, Blackie went house-hunting and started arranging for Martha, Betty, and Bob's move back East. Based on how a young airline pilot's life could occasionally be nomadic, Blackie decided to rent instead of buying a house. He found what he was looking for, a two-story stucco home, on a tree-lined street in Elizabeth, New Jersey, at 217 Glenwood Road which was located about three miles from Newark Airport.

On 21 December 1936 Captain Don Terry and Blackie, along with Hostess Gladys Entriken, enjoyed a few moments of fame for their record-setting speed on a non-stop flight between Chicago and Newark. They carved almost an hour off the scheduled duration of the flight and made it in two hours and fifty-eight minutes. It was not Blackie's only speed record; he would set several more in the years ahead, and like this first one, they all would promptly be broken by the next captain to enjoy a stronger tailwind. Blackie was getting to know Chicago like the back of his hand, and with a taste for varied cuisines, he rarely had dinner at any of the regular TWA hotels. Although Chicago was well known for good beef, Blackie's weakness for variety frequently took him to Chicago's Chinatown on the near south side. Here he met a young Chinese-American graduate of the Illinois Institute of Technology, working as a cook because he could not find an engineering job. His name was Ben Chan; his cousins would eventually be well known for Chinese food at their New York restaurant, The House of Chan, which became another one of Blackie's favorite restaurants. Blackie's regular flights into Chicago would come to an end in five years as would his enjoyment of Ben Chan's cooking. Unknown to Blackie and Chan at that juncture, they would resume their friendship in about twenty years.

The year closed well for TWA which carried 88,851 passengers for the year compared to 66,882 for the previous year. Air mail was up too, by over one-third. On revenue of $6,202,821, TWA made a gross profit of $397,348 and a net profit of $291,974. Billings on the TWA advertising account at the William Esty Agency suddenly went to zero; TWA signed on with Hanff-Metzger, an agency with a major presence in Chicago and Los Angeles. The agency came up with a theme that reflected a major TWA attribute: "First in Innovation." The quest for innovation became even stronger in the years ahead, but that was no guaranty that Hanff-Metzger could consider its having the TWA account in its back pocket forever.

By February 1937, Blackie had settled into his routine as an Atlantic Region co-pilot, usually flying roundtrips to Chicago and less frequently to Kansas City. His Air Corps Reserve flying was anything but routine after his O-46 checkout by Captain Wisehart on 1 February. Whenever he had a full layover day in Kansas City, Blackie was off on cross countries beginning with a flight to Scottsbluff via Lincoln to see his brother Charles. When he was not flying the O-46, he took up one of the old PT-3s for aerobatics practice. He frequently gave orientation rides to new enlisted men, including flights to municipal fields at Omaha and Wichita and military fields at Fort Riley and Fort Leavenworth. Occasionally Blackie practiced instrument flying with another lieutenant on board as safety officer, and then the roles would be reversed as Blackie became the safety officer. Sometimes, Blackie carried passengers to other bases; one such rider was Major Ennis Whitehead who later became an Air Force theater commander in the South Pacific during World War II. By far the most interesting Reserves work over the next four years was occasionally taking an O-46 or BT-9 from another Reserves Detachment in Illinois, Ohio, or Pennsylvania for periodic the approaches into non-military airports that the Air Corps frequently flew for training purposes. This "field checking" as Blackie described it in his logbooks took him into nearly any airport of any consequence on both sides of the airways from Philadelphia to Chicago. Some of these airports eventually were served by TWA, but in early 1937, places like Trenton, New Jersey; Lancaster, Latrobe, Allentown, Bellefonte, and Reading, Pennsylvania; Toledo and Dayton, Ohio; and smaller towns, such as Kokomo or South Bend, Indiana; and Joliet, Illinois, were not. First-hand familiarity with those airports was important to an aspiring co-pilot who could be assured he would be tested on knowledge of emergency locations at any point on an actual flight with passengers. In many cases, Blackie did not land at the airport – he just made a low approach and formed a mental picture of the field before heading onto the next objective.

Compared to most other reservists, Blackie was logging more than his share of hours in Reserve aircraft. Reserve commanders knew full well that very few of their pilots were getting the hoped for forty-eight hours aloft annually, but few if any of them, other than Captain Wisehart, knew that Blackie flew almost exactly twice that number for the full year of 1937.

On 12 February 1937, Dr. Charles Blackburn died of heart disease in Chicago. If Blackie felt any grief over his father's passing, it never showed.

On 15 March 1937, TWA lost another one at Pittsburgh, and it could so very easily have been Blackie. He and pilot Bill DeVries, who had joined TWA in 1934, were ferrying a DC-2 from Pittsburgh to Newark on what proved to be an uneventful two and a quarter-hour flight which was initiated in the morning while flying conditions were satisfactory. The weather went downhill as that Monday progressed, and Flight 15 from New York, scheduled to arrive at 5:58 P.M. that day never made it. The flight encountered such severe icing during its initial descent that the Two departed and crashed in a cornfield near the town of Clinton, about one mile west of today's Pittsburgh International Airport, with the loss of pilot Larry Dohnet, his crew, and all ten passengers. The ceiling at Pittsburgh was 1,500 feet, but the flight encountered such locally severe icing that even the ailerons were ice encrusted. The next day, Blackie was aloft again with Ole Olson on Flight 18 – to Pittsburgh and under a strict TWA rule that flights were to avoid any area where passengers could look down and see a crashed aircraft. Within a couple of weeks, when the investigation was complete, a large hole was excavated in the cornfield, the plane was dismembered, the remaining structures and engines bulldozed into the hole, and torched. The dirt was later 'dozed back over the hole, and within a couple of months the field was pushing up crops as though nothing had happened.

Although Blackie would primarily be flying DC-2s for the next few years, there were a couple of new experiences in the summer of 1937: on 1 July he made his first flight in the Douglas Sleeper Transport (DST). In September he added a new type to his logbook, the DC-3, at about the same time TWA added another city to its network, San Francisco. Like all DC-2 pilots, Blackie joyously welcomed one difference between the Two and the Three; he no longer had to pump the landing gear by hand (except in an emergency).

That Blackie was going to be checked out as a captain became obvious when he took his 3-B test on 9 October 1937 at Newark. That was essentially what is now known as an aircraft type rating, and there were two other pilots aloft on the flight which was to demonstrate Blackie's intimate knowledge of the DC-2 and how to fly it in a variety of situations. Hal Snead came along as the TWA observer, but the test was conducted by R.I. Hazen, the Bureau of Aeronautics inspector.

Beginning the next day, Blackie became the subject of twelve line checks involving nearly fifty hours of flying on the following thirteen days of flights between Newark and Chicago. That included all stops along the way: Camden (Philadelphia), Harrisburg, Pittsburgh, Columbus, Dayton, and Fort Wayne. The first four flights were monitored by Jack Zimmerman and TWA's assistant chief pilot, Otis Bryan. Blackie was well acquainted with Bryan from flying through Kansas City, although they had never flown a flight together previously. It was the beginning of an interesting relationship because Blackie's career and that of Bryan would become intertwined in just four short years.

Blackie and Bryan had much in common. Bryan came from a rural background in Indiana and had a lifetime interest in farming and ranching. He enjoyed ROTC and his engineering major at Indiana University, but left after three years because he loved flying even more. After March and Kelly, he also flew bombers, and after his active duty period he signed on with TAT as a co-pilot flying out of St. Louis, becoming a captain in less than two years. After TAT merged with Western to become TWA in 1930, Bryan was based in Kansas City. Six years later he became the chief pilot of the Eastern Division and concurrently the line's assistant chief pilot, working for Chief Pilot Harlan Hull. Among the older TWA pilots, Bryan was known as "Eagle Eye" because he was virtually unbeatable on skeet and trap ranges.

The remaining line check flights were monitored by Ole Olson who had been flying with TWA for over three years. Quiet and unflappable, Olson would enjoy a long career with TWA, and within a few years of checking Blackie he became TWA's chief pilot in San Francisco which was regarded as an absolute plumb of an assignment. Blackie passed that first series of line checks with flying colors, but was not immediately moved to the captain's seat full time. Instead, like all graduating co-pilots of that era, he was moved into a gray area with the title of "Reserve Captain." That meant that Blackie, for the better part of the next year and a half, flew some flights as the captain but many more as a co-pilot. Failure to pass the new captain's line check was quite literally a ticket right out of the TWA door. There would be other line checks in the upcoming months, but Blackie's services beginning in early 1938 were in demand as a co-pilot on some special "missions." Blackie's regular Chicago and Kansas City flight schedules were frequently upset by Jack Zimmerman, one of the first captains he had flown with upon moving to Newark from Los Angeles and the pilot who flew on Blackie's first line checks. Zimmerman, at that point chief pilot at Newark, took many of the charter flights that came to TWA, something every pilot wanted because it was a welcome break in the routine and a chance to fly in other areas of the country. Most of the charter flights were initiated by small groups of very well-heeled businessmen who wanted to go fishing in the Keys, attend the Rose Bowl, hunt big game in Montana, or attend some other special event. But when Zimmerman selected Blackie as his co-pilot for one client in particular, Blackie saw no fishing rods, deer rifles, or football pennants in the baggage. He may have seen some rods in shoulder holsters – the .38 or .45 type of rods that is. And adding to the exclusiveness of this particular charter client was its desire for plenty of publicity once it had completed a successful operation. The means, including transportation, leading to the successful outcomes were to be kept completely below the press and public's radar screens. The client was none other than J. Edgar Hoover and the Federal Bureau of Investigation. United Airlines had the original FBI charter contract, but some loose-lipped crews bragged about the flights, thus prompting Hoover to terminate the United relationship and to give it to TWA.

Zimmerman had flown the first secret FBI charter in 1936 when he flew Hoover and several other "G-men" to New Orleans where the notorious desperado Alvin Karpis was arrested. Karpis was bound up and taken to the waiting TWA plane which took off for St. Paul, Minnesota, where he was wanted for kidnapping. Turning Karpis over to authorities there, Zimmerman flew to Newark after dropping off G-men in Chicago and Cleveland. Soon thereafter, Karpis relocated to California, where he resided in Alcatraz for the next twenty-six years. The Karpis arrest had several ramifications. Hoover appreciated the cooperation received from TWA which went to great lengths to maintain secrecy. The nationwide news about the arrest fed Hoover's ego which resulted in his seeking additional publicity through more high profile successes, particularly in light of one of his agents, Melvin Purvis, having been the one to bring such high profile criminals as John Dillinger and Pretty Boy Floyd to justice. Hoover thereafter treated TWA as his special duties aerial unit.

By the late 1930s, the FBI was still involved in capturing or killing big time crooks, but was increasingly involved in surveillance at the request of the Roosevelt administration. FDR was becoming concerned about domestic security considering political and military trends in Europe. In either case, it made for some interesting flights for Blackie well into 1939. As the co-pilot, Blackie was the one who had to communicate with tower operators who never really knew they were dealing with a plane full of G-men. Blackie simply called the tower and acknowledged as TWA Flight 37, Flight 52, or some other innocuous-sounding number that was not in the current timetable. As the frequency of those charters grew, TWA avoided airports where the news media might typically lurk. For example, rather than be called to Newark, Zimmerman and Blackie might pick up the FBI agents at Floyd Bennett Field on Long Island instead and then deliver them and their captives to smaller airports near the ultimate destination under the cover of darkness. Some of the FBI charters were to out-of-the-way places such as Goshen, Indiana, while others were to cities eventually becoming part of the TWA route structure, including Miami and Detroit. Some of the flights were to places with which Blackie was very familiar: St. Louis and Chicago. Jacksonville, Florida, and New Orleans also entered Blackie's logbook for the first time with the FBI charters.

Blackie's story of his first FBI flight with Zimmerman became a classic as told and retold in TWA circles. Zimmerman called Blackie early one morning and admitted that he was a bit under the weather from partying the night before. Blackie inquired as to which sets of maps he should bring, although he expected the answer would be the usual Pittsburgh to Chicago route and possibly to Kansas City. Zimmerman told Blackie to bring maps for <u>all</u> routes in the continental United States. The two pilots had no idea where they were headed until Hoover and several G-men arrived at the TWA hangar in a big black sedan. Zimmerman was cobwebby, but told Blackie, "You take it the first leg, and then I'll be fine."

By the late 1930s, most Americans who followed foreign affairs focused on European matters, particularly as the growing militarism in Germany could hardly go unnoticed. Possibly because Europe was closer and it had been the main focal point of America's involvement in World War I, there was less attention paid to what was happening in faraway China, although several of Blackie's Kelly classmates were very much involved in a shooting war there. Japanese adventures on the Asian mainland had begun in 1931, and in December 1937, while Blackie was on two weeks of active duty with the Reserve, two events did capture American attention. The first was the Japanese sinking of the American gunboat, *Panay*, by bombing in the Yangtze River on 12 December. The other was the mass killing of civilians beginning on 13 December and lasting for several weeks in what became known as the "Rape of Nanking." Long before that, in 1935, three of Blackie's Kelly classmates were flying for the Chinese, primarily as bombing instructors. None of the men (Cecil Folmar, Sterling Tatum, or Roland Sansbury) had been in Blackie's inner circle of friends, but through the Kelly jungle telegraph Blackie was able to stay abreast of his classmates' adventures. Folmar, who had gone from Kelly to Mitchell Field on Long Island as an observation pilot, also had difficulty finding a job in aviation after active duty. He spent two years as a supervisor in a Civilian Conservation Corps camp before signing on with the Chinese. Fate was not so kind to Tatum who was killed, but Folmar returned to the Air Force and made a career of it while Sansbury became a test pilot in America after war was declared in 1941. Here and there, a few of Blackie's Kelly classmates were granted Regular Army commissions in the late 1930s as the Air Corps was strengthened, but Blackie, knowing his age was against him, never tried to gain a regular commission, particularly as his future with TWA seemed to be promising.

Blackie got back from a nearly six-hour sleeper roundtrip flight from Kansas City on 26 January 1938 and had the following day off. Several weeks earlier, a large disassembled object, completely covered up on a truck, had been moved under the cover of darkness from the Lockheed plant in Burbank to a remote hangar at March Field. The object, a curiously configured new aircraft was assembled and took off on its maiden flight on the twenty-seventh. Although Blackie was on a day off, he heard no news broadcast of the event, nor would the existence of this radically-shaped aircraft be acknowledged for a few more weeks. "It" was the prototype of the Lockheed P-38 Lightning, twin-boomed, twin engine fighter, and it more than met the Army's specifications; over 360 mph at 20,000 feet, better than 290 on the deck, and the ability to take off and land over a fifty-foot obstacle using a 2,300-foot runway. Blackie could only dream about flying a Lightning – after all, he was nearing forty years of age and the most modern fighter he had ever flown, the Boeing P-12, was a plane from the era of wing struts, bracing wires, fixed landing gear, and great round engines. His chances of flying a Lightning were just about zero. Or were they?

In November 1937, spurred on by growth in passengers and new routes linking Columbus, Dayton, and Fort Wayne, TWA ordered twelve new DSTs and DC-3s. The company also started experimenting with lower fares, including the announcement of winter excursion rates: available on any Monday or Saturday and good for any fifteen day excursions. Promoting such innovations could not save the account for Hanff-Metzger which lost it to the J. Sterling Getchell Company which had won the business by presenting TWA as the "Skyway of the Stars" who enjoyed "Scenery and Service." Getchell, whose showcase accounts included Chesterfield smokes, General Tire, and Chrysler, probably had no inkling of what was just around the corner when it came to Hollywood connections.

In early 1938, TWA suffered another national black eye when one of its flights crashed in California. In addition to temporarily slowing TWA's passenger growth which delayed Blackie's move to a full captaincy, the flight had some personal meaning for Blackie. He had flown with the pilot, John Graves shortly after joining TWA – and he had flown with Graves on the Eleventh Bombardment Squadron's mercy missions to snowbound Native Americans six years before. Graves' flight departed San Francisco bound for Winslow on 1 March with a 6,000 foot ceiling, but conditions quickly deteriorated as it flew over the mountains and into a major storm. Graves reported his plane was picking up ice and that he intended to make an emergency landing at Fresno. He never made it to Fresno, and the plane had been missing for over three months when it was discovered on 12 June in Yosemite National Park, considerably north of the route Graves had intended to fly. The plane had been completely buried by snow, thus preventing its earlier discovery. There was considerable storm static that night, and it was surmised that Graves had taken up a compass course rather than rely the on radio range to get into Fresno. But the storm's winds were much stronger than predicted, accounting for the DC-2 being so far off course.

On 12 March 1938, Blackie was again Jack Zimmerman's co-pilot on a night sleeper flight to Kansas City via Chicago. On that day, the first major step was taken by Germany on the road to World War II. It was the Austrian *Anschluss* in which Germany annexed Austria in direct violation of the Treaty of Versailles. The European victors of World War I, Great Britain, France, and Italy did little more than protest. The second step in the road to war was not long in coming.

Blackie was a co-pilot with Ole Olson on Flight 11 to Chicago on 14 April when a landmark piece of legislation passed Congress and was sent on to President Roosevelt. It was the Civil Aeronautics Act of 1938. Over the years, many airline managements have expressed displeasure with new rules and regulations coming out of Washington, but this legislation, signed by President Roosevelt on 23 June, was an exception. The old Bureau of Aeronautics was replaced by the new, independent Civil Aviation Authority which was responsible for air traffic control, the airways, and the safety thereof. The act also provided for a separate group called the Civil Aeronautics Board which enforced rules, conducted accident investigations, and dealt with airline routes and fares. Jack Frye welcomed the act because he felt that the old Bureau of Aeronautics had become a roadblock on his drive to open more routes which could feed into the company's main east-west artery and also allow TWA to take many more passengers to two major events planned for 1939: The New York World's Fair and the San Francisco Exposition.

On 7 June 1938 Blackie was on a routine flight out of Newark, bound for Chicago with Capt. Larry Trimble who in little more than fifteen years would be TWA's superintendent of international flight operations. Far to the west on that day at Clover Field, where Blackie had frequently landed his little Spartan C-3 biplane to build up his hours need for a private license, a new airplane took off for its maiden flight. The big new ship was the Douglas DC-4, a four-engine passenger plane for which design work was actually initiated before the DC-3 went into widespread use. The DC-4's launch customer was United Airlines, but all of the major airlines, Eastern, American, Pan Am, and TWA, had put up some seed money (nearly $92,000 apiece). Although the early test flights were uneventful, the original DC-4 did not live up to expectations which is why Pan Am and TWA decided to drop out of the program. With Douglas and four airlines having a say in the specifications and design, the new transport became a committee-designed airplane cursed with a heavier than planned weight (over 65,000 pounds) and unacceptably high operating costs which is why both TWA and Pan Am were hedging their bets; each was very interested in a commercial variation of the B-17 Flying Fortress heavy bomber. It was back to the drawing boards for the DC-4 which would re-emerge in four years as an essentially newly designed DC-4. United, however, took delivery on none of the improved DC-4s because all forty of the initial batch were promptly pressed into military service as Army Air Force C-54 Skymasters.

Having backed out of the Douglas deal, Pan Am and TWA placed orders for six of Boeing's new Model 307 Stratoliner four-engine passenger transport which offered several key advantages. The new airplane, which first flew on 31 December 1938, was, in essence, a modified Boeing B-17C bomber. The wings, tail, engines, and landing gear were stock B-17, but the fuselage was vastly different. The fuselage was quite round, much larger, and it was pressurized which offered an important advantage in addition to passenger comfort. Because it could fly higher, it could safely climb over most threatening weather. Among the major airlines, TWA had been conducting significant high altitude research and already had some experience with the jet stream, although the full scope of that phenomenon was not yet understood. Still, TWA was looking into what was emerging as pressure pattern flying, a subject with which Blackie was quite fascinated. That technique placed less emphasis on flying straight through on the published airways and instead emphasized harnessing prevailing winds around high and low pressure systems to reap the advantages of flying in more tailwinds (or minimizing headwinds). Blackie was aware of United's taking delivery of the first DC-4 and putting it into flight trials, but never assumed he would be flying one since TWA had opted for the Boeing design. Blackie was right – sort of. The original DC-4 design was heavily modified to create a slightly smaller, more efficient design, but by the time that was done and the new, improved DC-4 went into production in 1942, all new DC-4s had their names changed to C-54 Skymaster since United was no longer the customer. The Army Air Force was the customer and Blackie, although hardly apparent to him in 1938, did indeed eventually log many hours in the Douglas design.

At about the same time, another of Blackie's close friends from cadet days joined TWA. That was Elmer "Bud" Gorman, the ordained California minister who kept up his Reserve flying until landing an airline job. Gorman, who had been the cadet captain in Blackie's Kelly class, combined his pulpit presence with a strong penchant for practical joking. He pushed the comedy envelope right up to the point where his next caper might invite company discipline. On his occasional walks through the passenger cabin when he might pause to answer a question or point out an interesting landmark below, he was frequently seen carrying a soft cover book entitled *Learn to Fly in Ten Easy Lessons.* There was another favorite gambit, one that he frequently employed on unsuspecting new hostesses after he had made captain. He would press the cockpit button to request that the hostess come forward to the cockpit. The unsuspecting young lady would arrive to see both pilots staring forward, and just as she was about to ask why her presence was so urgently requested, Gorman suddenly turned around, leaving the hostess quite aghast. Bud had brought a most terrifying Halloween mask on board and had yet again used it to devastating effect.

Gorman and Magden flew uneventfully as airline pilots, right up to retirement. The same could not be said for all of Blackie's classmates. On 10 June 1938, Blackie was flying from Kansas City back to Newark with John Collings, a hard-driving captain who had joined TAT in 1929, when Samuel Stephenson, one of Blacie's Kelly classmates, was killed. The accident occurred near Asheville, North Carolina, after Stephenson could not recover in time from a low altitude spin in a BT-9. That was the only fatal accident among Blackie's Kelly acquaintances, but there were nine other classmates who were involved in non-fatal mishaps once they left active duty by mid-1933. Those accidents ran the full gamut of misadventures: minor mid-air collisions, mechanical failures, off-airport forced landings, and hard landings on airports. Blackie had avoided so much as denting a wingtip from his first days in the air and would continue to do so for the rest of his days as a pilot.

On 11 August, Blackie was on a sleeper flight from Kansas City to Newark when a significant aerial event occurred back in New York. A Lufthansa German airline Focke-Wulf FW-200 Condor four-engine aircraft landed at 4:00 P.M. after a twenty-five hour flight from Germany. At first glance, that news was not earthshaking as Blackie read about it the next day; after all, England, France, and Germany had all flown the Atlantic before. Those airmail flights were all made by seaplanes and had flown via the Azores and Bermuda. The aviation community was well aware that Pan Am would start its four-engine flying boat transatlantic passenger service not too far into the future. What made the Lufthansa flight notable was that it had been *non stop*, from Berlin and the FW-200 was a *landplane*, not a seaplane. That flight would not be forgotten by TWA or by a few clairvoyants in the Army Air Corps. Blackie, however, had his mind on other matters, including a forty-five minute instrument flight check with Hal Snead on the thirteenth. Eventually, Blackie would come to have a great deal of respect for, and even fear of, later versions of the German Condor. Although quite unimaginable at the time, Blackie would eventually be proud to say that he had made friends of quite a few Condor pilots, but that was still two decades into the future.

Blackie put in a long day on 30 September 1938, taking Flight 316 in a DC-3 from Harrisburg to Pittsburgh and then returning in the afternoon from Pittsburgh to Newark via Philadelphia. Neville Chamberlain, the British Prime Minister put in a less productive day as he and the representatives from France, Italy, and Germany signed away the Sudetenland, a German-speaking portion of eastern Czechoslovakia, to Nazi Germany. That grab, including heavy industries and significant defensive assets was officially sanctioned by the major European powers. The next grab would not be.

Earlier in the month, Jack Frye gave Otis Bryan another responsibility; setting up and running a new school for pilots of the new Stratoliner. Although it would be months before the first "Strat" would go into service, Bryan had accumulated sixty hours of flying the B-17 Flying Fortress and was therefore considered the most experienced four-engine pilot in the company. In moving into that job, he gave up a temporary assignment as chief pilot in the western region.

Blackie spent the balance of 1938 as an Army Reserve pilot or co-pilot flying Newark to Chicago roundtrips with an occasional flight to Kansas City. The only exceptions were the special charters with Zimmerman and one rather exceptional charter on the night of 7-8 November. The flight, lasting just one hour out of Newark, was in a DC-3 with Blackie acting as captain. The passengers all had one thing in common because they wanted the best seats in the house for an exceptional performance: a full lunar eclipse.

On 1 March 1939, Blackie's copilot and reserve captain days were officially over; he was rated as a captain, flying Newark to Chicago roundtrips. He was being watched, as were all new captains, and went through several instrument checks and line checks with Zimmerman, usually in a DC-2 but occasionally in a Three or DST. In early December 1939, Blackie's domicile changed: from Newark to LaGuardia. Finally, New York had an airport that it could call its own, and a major reason for the new field was the City's mercurial mayor, Fiorello LaGuardia, who had been an Army pilot during World War I. That the airlines, TWA included, printed timetables saying "New York" while those flights originated and terminated in New Jersey had always rubbed LaGuardia the wrong way. TWA became the focal point of one of the mayor's more blunt responses to the geographical misrepresentation. On one descent into Newark as a TWA passenger, LaGuardia became so livid that the TWA dispatcher and captain continued the flight to Floyd Bennett Field because LaGuardia said he was not deplaning anywhere but New York – as was indicated on his ticket.

Trans World Airlines

TWA, as the "Lindbergh Line," overflying Manhattan.

On 15 May 1939, Blackie was the captain on a DC-3 flight from Newark to Chicago when the descent into World War II became much steeper. Germany grabbed the rest of Czechoslovakia on that day. Only six weeks before, the Spanish Civil War had come to an end with a Nationalist victory. The Loyalist/Republicans had been losing ground steadily since the first of the year, and by early March its army had all but crumbled. With the war concluded, Germany was able to withdraw the aviation and armored assets that had fought on the Nationalist side back into Germany to thwart any Allied reaction to his occupation of the rest of Czechoslovakia.

As a line co-pilot Blackie paid little attention to board room politics other than casually reading about them in the company employee newspaper, *The Skyliner*. What was not reported in those pages dating back to late 1936 was the growing friction between top management (Jack Frye in particular) and TWA's majority stockholder, John Hertz, who was a partner at the investment banking firm, Lehman Brothers, which also held a substantial share in TWA. At the time, there were close to 30,000 separate, individual owners of TWA stock. Hertz, who had made his fortune in Yellow Cabs and later with General Motors, was able to load the TWA board with his cronies of whom none shared the vision of TWA's future as seen by Jack Frye and Paul Richter. Frye had seen the day when transport planes would need four engines to carry more people to altitudes where weather would not be such a problem. Frye correctly anticipated the major fight with the board when he announced his plans to talk with Boeing about building just such a plane. Richter was particularly frustrated by Hertz's lack of technical understanding when it came to innovations that would bring performance advantages to TWA. Frye's fundamental problem with Hertz was about profitability. Hertz, whose controlling interest dated back to December 1935, wanted a better bottom line and would veto any investments in the future. Jack Frye had known the eccentric Howard Hughes since the late 1920s when both were prominent members of the West Coast aviation community while Paul Richter's acquaintanceship with Hughes dated to late 1936. It was Howard Hughes who had bought the sole DC-1 from TWA, planning to use it in an around-the-world flight which was later scuttled in favor of a faster Lockheed twin. That started a dialogue in which Hughes became aware of TWA's top management having no confidence in the board of directors. In late 1938, Frye and Richter again met with Hughes, recounting all of their grievances with the two largest shareholders, John Hertz and Lehman Brothers, which combined owned about twelve percent of TWA's outstanding stock.

The conflict between Frye and the board of directors was heating up because Hertz had blocked the scheduled pre-payments to Boeing for the Stratoliner order. Frye, who saw the Stratoliner as the wave of the future much as he had seen the DC-2 as the industry's leading edge nearly ten years before, found the situation intolerable. Convinced that TWA was a good investment and having confidence in Frye and Richter, Hughes started buying TWA shares in January 1939 and soon had a controlling interest which was later expanded to over three-quarters of all shares by early 1940. The new ownership would prove to be a mixed blessing; Hughes was technically very savvy and had very deep pockets resulting from his other businesses, particularly in the petroleum industry. Counterbalancing his financial resources were several inescapable Hughes traits; he was autocratic, eccentric, and as time went on, increasingly neurotic.

TWA's top line was looking good for the first quarter of 1939 – it was the best quarter in the company's history for ticket sales. Air mail business was up fifteen percent and air express even more at thirty-five percent. The crack in the bottom line, John Hertz's fixation, was becoming another matter.

Blackie's contacts with Hughes were few and far between. In Blackie's last few years of flying with TWA, he considered Hughes a meddling, micromanager. That view would eventually change, but in the immediate future Blackie was definitely affected by Hughes decisions in many ways including the types of aircraft flown and the routes over which he flew those aircraft. In the meantime, disaster struck the Stratoliner program on 18 March 1939 when the Stratoliner prototype, which was originally destined for use by Pan Am, crashed during a demonstration to representatives from Royal Dutch Airlines (KLM), killing all aboard. Blackie was captaining a two-hour DC-2 flight that day from Newark to Pittsburgh and heard about the accident early the following morning. The accident, in which the wings and tail separated from the fuselage, was nominally caused by rudder lock during spin recovery in which stress caused structural failure. The Dutch pilots asking to see the plane flown with two engines out on the same side was the initial source of the spin, but the real issue was faulty rudder and aileron design. Those shortcomings were eventually corrected (note the addition of a large dorsal fin ahead of the rudder on B-17s following the C model and on production Stratoliners), but the loss of the prototype was costly to TWA in more ways than one. Harlan Hull, TWA's chief pilot and a friend of Blackie, was on the demonstration flight. TWA had originally ordered six of the Strats, but then reduced its order to five. The other launch customer, Pan Am, reduced its order to three aircraft which it intended to use on its Miami-Latin American routes. From a personal viewpoint, the tragic loss of Harlan Hull favorably impacted Blackie's career in one respect; Otis Bryan became TWA's chief pilot.

On those two fateful days in early September 1939 – on the first when Germany invaded Poland and the third when the Great Britain declared war – Blackie was off on the first and flying from Newark to Chicago on the third. After Poland collapsed, the war slowed until the following spring. Blackie was not directly impacted by the new European war, nor were most Americans on Main Street. That would take another two years of waiting and watching, but the American War Department was at long last starting to build up the military and the Army Air Force in particular, although it had a long way to go before American military aviation was the equal of the Royal Air Force or the German *Luftwaffe.*

Fiorello LaGuardia got his wish and his city's own airport located within the Empire State when New York Municipal Airport, better known as LaGuardia, was dedicated on 15 October 1939. It officially opened on 2 December, or perhaps more accurately, it was opened by Jack Zimmerman on a DC-3 flight from Burbank. Jack Frye had dearly hoped to open LaGuardia with the Straoliner, but delays in the program prevented that. Instead, Jack Zimmerman was inbound on a DC-3 flight that had originated in Burbank. The weather was atrocious and Zimmerman was faced with the real possibility of having to divert to an alternate which probably would have been Newark. Not wanting the honor to go to some other airline, Zimmerman predictably toughed it out and landed at several minutes after midnight just before the field was closed due to visibility going below minimums.

The year 1939 came to a close. Blackie was little affected by the opening of World War II. Nor was TWA – yet. In one sense, TWA was already at war, and the foe was itself. The company had enjoyed a banner year in terms of passenger revenue miles, closing out the year at $99,620,652 compared to $71,774,600 in 1938. Passenger revenue for 1939 was $5,195,780 compared to $3,734, 963 in 1938. Unfortunately, costs were rising fast; TWA suffered an operating loss of $188,827 for the year, and the financials would not prove very encouraging for the first three months of 1940 when the first quarter loss came in at $391,458.

Jack Frye attributed the losses to the cost of increased service, but optimistically believed that the losses were only temporary. Many TWA employees hurried to the bank with their paychecks, but time was on Blackie's side. Co-pilots would get pink slips before he would if matters slipped that badly.

On 1 February 1940, Blackie became a check captain. He welcomed the promotion since it meant more pay, nearly nine hundred per month, and the recognition that came with it. He was less thrilled with the work itself because he enjoyed flying more than critiquing other pilots. Years later, he attributed that promotion to his own training by flying the line with talented TWA captains and not from any formal ground schooling, although by that time TWA did have a world-class training facility in Kansas City. In the beginning, most of Blackie's check flying was devoted to instrument evaluations and airwork, or what Blackie called "transition" checks, particularly with co-pilots who were being upgraded to captain. The pilots being checked were of a generation which would go on to four-engine flying during World War II and eventually close out their careers in the 1970s flying the Boeing 747. Among those pilots Blackie checked was Bob Buck, later to become a TWA chief pilot, thunderstorm expert, and author of books on aviation weather. Another pilot was Marvin Horstman who eventually became TWA's chief international pilot.

Blackie also had to check some of his old Kelly friends, Bud Gorman and Johnnie Magden, both of whom were by then flying in TWA's Atlantic region. There was one other TWA pilot whom Blackie gave a check ride: Charles Lindbergh. Although not a regular line pilot with TWA, Lindbergh nonetheless had to undergo a periodic check ride because his association with TWA required his being current on anything TWA flew – DC-3s included. Blackie would eventually go on to become an international check captain and would also gain a reputation as a tough but fair taskmaster. He was well known for his cockpit discipline, being point-blank forthright, and not suffering fools gladly. Occasionally, check rides could bring out the more biting side of Blackie's normally very dry sense of humor. After World War II, when TWA's fleet included both the DC-4 and Constellation, Blackie had to give a down check on a Constellation qualifying ride to a man who had performed satisfactorily as a DC-4 captain. The man, whom we will call Captain Jones, was not having a good day and subsequently got busted on his check ride. Blackie commented, "Captain Jones was one of our best DC-4 pilots – and still is."

Most of those check rides were flown out of Newark and/or LaGuardia, but within a few months of the promotion, Blackie's check rides were almost exclusively flown out of LaGuardia which created a much longer commute to work. As soon as the lease on their residence in Elizabeth was up, the Blackburns moved to a home at 370 Congress Avenue in East Williston, on Long Island. The home was a well-shaded, two-story Tudor-style house with an arched stone entrance, a two-car garage, and a large fireplace. The home was the largest the Blackburns had lived in, and with good reason too; Martha was expecting a baby in mid-January. Because he was aware of how quickly TWA could move him to another domicile, Blackie again decided to lease instead of buying. The house had only one drawback; it was fourteen miles as the crow flies to LaGuardia, but Blackie did not have to drive through New York City to get there. The decision not to buy would prove most prudent.

Blackie was now averaging ninety flying hours per month which, with the exception of an occasional ferry flight between LaGuardia and Newark, was almost exclusively check flying, of which the majority was on night flights either in a DC-3 or DST. With the exception of Air Corps Reserve summer camp, he was averaging about five hours a month with the detachment in Kansas City. Nearly all of that time was in the O-46, although there were some local flights in what was a relatively new trainer, the BT-9, and the then antiquated fighter, the Boeing P-12. Blackie's last flight for the year with the detachment was flown on 11 October when he flew a BT-9 on a cross country to Wichita at the same time the titanic Battle of Britain was in its closing phase. The outcome of that epic struggle would have a direct impact on Blackie's responsibilities, his income, and his domicile within six months. Battered, but victorious, the Royal Air Force's need for aircraft vastly outstripped its production capacity which is why Britain looked to America for new weapons. Blackie's demanding check pilot schedule did not allow him enough time in Kansas City to fly the detachment's obsolete aircraft during much of 1940. But within another six months Blackie would be out West again and flying two of the Army Air Corps' frontline aircraft. He was quite unaware of that, however, as he landed back in Kansas City from Lincoln, Nebraska, on that eleventh day of October in an O-46.

In late October 1940, readers of American newspapers were learning that the Battle of Britain had just about wound down, but did not know that the months of night bombing which came to be known as the Blitz on British cities would be starting within weeks. On 26 October, Blackie was taking Flight 19 from LaGuardia out to Chicago, a day when the large Canadian troopship, RMS *Empress of Britain* was cruising just off the northwest tip of County Donegal, Ireland. The ship had been built in 1931 for the Canadian Pacific Steamship Company. At nearly 43,000 tons, it was once the largest and most luxurious ocean liner of the 1930s, but with the outbreak of war, it had become a troop ship and ventured as far as Suez. On that last Saturday of October, the *Empress of Britain* was attacked by a German Focke-Wulf FW-200 Condor, the bomber version of the four-engine transport that had flown non-stop from Berlin to New York two years before. The *Luftwaffe* pilot at the controls, Bernhard Jope, had been in training for service with Lufthansa, but had gone into the *Luftwaffe* instead and had flown bombers in Spain in support of the Spanish Nationalists. Jope, age twenty-six in 1940, hit the *Empress* with three bombs that disabled the liner which was then taken under tow by tugboats. As it was no longer able to maintain much speed or to zigzag, it became a sitting duck for any Nazi submarine in the area. The U-32 sank the Empress two days later, and Jope went on to become a highly decorated pilot who sank the Italian battleship *Roma* on 9 September 1943 as it was sailing to surrender to the Royal Navy in Malta following the Italian capitulation. Jope later severely damaged the British battleship, HMS *Warspite*, the cruiser HMS *Uganda*, and the American cruiser, USS *Savannah*. If told that he would be sitting side-by-side in a cockpit flying the Atlantic several times in less than twenty years with Bernhard Jope, Blackie would have been flabbergasted. But exactly that would come to pass.

TWA watchers during the first half of 1940 had a lot to observe and anticipate. The focal point was the Boeing Stratoliner of which TWA expected the first delivery in May. Financing the planes had become an issue until late in March when 119,154 shares of TWA stock were sold at $14 a share to Hughes Tool Company. By mid-June all five Strats arrived in Kansas City and were readied for their proving flights preparatory to going into revenue service. The early proving flights were not without a few scary moments; Otis Bryan encountered some very problematic carburetor ice that he could not overcome. He had to make a dead stick, off-airport landing in a practically unpopulated section of southern Colorado after breaking out of a 500 foot ceiling. After demonstrating that superior airmanship, Bryan flew it out of the prairie to Albuquerque several days later. The other dress rehearsal flights resulted in considerable positive publicity considering the complimentary tickets given to influential passengers, including the nation's leading aviation writers, magazine publishers, and aviation industry leaders.

The inaugural flights were flown on 8 July starting eastbound from LA to New York with Jack Zimmerman flying the first leg to Kansas City and then Waldon Golien from KC to New York which set a new record at twelve hours, thirteen minutes. The return westbound flight, with Otis Bryan flying NY to KC and Mel Bowen continuing to LA, also set a record: fourteen hours, nine minutes. Nobody ever called Golien by his first name, "Waldon." It was always "Swede" dating back to his days as a Western captain. Prior to that, Golien had been a Marine fighter pilot, serving in Nicaragua.

When the first regularly scheduled flight left LA for New York, it was star-studded, just as the advertising had promised; on board were Tyrone Power, the famous British producer/director, Sir Alex Korda, Prince Michael Romanoff, and Paulette Goddard. There were ceremonies and celebrities galore at the departures and arrivals: Mayor Fletcher Bowron and TWA Executive Vice President Paul Richter to welcome Mel Bowen and, in New York, Fiorello LaGuardia to greet Swede Golien.

Five days later, Blackie was on a flight from Kansas City back to La Guardia in which he gave the first leg, to St. Louis, to his young co-pilot, Dick Beck, who was about to get a major lesson in precision flying. Then Blackie took it to Indianapolis, but while entering the downwind leg talked young Beck through what would follow.

"We'll be landing to the west, so here's how to make your approach. When you are downwind and opposite the point on the runway where you want to land, you should be 1,000 feet above the ground. Reduce manifold pressure to twenty-three inches and descend to 800 feet. Then turn left for your base leg and reduce manifold pressure to twenty inches and descend to 500 feet. As you turn for final start reducing power and continue descending so that when you are over your intended landing point, 500 feet from the end of the runway, your throttles should be closed and your landing gear should be one-sixteenth of an inch above the concrete." Beck was incredulous, thinking he had the world's biggest braggart or joker sitting next to him – until Blackie's performance followed his instructions precisely.

At about the same time, Jack Frye, feeling bullish about the future ordered $1,100,000 worth of 1,200 hp Wright Cyclone engines, 111 of them, for the company's DC-3s and an additional fifteen DC-3s scheduled for December delivery. He did not get all of the Threes and engines in December, but had to accept deferred delivery because the Army Air Corps got thirteen of the Threes and all but thirty-four of the engines. Although America was not at war, Uncle Sam was gearing up for one. Many TWA reservists and National Guardsmen were called to active duty and others were enlisting. Many others had no choice in the matter; the peacetime draft went into effect in October. Paul Richter promised that those taking a leave of absence would return to better jobs than the ones they left. There was no doubt about the sentiments of at least one TWA office; employees in New York started donating blood for the military and civilian wounded in Great Britain which, at the end of 1940, were suffering under the German Blitz, particularly those living in London.

Blackie took a well-deserved, overdue vacation (unpaid) from New Year's Day through 15, January 1941. He became a father again with the arrival of daughter Bonnie on 12 January. Bonnie did not arrive at any Long Island hospital because Martha had decided that she wanted to have the baby back in San Antonio where her parents and brother Bill, an M.D., were close at hand. It was a very practical decision considering her already having two very young children at home and that fact that Blackie could be gone for several days at a time.

For the first six months of 1941, Blackie was averaging nearly ninety hours a month as a check pilot on DSTs and Threes, and the majority of that time was flown between LaGuardia and Chicago with only an occasional flight to Kansas City. In early June, TWA issued two press releases of which the lesser definitive one would have major impact on Blackie's career. The harder news of the two announcements was of Noah Dietrich's joining the TWA board. Dietrich was Howard Hughes' trusted right hand, financial man – one of few people Hughes really trusted. Eventually, the two would have a falling out, but that was far into the future, and in the meantime Dietrich would become a force to be reckoned with in the management of TWA. While John Hertz had been a nuisance to Jack Frye, Noah Dietrich would become an ever-present nightmare whose outlook was identical to that of Hertz. Dietrich and Hertz both thought that Frye's notion of TWA as an airline run by flyers was a good way to run an airline into the ground, and that the only sane way of managing an airline was with a businessman in the corner office.

The other announcement was of what was billed as a "Hughes-Frye" development of a fifty-seven passenger, four-engine plane with a performance that would leave the Stratoliner in the dust from both the financial and performance viewpoints. It was the Lockheed Constellation which, with seating for almost 60 passengers, could carry nearly twice as many people as the Stratoliner. That was a major issue because with seating for only thirty-three passengers, the Stratoliner could not carry enough people to become profitable, especially considering all of the costs attending its operation. The Connie's operating costs included a new, third man in the cockpit, the flight engineer, and a host of indirect costs such as publicity and advertising.

Hughes and Frye had a vision of new technology for a better future, and both were very much aware that Pratt and Whitney along with the Wright Corporation were in the final stages of developing a new generation of compound, supercharged radial engines – powerplants for which early models offered a thousand more hp than the Wright engines on the Stratoliner. TWA would be in the forefront of harnessing all of those new horses to power the next generation airliner. Blackie would be among the first to fly it, but the journey to putting the plane into commercial service would have its share of bumps, blind turns, and roadblocks.

Chapter Five
The Eagle Nest Flight Center

On 1 April 1941, Blackie accepted a new position, one that was among the most pivotal in his career. He had gotten a call several days before from TWA's chief pilot, Otis Bryan, in which Blackie was offered an important job in what was a radical departure from TWA's normal business as an airline. TWA was going into the pilot recruiting and training business in a big way as a result of Jack Frye's growing connections in Washington. TWA had signed contracts with two Washington-based organizations, the American War Department and the British Air Commission (BAC), to provide navigation and multi-engine training to two disparate groups of which the first was recent graduates of Army Air Force pilot training who would be assigned to ferry B-24 Liberator bombers. The second group was private citizens who would ferry bombers, particularly the Lockheed Hudson, across the Atlantic where they would promptly be put into service with the Royal Air Force. Great Britain could not begin to manufacture the number of warplanes it needed and had been placing large orders for American aircraft since the previous summer. Its first purchase of American military aircraft, however, dated back to the 1938 procurement of 200 Lockheed Hudsons.

Blackie would be in charge of all instrument flight training at what would become known as TWA's Eagle Nest Flight Center at the airport in Albuquerque, New Mexico. The airport was rapidly becoming known by another name: Kirtland Army Air Base. The Army Air Forces (successor to the Air Corps in mid-summer 1941 and usually referred to in the singular) had started construction of the Kirtland base on property adjoining the municipal airport in January 1941and would complete it in August. At the core of the operation was the advanced flying school which trained bombardiers; the base started with just one B-18 bomber, but its numbers would swell with the arrival of dozens of Beech AT-11s which were used in the bombardier school. Ironically, the municipal airport had been Albuquerque's second airport, and it was launched by Western Air Express, one of the two companies which merged to form Transcontinental and Western Air. Just how Jack Frye came up with the name Eagle Nest is uncertain, other than its being synonymous with an eagle eyrie. It is possible that he derived the name from the village of Eagle Nest, located just northeast of Taos, New Mexico, and not very far from the Vallejo Ranch in nearby Cimarron. The Vallejo Ranch, and other upscale dude ranches in the Southwest, hosted Conquistadores del Cielo (Conquerors of the Sky) a group founded by Jack Frye and Paul Richter in 1937 for what amounted to an annual, three-day executive retreat attended by other airline top managements, aircraft manufacturers, influential aviation editors, and publishers. The exclusive, secretive group, now composed of airline and airframe/engine manufacturers still meets, although New Mexico has given way to Wyoming as the preferred location for the annual wing-ding.

Jack Frye was able to land this contract because TWA was the only airline in America with extensive domestic service on the Boeing 307 Stratoliner. TWA also ran a substantial multi-engine training program in Kansas City for its own newly hired co-pilots because its route structure was growing rapidly. Most importantly, it had a cadre of co-pilots who had trained on the 307, and these men, along with those still in 307 training, were prime candidates to become twin and four-engine instructors.

The British were already acquainted with the Boeing 307. In late December 1940, several regular RAF bomber crews were sent to America for an orientation on the Boeing B-17, an aircraft type they would soon be delivering back to Britain. The crews took a side trip to Kansas City to see TWA's 307s before continuing on to Blackie's old base, March Field where they were to learn more about the B-17. Unfortunately for the Brits, they were not allowed to gain any flying experience with the resident Ninety-third Bomb Squadron due to the existing neutrality laws at the time. Thus TWA, a civilian concern, was seen as one reliable source of four-engine aircraft training.

TWA was to benefit from the Law of Unintended Consequences because it was certainly at the right time, at the right place, and with the right resources. The Royal Air Force and the Royal Canadian Air Force were not in a position, in late 1940 and early 1941, to release more than just a handful of pilots for ferrying or even training ferry pilots. The first flights of American aircraft to Britain had been flown by those with the most experience in long, ocean flights – pilots from the RAF Coastal Command. Only a few of those pilots could be spared since they were needed to fight the escalating German menace at sea. An unacceptably high number of planes flown by civilians from Canada to Britain were not getting through, and much of that loss was attributed to poor training. Concurrently, the United States Army Air Force was paying the price of emphasizing pursuit, attack, and observation training at the expense of bomber training during the interwar period. It had very, very few heavy bomber pilots, and heavy bomber instructor pilots were a rare species indeed. It was rumored that the Army Air Force had less than a dozen pilots qualified to fly the Atlantic in the spring of 1941. And it was only in the middle of June 1941 that the Army Air Force received its first Liberators.

The severity of the qualified pilot shortage was noted by the RAF's Air Vice Marshal John Slessor who was head of the Air Ministry Planning Department.

The present system where we bribe a few American pilots to fly machines over will not touch the fringe of the problem when we begin to get deliveries in really big numbers. Ultimately, we will want at least 1,000 pilots on this job, and that, as far as I know, is a commitment which we have never faced up to in our calculation.

The need for trained multi-engine pilots was desperate, and the number of airplanes being produced for the RAF was growing at a geometric pace. The pipeline through which growing numbers of American bomber aircraft started reaching Britain in early 1941 was the Canadian-based Atlantic Ferry Organisation (ATFERO) which had started recruiting new pilots from American, Canadian, and British airlines. Those sources could not fill the needs, so ATFERO had to recruit elsewhere. It is beyond the scope of this book to thoroughly trace the origins of ATFERO other than to provide a thumbnail sketch of the government, airlines, railroads, and characters who contributed to what is a circuitous and fascinating history of how the North Atlantic indeed became an aerial lifeline for the Allied cause early in World War II.

ATFERO was the publicly known name of the Canadian Pacific Railway Air Services Department. How could a railroad have such a dramatic impact on Blackie's life and the flow of warplanes across the North Atlantic to beleaguered Britain early in the war? In 1932, the Canadian government took steps to create an air route over the North Atlantic by signing an exploratory agreement with Newfoundland (not then a Canadian province), Ireland, and Great Britain. Initial planning focused on surveying potential routes and launching an airmail service before passenger flights could be established. The Canadian government felt that it should have a major role in developing the new air route instead of an American organization that was emerging as a threat – Pan American Airways.

National Archives

Consolidated B-24D Liberator of the Combat Crew Training School based at Kirtland Field, late spring 1942.

Three years later, a weather station was established in Newfoundland, and the Canadian government took steps to establish a single airline providing service between the Canadian coasts. That airline would provide the link to Newfoundland, the western terminus for the new transatlantic service. It was widely assumed that Canadian Airways would become the trans-Canada airline because it was the largest operator in Canada. It was also assumed that Canadian Airways would develop the North Atlantic route in partnership with the British Imperial Airways. The latter became British Overseas Airways Corporation (BOAC) in 1939 and later emerged as British Airways. The new Canadian airline would need much new capital for the expansion, and that was expected to come from the two railroads which each held a ten percent ownership in Canadian Airways: the government-owned Canadian National Railroad and the privately held Canadian Pacific Railway.

As it turned out, the Canadian Parliament did not favor Canadian Airways, but instead set up the government-owned Trans-Canada Air Lines with most of the capital coming from its state railroad, Canadian National. In 1939, the same year Pan Am established regular transatlantic service on the South Atlantic route, Canadian Pacific Railway bought Canadian Airways. Prior to that, construction of an airport in the Newfoundland wilderness was launched, particularly with support from Britain which wanted a modern airport located on the Great Circle route which would be flown by a revolutionary new land aircraft. That was the DeHavilland Albatross, a sleek, four-engine transport of plywood construction. The new airport was one that would appear frequently in Blackie's future logbooks: Gander.

On 11 November 1938, before Gander was operational, the first group of Lockheed Hudson bombers landed at Aldergrove, Northern Ireland, after a nine and one-half hour flight from Botwood, Newfoundland. Even with that, there was still very little activity over the North Atlantic where the weather completely shut down transatlantic flights in the winter. Infrastructure was still lacking, especially in radio navigation aids, although weather forecasting had been improved by the Canadian government which established more weather stations in Newfoundland. In fact, between 1919 and 1939, there had been less than 100 successful Atlantic crossings, but that would change dramatically as World War II erupted.

The conflict rapidly accelerated opening the North Atlantic air route in which many parties had a direct interest. The Royal Air Force needed more aircraft than Britain could build. It became the responsibility of the Ministry of Aircraft Production to build what it could and get the rest from America, which included seeing that the planes were delivered safely. Delivery by sea was hazardous at best. By the end of 1940, the Nazis had sunk over 1,200 British Commonwealth ships with half of those losses attributed to U-boats alone. In the early autumn of 1939, Sir Edward Beatty, chairman of Canadian Pacific Railway offered to put his company's air services at the disposal of the Canadian government. He was politely turned down although his aviation assets were already operating seven pilot and navigator schools for the Royal Canadian Air Force through the British Commonwealth Air Training Plan (BCATP). The latter was a program to rapidly boost the training of aircrews throughout the British Commonwealth of which Canada was a key resource. Beatty had friends in high places, and his offer did not go completely unheard. One of his friends was William Maxwell Aitken, better known as Lord Beaverbrook, the Canadian-born publishing magnate who became Churchill's Minister of Aircraft Production. The "Beaver" was a forceful and unconventional administrator who very quickly accepted his old friend's offer to provide services for ferrying aircraft from Canada to Britain. Beatty's organization inheriting this task was his Canadian Pacific Air Services Department, but its new, proper name was the Operating Division of the Ministry of Aircraft Production. That mouthful never flew and the new entity was simply called ATFERO (Atlantic Ferry Organization).

The Newfoundland airfield base was not without its problems, which prompted the British and Canadian governments to look for another base which offered more shelter from fog and the Atlantic's ferocious winter weather. That new base was Goose Bay, Labrador, where ships with construction equipment landed in September 1940. "Goose" was on its way to becoming a springboard for several more bases in Greenland and the established airfield on Iceland. All of those bases would eventually appear many times in Blackie's logbooks. The United States had actively been looking at North Atlantic real estate; that research resulted in the "Bluie West" bases in Greenland which could be used by shorter range aircraft before proceeding onto Iceland.

Where did ATFERO find pilots? Anywhere it could. It began by recruiting with the airlines and offering very tempting salaries, but that was not especially successful. The organization had more luck recruiting aerial adventurers, bush pilots, and what few multi-engine pilots it could find in the civilian sector. In late 1940, the Clayton Knight Committee, best known for recruiting potential fighter pilots for what became the three Royal Air Force Eagle Squadrons referred some men to ATFERO, but they were found to be largely ill-prepared for the dangerous flights over the North Atlantic. Long range ferrying was not without peril either – at war's end in 1945, over 560 lives had been lost in disappearances and crashes of aircraft ferried to British combat areas.

Shipping was realistic only for short-ranged, single-engine aircraft. Hope also was placed in the British Commonwealth Air Training Plan which was expanding rapidly, but at that point it was not beginning to turn out enough new pilots to make any real difference in transatlantic flying. Eagle Nest also was involved in recruiting for ATFERO and had transferred one of its traffic managers, Ralph Oursler from Los Angeles, to conduct final interviews in Albuquerque to screen applicants recruited from across the country.

In April 1941, Blackie, Martha, and their children, Betty, Bob, and the infant Bonnie who had been born in January, were headed away from New York

State for sunnier climes: Albuquerque, New Mexico. The decision to locate the Eagle Nest Flight Center at Albuquerque was a simple one. The location had to be in the Sun Belt and also on the TWA route structure; "ABQ" fitted the bill perfectly and was the final recommendation of Otis Bryan who had been dispatched in Jack Frye's personal airplane, a Lockheed Super Electra, to survey potential sites. The move to Albuquerque disrupted the Blackburn household; the Blackburns were now faced with leaving their comfortable Tudor-style home on Long Island and finding a new place to live in Albuquerque.

On 16 May 1941, Blackie was returning to LaGuardia on a check flight from Chicago, which is how the entire month was spent: one day out to Chicago or Kansas City and the next day back to New York. On that date, Sir Henry Self, head of the British Air Commission in Washington, directed that three Lockheed Hudson bombers, two North American AT-6 trainers (designated the "Harvard" in British service) and one Liberator bomber be diverted from ATFERO to TWA in Kansas City. This was just the beginning of TWA's fleet in Albuquerque; as the school grew, it would have eighteen B-24s in service. Its first Liberator was not strictly speaking a B-24; rather it was an LB-30 Liberator which was essentially the early export version of the B-24A. Unlike later models of the Liberator, the LB-30 utilized a mechanical supercharger instead of the turbochargers with which later models were equipped. Any airplane watchers hanging around the Kirtland Field fence may have thought the base had been invaded by a small foreign air force. The Hudsons were not painted in the typical olive drab and gray seen on Army planes, but were painted in two shades of gray: dark sea gray on top and sky gray on the bottom. The Harvards were bright yellow and the Liberator was earth-tone on top and black on the sides and bottom. All carried the early red, white, blue, and yellow roundel insignia of the RAF.

TWA had a massive job on its hands, and it had to have the new school up and running by mid-June. Otis Bryan was named superintendent of the school. Over fifty of the staff from TWA's training operation in Kansas City alone were transferred – and they were just the first of what was anticipated to be several hundred employees headed to the Land of Enchantment. TWA could not fully staff Eagle Nest from within, so many staff members were recruited from the outside. Much of that hiring was done by word of mouth and through personal contacts, especially through acquaintances in Air Corps Reserve squadrons and Air National Guard units. Blackie was not the only top gun pulled off the line.

The second in command, as supervisor of training, was Blackie's old friend from Air Corps Reserve flying out of Long Beach, Joseph D. Mountain who had been with TWA since 1937. After his Reserve duty in California, Mountain flew mapping and reconnaissance flights in Saudi Arabia for what later became the giant petroleum company, Aramco. His employment with Eagle Nest lasted for only a few months because he later returned to active duty and service on the air lift into China from India, "the Hump." Blackie was aware of his friend's colorful past, but had no idea in the summer of 1941 that his own path would, in the not so distant future, take him over the same deserts and peaks Mountain had flown over before and during the war.

Blackie's friend and senior pilot Swede Golien was the director of transition flying and concurrently recruiting ATFERO pilot candidates by conducting interviews in major cities from coast to coast. Additionally, Golien conducted the first staff checkouts on the Liberator. Although Blackie met Golien shortly after joining TWA and was well aware of Golien's reputation, they had never flown together prior to 1941. Frank Busch, a fellow co-pilot from the early Glendale-TWA days, was the director of twin-engine training, and like Blackie and Joe Mountain, was a Kelly graduate who went onto fly bombers before joining TWA. Among his close friends, Busch was called "Francisco." Busch always had a minor problem on forms where "place of birth" was required because in his case the location was best described by latitude-longitude coordinates. He had been born in the middle of the Pacific Ocean while his parents were traveling by ocean liner between the Sydney, Australia, and Vancouver, British Columbia. He graduated from the University of Southern California with a B.A. in 1927 and subsequently from Kelly Field with a commission and Army Air Corps wings. Busch focused on aviation early and had chosen not to go into the family business; his name was the Busch in Anheuser-Busch of brewing fame.

One of the twin-engine instructors was Joe Carr, a former Navy Catalina pilot who, along with another former Navy flyer, Bill Townsend, brought a rather unique distinction to Eagle Nest. In addition to having been rated naval aviators, they also were rated celestial navigators who had taught navigation by the stars at the TWA training center in Kansas City. It was Joe Carr who checked Blackie out on the Lockheed Hudson; it was not love at first sight. Blackie, having lived not so far from Lockheed's Burbank plant, was very familiar with the Hudson and its close relatives: the Lockheed Electra, as popularized by Amelia Earhart and other distance flying record setters, and the Lockheed Lodestar. In fact, the Hudson was essentially a militarized, first cousin of the Lodestar which was in regular service with Northwest Airlines. Like Blackie, Carr was very articulate in both verbal and written English which was remarkable because his parents were Slavic immigrants; Joe did not start speaking English until age six.

From talking with Northwest pilots, Carr was well aware of the Lodestar's problematic ground handling characteristics. The portly Hudson resembled a sumo wrestler, and like the wrestler who could easily toss an opponent out of the ring, the Hudson was just as likely to ground loop right off the runway with only the slightest provocation. The Northwest pilots had an advantage: their Lodestars were configured for carrying passengers and a co-pilot, not a bombardier. Eagle Nest got the bomber version which had only one set of controls for the pilot. Instead of a co-pilot, the Hudson carried a bombardier ("bomb aimer" in the RAF), and where the co-pilot would have been stationed, there was only a passageway into the nose and its bombsight. Therefore, instructing a student on the Hudson could be a trying experience, especially at the first sign of a ground loop. The aircraft was, however, quite successful on operations, particularly in use with the Royal Air Force and the Army Air Force which used the Hudson extensively for maritime patrol. The RAF also used it as a light, hit-and-run bomber for raids over France and the Low Countries early in the war. The Hudson was also the choice of RAF special duties squadrons for their nocturnal cloak and dagger missions with OSS agents operating to and from Occupied Europe. Blackie's introduction to the beer-bellied bomber went smoothly; and although his flying time with it in Albuquerque amounted to only several hours, it would not be his last encounter, with a pot-bellied, Hudson sporting foreign-looking colors.

Blackie had nearly a dozen people reporting directly to him, including the gentleman who was in charge of the school's long row of Link trainers. That was Hugh Davis who had packed a lot of instrument flying and education into his twenty-eight years. A native of the very picturesque village of Sonestown in northeast-central Pennsylvania, Davis attended Bucknell University before starting to fly. He had taken the instrument flying course at Randolph Field and had also been a sales engineer with the Edwin Link Company, manufacturers of the Link trainers which by the war's outbreak had gained wide acceptance in flight training. Davis had also taught Link trainer courses in Britain and had only recently returned from his work with the RAF.

Otis Bryan recognized that just because the flight instructors were some of TWA's finest pilots did not mean that their communication skills were at the same high standard. The articulate Captain Blackburn was a notable exception, but to dramatically improve overall teaching effectiveness, Bryan hired a Kansas City speech and debate professor, C.C. Fairchild, to serve as Eagle Nest's public speaking coach. In 1941, Blackie had absolutely no idea of how that training would become even more valuable to him exactly twenty years later.

On 2 June, Blackie flew TWA Flight 11 out of his base at LaGuardia and landed at night in Kansas City after the seven-hour flight from New York. He spent the next few days meeting his new instrument instructors and attended meetings for the purpose of getting Eagle Nest up and running by mid-July. On the twelfth, Blackie took Wilbur Knudsen, a TWA instrument instructor headed for Eagle Nest, aloft in a North American Harvard at Kansas City. There was no time-consuming check out in Blackie's transition to the Harvard; he had many hours in the Harvard's predecessor the North American BT-9 whose principal differences were a fixed landing gear, fabric covered aft fuselage, and a 400 hp engine instead of the Harvard's 600 hp engine.

Over the next week he flew with Hugh Davis and other members of his immediate staff on what amounted to instrument and night check rides between Kansas City and Tulsa. Then on 18 June, he flew alone to Albuquerque with only a brief stop at Amarillo, and spent the next two weeks on the ground, busy with the administrative details attending the Eagle Nest start-up. Four days after that flight came one of the war's greatest thunderclaps: the German invasion of

Russia on 22 June, a date on which one of the biggest nails in Hitler's coffin was rammed home.

By that time, Eagle Nest had already started up with a token staff and its first students. For the next six weeks, Blackie's flying was limited to the Harvard and consisted solely of quality control, i.e., instrument check flights with the twin and multi-engine flight instructor staff and his own instrument flying team which numbered eight men, about half of which had been hired from outside of TWA. The outsiders were not complete strangers because nearly all of them had contacts within TWA as the result of service in Air Corps Reserve or National Guard units. One of the outsiders was particularly well-connected; Robert Redding who owned a large farm just outside of Minatare, Nebraska. Redding and Otis Bryan were also business partners in a 7,500-acre cattle ranch just west of Wheatland, Wyoming, near Laramie Peak. One of the prominent features of the ranch was a large lake which offered particularly good duck hunting, a favored sport of both men.

Like Blackie, Redding was consumed by aviation, but unlike Blackie he was given a Curtiss Jenny by his wealthy father. The young Redding was giving flying lessons on the Jenny at age eighteen. Redding was an adventurer and had spent several years south of the border flying Fokker Trimotors for TAG, a Panamanian airline, where one of his misadventures included landing a flaming Trimotor in the jungle and walking out with his passengers. He had been one of the youngest airmail pilots in the United States, was a captain in the Air Corps Reserve Detachment at Richards Field, and was in Kansas City frequently enough to become well-acquainted with many other TWA pilots. In addition to his Reserves flying, Redding was the owner of a light plane, a high wing, all-metal Luscombe Silvaire, which was used at another of his joint ventures, a flying school in Scottsbluff.

Redding, about twelve years younger than Blackie, had grown up on a prosperous farm near Minatare – not much more than twenty miles from Blackie's home in Mitchell. Redding would become Blackburn's protégé very much in the manner that Blackie was becoming Otis Bryan's protégé. Following the Blackburn family's move to Albuquerque, Martha Blackburn and Redding's wife, Helen Jane, would also become good friends. Like Blackie, Bob Redding had a young family, including a son named Michael who was just two years old.

One might easily wonder why Blackie had so quickly accepted the offer to move to Eagle Nest; it was quite a departure from the family's pleasant life back in East Williston on Long Island. Blackie was a loyal TWA company man who was grateful for his hiring during the depths of the Great Depression. He had also been handpicked by the company president Jack Frye and the chief pilot, Otis Bryan, as the instrument flight supervisor. And finally, there was a substantial financial incentive – one that really could not be turned down. The salary levels were such that those instructors who had been pulled out of Boeing 307 co-pilot school, considerably disappointed, had their pay tripled from $200 a month to $600.

Blackie had to find a place to live and found a house not far from Kirtland. It was a one-story, Spanish style adobe home at 323 South Girard Avenue which was only about a mile from the field's north gate. The house did not have the charm of the Blackburn's former Tudor-style home in East Williston, but the commute was a lot simpler. Not knowing how long the Albuquerque assignment would be, Blackie wisely decided to rent again instead of buying. The Army turned over a brand new building for TWA's use at Kirtland Field: the new bachelor officers' quarters, and that was just the first structure in a complex that eventually numbered fifteen buildings. The facilities served many purposes including ground school classrooms, radio code practice, administration, aircraft parts and supply storage, and space for the Link trainers. The Army did not provide TWA with access to the base mess halls, so many of the staff drove their cars or rode bicycles home for lunch. Frank Busch was more interested in scenic surroundings and had rented a place up in the mountains on Sandia Pass; thus he and others took a lunch pail. Busch's upscale rental property also became a favorite target for buzz jobs. There was a supplemental benefit of living up in the mountains in late 1941; Busch was a deer hunter and he bagged a very nice mule deer buck during the first week of December 1941.

In mid-August, Blackie had to attend meetings back in Kansas City, but did not fly there in a Harvard. He changed out of his regular Eagle Nest uniform (Army trousers and garrison cap, steel gray shirt, and black tie) into his regular TWA uniform which was slightly different than nearly all TWA captain's uniform; it had a slender silver-bordered black band near the right arm cuff embroidered with the words, "Eagle Nest Flight Center." On 18 August, Blackie supervised Flight 378 from Albuquerque to Kansas City via Amarillo and Tulsa as a check pilot. There was astounding news waiting for him at TWA's headquarters. At the end of the month, ATFERO was cancelling the Eagle Nest training contract. Jack Frye related the circumstances to both Otis Bryan and Blackie, saying that from the outset, the British Air Commission had viewed Eagle Nest as a stop-gap measure.

ATFERO, BAC, and a highly placed member of the Royal Air Force had major objections to Eagle Nest's entire training and flight operations which they discussed internally in a meeting held in Washington on 31 July. Attending for the RAF was Air Marshall Sir Frederick Bowhill who had most recently been in charge of RAF Coastal Command. Bowhill had been a combat pilot in World War I where he had been decorated with the Distinguished Service Order and bar (roughly the British equivalent of the American Distinguished Service Cross). Ocean flying was a subject he knew well from having commanded coastal air units going back as far as the pre-World War I era. The other heavy hitters at that meeting included Mr. Morris Wilson who was president of the British Purchasing Commission (and as such the personal representative of Lord

Beaverbrook), Sir Henry Self and Mr. William Hildred, both of the BPC, and A.C. Boddis of the BAC.

The school's safety record was blemished in that a very hard landing had taken one of the school's Liberators offline and two of the three Hudsons had been damaged in landing accidents, including one where a collapsing landing gear strut came very close to puncturing a fuel tank. The Liberators had brake issues that were relayed back to the manufacturer which eventually did fix the problem, but it did not help get Eagle Nest's Liberator back into the air overnight. The British felt that Albuquerque was not a good choice because, in their view, TWA's Eagle Nest was being stifled by the Army Air Corps bombing school which was located on the other side of the field and that the steady parade of its many AT-11 twin-engine trainers created unacceptable delays. Blackie had one major objection to Kirtland, and that was the practically endless dust storms created by Albuquerque's hot, breezy climate and all of the earthmoving that accompanied the base's rapid expansion. Funds for the school and pay for the civilian pilots flying the Atlantic did not come from Lend-lease, but required cash payments from a very financially strapped British government. From a ferry pilot's viewpoint, flying the Atlantic, in spite of knowing that not all crews got through, was financially attractive; a first pilot could make up to $1,200 a month – a pay scale that was equally unattractive to the British whose coffers were nearly empty. Other complaints focused on the quality of training. ATFERO felt that it should have a check pilot on station in Albuquerque instead of having to blindly accept what TWA sent north. Both ATFERO and the RAF expressed confidence in being able to take up the slack with the Commonwealth flight training plan, but also recognized that would not happen overnight.

The next day, A.C. Boddis, who had previously been Assistant Director of Aircraft Contracts at the Air Ministry met with Jack Frye and three officers from the Army Air Force at the War Department Building in downtown Washington. The senior Air Force officer present was Colonel Robert Olds who was in charge of the Army Air Force Ferrying Command which had just been set up on 29 May 1941. Olds had flown in France during World War I and had just received a bronze medal and trophy from the International League of Aviators for his leadership of international goodwill flights flown by B-17s under his command in the late 1930s. Olds' adjutant was none other than Major William Tunner, Blackie's old flight commander in the Eleventh Bombardment Squadron.

Boddis explained what the British position was: that the RAF should control training and transatlantic flying directly instead of placing it in the hands of second and third parties. Olds tried to keep the ship afloat by suggesting that he could investigate moving the Eagle Nest operation to a more suitable location, such as Topeka, Kansas. Olds was also optimistic about the possibility of transferring a few B-24s from Barksdale Field in Louisiana to the school. ATFERO wanted more say in the school admissions process and felt that having had one of their own men on station in Albuquerque would have led to more qualified graduates. ATFERO also believed that time was on its side, particularly as winter was approaching, and with bases in Greenland still undeveloped there would not be substantial activity until the spring of 1942.

Unlike the British, Colonel Olds was very satisfied with the training of the American Liberator ferry crews. Otis Bryan had forwarded a letter to Blackie from Captain Herbert Chambers, one of the early graduates, who emphatically stressed that his training on the Liberator was far superior to that of the other Liberator ferry pilots he encountered who had been trained elsewhere. The Chambers crew had received twenty-two hours of in-flight instruction from Blackie about whom Chambers was glowing in his praise, "In my opinion, he is the most accomplished pilot I have ever flown with. I wish I were just half as good."

Winston Churchill and his minister of aircraft production, the formidable Lord Beaverbrook, were very sensitive to the problem of large numbers of badly needed warplanes coming out of American factories only to be parked in lines on both Canadian and American tarmacs waiting for delivery. The British decided that they could control the situation better and, accordingly, set up an entirely new command in the RAF – Ferry Command which absorbed ATFERO. Bowhill placed confidence in his and the RCAF's growing capability to train crews. He was also confident of at least temporarily recruiting more qualified aircrews from Canadian airlines, BOAC, and from his own Air Transport Auxiliary. The BCATP was turning out more aircrews while ATFERO had just started training crews at its base in Montreal. ATFERO had also initiated its Return Ferry Service using a Liberator bomber converted for passenger service. Ferry pilots were then returned to Canada by air instead of embarking upon a perilous, time-consuming ocean journey back. That alone increased the ferry pilot pool dramatically. And the RCAF, under RAF control, was launching the first of several operational training units for ferry crews.

The Roosevelt government also wanted to loosen the bottleneck and proposed that the Army Air Force assume responsibility for getting the Consolidated Liberators and Lockheed Hudsons to Montreal instead of the free-lance pilots that Lockheed and Consolidated had been using. The first contract pilots flying for Consolidated and Lockheed were very experienced men. Typical of those early pilots was Charles "Slim" Gropstis whose six-foot, six-inch frame fit very nicely into the Liberator. He was an Alaskan bush pilot with nearly 5,000 hours, had flown for Alaska Pacific Airlines, and had extensive experience in Caribbean flying. Before he ferried his first Liberator from San Diego to the East Coast on 22 March 1941, he had already flown the Consolidated PBY Catalina patrol bomber and Lockheed Lodestar (the Hudson's first cousin). Gropstis spent the spring, summer, and early autumn of 1941 ferrying Liberators and Hudsons to the East Coast and also flew Consolidated PBY Catalina maritime patrol bombers from San Diego to North Carolina and then to Bermuda. From there the

Catalinas were picked up by RAF crews and flown to Britain via the Azores. As far as ATFERO was concerned, there were not enough good men like Gropstis available, and the experience level quickly went down as new men were hired. Sadly, Gropstis was killed in a flying accident on Salal Island, British Columbia, on 21 October 1941.

The Roosevelt plan had pilots like Gropstis joining ATFERO and flying the aircraft to Britain under direct control of the Royal Air Force itself. Roosevelt's offer was accepted by Churchill, and thus RAF Ferry Command came into being on 20 July 1941 with Bowhill as its commander. Some Canadian noses were put out of joint since it was basically their effort that established ATFERO, but the transition to military control was eventually a smooth one. There were a few flying accidents, but the combination of ATFERO and RAF Ferry Command moved a total of 593 aircraft to Britain in 1941 – and that was just the beginning.

Jack Frye pressed for a prompt resolution of the differences brought up in his meeting with A.C. Boddis. The answer was only days in coming, and when the ATFERO contract was terminated Frye was neither surprised nor disappointed. His Plan B was going into place in September because General Olds still had a burning need for many more qualified B-24 ferry crews. By early September, cash-strapped TWA had a much-needed new contract worth $599, 706 and an expanded assignment for training strictly American Ferry Command crews on the Liberator. Eagle Nest's planes in RAF markings went away and the training of RAF Ferry Command crews became an RAF responsibility as administered through RCAF staff and facilities. The turmoil worked to Blackie's advantage because on 1 September, he became Eagle Nest's director of operations.

That new position, however, presented Blackie with a delicate problem, a tightrope of sorts and one that he would have to walk several times in the future. Was he a TWA pilot or was he a member of TWA management? Blackie was not the only one concerned about status vis-a-vis TWA/Eagle Nest and the Airline Pilots Association. Indeed, Blackie and the other pilots were apprehensive about their seniority because being absent from the line and the domicile/assigned station in excess of ninety days jeopardized seniority which was, and remains, very important to every airline pilot.

Fortunately, the immediate seniority issue was put to rest by Larry Fritz, TWA's director of operations, who sent a memo to all regular TWA pilots at Eagle Nest confirming their status as TWA pilots and that their seniority positions would never be at risk. They would be treated exactly as those pilots who were working at the training center in Kansas City. Fritz also commended the Eagle Nest pilots for their "patriotic and useful service in the interest of national defense...." Blackie, however, was more than a pilot; as the new operations manager, he had crossed solidly into company management. Blackie spent the first two weeks of September reorganizing the school for the influx of Army Air Force personnel which now included full ferry crews, not just pilots. In the midst of all of that, Blackie was occasionally called back to Kansas City for meetings and he traveled to and from there as a check captain on Flights 1 and 10.

Both Jack Frye and Otis Bryan occasionally passed along pearls of wisdom regarding supervisory techniques to all management employees. Blackie had no academic degree in business administration or any previous formal schooling in management, but he saved those memos and was starting to generate his own management philosophies which were usually passed along to his subordinates in the form of memos or pep talks at meetings. The first of these guidelines was presented at an instructors meeting on 8 August when Blackie discussed responsibility of instructors. Subsequent subjects would include authority, communication, full utilization of resources, and setting the example among many other hallmarks of leadership. That type of thinking would characterize his entire career and extend into his life well beyond his airline flying days.

Blackie's first recorded approach to personnel management was recorded in an internal memo requested by Otis Bryan in late July following Blackie's assignment as head of instrument flying. The document was Blackie's own problem-solution primer for use in a variety of situations, particularly in conjunction with the Link trainer. In reading this memo, Otis Bryan could quickly see why Blackie was so successful in dealing with students who displayed the full range of learning issues: the discouraged, lazy, slow, overconfident, rebellious, inconsiderate, and uncooperative. In the final part of the memo, Bryan asked for Blackie's self-evaluation – the responses were clearly what Bryan was looking for when he promoted Blackie to the operations director post. Blackie's self-perception was that of a technically oriented, perfection-driven optimist with ambition tempered by adaptability and concern for others. In the matter of self-control, Blackie replied, "Had a temper as a child, but believe I have it pretty well tied for a red-head." Blackie's youth was on his mind when he answered the question regarding industriousness, "Have practically been on my own since the age of twelve and have supported two families." When asked about hobbies, Bryan already knew the answers, but the sequence of Blackie's answers are revealing: family, athletics ("can still do hand springs at age 40), home movies, reading, hunting, and fishing. Of the latter two, Blackie lamented that his having been based in Newark and LaGuardia had pretty well minimized available time for those endeavors.

On 23 September, Blackie and several other senior pilots were checked out on the B-24 Liberator by Otis Bryan. Blackie was familiar with the Liberator, having seen the school's LB-30s and B-24s, but never mentioned the plane over the years when asked what his favorite airplane was. The prominent aviation writer Earnest K. Gann had a low opinion of the C-87 Liberator Express, the cargo version of the B-24 bomber, and claimed that the Liberator resembled a "harpooned whale." That may have been a bit unkind although a flight of Liberators could well be likened to a bloat of dyspeptic hippopotami, considering the plane's great slab sides, blunt nose, and burly appearance. Although the school had started with just one British Lend Lease LB-30, the school was now receiving American LB-30s and B-24As which was timely because the number of students was increasing significantly. Blackie spent the balance of September strictly as a B-24 instructor and check pilot. Beginning on 1 October, he was in the air at least once a day and sometimes three times a day on flights which averaged two hours each. All of that flying was in addition to his administrative duties as the operations director. That was quite simply his response to the management vs. pilot question. When anybody inquired about his flying, Blackie's response was that to be an effective manager required intimate knowledge of the school's operation and that expertise could never be gained by sitting behind a desk. That reply always worked well during the war, but the war would not last forever.

The Liberator had been conceived by Consolidated Aircraft in San Diego as a heavy bomber, but experience was showing that it had other roles to play. While not quite as fast as its more glamorous competitor, the sleek B-17 Flying Fortress, the Liberator could haul a heavier load higher and haul it much farther, although it certainly was nowhere near as responsive as the B-17 Flying Fortress. The 24's twin tails were almost a Consolidated Aircraft trademark because they had been used on several of the company's large amphibious aircraft. Most B-24 pilots simply thought that each of the two big rudders just fought with each other and neither came out a winner. As a result, the bomber's yaw stability suffered accordingly. Nonetheless, the big B-24 was a comparatively safe ship to fly. When the statisticians closed the books on wartime accident rates, the B-24 came in at 35 accidents per 100,000 hours of flying (the B-17 Flying Fortress was marginally safer at 30). Aviation cadets preferring life in the fast lane had their wishes granted if they were assigned to P-39 Airacobras where the crack-up rate was 245 per 100,000 hours of flying.

In March, a Liberator had flown the nearly 3,000 mile non-stop flight between Montreal and Prestwick, Scotland. The big ship had already gone into the Trans-Atlantic Return Ferry Service to bring ferrying crews back to Canada due to its capability of staying aloft for over nineteen hours. To its roles in heavy bombing, maritime patrol, weather reconnaissance, and VIP passenger transport, the Liberator, in its C-87 version, also became a cargo hauler. What the Liberator did not have was proper instruction manuals for new Eagle Nest crews. Consolidated Aircraft provided only the bare minimum of materials for training, and the pilot's manual for Eagle Nest's first Liberator was written for British readers. That is to say it was not a *Pilot's Manual*; it was instead *Pilot's Operating Notes*. Thus "airscrew" had to be interpreted as "propeller," "windscreen" as "windshield," "undercarriage" as "landing gear," etc.

Eagle Nest promptly set out to correct those shortcomings by translating the existing pilot's document for Yankee audiences and, much more importantly, creating a complete instruction syllabus for training Liberator crews. This heavy document had a long shelf life and was adapted for use in the Army Air Force Liberator transition schools which were expanded significantly in 1942. The new *crew* training syllabus included the responsibilities of pilot, co-pilots, flight engineer, and eventually radio operator. Although all departments, including Blackie's instrument flight group, contributed to the document, Royce Welliver, originally from the Kansas City engineering staff, wrote most of the technical sections. Blackburn and Welliver were good friends – both were from western Nebraska. The syllabus also outlined, flight by flight, the airborne instruction and testing sequences; it also included written tests and an answer guide for the ground instructors.

Blackie was averaging close to sixty flying hours a month as 1941 drew to a close. Nearly all of this time was taking Army crews on check flights after other instructors had taken a crew through its roughly 50 hours of flying. Not all of that time was in the left seat for the Army pilot trainees because standard procedure was to take two crews up in the same Liberator and have them switch places half way through the flight; thus only about twenty-five hours were available for a trainee at the controls, but that appeared to be the bare minimum to do the job. In addition to demonstrating competence in pre-flights, ground handling, takeoffs and landings, trainees had to pass a flight check which included night flying, instruments, and basic airwork.

However, a significant problem was beginning to emerge. The school was a seven-day a week operation, and the classrooms and Link trainers were in constant use, although staffing was not a problem there. The problem was not enough B-24 instructors. There were only two solutions: upgrade some of the pilots who had been working with the Harvards and Hudsons to B-24 instructor pilots and/or start looking outside TWA for talent that could quickly be brought up to Liberator instructor standards. Blackie started by upgrading his existing single engine-instrument flight instructors and logged many training flights with Cliff Dombrowski, Wilbur Knudsen, Marvin Sauers, Bob McReynolds, Jim Clifford, and Vernon Irons. He spent more time with Bob Redding, but not because Redding was a slow student. Quite to the contrary, Blackie was grooming his close friend for bigger things to come.

Blackie and Redding shared a misadventure on 19 November that was picked up in the *Albuquerque Journal* the next day. The two were flying a Liberator on a training flight that should not have lasted more than two hours at most. As they approached the field, the main landing gear stopped descending while half-way down. The Liberator had a back-up hydraulic system, but, even that failed to budge the gear. The flight engineer tried to crank the gear down manually, but inconceivably that also failed. Blackie was most reluctant to belly land the Liberator, although the wings were high enough to ensure no damage to the props. Pranging a Liberator would create a public relations issue the Army Air Force Training Command, to which the school technically reported. The command was pleased with the quality of instruction, but also very sensitive about the Hudson accidents and the first British LB-30 that had been taken out of inventory due to a very hard landing.

Blackie eventually deduced that the problem was not hydraulic, but that the culprit was in the closed loop cable system that operated the landing gears. The flight engineer had to disconnect the cable and temporarily create three separate circuits. The fix, including confirming that the gear locked when it did finally lower, took two hours and resulted in a dusk landing. When interviewed by the newspaper, Blackie, in his typically understated manner, described the action as "...just normal procedure." Otis Bryan did not see it that way and sent Blackie and Redding a warmly congratulatory note the day after the news appeared in public.

On Wednesday 3 December, Bryan had just returned to Albuquerque from a meeting with Jack Frye in Washington; Bryan then called Joe Mountain, Blackie, and Joe Carr into his office for a meeting in which Carr thought the discussion would focus on Eagle Nest operational issues. Next to nothing was said about what was going on in Albuquerque because the discussion was entirely about Carr's unique, pre-TWA experiences. Everybody was aware that Carr had flown Navy PBY Catalinas in the Caribbean and that he also held a celestial navigator rating, but the meeting quickly shifted into another, less familiar aspect of Carr's pre-TWA life. As it turned out, Joe Carr had been assigned to temporary, special duty flying as a navigator and co-pilot on Pan American Airway's big Boeing 314 flying boats. Bryan was very interested in hearing all about Carr's two transatlantic trips to Lisbon with Pan Am. In fact, those two trips represented the sum of all knowledge within TWA when it came to crossing the Atlantic Ocean. The very near future would reveal just how important Carr's ocean flying knowledge really was.

Bryan asked Blackie to see what he could learn about long distance, ocean flying from any of his wide circle of airline pilot acquaintances outside of TWA. Most of Blackie's contacts were at the decidedly land-bound United, American, and Braniff, but he did know several Pan Am pilots and made it a point to host two of them in the posh bar, Top of the Mark, at the Mark Hopkins Hotel in San Francisco when he was in California in early January. Blackie had been there before when he was living in Klamath Falls – and, he was on a solo tour of California's dance halls and night clubs trying to decide between heading for the big time in the music business or to bite the bullet and become a pilot. The Top of the Mark still had talented jazz bands, but the clientele had changed; now many of the patrons were Navy officers enjoying a few drinks and fabulous sunsets over the Pacific where they were shortly headed. The Pan Am Clipper pilots played it close to the vest and divulged little other than the fact that they were very experienced at plotting howgozit charts (with which Blackie was already very familiar). It was clear that if Bryan wanted to get up to speed on flying long distances over water where radio navigation was essentially non-existent, he was going to have to seek sources other than off duty Pan Am personnel. Still, he approved the not inconsequential expense voucher for Blackie's trying to pump the Pan Am flyers.

On 7 December 1941, Blackie flew three times; the first two were brief flights with Army crews in the morning and the last, by then hearing about the Pearl Harbor attack, was an hour-long flight with Bob Redding. The declaration of war intensified the staff shortage problem because the Army now demanded that the school take on larger numbers of students and reduce the amount of time they spent actually flying a B-24. To top that off, TWA itself was losing not only pilots but top managers who, as reservists, were shortly headed for active duty. Jack Zimmerman who had been chief pilot in TWA's Atlantic region was going into the Ferry Command. Larry Fritz, the director of operations, would be going to Ferry Command himself in a few months. The airline's board chairman, T.B. Wilson joined General Douglas MacArthur's staff as chief of transportation and Paul Richter, TWA's executive vice president, went into the Naval Air Transport Service. The engineering V.P., Tommy Tomlinson, went back to the Navy in a training function. Joe Mountain's days in Albuquerque also were numbered, and he was Blackie's immediate superior.

Many of Blackie's friends from Kelly Field already were in uniform. The first man, alphabetically, in Blackie cadet class, Addison Baker, had been flying with an Ohio National Guard observation squadron, but was soon headed to Barksdale Field in Louisiana to fly B-24s with the Ninety-third Bomb Group. He would be going overseas in just a few months. Clarence Shoop, also had been called to active duty from an Air Guard observation squadron in California and had started flying Army acceptance flights on Douglas A-24 dive bombers and A-20 light bombers at the Douglas plant in Santa Monica. He had more recently moved to Lockheed's Burbank operations to fly acceptance flights on Hudsons and P-38 fighters. Philo Rasmusen, who went to Mather Field as a fighter pilot, was in Hawaii, commanding the Ninety-eighth Bombardment Squadron. He was soon on his way to the South Pacific where he would receive a Distinguished Flying Cross for leading his B-17s in strikes against Japanese shipping and land targets. Richard E. Morehouse, who went from Kelly to Crockett Field as an attack pilot, was now a major, commanding a squadron in the Seventy-fourth Recon Group. Karl Baumeister, who started with Blackie as a cadet at March Field in primary, had just arrived at McDill Field in Florida to activate the 320th Bomb Group which would head to North Africa with its B-26 Maruaders before the year was out. Baumeister would eventually become a general officer. He was not the first of the old March Field '30 to attain high rank. The several West Pointers in Blackie's class got their stars while Blackie was instructing in B-24s in Albuquerque or very shortly thereafter. Walter Todd, then a colonel, was on the staff of the Fourth Bomb Wing at Westover Field, Massachusetts, and would eventually be heading to England as assistant chief of staff, Eighth Air Force. Thomas DuBose was commanding the Army's basic flying school in Merced, California, while Blackie was in Albuquerque, but in three years was off to the Pacific as commander of the B-29-equipped 316th Bomb Wing. Blackie had expressed no dissatisfaction with what he was doing, but as more friends and associates went on active duty, he started having reservations about not being where the action was. Eventually, that longing would become a much larger, festering problem.

Blackie was not flying on the day of Christmas Eve, but Otis Bryan was. Very early that day, Bryan was on his way back to Kansas City to spend the holidays with his family and get in a week of quail hunting in the Ozarks, but he was intercepted at the airport and told to report immediately to Jack Frye. On that very day, Frye would sign a monumental contract with the United States Army. That arrangement created TWA's Intercontinental Division (ICD) which was essentially an airline within an airline. Its mission was to provide passenger and freight service to the Army and manage all of the related functions including recruiting aircrews, training, maintenance, and the many details related to international flights, i.e., licenses, landing rights, etc. And this was to be done on a global basis, hence the name "Intercontinental." In effect, ICD would be an arm of the Air Force's now rapidly expanding Ferry Command which itself would undergo a monumental transformation in July 1942. The Army Air Force supplied the planes and TWA, along with the other major airlines, supplied the aircrews and everything else. The most urgent requirement was to get ICD up and running immediately with service to Cairo. Bryan was assigned as ICD's CEO and was to establish the organization's headquarters at the relatively new Washington National Airport. He got two days with his family and one day of bird hunting only; then he was off to a small office on the second floor of TWA's hangar at Washington National.

All of the major airlines were going through the same upheaval supporting the Army's new global requirements, but TWA was the best prepared to comply. Even as early as 1940, Frye had promised the Air Corps chief, General Hap Arnold, the use of TWA's Boeing 307s in the event of a national emergency. Every major airline CEO was also very familiar with what amounted to a mobilization plan that had been prepared by the Air Transport Association (ATA) chief, Col. Edgar Gorrell. He was no lightweight. The Baltimorean Gorrell was a West Point graduate, had won his Army wings in 1915, flown in the Mexican Punitive Expedition of 1916, and was a decorated veteran of World War I. As a pilot commissioned into bombers, Blackie was very familiar with Gorrell's treatises on strategic bombing. Gorrell was also a clairvoyant in the Churchillian sense; as early as 1936 he became alarmed at Japanese aggression in China and Hitler's saber rattling Germany. As ATA chief, Gorrell began preparing an airline mobilization plan for use in the event of war.

President Roosevelt was about to sign an executive order nationalizing America's airline industry, but rescinded it after consultation with Gorrell shortly after Pearl Harbor. Gorrell's plan essentially outlined the creation of a cargo/passenger airline reserve that could be activated on very short notice. And that is almost exactly what happened. The Army Air Force did take over many of the airline industry's DC-3s – in fact it commandeered 200 of the major airlines' 400-airplane fleet of which the majority were DC-3s. The first to become militarized were all five of TWA's prized Boeing 307 Stratoliners. Blackie had never flown a 307 before, but he would find himself in the left seat of one just a few months into the future.

On Friday, 16 January 1942, Blackie was not flying, a rarity for that month when he was averaging two hours almost daily with young lieutenants learning to fly the Liberator. That day was anything but uneventful for TWA because Flight 3 flew into Potosi Mountain shortly after takeoff from Las Vegas with all aboard killed instantly. That was bad enough, but the passenger list included Carole Lombard, a popular movie star whose husband was the super star, Clark Gable. Lombard had been on a war bond tour and took TWA out of Indianapolis bound for Burbank. There was a crew change in St. Louis where Captain Wayne Williams and First Officer M.A. Gillette took over and flew to Albuquerque where some passengers were bumped off the flight due to a heavier than average load of mail and priority cargo. Among the passengers were fifteen Army officers and men on their way to their base in California. The station manager asked three passengers to give up their seats in order to satisfy the flight's weight and balance requirements. Lombard preferred to remain on board with her mother and an M-G-M publicity agent; when TWA learned that Lombard and her party were traveling under presidential orders, they quickly persuaded three of the Air Force men to give up their seats. Williams normally would have flown directly to Burbank from Albuquerque, but with his heavy load and forecast of strong headwinds he decided to stop for fuel in Winslow and Las Vegas. After taking off, he received an amended flight plan calling for just one fuel stop – at Las Vegas. Twenty-two miles southwest of Las Vegas, Potosi Mountain rises 8,500 feet into the sky. Gillette filled out the flight plan, but Williams never endorsed it which was a breakdown in procedures. The flight plan called for an altitude of 8,000 feet and a heading of 218 degrees, i.e., a recipe for disaster since that would take the flight directly toward the mountain.

The investigation eventually blamed Wayne Williams, who otherwise was considered to be a very experienced pilot. Based on where the bodies were found amidst the wreckage, investigators believed Williams had put his co-pilot in charge, walked back into the passenger cabin, and was talking to Carole Lombard at the moment of impact. In trying to recreate the path to disaster, it became clear that Gillette was simply following a heading on the directional gyro and not using the radio range which would have instantly told him he was off course. The Amber Two Airway, used for flights from Las Vegas to Burbank, starts out of Vegas at 205 degrees and bends to 212 degrees when the Silver Lake beacon comes into range. Following the range would have taken the flight safely south of Potosi Mountain. There was also some thought that Gillette may have focused on the wrong airway light beacon flying out of Las Vegas.

Finally, the accident report concluded that inasmuch as the night of 16 January was very clear, the outline of the mountains ahead should have been obvious, assuming that the cockpit lights were dimmed. Not seeing the lights of two towns, Arden and Goodsprings, which were beside the airway, along with car headlights should have been another visual cue that the flight was off course.

Flight 3 hit Mount Potosi roughly 700 feet below the rocky crest, and among the victims were seven recent Eagle Nest graduates, members of the first Air Force graduating class. They and their crews were on their way back to the Ferry Command base at Long Beach and would have picked up more Liberators for flights to Canada. The crash was another high-visibility blot on TWA's safety record, but the tragedy hit closer to home for Blackie. Among the Army dead were Lieutenants James C. Barham, Charles D. Nelson, and Frederick J. Dittman – all of whom Blackie had personally checked out on the B-24.

By this time, the name "Eagle Nest Flight Center" was officially retired and Blackie became an employee of the "Jack Frye Four-Engine School," or JFFES, although "Eagle Nest" still hung on for many of the employees. The staff manpower requirements had mushroomed; many new first officers or "co-pilots" were needed for double-duty as ground school instructors as well as inflight assistants to the school's instructor pilots. Recruiting those co-pilots was done at the country's best known flying academies and aviation colleges such as Parks in Saint Louis. Some of the new co-pilots had been waiting on the outcome of their applications with the airlines, and just decided sign on with the Four-Engine School instead. All of these new hires had to go through the same pilot training on the Liberator as the Army students, so it took a month to bring them up to an acceptable experience level.

Typical of those young men was Harold "Hutch" Thurston, a young man with an instrument rating on his commercial license, a B.S. in Aeronautical Engineering from the University of Minnesota, and over 1,000 hours in his logbook, including a lot of time in the Beech Staggerwing, one of the heavier single-engine planes of the day. Once Blackie approved the application, Hutch resigned from working on autopilots at Honeywell in Minneapolis, withdrew his application with Northwest Airlines, and headed for Albuquerque.

Personal friendships resulted in hiring other copilots. Bob Redding was acquainted with a Californian who had been a friend while flying in Panama. The friend was Jonas Ruff who, at age thirty-two, was a bit older than the other copilots, but he had a commercial pilot's license with an instrument rating and was a qualified flight instructor. Ruff had considerable commercial experience, including having worked as a part-time pilot for non-scheduled airlines in Panama. He also was a part-owner of a Travel Air 6000 which he used for commercial flying in Panama, although his regular work there was as an instrument technician for the Panama Canal Authority. Ruff left his job as a technician with the Department of Commerce weather bureau operations in Washington and checked in at the school on 2 February 1942. His salary was $4,400 per year (less the cost of his Eagle Nest uniforms which he had to buy himself). Just the day before, Blackie had gotten some good financial news himself; Larry Fritz upped Blackie's salary to $1,100 per month – big money in early 1942. For $25, Martha Blackburn could easily do the weekly grocery shopping for her family of five, skimping nowhere, completely load one grocery cart to overflowing, fill up the better part of another cart too, and still have a few dollars left over. At about the same time as his salary was increased Blackie received some more recognition, although it was not tangible. Brigadier General Russell C. Charlton, Adjutant General of the State of New Mexico granted Blackie a commission as a Colonel, Aide-De-Camp on the staff of His Honor, John E. Miles, Commander in Chief and Governor of the State of New Mexico. Blackie remained, however, a first lieutenant in the United States Army, Air Force Reserve with little to no chance of becoming a colonel in that organization.

Three weeks later Otis Bryan completed ICD's first flight to Cairo via what would become the heavily traveled southern route, i.e. from the eastern tip of Brazil to West Africa. It was in one of the 307s which had been completely transmogrified into an Army transport. The plush seating was gone, the heavy pressurization equipment had been removed, and an olive drab/gray camouflage exterior coat had been applied. The plane ceased being a 307; it and the four other Strats had become C-75s and no longer prominently sported the individual and memorable Native American names that Jack Frye had bestowed on them: "Apache," "Cherokee," "Comanche," "Navajo," and "Zuni." Gone too were the slogan "The Transcontinental Line" and TWA logo. Now the Strats carried an Air Transport Command logo (a globe and futuristic abstract airplane), dull yellow serial numbers back on the tail, and the original, individual aircraft name (very understated) just above the passenger door. Bryan's high priority cargo consisted of proximity fuses which were desperately needed in the desert war which was not going well for the Allies in early 1942.

At about this time, Blackie was advised of a request from the TWA publicity department which wanted Eagle Nest photos for use in the company newspaper. Photography on military bases was tightly controlled so Blackie contacted the executive officer of the bombardment school, Colonel William Lewis, to see what arrangements could be made. Lewis, a lawyer and former U.S. attorney in civilian life, was a World War I veteran who had earned his wings in the interwar years. After getting his law degree from the University of Michigan he had briefly been a professional baseball player. He had recently arrived from Barksdale Field in Louisiana where he had been the executive officer of the Third Bomb Group. The photography was approved, but it consisted largely of personnel photos without any close-ups of the B-24s. Shortly thereafter, Lewis hosted Army Signal Corps photographers who did get some excellent air-to-air shots of the school's Liberators. The most striking photo was a head-on view of Blackie and one of his co-pilot instructors, Nick Wasil, flying one of the new B-

24Ds. That photograph was later picked up by *Flying Magazine* which used it on the cover of the April 1943 issue.

The most extensive photography which Blackie had to coordinate and fly for was the Kirtland location shooting of the movie *Bombardier* starring Pat O'Brien, Randolph Scott, Eddie Albert, and Robert Ryan. Blackie supplied one B-24 for use as the camera plane and led two other Liberators in the bombing of an enemy naval vessel. The "bombing" required some imagination because the enemy capital ship was a large wooden mock-up built on a flat section of the west bombing range outside of Albuquerque. With controlled lighting, a very soft focus, the right angle, and some imagination, a flat area of prairie could pass for the ocean at dusk. To compensate for the "ship" being smaller than the real McCoy, Blackie's three liberators bombed up with twenty, 100-pound bombs, but reduced the explosive charge to just six pounds per bomb. When Blackie was on the final bomb run, he had to abort and call a go-around because the camera plane was not in the right position. Unfortunately, one of Blackie's other Liberators misunderstood the call and instead of rolling up his bomb doors, salvoed his eggs. Fortunately, the bombs only tore up some less than prime real estate along the west bank of the Rio Grande River.

National Archives

Albuquerque-based B-24D Liberator with Rio Grande in the background.

Bombardier did not release until 1943 and was plagued by a few problems of which one reportedly related to a cast member; that was trying to keep leading actor Pat O'Brien on the wagon while the movie was being shot. The JFFES community had issues with the production because the school's B-24s appeared only in a few of the movie's opening sequences. The film editors wound up cutting all of the B-24 aerial photography and featuring the B-24's arch rival, the B-17 Flying Fortress, in all of the action scenes. *Bombardier* was produced by RKO Pictures which would be acquired in a few years by the same man who already had the majority interest in TWA: Howard Hughes. Fortunately, Blackie's next direct encounter with RKO would be one in which the outcome was much more favorable, but that was still a decade into the future.

Colonel Lewis, because his troubles were not over by far, claimed that he had doubts of ever seeing a brigadier's star based on what else happened that day. A bombardier trainee accidentally fell or deliberately committed suicide by jumping out of an AT-11 and plummeted to his death. One of Kirtland's B-18s ran its gas tanks dry, and the pilot baled his crew out. He then set the ship on autopilot and headed back to the bomb bay to jump, but saw how close he was to the ground and decided to ride it in. The B-18 hit the desert at a very shallow angle with the pilot hanging onto a bomb rack post for dear life. He walked away from it, but the B-18 had to be written off.

On 4 March, Blackie flew twice. On the first flight he gave instrument transition to two of his first officers and in the afternoon was aloft giving an initial orientation on the B-24D to two young lieutenants from the latest Army class which had arrived only days before. Half a world away, the Jiggs squadron was retiring to Melbourne Australia from the Dutch East Indies which the Japanese were about to completely overrun. Once in Australia, the squadron turned over its surviving B-17s to the Nineteenth Bombardment Group and prepared to return to the United States where it would re-equip with the B-25 Mitchell bomber before deploying to India and the war in China. News of his old squadron was greeted with mixed emotions. Had Blackie been able to remain on active duty, he would have been the squadron's commanding officer or, most likely, the commander of the squadron's parent organization, the Seventh Bombardment Group. On the other hand, not being commander of the Eleventh Bomb Squadron would have been providential. Its commander, Major Austin Straubel, had died of burns suffered in a crash near Surabaya, Java, on 3 February. In an interesting twist, many of the first Army bomber crews Blackie trained in Albuquerque were eventually assigned to the Nineteenth Bombardment Group, a very distinguished organization which ended the war flying B-29 Superfortresses on missions over Japan.

At about this time, Blackie made the first of several requests to the War Department to be transferred to active duty. He was promptly turned down on 20 February and advised that because of his important civilian position his mobilization was out of the question and that he was being retained in the War Department Reserve Pool without eligibility for promotion or transfer to active duty. The only circumstances that might prompt a return to active duty were a change in employment status and direct authorization by the Secretary of War. TWA certainly was not inclined to release Blackie who with considerable justification saw himself well-prepared to function as a heavy bomb group commander or executive officer and certainly a squadron commander at the very least. The matter turned into a gnawing issue that the normally stoic Blackburn would take up several more times before the war was over.

In mid-March, the Army Air Force personnel office contacted each of the major airlines and advised them to collectively provide a list of eighty pilots and eighty co-pilots who could be promptly released from airline duty to go on active duty in the Air Force for special duties. These men would be flying C-47s over the South Atlantic, across Africa, and eventually to Burma where they would be the first to fly supplies into China over the Himalayas, better known as the "Hump" route. Some of the more experienced pilots were diverted from that group, including Alexis Klotz with whom Blackie flew as a co-pilot from Newark in 1936, and went into the Army's Ferry Command to fly VIPs. TWA's top management made sure that Blackie's name was not on that list.

By 1 April 1942, the JFFES flight instruction staff included eighteen first pilots, twenty-eight first officers (co-pilots), and twelve flight engineer instructors. Swede Golien was no longer an Eagle Nest first pilot – he had just taken the job of acting chief pilot for the company after having spent the previous few months as chief pilot for the Pacific Division. Blackie's personnel problems continued, and the school started scouring aviation trade schools for flight engineer instructors. Several weeks before, Blackie got lucky and was able to hire a former Navy pilot, Howard Kincheloe, who also had Pan Am four-engine flying boat experience. His last Navy assignment had been flying Catalinas out of Coco Solo in Panama. "Kinch," originally from Wichita, had spent the previous year as an instructor teaching basic and advanced flying to Dutch Air Force aviation cadets in Batavia, Dutch East Indies, before being run out of the country by the approaching Japanese.

Men like Kincheloe had a real challenge on their hands as the school went through another change to accommodate the Army Air Force's mushrooming new needs for heavy bomber crews. Thus, the school became the Combat Crew Training School: Four Engine (CCTS), and it faced transitioning young pilots who had gone through multi-engine advanced flying schools flying single engine aircraft (North American AT-6s). The CCTS training was to include flight engineers and radio operators, of whom many were very, very green. Many years later when interviewed by Robert Serling for the book, *Howard Hughes' Airline*, Blackie shared a story about just how raw the new crews could be. Blackie was aloft in a B-24 and sitting in the co-pilot's seat which is directly in front of the radio operator's station. Blackie asked the young radio operator how things were going only to be told that the young man *thought* things were probably going okay. "Thought," being an inexactitude, was not good enough and never would be for Blackie who demanded absolute precision on the flight deck. When Blackie asked why the radio operator was so uncertain, he was told that the young man had never been up in an airplane before. It was about this time that "burp bags" were added to the schools B-24 minimum equipment lists. The number of B-24s on the CCTS flight line had expanded – now there were sixteen new B-24Ds to join the older B-24As and LB-30s.

Unfortunately, TWA would not be getting its deferred, new DC-3s from Douglas; the Army Air Force got them all. Something had to give and it was TWA flight schedules which were to be cut back by 1 April. And then the full year 1941 financials were reported; the company lost $487,615.

Blackie had to update the syllabus in order to accommodate the Air Force's demand for more graduates. At this point, an arriving crew in training consisted of two pilots, one radio operator, and one flight engineer. The ground instruction for pilots and engineers continued to be intense and focused on all of the B-24's mechanical systems. Most radio operators had been trained at Scott Field in Illinois, so their participation in the airborne instruction was largely as observers. When not flying they concentrated on "buzzer practice," as Blackie called it. Radio operators were among the most qualified and technically trained enlisted men. They had to be able to disassemble and reassemble a radio blind-folded and had to be able to create one working radio out of several others that had been disabled.

By now, the trainee pilots only got twenty-five hours in the air, including twelve hours in the left seat. Flight instruction began with a demonstration in which the student pilots (usually two or three sets of them on a single flight) watched the pilot and co-pilot as they went through their duties. Gradually, the student pilots moved to the right seat, and eventually got to the left seat where they learned how to taxi the Liberator, take off, and fly basic maneuvers such as steep turns and approaches to stalls. Landings were, of course, critical and when a student demonstrated three satisfactory landings he was considered proficient with the plane. Night, instrument, and formation flying were included, but just barely. Some of the pilot and engineer training modules were combined, i.e., engines, hydraulics, electrical, etc. When not flying or in ground school, the pilots were required to maintain instrument efficiency via the school's Link trainers. Later in the war, B-24 student pilots logged many more hours in the left seat, but their training syllabus, updated to include more flying time, had been created by TWA.

On 25 March, Blackie was aloft in an LB-30 with three young lieutenants on what was a rather short flight of two and a quarter hours, considering the fact that most of his flights, even local ones, usually lasted four hours or more. For lunch that day, Blackie had brought his own lunchbox – much in contrast to a lunch that was just breaking up at the White House in Washington where General Arnold, the Army chief of staff, had presented his plans for a cross channel invasion. President Roosevelt and his key advisor, Harry Hopkins, were non-committal about that invasion and another one, code-name "Sledgehammer," which was to take place on the Cherbourg Peninsula in the autumn in order to provide a stepping stone for the main offensive to follow in 1943. Sledgehammer was very dependent on German strength in the West which in turn would be influenced by who was gaining the upper hand on the Eastern Front. Roosevelt did not veto Sledgehammer but in effect told Arnold to sell his plan to the British before expecting Presidential approval. That precipitated the need to meet with Churchill and his staff – a meeting that the influential Hopkins would also attend. In his wildest dreams, Blackie could never imagine that he would be the conduit by which Arnold, Hopkins, and several other VIPs would reach London in just four months. Instead, Blackie was preparing for his next flight scheduled for the following day, a two-day roundtrip to LaGuardia and back in a B-24D with three more very young lieutenants.

By late April, management of the school changed. Blackie had been made superintendent of the school because Joe Mountain had gone back onto active duty with the Army Air Force and would soon be headed to the China-Burma-India Theater. Blackie promoted Bob Redding to operations officer, and both men were flying more than desks. Redding was not going to be available much longer; he already had his call-up notice and expected to be on active duty in July. Blackie was still flying check rides with new co-pilot hires, but was starting to spend more flying time as check pilot on the "graduation" flight with students. These flights were flown at high altitude, typically around 20,000 feet, to introduce oxygen procedures while on a long cross country. Thus Blackie was taking students as far as Minneapolis, Washington, New York, and Miami on two-day trips in addition to shorter, one- day cross countries to California or Kansas City. To maximize availability of the B-24s, one student pilot flew the outbound leg, and became the observer on the return leg which was flown by the other student. On the day he logged three hours on night check-outs, 31 March, 450 air miles to the northwest eighty-three officers were assigned to the Flying Fortress-equipped 305th and 306th Bomb Groups, then forming up at Salt Lake City Army Air Base. Seventy-three of them were second lieutenants who had just won their wings in late February as part of cadet class 42-B. Although the group came from all nine advanced flying schools then in operation, most had graduated from Stockton and Mather, both in California. The seventy-three were divided into two groups and sent on temporary duty to two of the early four-engine schools: thirty-six to the B-17 school at Davis-Monthan Army Air Base in Tucson and thirty-seven to the CCTS in Albuquerque. It was assumed that those going to Albuquerque could easily conduct their own transition to the B-17 after four weeks of B-24 training (such was the lack of training capability and coordination in early 1942).

The group arriving in Albuquerque became CCTS Class 42-4-E which consisted of forty pilots, twenty flight engineers, and twenty radio operators. As was typical of the early bomber pilot training classes, this group faced discouraging prospects of surviving the war. Of the forty pilots, fourteen were killed in combat or flying accidents and three others were shot down, but became prisoners who survived the war, i.e., a total casualty rate of forty-three percent. Among the survivors, two came home with Purple Hearts.

When the group of thirty-seven pilots reached Albuquerque from Salt Lake on 3 April, they found that three other new pilots assigned to the CCTS from other bases had gotten the last housing at Kirtland. Thus the thirty-seven settled for the alternative – motels along Route 66. Of the thirty-seven, only thirty-five would be returning to their respective heavy bomb groups nearly four weeks later.

On 18 April, there was national news about one of Blackie's acquaintances. Jimmy Doolittle had taken a flight of sixteen B-25 Mitchell medium bombers from the American aircraft carrier, *USS Hornet*, and raided Tokyo and several other Japanese cities. The damage was minimal, but the boost to American morale was incalculable. The news stirred Blackie's growing desire to contribute more to the effort than transitioning brand new second lieutenants to the B-24 Liberator. Blackie did not actually read about it until the morning after because he was on a long, high altitude cross country to LaGuardia via Chicago which also involved several hours of night and instrument flying for the five young lieutenants on board. The lieutenants, Joe Rutan, Edwin Patterson, John Brady, Louis Cunningham, and Lowell Walker all faced daunting prospects considering the fact that they all went on operations as heavy bomber pilots in late 1942, a year in which only about one-third of air crews could expect to complete a tour. Patterson was killed in a stateside flying accident later in 1942, John Brady was killed in action as a B-17 pilot with the 306th Bomb Group flying from Thurleigh, England, and Lowell Walker was shot down to become a prisoner while flying B-17s from Bassingbourne, England. Only Rutan and Cunningham survived the war unscathed.

Blackie kept a close eye on the school's flight schedules, particularly the cross country graduation flights and noticed that of the twenty student crews, crew number eighteen was scheduled for a graduation flight to Kansas City on 22 April. Quite by coincidence, Blackie had been summoned by Jack Frye to Kansas City for a meeting that day. Redding was told that Blackie would command the outbound leg to Kansas City, but would be remaining overnight, thus Redding would be the aircraft commander on the return flight. The return flight would result in one of the greatest disappointments in Blackie's life.

There were other last minute crew substitutions. Jonas Ruff, who had traveled extensively and was always interested in visiting new places, had never been any closer to Kansas City than St. Louis. He asked the scheduled co-pilot, Hutch Thurston, if they could swap flights on the twenty-second. Hutch had no problem with it, and was aware that Ruff was eager to fly with his buddy Bob Redding from their adventuring days in Panama. George Van Hoozer, who was a CCTS maintenance supervisor, had a couple of reasons for wanting to make the flight. Before moving to Albuquerque he had been a TWA mechanic in Kansas City, and was anxious to become a flight engineer instructor, especially since the Army had approached him about the possibility of a commission and becoming a maintenance instructor. Van Hoozer was yet another man with a Panama connection; he had been a seaplane mechanic there while in the Navy. He had not been back to Kansas City recently and thought the flight might provide him with an opportunity to see friends and relatives for a few hours when the B-24 would be parked in Kansas City before returning. Van Hoozer took the place of the regularly scheduled flight engineer instructor.

Student crew number eighteen came from diverse backgrounds. The flight engineer trainee, Philip Macomber, was a happy-go-lucky twenty-five year old from North Jay, in south-central Maine. His father had died young, so Philip became the man of the house early in life and was an auto-mechanic in addition to running the family farm. Very community-minded, he was quite active in the local Grange Hall and volunteer fire department. He had been away from North Jay before and had spent a year at the Sweeney Trade School in Cleveland learning automobile mechanics. He had a great Harley motorcycle which he rode to Florida and back in 1940, the year before he joined the Army Air Force as a mechanic and eventual assignment to the Thirty-fourth Bomb Group's Eighteenth Bomb Squadron. Philip was acquainted with the man in charge of group's Seventh Bomb Squadron, Curtis Emerson LeMay, who later became the 305th Bomb Group's first commander. Just prior to Macomber's assignment to the CCTS, he had been based with the Thirty-fourth at Pendleton Field in Oregon.

The radio operator was a quiet, slightly chubby twenty-two year old from Worthington in western Minnesota. Duane Peterson was from a family with a domineering, frequently absent father and was anxious to leave home early. After dropping out of high school he promptly joined the Army and had most recently been stationed in Salt Lake with the 310th Air Base Group. He was something of a shy, late bloomer and had volunteered for radio operator training, a specialty which required above average intelligence.

Of the nearly 1,100 cadets who graduated from Army advanced flying schools in February 1942, it would not be a stretch to say that the one flying the leg from Albuquerque to Kansas City was in the inner circle of the most unique cadets. He was Charles Reynard (pronounced "Rennerd"), a twenty-three year old from Hiram, Ohio. His cadet track had taken him from primary training at Mira Loma Flight Academy in Oxnard, California, to basic at Minter Field in Bakersfield, and finally advanced at Mather Field near Sacramento. He was a gifted baseball and basketball star, but so were scores of other 42-B cadets. His personality was outgoing and inquiring. Not surprisingly "Charlie" Reynard was a particularly talented writer and, like Blackie, a very good musician. He was from a very large family and had twelve brothers and sisters. He had been a big man on campus back at Hiram College, the small, but highly respected institution from which he graduated in June 1940. Reynard had an adventurous streak, but that was not unusual in a bunch of high achievers like the cadets in 42-B. Charlie had taken off right after high school and soloed Washington, New York, and Boston – hitch hiking and working at such odd jobs as a dishwasher and circus barker to pay his way. Reynard had a fiancée – a lovely young lady who was a student at Radcliffe College in Cambridge, Massachusetts. What really set him apart was his master's degree, obtained from Harvard University in June 1941.

The request of Kansas City as a graduation flight destination originated with the other young Lieutenant, Roland L. Jeffries. Kansas City was his hometown. Flying to Reynard's home was impractical because Charlie would never have enough time to visit Hiram if the Liberator had landed in Cleveland, Youngstown, or Akron – any one of which represented a nearly two-hour automobile roundtrip to Hiram. So, Charlie Reynard was content to visit Kansas City. "Jeff," as the nearly twenty-two year old Roland Jeffries was called, had no education beyond high school; his father had abandoned the family so Jeff had gone straight to work as a clerk in a downtown Kansas City insurance company. He got into aviation cadets without the required two years of college by taking the college equivalency exam or the "mental test" as cadets called it. To prepare for cadets, Jeff was tutored in math, and because he was concerned about his stamina, he launched a strenuous physical training program built around swimming laps at the Kansas City YMCA indoor pool.

Jeffries, an Eagle Scout, was no stranger to TWA which had been the sponsor of a Boy Scout unit called "Sky Scouts," the forerunner to the popular Air Explorers which followed a few years later. There were only twenty-one boys in the troop, and each one was an Eagle Scout, selected from different troops within the Kansas City Council, BSA. The magic number was "21," the number of required merit badges (in addition to other qualifications) to attain Eagle Scout rank. Many were members of the Tribe of Mic-O-Say, the Scouting honor society for campers in the council. Mic-O-Say had been created by the council executive, H. Roe, Bartle, who went on to eventually become the popular mayor of Kansas City. Bartle was always known as "The Chief," which was the source of the name for Kansas City's professional football team. The Sky Scouts met at the big TWA hangar at the airport and were treated to a DC-3 plane ride when the troop was chartered. The chartering partner was TWA, and the guest speaker at the troop investiture was Jack Frye, although the troop's weekly meetings were presided over by TWA personnel from all departments. By the time Jeffries graduated from Scouts, he already had a good introduction to aero engines, airframes, navigation, meteorology, and other subjects which would comprise his cadet ground school experience. The boys wore a unique neckerchief; it was made from parachute silk. TWA treated the boys well and gave them an expenses-paid vacation to Hoover Dam – via a DC-3. Not surprisingly, many of the young men who were Sky Scouts and members of the Tribe of Mic-O-Say who survived World War II came home to post-war careers with TWA.

The Liberator that Reynard and Jeffries would fly on 22 April was practically brand new, having less than 120 hours of flying time in its logbook. The big ship had been taken into the Army Air Force inventory at Long Beach in California and flown to Albuquerque on 16 March at an acquisition cost of $351,514. It was a standard B-24D with gray undersurfaces and dark olive drab on the sides and top. On each rudder, there were dark yellow numbers "11133" which stood for 41-1133, the actual serial number. Up front on all CCTS Liberators, large yellow numbers had been painted on both sides of the nose to facilitate quick identification of each ship. 41-1133 was sporting number "twelve."

Van Hoozer and Macomber had completed their part of the pre-flight just as the two young pilots arrived with their Form 23, the flight clearance obtained from the operations office. Macomber had previously pulled a recessed handle on the forward left side of the fuselage to retract the bomb bay doors so he could enter the ship to check the glass-column fuel gauges. Unlike most American bombers, the B-24D Liberator fuel quantity indicators were not on the instrument panel. Instead, they were nothing more than graduated glass columns; if accidentally broken through carelessness or other mishap, raw gasoline would spray onto the flight deck. That is why Liberator crews tended to be very fearful of electrical shorts, knowing that a spark could cause an explosion in which they would be incinerated in a flash.

Both pilots had gotten a weather briefing before submitting the flight plan which called for a straightforward cruise of 200 mph flown on a heading of seventy degrees. Lt. Reynard conducted the walk around pre-flight and then entered the bomber via the bay doors. The weather looked good outbound – just some scattered to broken cloud at 7,000 feet through which they would quickly pass anyway. Weather in Kansas City looked very pleasant all day, but the weather for the return flight looked ominous. After signing off on Form One, the crew settled into their stations while Reynard went through the pre-takeoff items as read off from a checklist card by Blackie. The flight would be unlike Reynard's previous Liberator flights because it would be flown at 21,000 feet since this graduation exercise included the oxygen orientation.

While Blackie was well-acquainted with his instructor crews and the mechanics working on the Liberators, he never learned much about student crews due to the limited time available and the fact that he was all business in the cockpit. Had circumstances been different, he would have found much in common with crew number eighteen. Nearly all of its members and the instructors on the flight had rural or farm backgrounds. Macomber and Ruff had both grown up on dairy farms, Van Hoozer's father was a vegetable farmer in Missouri, and Redding still owned a farm up in Nebraska and was part-owner of a ranch in Wyoming. The Reynard family home was on a large plot of land where they kept chickens and a dairy cow. Although Jeffries was a city boy, his Boy

Scout training immersed him in the great outdoors. Only the quiet Duane Peterson had no deep ties to the land. Blackie could very easily have started a band with that crew. Van Hoozer was a talented guitar and harmonica player, Redding played the trumpet in his high school band, and Reynard, like Blackie, could play several instruments. Most of them, especially Jeffries and Reynard, were very good singers. Jeffries was a star singer at Westport High School in Kansas City and Reynard had actually considered trying out for the Mormon Tabernacle Choir while he was based in Salt Lake City. Blackie was not the only former mechanic on board. Macomber had attended a trade school in Cleveland for a year before becoming a garage mechanic back in Maine, and Van Hoozer had graduated first in his aircraft mechanic training class at the Navy training base in Glenview, Illinois, in the early 1930s. Ruff excelled in mechanical courses in high school and became an appliance repairman after graduating. He later became an instrument repair technician with the Panama Canal Authority and had gone directly to the Eagle Nest Flight Center from his job as an instrument technician at what is now the Department of Commerce's National Weather Service headquarters in Washington, D.C.

Reynard was in the left seat for the flight to Kansas City. Excluding Macomber, he was the oldest member of the student crew, but his appearance was so youthful that he could easily have passed for a high school junior. The appearance was deceiving because he was a very capable young man and had served as a cadet company officer throughout his flight training in California. He was very confident and not at all intimidated by Blackie's authority and thousands of hours. Blackie was in the right seat on the outbound flight, functioning as the co-pilot in addition to being the check pilot. Thus, he was the one who read off the forty-eight items on the pre-start checklist. Once airborne, the roles were reversed as Blackie responded to Reynard's commands for landing gear, flaps, and power settings. After maintaining the runway heading until gaining several hundred feet, Reynard called for 43 inches of manifold pressure at 2,550 rpm and retrimmed the Liberator for 150 mph before starting a turn to his left and climbing at 750 feet per minute. His climb would take him over the Tijeras Pass beacon located to the northeast near Sandia Peak. Fifteen minutes were required to reach his cruising altitude of 21,000 feet which was attained after passing through a thin, scattered to broken layer of clouds. Reynard had already called for an oxygen check shortly after climbing through 10,000 and was then settled on a heading of seventy degrees magnetic. After the steady ascent, Reynard climbed through 22,000 before setting up the cruise. He had deliberately climbed through 21,000, but when reaching that 22,000, he pulled the turbos back for 32 inches and set rpm at 2,200. After shoving the nose down and the speed passing through 200 mph, Reynard retrimmed the big ship after leveling off at 21,000 very much to Blackie's approval. Although they were all wearing fleece-lined, winter flying suits, the cockpit was icy cold. On the other hand, the ceiling and visibility were absolutely unlimited. It was a wonderful morning to be flying, and the monotonous gray-brown landscape of eastern New Mexico was giving way to an increasingly verdant Kansas as they droned eastward. Unfortunately, those lovely sky conditions were poised to make a one-eighty by the time they were ready to return to Albuquerque late that afternoon.

They had taken off on Runway 17 in spite of the modest crosswind of seven mph from the east. There was some light rain falling east of the field, but it was moving away. At least it had put a damper on the blowing dust that constantly plagued takeoffs and landings at Kirtland. Until recently Runway 17/35 had been the longest runway in the United States, but the real reason for using it was the basically uninhabited real estate south of Kirtland. If something went wrong right after takeoff, there was not much beyond the end of the Runway 17 except the Rio Grande river bed. Unless there was a really terrific crosswind blowing, landings were usually made on 17/35 as well since its great length minimized the use of Liberators' brakes – something Blackie was very sensitive to considering the bomber's touchy brake system.

On the way to Kansas City, Redding had to be content to sit on the flight deck jump seat as a passenger. Nobody was allowed in the nose section on takeoffs and landings, but that was where Jeffries headed as they approached Kansas City. The greenhouse nose compartment offered the best visibility of any crew station in the plane. Only one person met the plane when it landed in Kansas City just before noon, and she pulled up to one side of the TWA hangar; it was Mary Casey, Jeff's lovely fiancé. They took off in her car and would not be back until 4:00 P.M. for the scheduled 5 P.M. takeoff. Charlie set off alone to solo Kansas City, and George Van Hoozer hung around the airport all day, phoning relatives and visiting with TWA mechanic friends. Ruff and Redding caught a taxi to Country Club Plaza which was Kansas City's extraordinary new shopping mall complete with many shops and enticing restaurants. Peterson and Macomber were headed there too.

Blackie met Jack Frye in Milleman and Gilbert's, the restaurant in the airport passenger terminal, for lunch. The restaurant was a Kansas City favorite, particularly for its dining rooms which offered great views of landing and departing planes. After lunch, Frye and Blackie moved to the TWA offices where Frye unloaded the bombshell – the Army would not renew the CCTS contract and that by mid-summer the CCTS operation would close. The Army was planning on taking the last two classes of CCTS graduates and turning them into B-24 instructors. There were more B-24 schools scheduled to open then, and when TWA left Kirtland, the Army Air Force would move in and take over the operation. The Army took not only the Liberators, but all instruction manuals and the school syllabus as well. Frye was not specific about Blackie's future, but it was made clear that going back to LaGuardia as a check pilot or to TWA's training operation in Kansas City were not options. The handwriting on the wall clearly pointed toward Washington. Because there were so many details to work out concerning closing down CCTS, Blackie would not be flying back to Albuquerque that afternoon on Liberator 41-1133 – Providentially.

By four P.M., the rest of the crew had gotten back to the airport along with Mary Casey and some very proud members of Jeffries' family. Charlie would not be flying the return leg, but instead worked up the flight plan which had been preceded by a visit to the weather office on the second floor of the terminal building. What he saw on the weather map was anything but encouraging. The massive high pressure system that made 22 April such a beautiful day over most of the entire eastern half of the country was also pumping moist air out of the Gulf of Mexico and sweeping it northwest where it would collide with Southern Rockies and a low pressure system descending in force from the northwest. Although flying conditions in eastern Kansas were still satisfactory, stations in northern and eastern New Mexico were reporting rapidly falling barometric pressure, very strong winds, significant precipitation, lightning, and declining visibilities and ceilings. A contemporary analysis of the national weather map for 22 April 1942 strongly suggests that a southern bend in the jet stream had looped down into southern Colorado and northern New Mexico, making matters that much worse. The eastern slopes of the Southern Rockies, including the Cimarron Mountains in the Sangre de Cristo range, are notorious for violent mountain weather. In fact, only a crescent along the Gulf of Mexico running from Louisiana to Florida receives more lightning than the Rockies running from New Mexico up into Colorado.

'1133 took off at five with Jeffries flying and Redding in the co-pilot's seat. Peterson called in position reports on time while over Kansas; and at 7:35 P.M. the crew checked in with the radio range station at Las Vegas, New Mexico, advising that they were twenty-five miles east of that station. Eighteen minutes later a garbled transmission was heard in which Redding said that they were at 14,000 feet, on instruments, and requested an instrument clearance into Albuquerque. That transmission was fuzzy, due to lightning, and the operator at Las Vegas could not understand what else Redding said, but interpreted it to be either 75 miles northeast or northwest of Las Vegas. It was a terrible night in northeastern New Mexico with high winds, torrents of rain, and Biblical flooding that took out not just roads, but railroad track beds too. The Santa Fe Railroad could not make it from Colorado into New Mexico because the tracks over Raton Pass had been completely washed away.

The crew was never heard from again. It was declared overdue forty-five minutes after failing to arrive on its ETA. Eight hours after takeoff, the plane was declared missing. What made matters worse was the weather over the next few days which severely limited search flights, and when the searching bombers were able to fly they concentrated in the immediate Las Vegas area. Blackie did not get back to Albuquerque until late on the twenty-fourth, but did combine a search with a regular student check ride on the twenty-fifth, a pattern he would repeat each day for the rest of the month, logging twenty-two hours in the process. One of the young B-24 students on those search flights was Lt. Russell Schleeh, a very good friend of Charlie Reynard's. Schleeh defied the law of averages and survived his combat tour with the 305th Bomb Group early in the war when sixty-five percent of Eighth Air Force crews did not. Schleeh later went on to a distinguished career as a test pilot, flew bombers in the Strategic Air Command, and after retirement held management positions with Douglas Aircraft.

On the twenty-seventh, Frank Busch, who had now become the CCTS operations manager, sent telegrams to the missing crew's anxious families saying, in effect, that there was little or no chance of finding the crew alive. Many years later, Betty Blackburn, Blackie's eldest daughter who was nine at the time of the accident, still recalled that her normally stoic father was shaken, especially by Redding's loss. Martha Blackburn's reaction was much more immediate; she was in the kitchen when she heard about it and became violently ill. Then at 10 A.M. on 1 May, Howard Kincheloe and Hutch Thurston, flying with a regular Army crew, noticed a disruption in the tree pattern on the southwest side of a 10,000 foot mountain nearly ten miles west of the village of Cimarron which is roughly seventy-two miles north-northeast of Las Vegas. CCTS crews had not been looking far enough north; but sure enough, it was '1133. Kinch promptly radioed the news and soon there were several Liberators circling the mountain which was on property owned by the Boy Scouts of America (BSA). Cimarron locals still called the property "the Philmont" which referred to the huge ranch which until recently had been owned by Waite Phillips, a Tulsan with vast ranching operations in New Mexico

Phillips had become fabulously wealthy in the oil business and bought tracts of New Mexico ranch lands, including large holdings in Colfax County. His wealth was matched by his extraordinary philanthropy, and in 1938 he had donated 10,000 acres in the northern sector of his Philmont Ranch to the BSA. In those days, the "camp" was called Philturn Rockymountain Scoutcamp. In 1942, Phillips granted another 117,000 acres to what would soon be called Philmont Scout Ranch; attendance picked up dramatically after the war, making Philmont the world's largest youth camping and backpacking operation. The gifts also included Phillips' palatial mansion, Villa Philmonte, which huge though it was, is paled by comparison with the Phillips home back in Tulsa which is now the Philbrook Museum of Art. Today Philmont Scout Ranch employs roughly 1,000 seasonal staff members in the summer to serve the roughly 22,000 Scouts and their leaders who backpack Philmont's many mountain trails. Philmont also serves as the BSA's national training center through which several thousand BSA volunteers and professionals pass every year.

In the summer of 1942, the Ranch served only a handful of Scout campers because its facilities were completely undeveloped and wartime gas rationing made travel difficult. The aftermath of the torrential rains would make a ground search difficult, but nonetheless Col. Frank Hackett, the Kirtland base commander, organized a New Mexico police-assisted ground search effort for Friday

2 May. Blackie, Col. Lewis, and two other Kirtland officers flew up to Philmont in a pair of BT-13 trainers and landed at what is now Philmont's buffalo pasture which is only several miles from the Vallejo Ranch, a frequent venue for Jack Frye's Conquistadores del Cielo. In the 1930s the relatively flat buffalo pasture served as a small, private airport. Among Waite Phillips' friends who had used the landing strip were Will Rodgers and Wiley Post who had visited Philmont immediately before their ill-fated flight out of Point Barrow, Alaska. Unfortunately, the search party, running into washed out roads and snow drifts, got a late start, and did not discover the wreckage until mid-afternoon. When the searchers reached the wreckage, they encountered a local cowboy, Bruce Bull from nearby Black Lake, who had ridden his horse up the mountain to discover why so many planes were circling over Trail Peak. Because the horses were having problems getting through some of the deep snowdrifts, the recovery mission was put off until very early the next day. Blackie and the Army officers spent the night at the St. James Hotel in Cimarron. The hotel had been launched by Henri Lambert, a man who had cooked for President Lincoln while working as a chef in Washington. The hotel, still going strong, has a guest register filled with the Wild West's most colorful characters including Buffalo Bill Cody, Annie Oakley, the Earp Brothers, Jesse James, Black Jack Ketchum, and Clay Allison along with media luminaries including Lew Wallace and Zane Grey. Through the swinging doors to the hotel's bar, many patrons passed in while others, lingering at the bar too long, passed out. Through those same doors, others emerged horizontally feet first as the bullet holes in the pressed tin ceiling bear silent witness.

That part of New Mexico is still sparsely populated and was even more so in 1910 when a youthful bookkeeper came to work as a cashier at a bank in nearby Maxwell, a village about the same size as Cimarron and located about thirty miles to the east. The young man from Wisconsin and his new Mississippi bride did not linger long in New Mexico, but like Blackie, sought a new future in Los Angeles. That good fortune did come to him in California at about the same time Blackie was struggling to become an aviation cadet. The former bank cashier would also have a major impact on Blackie's future. His name was Noah Dietrich.

The bomber had crashed into Trail Peak, elevation 10,242 feet, which is one of Philmont's ten mountains reaching 10,000 feet or higher. Blackie and the Army officers had a very early breakfast, joined other searchers, and then drove up the road to the Philmont Ranch where the group was again headed by Elliott "Chope" Phillips, Waite Phillips' son. When the searchers reached the wreckage in brilliant sunlight, they found that the Liberator had hit the mountain 200 feet below where it finally came to rest just under the mountain's southwest ridge where the snow was not quite as deep. Had 41-1133 been flying 250 feet higher, it would have just cleared the mountain. Due to the cold, the bodies were well preserved. Redding, Jeffries, and Macomber had all been catapulted through the cockpit windshield; they were later found just ahead of the completely demolished nose section. Van Hoozer and Peterson were found under the left wing root and Ruff was located just below the left waist gunner's window. Reynard's body was the least disturbed of all and was found tightly strapped into a waist compartment seat. The bomber had broken in half just aft of the wing's trailing edges, and the tail section had rolled inverted. The engines had all been ripped loose, were hurled over the ridge, and tumbled down the north side of the mountain. Col. Lewis gathered the possessions of the Army fliers while Blackie collected the remaining personal effects of Redding, Ruff, and Van Hoozer.

It was Blackie's sorrowful duty to assist packing out his good friend's body which, with the others, reached the Strong Thorne Mortuary in Albuquerque on Sunday, 4 May. From there, the bodies were prepared for rail shipment to hometowns in which funerals were held later that week. Blackie's involvement with the accident investigation board prolonged the matter and raised more questions than it answered. Liberator 41-1133 had been on an east-northeasterly heading when it hit the mountain around 8:30 P.M. on 22 April. The bomb bay doors had been retracted, the crew was on oxygen and had their parachutes attached, and the number four engine had been feathered before hitting the mountain with the left wing low at the moment of impact. That the parachutes were attached suggests that Redding had prepared the crew for the worst while open bomb bay doors pointed to the possibility of fuel leaks. Early Liberators were notorious for fuel leaks; the slightest electrical spark could ignite the vapors and instantly create a flying fireball. Prudent Liberator skippers would open the bomb bay doors to clear out fumes in a hurry – some pilots flew with the doors slightly rolled up all the time just to be on the safe side. Even with the number four engine shut down, it was estimated that the bomber had been flying at 200 mph, or almost exactly the recommended maneuvering speed for flying in thunderstorms.

The landing lights had been on suggesting that the crew had been looking for snow or the possibility of ice building up on intakes and the wings' leading edges. One might question why Redding had taken the ship to the north of the planned course. There was lower ground to the south and higher terrain to the north. The weather was actually more threatening to the south and heading north could result in putting the winds behind the Liberator since winds tend to flow counterclockwise around major storms. The radio compass card had been set for 250 degrees or roughly the original course that would have taken them to Albuquerque from the Las Vegas range station. Blackie concluded that Redding had run into more storm than he was comfortable with and turned north to fly around the storm because the most recent weather for Santa Fe was not as threatening as that to the south. Unfortunately, Redding ran into weather worse than that predicted to the north and was caught between a couple of aviation's most forbidding rocks and hard places – mountains immediately below and the freezing level not far above. He did not have much to work with; there are numerous peaks in the 11,000 to 13,000 foot range in the immediate area and the freezing level that night at his position was recorded at almost 13,000 feet. New Mexico's highest mountain, Wheeler Peak, at 13,167 feet, is only twenty-two air miles northwest of Trail Peak.

It would be a couple of decades before aviation science described the "downburst" phenomenon but that is the nearly certain cause of the '1133 accident. A downburst's rapidly descending column of cold air can easily push a 50,000 pound bomber down thousands of feet per minute and there is rarely little a pilot can do to break out of the downburst's grip. The energy in the thunderstorms recorded on the night of 22 April was more than powerful enough to generate strong downbursts. The most likely scenario was that Redding was finally able to break out of the downburst at about 10,000 feet but within seconds flew into the side of 10,242 foot Trail Peak. Nevertheless, the accident board report suggested pilot error on Redding's part because he could have turned back, landed somewhere in Kansas or Texas for the night, and flown down to Albuquerque the next morning. Army Air Force policy was explicitly clear on how to deal with thunderstorms: complete avoidance. The final truth is that the bomber was just one more to be done in by the four-jawed vise of aerial doom: pilot error, mechanical failure, terrible weather, and bad luck.

Blackie's sad involvement with his friend's passing was prolonged by Redding's funeral which was held back in Minatare. Redding's ashes had been delivered to Scottsbluff where Blackie would pick them up in a Liberator. He would then drop them over the Redding farm two and one-half miles north of Minatare right at the conclusion of the funeral service which was held on 9 May. Blackie was almost late over the farm, having briefly gotten stuck in some mud at the Scottsbluff airport while picking up the ashes. Several weeks later after the snow had melted, the Army sent a salvage team up to Philmont, but there wasn't much to salvage – just some of the instruments and engine components which were easily carried off the mountain by horses.

By 1942, Blackie was becoming hardened to the deaths of his friends in flying accidents. Greeson and Redding would not be the last, but it would be those two that hurt the most. In the case of Bob Redding, Blackie would be haunted until the end of his days about what might have happened if he had not spent two days in meetings at TWA headquarters. How would the flight have turned out if the meetings wound up early and he had been able to rejoin '1133 on that flight back to Albuquerque? Would his skill and intuition have brought all eight of them back to Albuquerque safely? Blackie always thought his being on board would have carried the day, but there was no way of turning the clock back.

Ironically, one year, almost to the day after CCTS Liberators circled Trail Peak, more Army Air Force planes were zooming down Philmont's canyons and rolling above Philmont's majestic landmark, the Tooth of Time which had been a waypoint on the western spur of the Santa Fe Trail. Those new aircraft were F-4s, the photographic versions of the Lockheed P-38 Lightning, and they were from the Twenty-second Photographic Reconnaissance Squadron. They had come down from their training base in Colorado Springs to Raton, New Mexico, and its tiny Crews Field for their bivouac exercises just before heading to England. Blackie was by then long gone from New Mexico and had no idea of how these hot fighters turned into photo ships would form a remarkable aspect of his wartime flying in just another year.

The balance of Blackie's flights in May 1942 was spent as a check pilot on cross countries, including several to Washington. While on an overnight layover at Washington National following a check flight on 21 May Blackie borrowed a Cessna UC-78 (twin engine trainer/utility aircraft) and flew over to Bolling Field where he enjoyed dinner with the now Colonel Joe Mountain. The flight to Washington included what was originally planned as a very brief stop in Kansas City. The stop lasted much longer than planned because it involved a landing accident in which the nose gear failed and the left main tire blew out. Blackie was on top of it and quickly and kept the ship, 41-1106, on the runway and rolling for another 2,000 feet before stopping. Unable to obtain parts at nearby Sherman Field in Fort Leavenworth, another Eagle Nest Liberator flew in with a complete nose wheel assembly and new left tire the same day. Blackie's training crew took that Liberator for the continuation to Washington while a TWA maintenance crew repaired '1106. The return flight to Albuquerque on 23 May was without incident.

Other than the Doolittle raid in mid-April, there was little encouraging war news for the media to report in the early summer of 1942. The major exception was the Battle of Midway which was fought 4-7 June and concluded with Japan's losing the core of its aircraft carrier fleet. The day before the battle opened, Blackie had been in Kansas City for meetings and took Flight 3 as a check pilot on the trip back to Albuquerque with intermediate stops in Wichita and Amarillo. Offsetting the news about Midway was the ongoing Battle of the Atlantic which the Allies were losing. Brazen Nazi submarine commanders were sinking American ships practically within sight of the American East Coast. Exactly one month before, the Philippines fell to the Japanese. General Joseph Stilwell had been run out of Burma by the Japanese who occupied enough of the country to sever the main supply road to China whose army was essentially so under-equipped and poorly trained as to be completely ineffective in its land war against Japan. The Germans still had the upper hand in Africa, although that would change in mid-autumn when the British were victorious at the second battle of El Alamein which coincided with the Allied invasion of French North West Africa.

Most of the men Blackie checked out were lieutenants who would go on to do their duty and come home to lead relatively anonymous lives – that is the small number of those early graduates who did come back. There were a few exceptions to the anonymity, and one was Norman Appold who was in the left seat, as a lieutenant, on the way back to Albuquerque following the landing mishap in Kansas City. A year later Major Appold was a flight leader in the 376th Bomb Group (The "Liberandos") on the epic mission to bomb the Rumanian oil complex at Ploesti. He would go on to distinguished careers in both the Air Force and Lockheed.

Probably the best known aviator to benefit from Eagle Nest's expertise was a Blackburn acquaintance of several years – Charles Lindbergh, who spent the better part of 16 April 1942 with Blackie and Colonel Hackett. Lindbergh was particularly interested in the B-24's performance because he was a consultant to the Ford Motor Company which was about to start manufacturing the B-24 at its massive Willow Run, Michigan, plant. He talked with the pilots, asking how the B-24 might be improved and got an earful about the brakes, fuel system, hydraulics, nose wheel issues, and improved visibility for night landings.

Lindbergh already had several hours in the Liberator and was not impressed, concluding that its controls were unusually heavy. He did concede that the big bomber handled well in a heavy wind – it was blowing a gale with plenty of dust in the air when he flew one of the CCTS Liberators for an hour after lunch. What did impress Lindbergh was the B-24D training document Blackie gave him, noting it was better than anything he had received from the Army Air Force. He was appalled, along with many others, at how 200-hour second lieutenants were given such a condensed course on the Liberator and then expected to function as well-trained four-engine pilots. But that was the state of affairs in early 1942, and it would be months before America had the resources to thoroughly train heavy bomber pilots.

Although he knew that the CCTS was going to wind down that summer, Blackie logged more flying hours in June, upgrading his instructor pilots instead of supervising cross country check rides with students. Those promising pilots being upgraded included Nick Wasil, Jim Wheeler, Merrill Shurtleff, Hutch Thurston, and several others. Some of Blackie's senior instructor pilots, Joe Carr among them, had already been transferred to TWA's new Intercontinental Division (ICD), and considering how many of TWA's domestic line pilots, as reservists, had gone back into the military, promotion from within was about the only option left. By 1 July, Eagle Nest was in close down mode. All of the Liberators would soon be flown to Smyrna Army Air Base in Tennessee where the Air Force had set up a new B-24 school. Some of the first officers who had joined Eagle Nest in 1941 went to TWA's domestic operation, including Hutch Thurston. A few, like Nick Wasil, were recruited directly into ICD to the great consternation of those flying domestic routes. Several pilots were offered Army commissions and went into the Training Command. Other reservists like Harry Campbell were called up and went on active duty. Howard Kincheloe, who had found the wreckage of 41-1133, declined an Air Force major's commission and went to Consolidated in San Diego as a test pilot and later finished the war as an F4-U Corsair test pilot with Chance-Vought in Connecticut.

Blackie flew his last flight in Albuquerque on the evening of 26 June, flying night check rides after getting back from a roundtrip graduation flight to Long Beach, California, the day before. On the morning of the twenty-seventh, he got a call from Otis Bryan. Blackie was given a couple of days to wind up affairs in Albuquerque before moving to Washington and TWA's Intercontinental Division. Somebody else was moving to Washington – Jack Frye. With a growing part of TWA's business coming from Washington, Frye decided he could be closer to the government while still managing manage much of TWA's business just as well from Washington as Kansas City. There was another reason; he knew the airline business would be changed forever by the war and that he needed to be close to the center of forces that would shape the post-war future of the major air carriers.

Ironically, one of Blackie's last assignments while at Eagle Nest was to critique a proposal that TWA furnish instructors for the new Army Air Force Ferry Command transition school at St. Joseph, Missouri. The plan called for a TWA flight instruction staff of ninety-two, including nearly forty instructor pilots and a dozen navigation instructors in the school which would have an inventory of nine four-engine aircraft (B-17s, B-24s, and C-54s) and twenty-one twin-engine aircraft. Blackie's favorable conclusions related to TWA's experience at Eagle Nest and the continuing opportunity to support the war effort. The unfavorable conclusions far outweighed the positive: friction between military and civilian operation, wartime exigencies preventing the Army from keeping commitments to supply planes and parts, and a shortage of qualified TWA instructors. Blackie had no reservations regarding the runways at St. Joseph Airport for use by experienced four-engine pilots but had little confidence in their use for instruction – certainly in contrast to Albuquerque.

As Eagle Nest was winding down, TWA had landed another government contract, and this one would be a long term success; it was going into the aircraft mechanic training business and spending $250,000 of its own money to build the training facility in Kansas City to which it would divert 100 of its regular employees. The training, which ran eight hours a day, seven days a week, would cover every phase of aircraft maintenance from balancing propellers to patching bullet holes.

Financially, Eagle Nest was a modest success during the roughly ten months of its existence as a contract school for the Army Air Force. When the books were closed, Jack Frye enjoyed a profit of nearly $90,000 on the nearly $600,000 training contract. Although it was in existence for little more than a year, the fact remains that TWA's Albuquerque operation graduated over 1,000 pilots and aircrew in its year of operation, and that was at a time when ATFERO and the Army Air Force very desperately needed trained ferrying and bomber crews.

Blackie had his new assignment: he was going to Washington National Airport as the Intercontinental Division's supervisor of flight operations. He journeyed to Washington via San Antonio where Martha and the kids would be staying with the Bondurants until Blackie could find a suitable home in Washington.

Chapter Six
TWA Intercontinental Division (ICD)

Blackie did not arrive in Washington from San Antonio until the end of the last week of June and would not fly again until 11 July which resulted in a flight that he knew was bound to happen sooner or later. At that point, all of TWA's ICD operations had been moved to Washington National Airport's Hangars One and Two, although the first few flights had been flown out of Bolling Army Air Base just to the south along the Potomac River. In the meantime, Blackie was very busy learning about his new job, assisting Cliff Mutchler, his immediate superior who was the assistant director of ICD, and looking for a place to live. Mutchler was not a pilot; but he had most recently been assistant to Larry Fritz, who was TWA's vp/director of operations until going on active duty with the Army Air Force. Blackie, as a new co-pilot in 1935, had flown with the mercurial Fritz whose hair trigger temper was legendary. Fortunately, Blackie never aroused the volcano which typically cooled almost immediately after any irritant was withdrawn. Fritz had flown with the Royal Air Force as a gunner in the Great War and then got his wings in the early 20s as a graduate of Kelly Field. His route to TWA was through Maddux and Western; when Blackie flew with him in 1935, Fritz was the regional superintendent for southwest operations.

Mutchler's aviation experience dated back to 1929 when he had been with the Bureau of Aircraft Production; he was later involved with quality control on the production of Liberty aircraft engines. Mutchler had more recently been in charge of service operations within TWA, including managing restaurants at TWA airport terminals and the hiring and training of hostesses. From the outset, General Hap Arnold decreed that there would be no women acting as hostesses on Air Transport Command (ATC) flights. So, TWA went back to the days of the bull hostess and recruited men for jobs as pursers. Considering the fact that the draft was going full blast, Mutchler had his hands full recruiting pursers but found enough of them by raiding Eastern Airlines (the only line not to employ women hostesses) and the nations' hotel service staffs.

Blackie's job as supervisor of flight operations carried a lot of responsibility, particularly as he was charged with managing all flight operations which thus necessitated close cooperation with engineering, maintenance, and training. He not only assigned crews, but also briefed them before departure and also debriefed them upon return. He was also responsible for safety and efficiency; thus he was also a quality control manager. This applied not only to Washington National, but also to the foreign bases that came on stream as ICD gathered momentum. Just getting a crew ready for a foreign trip took well over a full day. Blackie was also to function as the assistant division manager when Mutchler was not in Washington. Finally, there was one more responsibility that Blackie had to reluctantly accept – he would be in charge of flying accident investigations. Blackie would take the same approach to his job that he took at Eagle Nest: "I can't do my job unless I know what pilots are doing and the best way to do that is to take some of the flights myself." After all, Otis Bryan had set the example as a flying top manager. It would not be too much longer before Blackie's position with one foot in two camps came to the attention of the Airline Pilots Association.

TWA had entered into an arrangement with the Hotel Cairo near DuPont Circle on behalf of transient ICD crews since most of them had no intention of moving to Washington permanently. The Cairo was no longer one of DC's more fashionable hotels, but Blackie cared naught – he was house hunting and stayed there only briefly. Although the Cairo was only about four miles from National Airport, Blackie was looking for something closer and found it at 4812 North Seventh Street across the river in Arlington, Virginia. Again, it was a two story brick house well shaded by tall trees on a quiet street. TWA allowed Blackie $15 a day living allowance for a maximum of two weeks while he was looking for a house which, even in the summer of 1942, was becoming difficult considering the acute housing shortage in the city. TWA also provided his family's railroad fare from San Antonio and agreed to pay his household goods moving expenses too.

Otis Bryan had been in Washington for six months when Blackie arrived and had quite literally been building an airline within an airline at the same time the Ferry Command was evolving into the Army Air Force's Air Transport Command. That evolution was nearly complete just as Blackie came to Washington, and when finished the old Ferry Command became one of three functions within its parent organization. The other responsibilities of the new ATC were training crews for long range ocean crossings and flights over inhospitable terrain along with the establishment of air routes across the world complete with support facilities along the way. The airlines' contractual arrangement with the War Department was on a cost-plus basis; neither party was really prepared for the immense job at hand; and, there was some early friction between them. The airlines involved, TWA, Pan Am, American, United, Eastern and a few of the strongest regional airlines, felt that the military should have not only seen the need for more transport aircraft but also should have pushed for earlier development of aircraft such as the latest DC-4 instead of trying to rely on the nearly ten year-old DC-3/C-47 aircraft with its relatively limited capabilities. True, the big new Curtiss C-46 Commando answered some of that criticism, but it was a troubled airplane at first and was never available in large numbers.

Otis Bryan had neither staff nor airplanes when he came to Washington, but within a few days work on converting the Stratoliners began. On 7 January, the first Stratoliner was hauled off the line for modification. The pressurization system was removed along with the plush interior. The seats and berths in the two front compartments were ripped out and replaced with five 212-gallon gravity feed fuel tanks and a manually pumped, forty-five gallon oil tank. The fuel tanks had been flown in by Boeing under armed guard. The ladies' room, better known as the "charm room" went away along with its heavy mirror. Passenger capacity varied depending on how much freight might be carried, but never reached the regular 307 Stratoliner limit of thirty-three people. Crew capacity was increased because room had to be created for a navigator and a radio operator which created additional, structural work. The navigator needed an astrodome – that was quickly supplied by Pittsburgh Plate Glass and located in back of the front bulkhead where the navigator had his desk. The latest driftmeter was installed, making it possible to determine both drift and groundspeed, but because it was gyro-based, care had to be taken to make sure it was caged on takeoff or in rough air. Accommodation for passengers was flexible: four berths (or regular seating for twelve) plus four reclining seats. The converted Strats were noisy because effective sound-damping materials had yet to be invented.

New tunable radios were required and subsequently supplied by Bendix because the older crystal sets would not necessarily work everywhere ATC flew. Finally, there was the hush-hush "Identification Friend or Foe" radio system that told a friendly radar operator that the plane sending the signal out was also friendly. To create the best reception and transmission over some frequencies, a 400-foot trailing antenna with a five-pound weight at the end was installed. Its operation was not without some hazard for the forgetful radio operator who failed to crank it in. Landing with one extended invariably turned the weight into a five-pound missile that could hit anything near a runway, including people and cattle. The olive drab paint added a few pounds and by the time emergency gear was stored on the Strat, its performance was reduced. The Stratoliner was, however, the only game in town, and the five of them were all Otis Bryan had to work with until getting a few C-87s (the cargo version of the B-24 Liberator) later in 1942 and when the first C-54s (military DC-4s) started to arrive in the autumn. Eventually, ICD would have over twenty of the C-54s in addition to its Stratoliners and C-87s.

While the first crews arriving at ICD were well-qualified, there were not enough of them in the autumn of 1942 which left ICD with recruiting and training problems. For co-pilots, it was pretty much training on the job which was of little consolation to the captains. These problems would be eased in 1943, but in the autumn of 1942 Blackie described the recruiting as one of "shaking loads of hay looking for pilots and then looking under the wagon." Typical of the new co-pilots recruited was Robert Guss, a twenty-one year old flight instructor with 1,600 hours in his logbook. He hailed from Reading, Pennsylvania, and had gotten his pilot's license while still a teenager. He had a two-year technical degree, but could not quite pass the physical exam to become an aviation cadet in the Army or Navy. He took his flight instructor's rating to Spartan Aviation and worked as a flight instructor, ironically training some young Americans who eventually flew in the Royal Air Force Eagle Squadrons. By late 1944, he was twenty-three, had made captain, and also had thirty roundtrips to Scotland in his logbook.

ICD's manpower then precluded spending much time on training, so the new pilots like Guss were sent to the Army's B-24 Liberator transition school at Smyrna, Tennessee, which ironically is where most of Eagle Nest's Liberators had been assigned following the school's closing on 30 June. Harry Campbell, one of the Eagle Nest instructors and a Kelly Field '29 graduate, had gone back on active duty and was assigned to Smyrna as a captain and class instructor. He commiserated with Blackie about the lack of airmanship among the young pilots ICD and the other airlines sent to Smyrna, looking back on those days in Albuquerque as a time of plenty. The Smyrna training was even more condensed than had been the transition course taught at Eagle Nest, and hardly prepared young pilots for the full host of Liberator eccentricities. These shortcomings included major problems in the fuel system (dangerous leaks), inadequate lighting, runaway electric props, gremlins in the hydraulic system, a disinclination to leave the ground even when within the proper weight/balance limitations, and flight controls so heavy that pilot and co-pilot became a tug of war team. Once in the air, the Liberator Express did have one important virtue; it could fly tremendous distances without refueling. There was one major exception to the superb range; if a Liberator wandered into icing conditions, the range was in doubt because the Liberator's much vaunted, high-lift air foil, quickly gave up the ghost after accumulating even the most modest amount of ice.

On the other hand, ICD was considered a plum of an assignment because for a pilot who longed for faraway places, the pay was very good. Check pilots and captains made $1,100 per month while reserve captains and first officers were given $800 per month – that was good money during the war years. Navigators and engineers earned $500 to $600 a month. Otis Bryan personally selected the original pilot cadre for ICD and placed more emphasis on men he knew to be good in an emergency instead of seniority lists. Men with Stratoliner experience were among the first selected, but Bryan also wanted pilots who were known for initiative and resourcefulness when it came to getting out of trouble. If such men had occasionally been a nuisance to management, so much the better because ICD was going to be anything but routine, and Bryan did not want people who were inclined to call a dispatcher or start checking the company operations manual index when a problem suddenly appeared.

Blackie was in very good company with the original captains when he came to ICD. The pilot roster included men of whom several had supervisory or instructing responsibilities as well, including former Eagle Nesters Swede Golien as chief pilot and Joe Carr as director of training. The other names read like a "Who's Who" from TWA's early days: Doc Mesker, Chick Fredericks, Frank Busch, and navigator Pete Redpath. Other early captains in 1942 included many more of the men who had trained Blackie, including Mel Bowen, Milo Campbell, Larry Chiappino, Earl Fleet, Larry Trimble, Fred Richardson, Howard Hall, and Don Terry. The other early arrivals were senior pilots with considerable experience: Alton Parker, Dick Hanson, Stan Stanton, Charlie Kratovil, Gene Klose, Frank Niswander, Ben Dally, and Bob Buck. By the late summer of 1942 Blackie was also able to bring a few promising younger captains, like Nick Wasil, into the ICD fold from Eagle Nest. Sadly, not all of ICD's early captains would live to see the end of the war.

Otis Bryan faced staffing issues as the war went on, particularly when some of his pilots were called to active duty and especially when his aircraft inventory began to expand later in 1942. Some men, like Joe Carr who technically was a first officer, were moved up to captain, but were expected to have at least 2,000 hours and an instrument rating before being checked out. Navigators, for the most part, were recruited from the Merchant Marine, nabbed from Pan Am, or recruited from Parks Air College in Missouri while many radio operators were picked from ground operators who wanted to fly. Even a few talented amateur ham radio operators found their way into the Stratoliners' radio cubicles.

Bryan himself did not have to pound the pavement as a recruiter; he had a former Army pilot who had become a newspaper editor serving as his personnel department at first. The man, Kemper Jacks, had been a psychology major in college and usually put some practical psych to work when interviewing prospective pilots. Knowing how important multi-tasking and being attentive were to successful pilots, he had a screening technique that became legendary. At a pre-arranged signal, Jacks' curvaceous secretary came in and tried to distract the candidate. The encounter was much more subtle than her small talk and blatantly showing off her alluring cleavage and ample bosom, but she would be in the office long enough for Jacks to get a clear idea of how well-focused the candidate was on the interview.

Trans World Airlines

Blackie in Air Transport Command uniform, summer 1943.

The men of ICD in a few ways were neither fish nor fowl. ICD was somewhere between a "quasi" and a "para" military organization. Technically, they were not uniformed combatants, although every one of them carried a letter from the War Department at all times in flight – the letter clearly stated that the holder was not a combatant. Whether Germans who captured a crash landed ICD plane full of fighter aircraft engines, machine guns, or proximity fuses would buy that was, fortunately, never tested. Yet while on duty ICD crews did wear uniforms and had badges of rank along with their ATC pilot insignia. The wings were an interesting blend of Army Air Force and Royal Air Force designs, more closely resembling the former for pilots and the latter's wings for non-pilots, i.e., just one wing instead of a wing on each side of a central disk. Thus navigators, engineers, and radio operators wore a somewhat lopsided single wing insignia as did RAF navigators, bomb aimers, and gunners. Unlike RAF wings which were embroidered, Air Transport Command wings were available in metal or embroidered cloth. Senior pilots had a star above the central disk, and supervisory pilots such as Swede Golien, wore wings with a wreath around the star similar to an Air Force command pilot wings. The uniforms were stock Army Air Force issue – light khaki for summer and the woolen brown for winter. Blackie and all other eligible personnel each received a letter from Army Air Forces Headquarters authorizing them to buy officers' uniforms with only several restrictions. There was to be no black and/or gold piping on garrison caps and the uniforms could not be purchased with U.S. Army buttons or braid. The ATC rank badges were minimal and were essentially the same as their airline stripes which certainly did not resemble the insignia for the six ranks ascending from second lieutenant to full colonel. Thus Blackie, as a captain wore full stripes while his first officer wore one and a half. Inasmuch as the vast majority of early ICD pilots were military trained, the transition was an easy one, especially while enjoying the right to enter any military post exchange and purchase what he wanted to with only the limitations as noted above. They were also welcome in any officers club and while on a foreign station could expect to eat at the officers' mess and be quartered in officers' barracks. Such amenities, however, were not always available at every ATC foreign field in 1942.

Blackie did not actually receive his ATC command pilot's wings until 19 February 1943 when the very first wings were issued. In fact, Blackie's were the first command pilot's wings to come through, and he received them directly from ATC's commander, General Harold George.

At first, ICD flights were scheduled on an as needed basis, considering how few planes TWA really had. When the airline got its full complement of aircraft it was flying what amounted to regular schedules to Europe, Africa, and Asia. It would travel a long way from its very first flight when Otis Bryan took the Army Air Force's C-75, serial number 42-88627 (previously TWA's "Navajo"), to Cairo with a heavy load of desperately needed proximity fuses for the British Army's artillery. At the time, the see-saw battle between the German *Afrika Korps* and the British Eighth Army was not going well for the British and would not go their way for another eight months. Bryan had a full crew including Don Terry and Milo Campbell, both very experienced pilots and good friends of Blackie's. Bryan's Stratoliner took off on 26 February with a crew of eight bound for Cairo via the South Atlantic ATC Ferry Route. It would be a long roundtrip (over two weeks) because Bryan had been charged with a fact-finding mission as well as the delivery of enough proximity fuses to raise eyebrows when the weight and balance numbers were calculated.

After taking off from Bolling, Bryan cruised at 8,000 feet headed for Belem, Brazil. The South Atlantic route had opened in 1941 using a combination of former Pan American Airways' refueling bases and the development of Caribbean airfields which became available to America as a result of the 1940 "destroyers for bases" deal worked out between Prime Minister Winston Churchill and President Roosevelt. Thus American crews could land, in addition to Puerto

Rico, at former British bases on Antigua, Trinidad, and also in British Guiana. Bryan's was not the first American flight over this route; Pan American, through one of its subsidiaries, had ferried aircraft to the Royal Air Force in mid-1941 and the Ferry Command, using a B-24 Liberator named "Old Nosey," had flown a proving route in early 1942. The Old Nosey flight was very much a joint venture since the nine-man crew was composed of men from the Ferry Command and several airlines.

After making landfall on the African Gold Coast, Bryan refueled and set off for Khartoum in the Sudan where he became part of the clash between the twentieth century and Biblical times: he had to circle above the landing field until camels were driven off the runway. The Stratoliner was not greeted by a gasoline tanker or even a "follow me" Jeep. Instead it was directed to a huge pile of five-gallon cans with which the crew refueled the C-75 by hand, using a chamois cloth as a filter. Most of the refueling was done by locals who could cope with the blistering heat shimmering from the upper wing surfaces. Lewie Proctor, a flight engineer, could tolerate the heat on top of the wing for only a few minutes at a time. Returning from Cairo was essentially uneventful other than a stuck piston ring which caused a drop in manifold pressure between Monrovia and Accra. With that problem fixed, Navajo arrived back in Washington, completing the 5,683-mile return leg on 14 March which coincidentally was the same date that the last of the other four Stratoliners were converted to C-75s. Unlike some of the other planes entrusted to TWA, the five Straoliners would survive their war duty and eventually be de-mobilized and converted back to Boeing 307s for post-war service.

Another Cairo flight was scheduled very shortly thereafter when Dick Colburn and Mel Bowen took a C-75 across the South Atlantic, helping to pin down the procedures subsequent flights would use. Among the early practices were to fly the South Atlantic strictly at night. That way, navigators had more visible heavenly bodies aloft upon which to base their calculations. The early flights also deliberately tried to establish landfall about 100 miles north of the original objective and then turn south. That certainly beat arriving just south of the goal and choosing to fly south to try to find the original objective.

Freight was not the only cargo; even in the early days VIPs were carried in TWA's Stratoliners. On 23 April, a Stratoliner with three co-captains (Stan Stanton, Howard Hall, and Gene Klose) flew from Bolling to West Palm Beach for refueling and arrived in Natal, Brazil on 26 April after stops in Trinidad and Belem, Brazil. They hopped the Atlantic and arrived at Roberts Field in Liberia and eventually reached their primary destination, Karachi, India, on 4 June. There followed more shuttle flights between India and Africa until 10 June when Stanton took off from Karachi with a VIP on board: Jimmy Doolittle who, just weeks before, had bombed the Japanese homeland. When Stanton got home, it had been a twenty-six day odyssey involving 32,560 miles of flying, and almost 137 hours aloft, including over twenty-five hours on instruments.

Another ICD flight arriving back in Washington with pilots Frank Niswander and Charlie Kratovil brought home the wounded survivors of the Doolittle raid. They included Capt. Ted Lawson who would go onto write the bestseller, *Thirty Seconds over Tokyo* which was later turned into the popular 1944 movie of the same name. In the book and film, one of the more poignant moments is the one-legged Lawson coming forward to the cockpit where Niswander put him in the right seat, disconnected the autopilot, and let Lawson fly for an hour or more. It was a thrilling moment for Lawson to fly again and worthwhile for co-captain Kratovil too who managed a couple extra hours of sleep on what had been a very tiring trip.

The South Atlantic route had its share of challenges, particularly thunderstorms, but the North Atlantic was peril-fraught on the best of days. ICD's first trip on the North Atlantic was flown in March with Bryan as the pilot of the flight which was top heavy with Army generals. Bryan departed from Gander, Newfoundland, and ran into such atrocious weather just south of Greenland that he turned back, planning to wait it out and fly another day. Although Gander was still a relatively new base, it did have a skeet shooting range where Bryan was having a bad day, breaking only twenty-four out of twenty-five, in a competition with the Army brass.

Bryan's ICD organization was beginning to gather critical mass, even with its limited inventory of aircraft and *ad hoc* origins. There had only been that one flight in February, ten in March, eighteen in April, twenty-one in May, and thirty-two in June, the month before Blackie arrived. Still, there were many problems of which most were related to the demands placed on the Stratoliners. The takeoff weight had been raised to 50,000 pounds or about 5,000 more than originally certificated. In practice, it was not unusual for pilots to try to take off at 60,000 pounds and when they did they tried to stay on the deck for a while before venturing out over open water. The excess demands had repercussions from asking more of the 1,100 hp engines than they were designed to withstand: burned pistons which the pilots described as "piston ring feathering." The piston rings would fail which caused loss of power and excessive oil consumption. That was not a major problem when a Stratoliner was close to Washington, but if it occurred in Accra or Kano it was a big problem. For that reason, the Stratoliners carried a weighty little kit consisting of spare rings, pistons, and a few cylinders. The problem was never completely solved, but engineering pilot Bill Hanley devised a means of minimizing emergency repairs by making sure all flight engineers kept a very close watch on each engine's oil consumption. When consumption started exceeding three gallons per hour, the cylinder was changed.

Crews had to get used to being away from home for extended periods. Radio operator Bob Middlekauff 's first trip with pilot Dick Hanson lasted a full month out of country and amounted to nearly 230 hours of which most of the time was shuttling between Khartoum and Cairo. Milo Campbell also had a thirty-day foreign trip involving even more flying hours: 271. Hanson and Campbell were lucky as pilots; they could snooze while the other pilots flew. Navigators could pile up some Zs between fixes, but radio operators had to be on station listening for radio reports and sending position reports for extended periods. Some of the radio activity had its more amusing aspects: when going to Scotland, it was not unusual to receive a coded message from the ground radio operators asking how many silk stockings the flight had on board.

When he accepted the position of supervisor of flight operations at ICD, Blackie had already made up his mind that he would not spend all of his time flying a desk, even though the job description said nothing about his actually flying. His 11 June flight was a checkout on "Apache" or N-19907, one of the Boeing 307 Stratoliners. "Casual" best describes the checkout, especially considering there had been a TWA Stratoliner school in Kansas City until the planes had joined the Army Air Force. Blackie was checked out by one of the old hands Otis Bryan recruited into ICD – Doc Metzger who had been the chief pilot in Kansas City. Blackie was well-acquainted with Metzger whose wife represented a tragic link to Blackie's late, close friend Kenny Greeson. Doc's wife June had been aboard the DC-2 that crashed at Kirksville, Missouri, but had survived. The only reason that she survived was that she was so drowsy that Greeson granted her request to move from her seat next to the front bulkhead to a seat in the rear of the plane where there was more room to recline. She was sound asleep when the plane hit the dike – thanks to the fact that she was back near the tail where all survivors had been seated.

Blackie never mentioned the Stratoliner when asked about his favorite airplanes. Most TWA pilots transitioning to a Stratoliner did so from a tail-wheeled DC-3, but Blackie was coming from a tricycle geared B-24 which, although not a ballerina, behaved better on the ground than a tail-wheeled Stratoliner. Visibility from the cockpit was not one of the Strat's more pleasing qualities due the canopy design and less than optically perfect glass. That was not a major issue with Blackie because he would not log that many hours on the Stratoliners, but when he did they were memorable flights, especially the first one which was scheduled for 16 July. Blackie's first passengers on a Stratoliner never knew that he had a grand total of one hour and twenty-five minutes on type when he started the engines.

ICD got its marching orders directly from the War Department and more specifically from Colonel Robert J. Smith who was commanding the Air Transportation Division. Smith's office also issued orders to Army Air Force pilots flying in the Air Transport Command. As supervisor of flying, Blackie selected the pilots, first and second officers, navigators, flight engineers, and radio operators. Air Transportation Division decided who and who did not have priority and then supplied ICD with final passenger (or cargo) lists and the ultimate destination. While TWA, United, American, and Pan American were the primary international sub-contractors to ATC, the nation's regional airlines were also a very important part of the overall picture. Some, such as Delta hauled freight and passengers in the continental United States while others flew domestic routes and those outside continental America as well. For example, Braniff and Eastern flew to Latin America, Northeast to Labrador, Newfoundland, and occasionally beyond, and Northwest Airlines to Canada and Alaska.

ATC also assigned engineering officers to each of its airline contractors, and Blackie had no choice but to work with them, of whom many had no background in airline maintenance or aeronautical engineering. If TWA wanted to upgrade its C-54s, it had to go to ATC to get an approved technical order to do so.

Blackie was in Kansas City for meetings on 14 July and just happened to be handy when there was a need for a quick engineering flight early in the morning on a DC-3. He was on his way back to Washington later that morning and headed for the greatest adventure of his life to that date just two days later. The passenger list for his 16 July flight to Scotland consisted of several very important VIPs and their staffs. Leading the list were Harry Hopkins (President Roosevelt's right hand man), Army Chief of Staff General George Marshall, and Chief of Naval Operations Admiral Ernest King. The total passenger list amounted to nearly twenty people. Blackie had a large crew consisting of three other pilots; the lead navigator, flight engineer, and radio operator and back-ups as well. Among the other pilots were Stan Stanton and Robert Loomis. Stanton had originally flown with Braniff but left and joined TWA due to Braniff's periodically severe cash flow problems. The financial situation was occasionally so bad that pilots had to pay for aviation gasoline out of their own pockets and hope that they would eventually be reimbursed when the cash flow situation improved. The first leg of the flight to Prestwick, Scotland, was a nearly seven-hour flight to Gander, Newfoundland, where they remained overnight. Blessed by very favorable weather and tailwinds, Blackie flew across the Atlantic without stopping at Greenland or Iceland on a flight lasting over thirteen hours, including six hours of night flying. In doing so, he qualified as a full member of the Court of King Neptune's Airmen, the wartime aviator's reward for flying across the Atlantic and something of a parallel to those who cross the equator on a ship. Blackie would cross the equator in due time; again, it would be as the skipper of a Stratoliner. TWA itself would eventually award certificates to aircrew crossing the oceans and affix to them gold seals, one for each crossing. Blackie's certificate from Neptune, Rex, Court of the Dawn, the Sun, and the Moon would grow cluttered indeed by the many gold seals.

As a result of the crossing, Blackie signed Harry Hopkins' "short snorter," although Blackie never got carried away with short snorters as many servicemen did. The short snorter tradition dated back to the early days of aviation and involved co-participants of a flight placing their signatures on any banknote, typically a dollar bill. Failure to produce the short snorter when challenged resulted in buying a partially filled shot glass, i.e., a "short snort," for the person making the challenge. Some aircrewmen were so enthusiastic about short snorters that they carried bills that had been stapled together to contain all of the signatures. Blackie's friend Don Terry was a great enthusiast of short snorters – by the end of the war his stapled short snorter bills exceeded three yards in length. New first and second officers did not know quite what to make of Terry at first because he usually wore a white silk neckerchief and occasionally played his violin on the flight deck when not at the controls himself. Some short snorters became collecters' items, and the Harry Hopkins short snorter was no exception, considering the signatures it contained. In addition to Blackie and some of his crew (Stanton, Bob Loomis, and J.D. Love), Hopkins gathered a few other signatures on wartime flights: President Roosevelt, Louis Mountbatten, General George Patton, Anthony Eden, Averill Harriman, Dwight Eisenhower, Winston Churchill, and another Roosevelt Blackie would meet in two years: Elliott Roosevelt, the President's son who was an Air Force officer.

Prestwick was the principal landing field for ATC arriving flights because it was not prone to quite as much fog as were most locations on the English and Scotch west coasts. It was also assumed that it would not attract that much attention from marauding German bombers as places like Glasgow, and it was not festooned with barrage balloons which were definitely a hazard in high density airspace. It had been a smooth flight for Blackie and his VIP passengers. Indeed, the C-75s did retain a few of their previous creature comforts for such important passengers whose travels were about to become seriously bumpy after deplaning from the C-75, "Apache." Prime Minister Churchill had sent a special train to Prestwick to bring the American delegation south where he planned on entertaining them overnight at his beautiful estate, Chequers, in the Kent countryside southeast of London. General Marshall nixed the visit to Chequers and insisted on being taken straight to London where he set up an operating headquarters in sixteen rooms at the Claridge Hotel. Churchill quite naturally went ballistic at Marshall's snub which did nothing to promote Anglo-American relations at what was going to be a contentious conference to begin with. On the table were plans for the invasion of Europe, including timetables and preceding activities. Churchill called Hopkins and excoriated him about Marshall's brushoff to the extent that Hopkins promptly visited Churchill for a day before the conference opened. Churchill and Hopkins were friends who had first met while Churchill was a guest at the White House in December 1941, shortly after the Pearl Harbor sneak attack. Hopkins defused the situation, historians believe, by leaking Roosevelt's agreement in principal with Operation Gymnast (called Torch by Americans, the code name for the invasion of French Northwest Africa). That and the fact that Roosevelt, anxious for American troops to go into action somewhere to shoulder the burden, had decided that the European war would get the lion's share of America's manpower and war production in contrast to the Pacific war. His decision placed Churchill on the high ground in opposing Marshall's goal of a cross-channel invasion in 1943 or even later in 1942. It also put

Churchill and his generals in a position to veto Sledgehammer, Marshall's plan to occupy France's Cotentin Peninsula as a base from which to launch the full invasion after building up sufficient forces in Britain to sustain a successful drive into Europe. Thus, the way was paved for an Allied invasion of North West Africa to force Rommel into a two-front campaign that he could not possibly win, especially considering the way Malta-based RAF aircraft were bleeding his supply lines. Marshall was depressed; he felt that nothing had been done to take the heat off the Russians or had anything been done to build a foundation for invading the continent. As time would tell, the Allies were in no position to launch a major invasion of France in 1943, a year when the Afrika Korps ceased to exist, but a year when the complete success of taking Sicily evolved into the meat grinder that mainland Italy turned into. Marshall would have his day in June 1944.

Blackie, however, was still in Scotland and very content with his situation which at first was dominated by servicing the Stratoliner, albeit he was very tried from the flight over. He would, of course, eventually be influenced by the talks in London as would all Americans, but had his mind on other matters closer to Prestwick. Blackie and his crew stayed at the Saint Nicholas Hotel on Ayr Road in downtown Prestwick. The "St. Nick" was a three-story stone structure with tall chimneys on each end and great bay windows in the front. Appointments were Spartan by American hotel standards, but after what was essentially a sleepless night crossing the Atlantic that made little difference. Blackie was anxious to experience as much of Scotland as possible because he knew that in a couple of days he would be taking the Stratoliner down to London. The major attractions of Prestwick and Ayr held little appeal for Blackie since he was not a golfer. Thus the proximity of Royal Troon, Saint Nicholas (literally across the street from his hotel), Prestwick Golf Club, and Saint Cuthberts, all legendary British golf courses and all visible on the approach to Prestwick Airport, made little impression on him.

Food and drink were different matters for Blackie whose interest in cuisine could only be described as "adventurous." British cuisine by mid-1942 was very limited, not that it had ever established a reputation for excellence before that date. Britain was a net importer of food even before the war and with German U-boats sinking so many Allied ships, the variety and quality of food had declined substantially by the time Blackie arrived at Prestwick. Little children in rural Ayrshire grew up not knowing what an orange or banana was. They would eventually learn, of course, but for adults the title of a popular 1920s song, "Yes, We Have No Bananas Today," was a discouraging fact of life. Nevertheless, restauranteurs were under fewer restrictions than the average housewife, and many of the dishes common to Scots tables were still available, including haggis and neeps (the latter being mashed turnips and the former being an acquired taste best preceded by several drams of highland whisky), world-class oatmeal ("porridge"), and some seafood, considering Prestwick was in coastal Scotland.

Culture shock was most apparent at breakfast when Blackie was greeted with tattie scones (potato biscuits), black pudding (pork blood and oatmeal sausage), and kippers (smoked herring), but no orange juice or eggs. TWA personnel stationed at Prestwick were lucky to have one egg per month. There were advantages to having friends flying planes in from distant places. Later in the war when there were flights coming in from North Africa, Prestwick staff could look forward to fresh fruit, including oranges. Blackie had always been partial to his mother's biscuits, especially when accompanied with wild buckberry preserves, thus the various Scotch marmalades and shortbread were very welcome aspects of Scotch cuisine. Yet other special items were not in short supply and of those the most notable were woolen fabrics. Unfortunately, there was nary a Blackburn tartan fabric to be found.

Scotland changed Blackie's life in one important way. Countless other pilots, along with soldiers and sailors, found themselves on distant shores and were temporarily fascinated by their exposure to all things foreign during the war. For the vast majority of them, the experience of other cultures would eventually be consigned to the confines of scrapbooks along with other mementos of the war years which, in some cases, were best forgotten instead of fondly remembered. Blackie embraced the new experience which he, as an international pilot, was able to enjoy to the fullest. The flight to Scotland and, in a few days, to England was the taproot of a lifelong interest in other cultures, languages, cuisine, history, archeology, music, and art. This new curiosity would be expressed in many different ways throughout his life and would be passed along to his children and grandchildren. One would only have to look at the walls of Blackie's home to see how important having an international outlook was to him.

He had one full day on his hands before he was scheduled to take the Stratoliner down to London, and he was going to put that time to good use. It proved to be a very full day indeed. When he landed the day before, there were a number of aircraft types parked at Prestwick with the Liberator and Hudson being the most common since Prestwick was also heavily used by the RAF Ferry Command. Several Liberators were permanently based at Prestwick for the RAF's Return Ferry Service which returned ferry pilots back to Canada to initiate more eastbound deliveries. The number of multi-engine aircraft landing at Prestwick would begin to increase, eventually to geometric proportions, the following month when the first American heavy bomber groups began to arrive. Indeed, later in 1943 and 1944, it was not unusual for upwards of 300 aircraft to land at Prestwick in a day. Only several weeks before, three squadrons of Spitfires had landed at Prestwick and then were transported to the nearby River Clyde where they were placed on the American aircraft carrier *USS Wasp* for their voyage to Malta.

Word spread on the field that the pilot of the huge Stratoliner that landed on the morning of the eighteenth was one of America's most experienced Liberator instructor/check pilots. Thus Blackie was prevailed upon to give an American style check ride to a ferry pilot, Flight Lieutenant David F. Leggett who was new to the RAF Ferry Command. Blackie complied with the request and gave Leggett a two and a quarter-hour flight check in a new B-24D.

Unfortunately, the impromptu check ride with Leggett limited the time for visiting several destinations that were high Blackie's travel wish list, but he would save them for a later trip. Among them was the village of Blackburn between Glasgow and Edinburgh along with visits to those two cities too. By mid-morning on 19 July, Blackie and his crew made the pilgrimage to the two most important local attractions: Brig o' Doon (the medieval bridge immortalized in Robert Burn's poem, "Tam O'Shanter") and the nearby Robert Burns monument. Part of the day was spent locating something Blackie was vaguely aware of from past reading about Scotland and the origins of his Blackburn ancestors – the Blackburn Rocks on the Ayrshire coast. He did not have far to go to locate them while walking to Ayr over the bridge that separates the two towns. The rocks turned out not to be jagged spires with cascading waves smashing into them, but were instead small, rounded and much less impressive although still a minor offshore hazard to local fishing boats. The Blackburn Rocks are located southwest of Ayr harbor and about a quarter mile offshore. They can be seen, at low tide, within sight of Blackburn Road which is only a couple blocks away from the sea. Walking through the two towns which are separated by the River Ayr was another eye-opener for Blackie and his crew as they passed shops along the street. There were a few semantic issues at first, but the green grocer, fish monger, and iron monger (hardware) were quickly sorted out. There was not much in the butcher's window display case on a Monday morning, not that there ever was after rationing started, but the pilots were initially puzzled by some curious sausages resembling misshapen, overgrown grayish-brown softballs which turned out to be a display of haggis.

The Scotland in Blackie's mind's eye was gone. The church spires, parks, bridge over the Ayr River, and monuments were still there and the local people spoke with Scottish accents. The two towns in the summer of 1942, however, were essentially an armed camp and had been for almost two years. Nowhere could Blackie see a Scot wearing a kilt, sporran, Argyll jacket, or Glengarry. Instead there were the uniforms of the British Army and the Royal Navy – both services had installations in and around Prestiwck and Ayr. In his travels that day, Blackie discovered that the Royal Air Force had provided for the defense of Ayr and Prestwick. Just to the south of the main airfield at Prestwick was RAF Ayr, home to the twelve gray-green camouflaged Spitfires of 165 Squadron which was just one of several fighter squadrons located along the east, north, and west coasts of Scotland. Late in the afternoon, there seemed to be an unusually large number of school children in Ayr. Many of them could not, however, call Ayr their real home because they were kids from Glasgow who had been evacuated due to the threat of German bombing. It was safer to be in Ayr than the principal *Luftwaffe* targets in southwest Scotland: the port at Greenock just to the northeast and the manufacturing facilities at Clydebank and Glasgow to the east. There were bomb shelters in Ayr and Prestwick, although the *Luftwaffe*, surprisingly, never deliberately targeted either town. Ayr was bombed at the end of the first week of March 1941, but only as a secondary target. The attacking three waves of bombers were bound for Greenock to the north, but the Germans were intercepted by night fighters from RAF Ayr. Many of the *Luftwaffe* pilots broke formation and bombed targets of opportunity, including Ayr, instead of continuing to the important port of Greenock.

On Tuesday the twentieth, Blackie took off for the two hour and twenty-minute flight down to London and landed at RAF Northolt just to the west of the city. There he saw many more Spitfires because Northolt was home to 303, 306, and 307 Squadrons; each squadron was composed of Polish pilots who had fled Poland in 1939 to carry on the fight another day. Northolt had been London's principal civil airport before the war but was now, along with Stansted, Gatwick, and Croydon airports, part of London's defensive ring which included such prominent original fighter bases as Biggin Hill, Hornchurch, and Kenley. Blackie and his crew became tourists, although their hotel was not as upscale as the exclusive Claridge's.

London in the summer of 1942 was relatively free of *Luftwaffe* threats because the German Blitz bombings which began in late 1940 were over by the end of May 1941. The German V weapons, the V-1 flying bombs and V-2 ballistic rockets, would not start falling until late 1944. The latest German menace, the Baedeker raids, had pretty much run their course by the end of June 1942 and targeted neither Prestwick nor London. Those raids were named after the popular, pre-war German *Baedeker Guides* to popular British tourist destinations and had been ordered by Hitler as reprisals for recent RAF raids on Rostock and Lübeck. Blackie did not visit Bath, Canterbury, Exeter, Norwich, York, or the other towns that the Germans attacked on the Baedeker, small scale night raids. Nor did Blackie visit any south coast towns which were occasionally hit by fast moving, low-altitude German fighter bombers. Just walking on the street made Blackie quickly realize that London in July 1942 was at war: barrage balloons lolling about in the sky, the night black-out, anti-aircraft and searchlight batteries here and there, air raid shelter signs, other signs limiting access to restricted areas, a proliferation of uniforms (but American uniforms still in small numbers), bombed out skeletons of buildings throughout the city, and vegetables being grown in what once were formal flower gardens. There were signs in many shop windows reading: "Coupons Required." Few women were seen on the street and of the ladies that Blackie saw, the vast majority were in uniform. The other women were busy at work in war production plants. Their young children, the unlucky ones who had not been shipped to Canada or rural Britain, spent the day in special nurseries. The Ministry of Food established a chain of its own restaurants and canteens to feed the influx of war workers and military personnel in the cities. The food was served at cost although the menu was limited to such items as Woolton pie (carrots, turnips, onion, and other vegetables baked in their own broth and thickened with oats, topped by a potato crust), liver sausage, or mutton with sponge cake or rice pudding for desert – all sold for just a few pence per dish. It was discovered that bombed out buildings, with their basement walls still intact, made wonderful places to keep pigs. Thus the occasional snarkle, gork, and squeal so common to a Yorkshire pigsty could be heard in close proximity to Cockney, BBC English, and the many accents of the British dominions heard on London's wartime streets. Otherwise London was business as usual. People still went to work and to church. They went to the parks during lunch and listened to free concerts. Little boys floated sailboats on ponds in the city's parks, and people still flocked to theaters and cinemas.

In his weather briefing at Prestwick on the twenty-sixth, the winds aloft forecast clearly showed Blackie that a non-stop return was out of the question, so he submitted a flight plan with an intermediate stop at Reykjavik, Iceland, which he reached six and one-half hours after leaving Prestwick. The Stratoliner was out of Iceland early the next day on a fourteen and one-half hour flight to Presque Isle, Maine, where they overnighted before flying onto Washington the next day. Had Blackie been in Iceland just two weeks later, he would have gotten some exciting news firsthand. The first German plane to be shot down by an American Army Air Force pilot in the European theater of war happened on 14 August when a P-40 flown by Lt. Joseph Shaffer of the Thirty-third Fighter Squadron stationed at Keflavik attacked a Focke-Wulf FW-200 Condor. Shaffer's attack only damaged the four-engine reconnaissance aircraft which was then set upon by two P-38 Lightnings flown by Major John Weltman and Lt. Elza Shahan of the Twenty-seventh Fighter Squadron of the First Fighter Group. Weltman's Lightning was damaged by return fire from the Condor, but Shahan's guns destroyed the bomber which fell to earth on Seltjorn Point, not far from Reykjavik. It was not the last *Luftwaffe* weather or maritime reconnaissance that would be bagged over or near Iceland. Indeed, the *Luftwaffe* was very active in the area, not only with ultra-long range Condors, but also with twin engine Heinkel 111s and the fast Junkers JU-88s. All three German types were armed and therefore were a threat to ATC crews, although the Germans for the most part were operating that far west primarily for weather reconnaissance and acting as the eyes for German submarines. Still there was another threat looming in the immediate days ahead, and it came not from the *Luftwaffe*, but rather from the Army Air Force, U.S. Navy pilots, and the Royal Air Force. ATC was about to start flying the Douglas C-54 Skymaster, or more accurately the military version of the DC-4s that United Airlines never got. In the excitement of spotting a bogie and closing in with machine guns armed, an Allied fighter pilot could very easily fail to properly identify a C-54 since it was practically a dead ringer for the FW-200.

After getting back to Washington on 27 June, Blackie did not fly again until 2 July when he checked out Charles Kratovil, the slow-talking, deliberate, former DC-3 captain on the Stratoliner. He also spent the following two days checking Larry Chiappino out on the Stratoliner, but on 10 July, it was Blackie's turn to be checked out – on the C-54 Skymaster. The check pilot turned out to be a major surprise. He was an old friend and none other Benjamin O. Howard, better known by Blackie and thousands of others as "Benny." Howard and Blackie had first gotten acquainted in 1929, just before Blackie went into cadets. At the time, Howard was a famous air race pilot and designer of the well-regarded Howard DGA series of high-wing monoplanes of the 1930s. His prowess as a race pilot combined with engineering talent led him to Douglas Aircraft where he became a test pilot on many aircraft including the DC-4/C-54. As with the Stratoliner, Blackie proved to be a quick study and was signed off on the C-54 after only two hours of instruction. Paul Richter was in town the following week, so Blackie and Swede Golien, whom Blackie had checked out on the C-54, gave Richter his first orientation ride in the C-54 – all of thirty-five minutes worth. Over the following two days, Blackie gave more-in depth C-54 instruction first to Joe Carr and then to Bob Loomis who had, like Carr, been a Navy pilot before joining TWA. The first production run of C-54s was divided by TWA and Pan Am – each got twelve of the new Skymasters while the Air Force got just one. Golien and Joe Carr immediately set about to create an orderly check list.

Several days after the Carr-Loomis checkouts, Blackie flew a Skymaster full of senior ICD pilots on what amounted to a multi-purpose mission: delivering Army personnel to Greenland and Arctic training/orientation for a few pilots, including Blackie who was essentially on what amounted to an inspection tour of the two newly opened bases on Greenland. Among those aboard were Otis Bryan, Pete Redpath, Al Brick, and several new men along for the orientation. Blackie had taken off from Washington on 19 August and landed at Presque Isle, Maine, for fuel and then continued to Goose Bay. Fortunately, it was a clear day, the type of day when navigating to Goose was not a problem. Blackie had only to take up a northeasterly heading and follow that until encountering the first principal landmark approaching Goose which was the Hamilton River (since renamed the Churchill River). The river could be followed right to the airfield, itself little more than a patch of gravel hacked out of the woods and populated at first by a few men of the Royal Canadian Air Force. The airfield was located just north of the river and west of Lake Melville's Goose Bay and was destined to grow as the skies would soon swell with legions of American warplanes on their way to Britain. The deeply forested terrain was not friendly, i.e., mountainous and rivalling Minnesota for countless unnamed lakes. Even the aeronautical charts carried the warning "Unexplored" over vast sections around Goose Bay. It was friendly territory for hunters of caribou, fur trappers, and salmon fisherman, and when Blackie flew in there was still an outpost of the Hudson Bay Company serving the needs of a few fur trappers in the area. While idyllic for outdoorsmen, pilots found it far from that in bad weather particularly because Goose Bay's range station was both weak and frequently unreliable.

On the twentieth, Blackie touched down after a nearly five-hour flight in clear weather at Bluie West One, the base built on a glacial moraine at the southern tip of Greenland near the village of Narsaruaq. After deplaning several passengers the Skymaster was again aloft for the three and one-half hour flight to Bluie West Eight, a base built north of the head of Kangerlussuaq Fjord on the western side of Greenland, just above the Arctic Circle. Bluie West Eight had been opened as an alternate Greenland base for Iceland-bound flights when the weather made flights over southern Greenland dangerous. Actually just getting to Bluie West Eight was dangerous and fraught with peril. The western coast of Greenland is punctuated by many fjords, all of which closely resemble each other. Mountains rise along the coast and the initial point on the approach was a range station on Simiutaq Island, located at the mouth of the fjord, that was subject to all of the foibles of radios of that era. The standard approach was to enter the fjord and fly visually up to the base which frequently was more easily said than done if visibility happened to be poor. With poor visibility, a premium was placed on getting the right altimeter setting from Bluie West Eight since there were mountains on both sides of the fjord and icebergs, some quite high, were just below. Although finding the range station was comforting, there remained the problem of ascending the correct fjord since there were three major fjords and several smaller ones in the immediate area. Ascending the wrong fjord was essentially a dead-end proposition in poor or declining visibility. Fortunately, Blackie's approach to Bluie West Eight was made in the clear, and he had no problem identifying the next fix – a half sunken freighter along the north shore, about thirty miles up the fjord.

Otis Bryan was the guide on this flight; he had taken the Stratoliner Apache up to Greenland and was the first to land at Bluie West Eight, the first of the Bluie West bases to be completed. On 20 April, a picture was taken of his Stratoliner Apache's crew and the base commander, Colonel Bernt Balchen, the famous Arctic pilot and explorer. It was twenty-two below zero when the picture was taken. The plan was to continue on to Scotland, but since the northern lights were playing such havoc with the radio operator's radios, no weather reports were received. So Apache sat at Bluie West Eight for two weeks before it was safe to proceed to Scotland. The second plane to land at Bluie West Eight was a C-87 Liberator Express flown by Lt. Col. Milt Arnold and it was still there at Bluie West Eight long after Bryan took off for Scotland. The C-87 had not been winterized, and all of the oil and hydraulic lines froze solid. At least the crew got a certificate stating that they had flown north of the Arctic Circle, but it was nearly a month before the Liberator reached Scotland via Iceland.

Iceland had been neutral in 1940, but was occupied by Royal Marines in May of that year and then garrisoned by American forces the following year. But even so, there was a burning need for bases between Iceland and eastern Canada, particularly as there would be more planes flying eastward in the near future, and they did not all have the long legs of a Liberator or Stratoliner. With the first two bases established on Greenland, there would be three legs to cross the Atlantic instead of two legs, and no leg would be longer than roughly 900 miles. In late 1942 construction began on Bluie East Two on the eastern coast of Greenland and about 400 miles northwest of Iceland, although its purpose was primarily as a weather and radio station. Providing the means by which more aircraft could safely fly the aerial highway from Canada to Scotland was not the only motivation in establishing a stronger presence in Greenland. Early in the war, the Germans had a strong interest in Greenland because it is the source of much of Europe's weather, and having weather and radio facilities in Greenland would be important assets to Germany's intelligence machine. The Germans, whose long range recon aircraft periodically visited eastern Greenland, successfully established several weather reporting stations on the remote northeastern coast. It was not until 1944 that the last German base was neutralized by Allied forces staging from Bluie East Two. On the southwest coast of Greenland, there was a cryolite mine at Ivigtût, and it had to be defended by forces supplied from the Bluie bases. Cryolite itself could be used as aluminum ore and it also functioned as a catalyst when refining bauxite into aluminum which, of course, was essential to the production of aircraft.

The hop from Goose Bay to "BW-1" took just under five hours, and on the return from Bluie West Eight Blackie skipped Goose and flew direct to Washington in just under thirteen hours, blessed by a favorable tailwind. Although the reality was that stage lengths had been shortened, the weather was as atrocious as ever, particularly in winter, as in the approaching winter of '42-'43 when the northern route was shut down until the following spring. Blackie would not fly the full northern route until well in 1943, but he heard about its perils from his debriefings of ICD pilots who were flying it on a regular basis.

For the most part, ICD crews flying from Scotland back to Washington could plan on substantial headwinds, much in contrast to Blackie's recent return from Blue West Eight. There was one extraordinary case of a TWA captain enjoying even better tailwinds, but his management of the flight had some comedic overtones. Frank Niswander was the captain of a roundtrip to Scotland in early July on which the outbound segment carried a VIP in the person of the exiled King Peter of Yugoslavia along with several British diplomats. The purser on the flight, Bob Stevens, was normally a dispatcher, but hands were short, so he was pressed into service in spite his lack of a culinary background. On the return flight, fuel and provisions were taken on at Reykjavik while Niswander got his weather briefing which turned out to be so very favorable in the wind department that he planned to bypass Goose Bay and fly directly to Presque Isle,

Maine. Stevens signed for the food, but failed to notice no utensils were included. The uncharacteristic tailwind proved so strong that Niswander revised his flight plan and decided to land at Boston which also proved to be unrealistic. By then it became apparent, that Washington was a possibility if Niswander could lighten the ship and reseat the passengers and keep them in their seats for the best aerodynamic profile. Out went all of the food along with some garbage. The passengers sat tight and Niswander made Washington, setting a record of twenty-one hours and 16 minutes not to mention snarling stomachs and numb butts.

On 8 August, Blackie got a questionnaire from the Seventh Service Command of Air Force Reserve Headquarters in Omaha. He had to fill it out and return it promptly but absolutely did not want to. Most of the form asked routine questions such as graduation dates from any service schools or specialized training, foreign assignments, World War I service, ratings, current Reserve unit an assignment, and general physical fitness questions. There was one question that would put a shadow over any attempts to get back on active duty in an operational squadron or group. The form called for a check mark to be placed in a space marked "over age in grade," and in Blackie's case he was beyond the thirty-five years of age limitation for first lieutenants. He stoically returned the form, assuming that where there was a will there was a way, especially with his growing number of connections in high places. The irony was not lost on him – as a mere first lieutenant, he, as commander of what amounted to two squadrons of specialists, had flown an admiral, general, and presidential diplomat across the Atlantic at a time when there were still not many pilots who had flown roundtrip to London from Washington. He had commanded a heavy bomber school that was more sophisticated than any training facility in the Army Air Force at the time (the CCTS flew more airplanes than any bomber squadron in the country). And he was physically as or more fit than any second lieutenant getting his wings and lieutenants bars at any advanced flying school in the country. The irony was compounded by the results of his most recent flight physical, taken three months earlier on 30 May 1942, which revealed a twenty-one year old pilot living in what appeared on the outside as a forty-one year old, slender, 155-pound man with a five-foot, ten-inch tall body with blue 20-20 eyes and sandy-red hair. His enviable blood pressure was measured at 120/68 with a pulse rate of 60 which returned to 60 after two minutes of hard exercise. Everything on the 42-point physical was normal or better than normal. But Blackie was too old to qualify as a commander of troops.

Not long after Blackie's inspection trip to Greenland, he debriefed his old captains, Don Terry and Larry Chiappino and thereby learned that his own first flight over to Scotland in July was about as good as it could get. Very typically, ICD east-bound crews would run into at least one major front and several secondary fronts. Blackie briefed both Chiappino and Terry who were scheduled to take separate flights on the same night to Prestwick. The weather maps showed that they would encounter a high en route. Chiappino opted for a direct route to

Scotland after departing Goose Bay. Terry chose to take a more northerly route via Iceland where better weather was forecast. Both pilots knew where the weather reporting facilities were, but that the forecasts were never a sure bet. Chiappino was off first and encountered a mild front that was not forecast. He encountered some rain, moderate turbulence, and icing above 3,000 feet. When he broke out on the other side, there was still an overcast; but it was breaking up, and his navigator was able to get some good star shots. The remainder of his flight was without difficulty.

Don Terry also ran into a cold front, one that was not included in the forecast. He encountered severe turbulence and freezing rain down to 400 feet, but still made it to Iceland where he could not linger for very long because the front was about to catch up with him; and by then the freezing rain came right down to sea level. Terry got away in time only because there was a temperature inversion between 200 and 500 feet; he encountered gradually improving conditions for the rest of the way to Prestwick. His eastbound problems proved minor compared to those of the return flight after he had passed the point of no return. He ran into an occluded front and had to descend to 400 feet to shed accumulated airframe ice. The turbulence was epic, and both Terry and his first officer had to manhandle the controls, occasionally dropping below 200 feet to steer clear of the freezing level. Terry's hell continued for nearly three hours – in turbulence with the freezing level just 100 feet above and a storm lashed sea less than 200 feet below. The C-54 was drenched in salt spray from the cauldron below.

In Terry's own words at the debriefing, "It was too tight a squeeze. For two and a half hours I thought I was making my last flight. It was the worst experience in my twenty years of flying." For Blackie, the debriefing was prophetic. More than once in the three years ahead, he would be caught in the same vice, with the churning sea immediately below and strangling ice just above. And after landing, Blackie would have to explain to the deplaning VIPs why the plane was being hosed down with fresh water.

There were hard lessons to be learned from the early flights in 1942 and many precautions for Blackie to share as he briefed outgoing crews. Chiappino stressed the importance of relying on much more than the forecasts when he urged pilots to pay more attention to weather in flight, particularly the wind shifts and freezing levels. He watched changes in cloud formations, turbulence, and temperature to locate the beginning of frontal conditions. Fred Richardson warned pilots not to be fooled by a lack of clouds when trying to determine frontal locations, noting that secondary fronts often lose their clouds when they move over the ocean. In addition to the wind shift and temperature change, he suggested looking at the waves below to determine the wind shift direction. Pilots quickly learned that they were going to encounter ice somewhere on nearly every trip across, and the ice layers could be 15,000 feet thick. Flying on top was the best way to for navigators to plot a plane's track, but if trapped below. The driftmeter was the only option. The worst of it tended to dissipate as planes flew eastward on the crossing to Scotland and worsened as west-bound planes neared Greenland and the East Coast.

There were other frustrations. All airfields in the United Kingdom were camouflaged, thus it paid to be absolutely certain one was landing in the right place. During the war, some ATC crews did land in one notably wrong place, Ireland, inadvertently or on purpose. German pilots landing in Ireland were held for the duration. RAF crews were interned but were released in 1943. After that general release RAF crews were promptly interrogated and then transported to Northern Ireland. Americans were usually escorted to the border with Northern Ireland if their planes had crashed while the rest were allowed to make repairs and fly on to the UK. Interestingly, of the sixteen known cases of ATC crews landing in Ireland, none was a TWA C-54 or C-75; they were all Army ATC crews ferrying heavy bombers with the single exception of one C-47. The overwhelming majority of the approximately 180 incidents of foreign aircraft crashing or landing in Ireland during the full course of the war were from the *Luftwaffe*, Royal Air Force, and United States Army Air Force. For the latter service, the majority of landing incidents were bombers very much off course on flights from North Africa or from one of the bases on the North Atlantic Route.

One TWA crew, headed by Howard Hall, one of Blackie's captains from 1935, was flying a specially equipped C-54 on a research project conducted by the Army Air Force. Others on Hall's research team included pilots Dutch Holloway and Stan Stanton. The Army had taken special interest in TWA's research into pressure pattern flying and loaned a C-54 to TWA; that Skymaster was specially equipped with the latest drift meter, an advanced radio altimeter, and LORAN (Long Range Aid to Navigation). When LORAN worked well, a navigator could fix his position with great accuracy. On 25 February 1944 Hall took off from Newfoundland for Prestwick planning to fly the southern border of a large low pressure system where he could reap the benefits of strong tailwinds. Unfortunately, the weather turned sour; they had to fly between cloud layers which ruled out use of the drift meter and the possibility of astral navigation. The radio altimeter was not working well, considering ocean surface conditions that day and the flight was in local conditions not conducive for LORAN. Thus, Hall had to wing it until the radio operator, Earl Korf, another one of Blackie's good friends, finally got some reliable information from the LORAN set. Their newly established position was far south of where they thought they would be and even worse, they were bucking a sixty-mile an hour headwind. Hall knew at once that reaching Prestwick was not very likely and that making Ireland was seriously in question. He was anxious to avoid the Republic of Ireland fearing the possibility of internment; so he headed for Northern Ireland and just made it into an RAF base, Nutts Corner (actual name, not a field for daft flyers). Nutts Corner is near Belfast, and there were other bases, but all of them were closed due to fog. Hall was cleared for a circling approach, but promptly told the tower that he was making a straight-in approach due to his perilous fuel state. Just as he cleared the runway, two engines spluttered and quit due to fuel starvation after nearly sixteen hours of flying. Hall's flight plan estimated eleven hours aloft. Bob Loomis was on a regular flight two hours behind Hall and also landed in Northern Ireland at the RAF base near Belleek and was immediately followed by an entire squadron of B-25 Mitchells which never could have made it to Prestwick.

Compounding problems was the blackout so common in or near war zones. Many airfields along the ATC route were blacked out in the pre-instrument landing system days, and one of the Stratoliners landing at Belem got off easy in what could have been a bad night landing accident. Belem would light flare pots along the edge of its runway when radioed in advance. The pots did not illuminate a small hill just short of the runway, and one night the pilots bringing Stratoliner Apache in felt a hard jolt on short final. That was followed by the sudden realization that the left main gear was gone at the moment of touchdown. The pilot, Bruce Pettigrew instantly knew what had happened and kept the left wing high for as long as possible. Beyond the sheared off left landing gear, the Strat suffered only damage to the left flap and wingtip along with two bent props.

If there was any consolation about British weather, it was that the entire place was not a dreary, foggy island. Usually if one section of the country was below minimums, fields might be open in other parts – assuming there was enough gas to reach the alternate. Larry Trimble had another hard-learned lesson to relate to Blackie concerning weather and radio reception. Trimble was westbound after having heard a promising weather briefing. Due to freak conditions, Trimble's radio was essentially inoperative for most of the eleven-hour leg he was flying. Fifty miles from his landfall, he was disappointed to learn that his destination was reporting a 100-foot ceiling. Trimble approached the field which by then was down to seventy-five feet. After bucking headwinds all the way, he diverted to an alternate and just squeaked in. Just after he landed, another ICD plane touched down, also almost out of gas, but with pilots aboard who were also glad they were familiar with the alternate.

Blackie's debriefing of returning crews served as a vivid recreation of many other dangers facing ICD crews on their way to Prestwick. Larry Trimble, with whom Blackie first flew in 1938, returned from a flight to Prestwick and related what had happened on the flight over. He was flying low trying to stay well under a 1,500 foot ceiling which he suspected promised icing above that level. West of Scotland, Trimble spotted what he thought was a large ocean liner with some sort of camouflaged escort vessel between his position and the liner. Trimble, at this point well within range of the escort's deck guns, altered course to steer away from the escort. Noticing flashes from the escort's deck, Trimble ordered the crew to shoot off the colors of the day to confirm his Stratoliner as a friendly aircraft. At about the same time, there were two huge geysers of water rushing skyward just ahead of the Stratoliner, and that was followed by several bursts of nearby flak. Trimble advanced the throttles and dove down to the deck in a right turn. About two hours later, and much closer to Scotland, Trimble was again bracketed by flak. He told Blackie he felt like a duck over a pond on the opening day of the hunting season.

Flares were used when approaching convoys simply because ATC did not want radios to be used for identification because the Germans always eavesdropped on radio transmissions. For a while, ICD planes used Aldis lamps for identification, but that proved to be a mixed blessing. Navy anti-aircraft crews assumed that the winkling and blinking lights up in the sky were coming from a four-engine German patrol bomber which meant they were being fired upon. So naturally, the Navy gun crews opened up on the signalling plane. Eventually, the friendly fire Aldis lamp incidents became such a nuisance that ATC retired the lamps altogether.

Trimble, along with Otis Bryan, was among the first to fly roundtrips on both the South Atlantic and North Atlantic routes. He was therefore among the first to appreciate loading every last gallon of gas until the weight and balance check said no more, especially when heading eastbound on the North Atlantic. When he first arrived at Natal, the jump-off point in Brazil for the flight to Africa, a B-17 pilot paid the ultimate price for sloppy fuel management. The pilot radioed that he was ditching in the sea about two hours after takeoff. The water landing apparently did not go well since the crew was never heard from again. A review of the B-17's paper trail showed that the pilot had only assumed that the '17 had been properly serviced and the tanks topped off. They had not. At about the same time, there was another story circulating around Natal about a pilot who committed the same blunder. In the latter case, the plane was a PBY Catalina, and the pilot enjoyed the extraordinary good fortune of alighting on the ocean after he ran the tanks dry – only to discover that a small convoy was approaching his position. There was even better news: one of the ships happened to be a tanker loaded with aviation gasoline. The tanker filled up the Catalina which continued on to the Azores. As it turned out, the pilot was one of those rare birds who never seemed to learn a lesson or use common sense. Trimble later heard that the same dense pilot continued to Africa where, although on instruments, he had maintained the proper course but paid too little attention to his maps. He had flown directly into a 4,000 foot ridge on the side of a 6,000 foot mountain.

Earl Fleet had a similar problem to Trimble's which he mentioned to Blackie when returning from Scotland. He was in his initial descent to Prestwick when he broke out of the clouds only to discover he was over what he took to be a flotilla of Royal Navy ships. He ordered the flares shot off immediately, but was dismayed to find that the Very pistol they were carrying did not fit the Stratoliner's flare chute. What to do? Fleet was convinced that the ships below had him in their rangefinders and the anti-aircraft gunners were all set to let fly. Hurriedly, the flight engineer opened a cabin door and shot the flares directly into the blue. Unfortunately, there was a flash back; the engineer was burned.

Thus an important lesson was learned about pre-flighting everything, including signal equipment.

If Blackie had not made it clear in his briefings about avoiding convoys, he got a good example to use on 12 November when Captain Stan Stanton and First Officer Randy Churchill were on a return leg from Scotland in the Stratoliner, Cherokee. They were cruising at 1,000 feet and were roughly between Iceland and Gander when Stanton decided to go aft and use the head which was located back near the tail. Stanton had just vacated the head when there was a sudden flash and burst from an anti-aircraft shell immediately in front of the Stratoliner. Churchill applied full power and turned immediately, but just as he did so, there was another bang which shook the Stratoliner violently. Stanton got back to his seat and ordered the trailing antenna hauled in, assuming that might have been the source of the big bang. It was not, and the rest of the flight was routine, including the landing at Gander where a quick inspection revealed no major, structural damage. There were many holes in the rear of the plane, including a few close to the john where Stanton had briefly been reposing. A thorough inspection once Cherokee was back in the TWA hangar at National revealed that the rudder cable had almost been severed and was held together by just one strand of wire. That had not been picked up at Gander because where the cable was severed just happened to be on the far side of a pulley and was therefore not visible to the inspector.

Blackie was never one to lavish praise for anything except major achievements which was adding to his reputation as a tough but fair taskmaster. In the case of Stanton's shot-up Stratoliner, Blackie thought the flight engineer, Raymond Darst, had done a superior job on the flight and in preparing the post-flight investigation. Blackie sent a congratulatory note to Darst saying that ICD expected exemplary work of its flight crews and that such enterprise was not worthy of any special praise. Blackie noted that Darst's performance during the endangered flight and the analysis that followed was clearly something special, but was surprised to see the congratulatory letter reproduced in the company newspaper for December 1943. The introduction of the letter by the *Skyliner* editor portrays Blackie as a superb public speaker and one not given to laudatory comments at the drop of a hat. Blackie had received a crash, post-graduate course in public speaking at Eagle Nest, but neither he nor the editor had the foggiest notion of just how polished a speaker Blackie would become in less than 20 years.

At about the same time, Larry Trimble was nearly in trouble again on his way to Prestwick. He had flown between layers which saved him from collecting a load of ice, but that also robbed his navigator of any opportunity to shoot the stars or use the driftmeter to determine the Stratoliner's position. After eleven hours of instrument flying, Trimble broke out and asked his navigator to use the driftmeter. It revealed that there had been a tremendously strong wind out of the south, thereby putting Trimble well north of his desired position. He assumed that he was within radio range and then asked for a QDM (heading to a radio station). Surprisingly, the response came almost immediately. It told Trimble to take a more northerly heading, but Trimble did not buy it; instead he took up a more southerly heading. Thirty minutes later, he asked for another QDM and got the same advice to turn north, i.e. a heading that would take him north of Scotland and to Nazi-occupied Norway if his gas held out that long. Then another voice came up on the radio, "Don't use radio bearings. Enemy transmission." By that time, Trimble could just pick up the beacon he wanted and completed the flight safely. That is something that could not be said for two other (non-TWA) flights from the night before which were missing. Here was another lesson learned: verify radio stations and always have a list of frequencies on board.

Cooperation with the Navy was improving, but the bottom line was that the ship's gunners always shot first and asked questions later. Still, they were interested in any intelligence ICD could offer. As a result Blackie always made sure TWA crews were fully briefed on procedures when and if an ICD captain encountered what he thought was an enemy ship. In 1942 and the first half of 1943, that was a very real possibility. His briefings included a one-page handout headed with acronym, "CIRES" – for Communications Instructions for Enemy Sightings. The procedures called for a radio message in the clear if there was no question about the sighting and no immediate danger to the TWA crew. Otherwise messages were to be sent in code, especially if there was a perceived danger or there was some doubt as to the nationality of the sighting. There were code letters for enemy submarines, naval vessels, aircraft, surface raiders, and suspicious activity such as burning ships, life rafts, etc. Friendly forces could also be notified by flare or Aldis lamp instead of a radio. At the bottom of the CIRES checklist there was a stern warning not to approach an enemy submarine for the purpose of getting more accurate information because "Submarines are equipped with high-velocity machine guns."

Although the ICD crews had good radios for those days, there were problems other than German jamming and storm static. Ross Weaver, who was hired in the second group to join ICD, was among the first to experience what the aurora borealis could do to the radios – the effect was about the same as Germans jamming the same frequency. Terrain and weather also played havoc by creating natural skip zones where no radio signals could be received.

To a certain extent, Blackie was in a "learn on the job" mode as was the Air Transport Command in general. One aspect of his job as the operations manager was conveying useful information that would at first seem out of place in wartime; and that was a list of American and British consulates that could be contacted while out of the country. War or no war, TWA crews were flying to and operating in foreign countries in Latin America, Africa, the Middle and Far East, as well as the United Kingdom. Therefore they had to comply with laws relating to passports, visas, and inoculations. The pilots all had to have current licenses and medical certificates which therefore required periodic flight physicals, all of which Blackie would eventually control.

Blackie shared many TWA-generated materials and TWA-edited materials with ATC. In late 1942, much of that shared material included TWA's Air Route Manuals which were refined from other, previously supplied government maps. These route manuals covered all possible routes from Miami to points south and as far east as Calcutta and Assam, India. Blackie's contact at ATC was its liaison officer, Captain Robert B. Hotz, a 1936 graduate of Northwestern's journalism school and by 1942 an experienced newspaperman. Hotz would later go to China and become a member of the Fourteenth Air Force headquarters staff under General Clair Chennault. After the war Hotz became editor of *Aviation News* which soon became McGraw-Hill's prestigious *Aviation Week and Space Technology ("Av Week")*. Hotz was such a law unto himself that he became concurrently editor and publisher of *Av Week* in a company and industry where editor and publisher were normally kept very separate. Blackie and Hotz would collaborate on other projects many years later. While Hotz was at the top of the *Av Week* masthead from 1955 to 1979, he made sure that his closest friends were placed on the magazine's complementary subscription list, Blackie included.

From late August until mid-October, Blackie flew very little, logging only a few hours on C-54s in order to give check rides for new captains and such mundane matters as compass calibration. That all changed when orders came through from the ATC for Blackie to take some VIPs to India, leaving on 15 October 1942. The orders had come through only the day before from ATC's Air Transportation Division; Blackie was to take the Stratoliner Apache with a crew of fourteen and ten passengers to Karachi. He had a full crew: one first officer, three second officers, four navigators, two flight engineers, and three radio operators. Not all of those men would make the full trip because ICD was moving toward temporary, foreign domiciles in which crews would be stationed abroad on a temporary basis before rotating back home. Blackie's passengers would deplane in Karachi and subsequently be flown to China by the newly formed India-China Ferry Command which was assigned to the Tenth Air Force headed by Major General Clayton Bissell. The orders also called for Blackie to fly to Accra on the African coast and fly shuttle trips between there and Natal, Brazil, thus making the effort a three week mission. Before undertaking this trip, Blackie was well and truly punctured. His tetanus and small pox immunization was up to date, but prior to flying south he was also jabbed for typhoid, cholera, and yellow fever as were his entire crew.

The VIP passenger list was headed by T.V. Soong, China's foreign minister and, for all practical purposes, the second most powerful man in the Chinese government. Soong was a Christian and particularly well-educated: B.A. in Economics from Harvard in 1915 and then a Ph. D. from Columbia. He had previously been minister of finance and on several occasions was China's acting premier. His father, Charlie Soong, was a native Chinese and an orphan who returned to China after an American education. He became quite wealthy by printing and selling Bibles and subsequently making even more money in banking and finance. Charlie Soong's three sons went into finance, although T.V. ("Tse-Ven" in Chinese) was by far the most prominent. T.V. rarely used his Christian name, Paul. Saying that T.V. Soong's three sisters married well is an understatement. Ai-ling married Harry Kung, a prominent financier and finance minister, Ching-Ling married Sun Yat-Sen, the father of modern China, and May-Ling married Chiang Kai-Shek, the Chinese premier. T.V. Soong had just wrapped up a nearly two and one-half year stay in America during which period he secured financing for the American Volunteer Group ("The Flying Tigers") along with several other loans totaling nearly $1 billion.

Air Transport Command was anxious for some photo documentation of this trip; and knowing of Blackie's interest in photography and capability with a movie camera secured permission from the War Department for him to take along a camera. On 27 July, the War Department through the Adjutant General's Office had sent a restricted directive to quite literally every commanding general in all commands of the Army (from Air Forces, to Ground Forces, to all Defense Commands, Service Commands, Ports of Embarkation, etc.) outlining procedures for foreign photography. In effect, the directive said that approval was required, that releasing the film to any party of other than the War Department was prohibited, and that the completed photography became the property of the U.S. Army. The request was made by Lt. Col. Robert Love, Deputy Chief of Staff of the ATC, which was supplying Blackie with movie film (a pricey item for civilians at the time). The approval came through quite literally at the last minute on 15 October just as Blackie was preparing to depart.

Blackie took off from Washington with his VIPs and landed almost six hours later at Morrison Field, West Palm Beach, Florida, on what was his first South Atlantic crossing. He had been preceded six months before by a group near and dear to his heart: the Eleventh Bombardment Squadron. The "Jiggs" squadron had completed transition to the B-25 Mitchell and flown to China where it had already completed its first mission from Kunming, bombing Japanese shipping around Hankow, China. Blackie remained in West Palm for the night and was off in the early morning hours the next day for Belem, Brazil, after stopping for fuel at Waller Army Air Field in Trinidad. Fortunately, Blackie was in and out of Trinidad in daylight which was good. The field had been carved out of the jungle and was not at all well lighted. Takeoffs were assisted by a strong searchlight which shone from well down the field to the tree tops at the far end. If a pilot was above the shaft of light, he was all right – below it meant big trouble. Technically, Blackie was in a foreign country, and that being the case, he followed through with a standard procedure when stopping at distant bases; he posted one member of the crew by the Stratoliner on guard duty. It was a long day in the air before Apache touched down at Val de Caes Field in

Belem, Brazil. The field was originally developed by Pan Am, but had been taken over and expanded by the Army Air Force and built into a large facility which included two concrete runways and asphalt taxiways. Belem had grown into the largest city in northern Brazil due to its position at the mouth of the Amazon River and by trade in sugar, rubber, and the Brazil nut which assured Belem's future as a port city. For Blackie and his crew and passengers, it meant a cot in an Army style barracks after nearly eighteen hours of flying over ocean and jungle punctuated by towering cumulous clouds typical of tropical flying. The group caught little more than forty winks and a hurried breakfast at a screened-in mess hall. Breakfast was fresh, local fruit: bananas, melons, oranges, and pineapple. Then it was off to Natal, Brazil, which they reached after covering the 965 miles in nearly seven hours of flying. The flight down to Natal was via Sao Luiz and Fortaleeza along the coast which was much in contrast to some of the previous day's flying over miles and miles of dense jungle. As the Straoliner cruised over the coast, Blackie and passengers could see the broad beaches and fishermen below. Because the beaches were so wide and the tides so high, the native fishermen merely arranged their nets in the shape of a big V with the bottom pointed toward the sea. As the tide went out the fishermen had only to wade in and remove their catch from the nets. Although Natal's Pamamarim Army Airfield was started by Pan Am and had become a major base, it was just a refueling stop for Blackie. He took off immediately for Ascension Island, 1,438 miles distant and half way to the African coast. Getting acquainted with the cities of Belem and Natal would have to wait for another day.

There was some weather on the way which forced Blackie to fly nearly two hours on instruments before reaching Ascension's Wideawake Airfield in darkness. Given a choice, most pilots preferred to arrive at Ascension just after sunrise and to confirm their navigation with sun shots. Daylight arrivals had their own problems too; it was very easy from a distance to mistake the shadow of a cloud on the water for an island. Not immediately visible until on short final was one of Ascension's hazards, large flocks of wideawake birds (sooty terns whose raucous cries kept everybody awake, hence the "wideawake"). Hundreds of the birds were always driven to the skies by the sound of planes departing from the single 6,700 foot runway. It had been an exhausting journey that day – fifteen hours of flying. There was a radio range station at Ascension whose only occupants before the war, other than the legions of seabirds, were a British crew which operated the cable station at Georgetown, the island's only port. The station could not be detected from very far away which put a premium on superior celestial navigation. The volcanic island was not that large to start with, just thirty-four square miles, and the only way a reasonably long runway could be put in was by bulldozing through part of the smaller of two mountains. The approach was over a 300-foot seaside cliff. The runway, although well over a mile long had a hump before the middle after which the runway eventually sloped downward. At the far end was a rock fall which attracted the wideawake birds; thus the prudent pilot landed as short as possible, preferably well before the runway high point, and stood on the brakes. While the runway length was quite adequate, the width was alarming. Whereas a pilot who ran a Stratoliner off a runway in Kansas City did not have a lot to worry about, the same shortcoming at Ascension would end in disaster. There were still jagged rock formations very close to both sides of the runway from which a path had been blasted through the smaller mountainside to create a landing surface. Thus takeoffs and landings at Ascension tended to be tense affairs. Although there were several beaches which could be used for swimming, most of the coastline was jagged and unsafe for taking a dip. The landscape was essentially rocky, and although there was some greenery located higher on the mountains, most of the Army men based there just called the place a "rockpile." It was a lonely place to be stationed, leading some soldiers to believe they had been exiled. That notion was very much in keeping with Ascension's nearest neighbor, St. Helena (730 miles to the southeast) where Napoleon Bonaparte had been banished.

The war had come to Ascension two years before in the shape of a German submarine which intended to sink British supply ships and shell the cable station. The island's garrison opened up on the sub with the only cannon on the island, and that convinced the U-boat to leave. The U-boats did sink enough supply ships at sea to nearly keep the early cable crews in a state of semi-privation although that did create one good use of the annoying wideawakes. The crews took to gathering the birds' eggs and using them to make omelets instead of using dehydrated eggs.

Nazi U-boats were still active in the area and were known to broadcast signals that might be confused with Ascension's weak radio homing range. Opening Ascension made it possible for light and medium bombers to hop the South Atlantic. Even fighters with drop tanks could make it, although they typically followed an ATC mother ship which performed the navigation. The hazards of poor navigation in the South Atlantic were popularized in one wartime song of which some of the lyrics read, "If you miss Ascension, your wife gets a pension, so cheer up my boys, bless them all."

Wideawake Field had not been opened that long – Army engineers arrived in March and by 10 July the runway was in business. Then they set about building the base, mess hall, machine shop, barracks, and most importantly, the desalinization plant. When Blackie arrived, each man was allowed four gallons of water a day for drinking and washing. At first it had been five gallons per week. Because Blackie had VIPs on board, they spent the night in the barracks unlike most of the airfield personnel, all 2,000 of them, who still lived in tents built over wooden platforms.

There was time for sleep in the Army barracks and breakfast in the mess hall because Blackie had accepted some good advice about arriving too early in Africa where coastal fogs in the morning were very typical. Thus, the Stratoliner took off for Accra after sun-up for the eight and one-half hour, 1,400 mile flight to British West Africa's Gold Coast (today's Ghana). On takeoff there were sounds of thuds and splatters coming from the leading edges as a few of the sooty terns passed onto the avian Great Beyond. Fortunately the birds' final ascent took place without doing any damage to the Stratoliner other than a few bloodstains on the deicer boots and some feathers inside the engine cowlings. Blackie reached the African coast in mid-afternoon and was greeted by a dark green landscape over which cumulous clouds were riding high. Below, waves were breaking on what looked like pristine beaches. Approaching Accra airport, the pilots observed curious little red bumps scattered below – almost like upside down ice cream cones. Landing was at Accra Airport which was under British control; it had been on the British Overseas Airways route to Khartoum and was used by the RAF Ferry Command to move aircraft to Egypt and India. Once on the ground, Blackie discovered that the red, inverted ice cream cones were anthills as much as eight feet high. Blackie's limited command of Spanish, along with one trip to Mexico, had provided him with only minimal preparation for Portuguese-speaking Brazil, but Ghana was a new experience in the extreme.

Having arrived at a British base, and being officers with VIP passengers, the group was accorded VIP treatment, i.e, mobbed by porters whose boss spoke English with a noticeable King's English accent. Blackie's crew and the passengers were taken to what amounted to a large concrete hotel right on the airport, and each man was given a spacious room complete with houseboy. The natives were found to be smiling and friendly, but also inclined to walk off with anything that was not nailed down. Ghana had been a British Colony, and the most visible signs of civilization were paved roads and a few government buildings. Accra's claim to fame was built on the cocoa business of which it was the world's largest exporter. Otherwise, Accra struck Blackie as a study in contrasts. In town, the locals were dressed, but as one approached the countryside clothing became more rustic. Bob Loomis snapped a photo of Blackie and Stan Stanton at a public market where the women were topless as was their custom. With his usual dry wit, Blackie captioned the photo, "Recruiting new TWA hostesses." There were a few badly-worn cars on the red dirt roads, but many more people were riding on burros.

The market was loaded with fresh fruit, particularly bananas, coconuts, and pineapples. There were vegetables including the Ghanian staple, a sort of sweet potato called *fufa* which was ground much like corn meal, mixed with water, and eaten by finger dipping just as Hawaiians eat *poi*. Blackie paused in the market and bought himself a cork pith helmet, a pair of wooden bookends, and a few gifts for Martha and the kids. The beaches were beautiful, but Blackie had no time for that on this particular trip. Nor was there time to watch the local fishermen who dragged their nets, suspended between two canoes paddling shoreward where helpers would remove the catch from the nets. For all of the beauty, there also was poverty, filth, oppressive heat, and a background odor of foul-smelling smoke from wood cooking fires and rotting vegetation typical of the tropics. Unlike the facilities on base, open sewers ran in the middle of Accra without the convenience of modern restrooms, thus answering nature's call was never a private matter for the locals, much to the chagrin of visiting westerners like Blackie. The humidity was so ever-present that closets back at the hotel were equipped with light bulbs which ran around the clock – to keep clothes from quickly mildewing. The airfield beckoned after a couple hours in town – there was cold beer at the officers' club and the mess hall served surprisingly good, fresh food.

The Stratoliner was off the next morning well before sunup on a two-leg flight to Khartoum via Kano, Nigeria, another former Imperial Airways base. Kano, with its surrounding city walls, was supposedly the oldest city in Africa; its only enterprise, other than servicing itinerant aircraft, was in cotton and dying it indigo blue. Kano was essentially just a pit stop after 1,242 miles and five hours of flying over bush country, and Blackie stayed just long enough to top off the Stratoliner's tanks before the taking off for the long ten and one-half hour, 1,631-mile haul to Khartoum. As the tides of war changed, Kano would recede into the past because once Rommel was defeated, ATC largely abandoned the Central African route in favor of the North African. The route to Khartoum took Blackie over Lake Chad, many herds of elephants, and miles and miles of desert and flat farmland, all of which created a bumpy descent into the airport.

Blackie had two days in Khartoum while the Stratoliner was serviced. Khartoum was the capital of Anglo-Egyptian Sudan where many cultures and nationalities collided over the centuries. The Arabs, British, Coptics, Egyptians, Ethiopians, French, Greeks, Jews, and Syrians had all left their marks on this cosmopolitan city along the Nile, and now Blackie joined the Americans who would leave theirs. Khartoum means "elephant head" which is roughly the topography of the land where it is located on the left blank of the Blue Nile near the junction with the White Nile. The city had been fought over many times down through history with the most notable conflict before Blackie arrived being that between the British Army and the Mahdi and his Arabs. More recently, Mussolini had threatened and bombed Khartoum before the British drove him out of the Sudan. Until the allies triumphed in North Africa, Khartoum was a key stop on the way to the Middle East. The airfield was located well outside of the city, surrounded by barbed wire, and heavily guarded by the "donkey fence patrol." Blackie observed that the name was apt, but not entirely descriptive of the soldiers riding the donkeys. The guards were ferocious-looking, tall men, well-armed, and were uniformed in sand-camouflage uniforms, over which they wore leather cartridge bandoliers. The impression of height was accentuated by the peaked caps which they all wore.

Unlike the facilities on barren Ascension Island, the facilities at Khartoum were excellent; they were in fact a college campus which was scheduled to open just as the war came to Khartoum. Classrooms and dorms gave way to living quarters, mess halls, barracks, and recreational facilities. Blackie and company left Wadi Seidna, the ATC base, for a visit to downtown Khartoum, finding it to be very cosmopolitan, but scorching hot. It was hot getting there on a very bumpy road; the base was located thirty miles away from downtown. The strong British influence was obvious, particularly at the Grand Hotel's British Club and its Starlight Roof where some of Cairo's best entertainers (fearing Rommel would make it that far) had relocated. Every city the crew visited seemed to specialize in some sort of souvenir, and in Khartoum it was carved ivory. Blackie had little time for extensive shopping because the next day, the twentieth, was going to be a long one indeed – taking off for the 859 mile flight to Aden before the long flight over the Gulf of Oman to Karachi. There was one blessing in such an early takeoff; the Stratoliner had been sitting in the dark and was not blazing hot inside.

Blackie never made it to Aden because his number four engine was running very roughly; so he decided to land at Gura in Eritrea instead of pushing on to the Royal Air Force Base in Aden. Gura Army Air Base was a wise choice because it was a major repair facility. Eritrea had been an Italian protectorate and was used by Mussolini as a springboard for his invasion of Ethiopia. Part of the Italian pre-invasion build-up was the establishment of a major air base which the Royal Air Force took over after Mussolini's forces had been routed in the East African Campaign in 1941. Fortunately for Blackie, the runway was lengthy because he needed a long one, considering the facts that the Stratoliner was heavy with fuel and the aerodrome was located on a high plain nearly 7,000 feet above sea level. Fortunately his landing was not made in mid-afternoon when the eternal sun, baking the earth, would have thinned out the air and made the landing even more demanding. The mechanical issues with the Stratoliner delayed Blackie's takeoff for Karachi until the twenty-second. Crew and passengers, except for the flight engineers, had a full day on their hands with little to do other than visiting the nearby small city of Asmara which boasted the Eritrean National Museum, an opera house, and other buildings influenced by Italian architecture.

At Gura, the Royal Air Force had employed an American subcontractor which operated a comprehensive repair facility that serviced damaged and shot-up aircraft from the North African war. Gura was chosen because its location was far away from marauding *Luftwaffe* planes and the fact that the defeated Italians had surrendered an intact airfield. The damaged RAF planes, including many P-40 fighters, were barged down the Red Sea coast to Massawa, trucked to Gura, repaired, and then flown directly to Cairo where they were put back into service. Gura had none of the charm of Khartoum, but did have more than a dozen, corrugated metal repair buildings, and considering its origins as a British base, a very good officers club and transient officers' quarters. The original repair subcontractor was none other than Douglas Aircraft of Santa Monica, California.

Blackie did not linger at the airfield but took off in darkness on the twenty-second for the fifteen-hour flight across the Red Sea, Yemen, southern Saudi Arabia, Oman, and the Gulf of Oman to Karachi. Blackie had not flown every hour of the journey, of course, but as the captain he had 100 percent of the responsibility, and soon part of the burden would be lifted when his VIP passengers were turned over to other pilots who would take them on to Kunming, China. The landing at Karachi was, after dark, providentially free of two well-known local hazards: dust storms and thunderstorms that produced world class hail stones. Fortunately for Blackie, Soong's flight had not been scheduled for a month or two earlier which would have coincided with the last of the monsoon season. The TWA crew knew what to expect in India (although Karachi is now part of Pakistan). The Army had prepared the pamphlet, "A Pocket Guide to India," which warned them about rampant disease, the necessity of staying away from rivers (crocodiles), seeing that all food was properly cleaned and cooked, disease-carrying insects, and other hazards of eastern countries. The Stratoliner and crew were inspected and fumigated (a wartime indignity that all ATC crews were subjected to at most foreign bases). Finally they were taken to the transients' quarters, except for T.V. Soong and his entourage who were whisked away to Karachi itself. With his inquisitive mind and gift for conversation, Blackie became a favorite of Soong's on the long flight which also revealed another Blackburn characteristic and one that occasionally could have fatal consequences. The crew noticed that Blackie spent very long hours at the controls and took only infrequent breaks for a quick nap. A longer period of restful sleep was the customary procedure on such lengthy hauls. Blackie was always unbending on cockpit procedures, such as getting adequate rest, but he may well have been one of those rare individuals who could function perfectly well on substantially less sleep than most people.

Blackie's hopes for spending a day or two in Karachi were dashed; he had only time to grab a snack, hit the sack, and wolf down a quick breakfast because the Stratoliner was scheduled for an early takeoff and return to Gura. The city fathers had billed Karachi as the Paris of the East, and for an explorer like Blackie touring it would have been a welcome adventure. They also promoted their nice beaches and that Karachi's location on the coast made it cooler than the hot, suffocating cities of the interior. There was fine Chinese cuisine at the city's two oriental restaurants, Chungking and Victory, which would have appealed to Blackie, but there was no time for leisurely meals. More conventional, occidental tastes could be satisfied at such British favorites as Café Grand or Café Oxford. There was a noticeable lack of taxis, or any cars for that matter, although there were trams and very British double-decker buses for those who wanted a less than leisurely tour of the city's racetracks, palaces, theaters, museums, and dance halls. For every tram, there were many more horse-drawn carriages, donkey carts, and camels towing larger wagons. Most traffic in the center city was pedestrian and likely to be wearing a garrison cap, pith helmet, or turban. Early on the twenty-third, Blackie took off from Karachi Airport and headed out over the upper Arabian Sea – and trouble.

After several hours of flying, the southern end of the Arabian peninsula came into view off to the right, and the next four hours were spent flying along the southern Arabian coast and its three ATC bases Masiram, Salalam, and Aden. With the Red Sea in sight, the number four engine failed and was promptly feathered, prompting a command decision: land at the ATC base at Aden which was practically below his right wing tip or push on with three good engines and fly the remaining two hours to Gura. Blackie's decision was for the latter based on the fact that the ATC station at Aden was on a Royal Air Force base and that the chances of better maintenance were available at Gura, considering the fact that it was a major repair depot. After thirteen and one half hours of flying, the Stratoliner landed without further incident. The problem turned out to be a major one, and required three days to fix, leaving the crew and its thirteen westbound returning ferry pilots with little to do except pass hours away at the officers' club or walk the streets of Asmara. Unlike the ancient land where it is located, Asmara was a very modern city in late 1942 because it had been built up by the Italians when Eritrea was an Italian protectorate. The Italians had tried to create a "little Rome" as Asmara was being planned by architects, and they succeeded which took some of the edge off being grounded for mechanical reasons. As an essentially modern Italian city, Asmara enjoyed tree-lined boulevards, many restaurants, street side cafes, bistros, ice cream parlors, and restaurants. And at nearly 7,000 feet above sea level, it was cooler by more than a mile than any ATC base in southern Arabia.

Work on the number four engine was completed late on the twenty-sixth, so Blackie took the Stratoliner up for a half-hour test flight very early on the twenty-seventh. Satisfied with that, he was off promptly off to Kano via Khartoum which he reached after nearly twelve hours of flying. The rest of the trip home was essentially a reverse playback of the trip out: Kano to Accra, Accra to Asencion at night, and Asencion to Natal in daylight on 30 October. Prior to the crossing, Blackie accomplished another of the flight's objectives for which the movie camera came in handy. While in Accra, he conducted a survey of the coast west and northwest for a potential airfield location to handle additional traffic in the event that Accra, Takoradi, and Monrovia became too crowded. As it turned out no additional bases in Liberia and what is now Ghana were ever established because the North Atlantic was eventually opened to winter traffic and North African ATC bases such as Dakar and Marrakech soon came on stream.

On 1 November, Blackie was invited to ferry a Lockheed Hudson from Natale to Recife by Lt. Col. Jaffe N. Bell, a member of the American air mission that was equipping the Brazilian Air Force which had been founded only the year before. The Brazilians were active participants on the Allied side during World War II, not only because they supplied a squadron of Thunderbolt pilots who flew in Italy, but very importantly because they were Nazi submarine hunter/killers. That Hudson was among the very first to arrive at the Recife base where the Brazilian Sixth Aviation Regiment had been constituted to conduct maritime reconnaissance and convoy escort duties. Blackie remained in Brazil for two days shuttling pilots between Natal and Belem before taking off on the final legs of the journey by flying from Belem back to Washington via a different route. This time, the Stratoliner flew to Atkinson Field in British Guyana which was more than just an airfield hacked out of the jungle for quick servicing of ATC planes. It was home for a light bombardment squadron which patrolled the southern Caribbean in search of German U-boats or other vessels that could launch an attack on the nearby bauxite mines so important to the aluminum industry. Atkinson was just a pit stop as was Blackie's next destination, Borinquen Field on Puerto Rico. After another six hours, Blackie was back in the United States, where, at Miami, he became a passenger on a regular commercial flight to Washington. He had been gone for three weeks – a long time for Martha and the kids to understand, but fairly typical for a pilot on the South Atlantic ATC route to India.

On 5 November, Blackie was back at his office when the TWA community was rocked by a news release issued by the War Department. Lt. Colonel Jack Zimmerman was missing and feared dead following a flying accident in the Gulf of Saint Lawrence. Ironically, Zimmerman's position in the Air Transport Command, North Atlantic Division, was chief of flying control, a job which put him, in effect, in charge of all safety matters. On 2 November, Zimmerman had flown an OA-10 Catalina (Army version of the Navy's PBY Catalina as used for air-sea rescue operations) from Presque Isle to Mingan, Quebec, to inspect a recently completed emergency landing field. That field, located on the north shore of the Gulf of Saint Lawrence, was roughly half way between Goose Bay and Presque Isle and was simple to locate since it was situated about twenty miles directly north of easily identified Anticosti Island. Zimmerman, with his eight-man crew, had flown to Mingan late in the morning of 2 November and completed his inspection by late afternoon. He had a photographer on board and planned on getting some aerial shots of the emergency field while there was enough daylight. The weather had worsened considerably during the few hours he was on land, but he decided to return to Presque Isle anyway. The Catalina had been moored at a dock where it was rocking in seas that were running as high as six feet. The first attempt to take off failed and resulted in the nose wheel compartment springing a leak from the pounding it was taking after hitting wave tops. The leak was temporarily checked by plugging it with a heavy winter flying jacket. On the second attempt, Zimmerman chose a crosswind takeoff to minimize the chance of worsening the leak. Even with both engines at full throttle and rpm, the Catalina failed to gather enough speed to lift off. Then it hit a rogue, seven-foot wave prompting Zimmerman to close the throttles. At this point, water was flooding in and the Catalina was sinking rapidly, nose first.

Only four men got out; the co-pilot wiggled through the window on his side and three men got out of the right side machine gun blister between the wing pylon and tail. They were able to grab a life raft, but it was carried off by the wind and current. Fishing boats reached the survivors in ten minutes just as the tail went under and the plane and five unfortunate flyers headed toward the bottom 108 feet below.

At the time of his death, Zimmerman had over 11,000 flying hours, but less than a dozen in Catalinas. His demise was probably avoidable had he been more prudent. Although he was blamed for the accident, some good did come from it. The accident board's recommended that all Catalinas be inspected for cracks or abnormalities in the wheel wells and that water takeoffs be limited to conditions where the ocean swells were three feet or less.

Blackie and Swede Golien flew a C-73 (originally a Boeing 247 passenger aircraft) full of ICD personnel to Zimmerman's memorial service in Fremont, Ohio, on 15 November and returned later that day. Almost exactly seventy years later, an underwater archeology crew from Parks Canada re-discovered the sunken Catalina quite by accident. The few human remains discovered in the wreckage were subsequently sent to the Joint POW/MIA Accounting Command's Central Investigation Laboratory for identification. Zimmerman had survived five previous flying accidents, including two that were nearly fatal. He was only in his mid-thirties when he died and had been asked why it was that he had so much responsibility at such a young age. His reply was prophetic, "Flying has never been an old man's game, anyhow."

Most of the flying done by ICD involved flying freight and passengers on what were rapidly becoming the final North and South Atlantic ATC routes. Less frequently an assignment would come to ICD involving ferrying military aircraft because Air Transport Command's Ferrying groups were still understaffed. On the night of 10 November, Blackie and Joe Carr ferried a B-24D down to Palm Beach, Florida, and flew maintenance test flights before an Army Ferry Command took the plane for a South Atlantic crossing. Blackie and Carr returned the following day on Eastern Airlines for the first of what would become an unpleasant aspect of their jobs as the flight supervisor and training manager: accident investigation.

Several days after getting back from Miami, ICD suffered its first flying accident when a Boeing C-73 landed much too long, overran an embankment, and was destroyed at Pittsburgh's Allegheny County Airport on 17 November. The C-73 was one of several given to TWA for navigation training and, fortunately, there were no fatalities, but one of Joe Carr's navigation instructors, H. B. Jobe, suffered a back injury.

The administrative aspects of Blackie's job, along with C-54 teething issues and training needs, caught up with him in force and he would not fly again until late January. One of the first problems with the C-54s was in the main landing gear. That was discovered when one of Blackie's friends, maintenance chief

Johnny Guy, was called down to Washington National in the middle of the night. A crew was running a test flight on a Skymaster following a double engine change and could not get a green light on the left gear when preparing to land. The usual remedy, a quick dive with a sharp pull-up failed to produce results, and cycling the manual pump still produced no light. Air Transport Command was notified and the order was given to fly the C-54 up to ATC's overhaul base at Middletown, Pennsylvania, (near Harrisburg) and belly it in. Swede Golien decided to intervene since the troubled Skymaster had plenty of fuel on board – enough for seven hours of flying. The longer the Skymaster was burning fuel the safer the landing would be anyway. Swede took off in another C-54 to observe from below and report any progress. Constant cycling of the manual pump was starting to produce some slow results, and after four hours of pumping, the crew finally got three in the green whereupon they made a safe landing at Washington. An immediate inspection showed that a spring clip was missing, a condition found on several other Skymasters when the entire fleet was inspected. The clip was replaced with a bolt, and the TWA fix saved ATC of potentially more landing incidents.

On 9 December, TWA suffered another accident, and this one hit Blackie hard since he lost another good friend. Nick Wasil had gone from Eagle Nest first officer to ICD captain in less than five months and had taken off from Natal late on 8 December for a flight across the South Atlantic in a C-87 Liberator when his plane simply disappeared. With him went his crew of four and five TWA mechanics who were being flown to ATC bases in Africa. A following flight reported flotsam and an oil slick 225 miles west of Ascension. At 3:37 A.M. Wasil had checked in with Natal radio and noted that he was cruising at 9,000 and that all was well. Weather along the route that night was completely benign, so any in-flight breakup due to storm-induced structural failure was unlikely. Speculation focused on the possibility of Wasil being downed by a German submarine, possibly being hit by friendly fire from Allied ships, or mechanical failure. While it was far from commonplace, TWA crews occasionally did spot Nazi submarines in the South Atlantic but had orders not to venture within range. In truth, being shot down by a sub would have been a unique distinction considering the fact that U-boat deck guns, while having the range, could not be elevated to any significant extent. Although the machine guns could be elevated, they did not have particularly long range. Many years later, Ed Betts, a TWA pilot and fine historian, cross referenced the Wasil incident with U.S. Navy ships' logs in the South Atlantic and found that no ships had fired at aircraft on the day in question. Betts checked other ATC reports of anti-aircraft fire in the Atlantic and cross referenced them with Navy ships' logs, learning that the fire did not come from American ships. That largely eliminated the friendly fire possibility, but Blackie had already made up his mind what caused the problem; it was the Liberator's poorly designed fuel system, and he was far from being the only pilot who thought it was dangerous at best. Wasil's C-87, serial number 41-

11707, had a history of in-flight fuel leaks, and all it took was a spark from an electrical short to create a fiery explosion in the sky. Naturally, smoking was forbidden in a Liberator, but that never stopped the most famous Liberator passenger, Winston Churchill, from lighting up one of his trademark cigars. Churchill's Liberator actually was a converted LB-30 featuring RAF night camouflage (black bottom and sides, dark earth upper surfaces), a bomb bay modified with seats and a berth for Churchill). Prime Minister Churchill only made several long range flights in his Liberator, named "Commando," before moving to a modified version of the RAF's Lancaster bomber in 1943. It was said that Commando, with its VIP passenger lighting up constantly, was quite simply too afraid of Churchill to blow up. One of the most engaging photos ever taken of the Prime Minister is the one where he was in the left seat of his LB-30, flying it (he had qualified as an RAF pilot years before) with a monstrous stogie in his mouth.

Blackie, although he never met Churchill in person, did have an interesting connection with Churchill's flights in Commando whose pilot was William J. Vander Kloot, Jr., a former TWA co-pilot. By the end of the war, Vander Kloot flew Churchill well over 200,000 air miles. Vander Kloot had last flown as Blackie's co-pilot in early December 1940 on the New York-Chicago-Kansas City route but left TWA to ferry Hudsons and Liberators to Britain because the pay was so far superior. Churchill was in the process of saying goodbye to Commando in favor of the AVRO York, a passenger/freight adaptation of the RAF's AVRO Lancaster heavy bomber, whose performance substantially eclipsed both the American Flying Fortress and Liberator. Interestingly, the York had one notable feature in common with the Lockheed Constellation: a tail with three rudders.

It was Otis Bryan's sad duty to inform wives and parents of the Wasil flight's loss at sea to which Blackie added his personal condolences to Nick Wasil's family in Cleveland. There were three outcomes related to the Wasil crew's disappearance. Blackie gave up smoking at about this time. Like many men of his generation, he smoked cigarettes, although he had never been a heavy smoker. The initial impetus for that move was concern about smoking and fire danger on airplanes but more importantly how he had become disgusted by the habit in general. His recent trip to India, with all of its stops and seeing people begging for cigarettes or lunging for a butt flicked by an American convinced him to have nothing to do with smoking.

At that point, TWA crews had really never received formal training in how to ditch an airplane or how to survive the aftermath if forced down in rough or hostile territory. To address that issue, the ATC created and distributed a 116-page brochure on survival techniques for use in a variety of situations, including the Arctic, jungles, deserts, and oceans. The book's introduction was written by World War I ace Eddie Rickenbacker who knew something about survival; he had survived a crash landing in mid-Pacific in late 1942 but survived in spite of spending several horrific weeks adrift on a life raft. The Rickenbacker experience had shown that a more orderly approach to ditching with individual assignments would have resulted in the crew's getting into life rafts with a maximum of survival gear instead of people just grabbing what they thought would be useful at the last minute before a plane sank.

To provide ICD crews with a step-by-step guide on water landings, Blackie, Joe Carr, and Swede Golien wrote TWA's first water landing survival guide for use with four-engine aircraft. The guide was broken down into two sections: before landing and after landing. The guide was perfectly explicit in that it told exactly what had to be done, when, and by whom for a full crew consisting of the captain, first and second officers, navigator, radio operator, flight engineer, and purser. The guide, which also indicated where the life rafts were to be stored on-board, was based on experience gained in the Potomac River using a mock-up C-54 fuselage. Blackie had the engineering department contact Douglas to determine how long, on average, a C-54 would float after ditching and which part of the aircraft would settle into the water first prior to actual sinking. Naturally there were variables such as gross weight, wind velocity and direction, wave heights, etc. Douglas engineers calculated that a C-54 whose fuselage and wings were essentially intact would remain afloat long enough for crew and passengers to quickly exit and paddle away before the plane sank. The engineers also indicated that the plane would float roughly ten degrees nose down before the angle increased to the point where the ship sank nose first. TWA never suffered an emergency water landing with its C-54s which would have been a rather academic matter in the rough North Atlantic anyway, but Pan Am did have a crew that suffered a water landing only to find that the Douglas engineers had missed the mark. The Pan Am water landing was in the profoundly balmy Gulf of Aden where their lightly loaded Skymaster remained perfectly level on the surface for an hour before submerging. That alighting (as water landings are correctly called) was not an emergency, but resulted when the crew brought their Skymaster too close to the water at night on final approach into Djibouti. The crew's depth perception failed which is not unusual when flying over calm water and simply touched down not that far from the shore which they soon reached in life rafts.

The third outcome of the Wasil tragedy was a while in coming, but TWA and the other airlines flying C-87s banded together and presented a grievance list to ATC with the fervent request that something be done about the Liberator's fuel system – an effort that would extend into the first quarter of 1943. In the meantime, ICD was about to complete its first year of operation, which held a pleasant surprise for Blackie. On 23 December, Jack Frye took a phone call from the National Aeronautics Association informing him that the Army Air Forces and the airlines of the United States had been selected as the joint winners of the Collier Trophy for 1941 for pioneering worldwide air transportation so essential to the eventual victory of the Allies in World War II.

The prestigious award, named for Collier Publishing heir Robert Collier, was invariably given to individuals for the greatest achievement in aeronautical performance, efficiency, or safety. The Collier Trophy had been awarded to such people as the Wright Brothers, Glenn Curtiss, and Glenn L. Martin although organizations such as the U.S. Air Mail Service and Pan American Airways had also won the trophy. Even a TWA insider had won not that long ago – Howard Hughes in 1938. Frye promptly advised Clarence Fleming about the details and that Fleming, who was then an assistant vice president, would be in charge of coordinating TWA's representation at the presentation which was scheduled for 29 December. Fleming was also the secretary of the company whose principal responsibility was regulatory affairs – hence his location in Washington. Frye was most pleased, and also saw the event as a great public relations opportunity which would be enhanced if the TWA representative were a pilot in uniform. Subject to the concurrence of Paul Richter (about to go on active naval duty), Lee Tallman (Paul Richter's replacement as executive vice president), and Otis Bryan, Jack Frye proposed that Blackie be the TWA representative since Frye himself would be leaving Washington shortly and remaining in Kansas City for the balance of the year, and thus not able to attend the presentation. Frye, seeing how important Washington had become to TWA's future fortunes, would eventually lease a grand suburban mansion to use as his Washington base. That was in addition to TWA's already large sales and public affairs offices in Washington. Arch rival, Pan Am, already had a strong Washington presence in the form an office building downtown.

A uniformed Blackie reported to the second floor of the Senate side of the Capitol Building at 11:30 A.M. on the twenty-ninth and was ushered into the office of Vice President of the United States Henry Wallace. Gil Rob Wilson and William P. Redding, president and executive vice president respectively of the National Aeronautics Association also waited there. Wilson had a remarkable resumé: World War I ambulance driver followed by service as a fighter pilot with the French, ordained minister, first president of the Civil Air Patrol, first member of the Aircraft Owners and Pilots Association, and one of the nation's leading aviation writers and promoters. Soon, the representatives of other airlines also arrived and were joined by the Air Force representative, Chief of Staff General Henry Arnold (Blackie's old base commander from March Field) and Colonel Edgar Gorrell of the Air Transport Association, representing the airline industry. The award was made en masse to the airlines and military which was followed by photographs of the individual airline recipients standing with Wilson, Wallace, Arnold, and Gorrell. It was a very distinguished group, but among the men present, just two enjoyed a unique distinction as the result of being named a co-winner of the 1941 Collier Trophy. The special two were General Hap Arnold and Captain Harold F. Blackburn, who were in that exclusive circle as winners of both the Mackay Trophy and the Collier Trophy. General Arnold had won the 1934 Mackay Trophy as the result of his leading the Martin B-10 bomber flight on the record setting roundtrip between Washington and Fairbanks, Alaska.

The Collier Trophy was a pleasant experience for Blackie after suffering several days of irritating downtime at the Liberator transition school in Smyrna, Tennessee. ATC wanted to be sure all pilots were instrument qualified and required Blackie to go through the blind-flying school. There were no exceptions – not even for an airline instrument check pilot, not even for an instrument instructor, not for anybody. It was a waste of Blackie's time, but ATC had him on a technicality; in his logbook, he could only scare up four or five instrument hours in as many months including his trip to India. According to the authorities that was not enough to be current. Blackie had no choice and completed the Link trainer-dominated course and received his certificate from the "Institute of the Blind" on his birthday, 25 December.

On 1 January 1943, ICD could look back at 1942 with a sense of a job well done. A year before, there had been one employee – Otis Bryan located in a small office at the Washington National terminal. Now ICD was, with over 1,500 employees (including roughly 225 pilots), a larger organization than TWA was when Blackie joined it in mid-1935. ICD's fleet had grown from its first converted Stratoliner, to five Stratoliners, a dozen C-54s with more coming, several C-87s, and some C-73s. It had flown 483 intercontinental flights over 2,449,000 miles requiring nearly 15,200 flying hours. Even though the government had claimed about half of TWA's domestic DC-3s, utilization went skyward and the Company's net profit for 1942 was $2,176,035. Although the number of flights was down, the seats filled rose to an all-time high, even with seating restrictions in effect. Top priority went to people traveling on orders from the White House or War Department in that sequence which was followed by Ferry Command pilots on their way to or from aircraft deliveries. Valuable cargo came last.

TWA had played a major role in shaking the pre-war belief that only Pan Am and British Overseas Airways could fly across the oceans with their great flying boats. Part of the progress was directly tied to how quickly crews learned the tricks of the trade and passed them along in their post-flight debriefing sessions with Blackie. For the first quarter of 1943, Blackie flew very little, having his hands full with administrative details, particularly assigning crews, briefing them, and debriefing upon their return. It was during this period that ICD started to mature; it had started as essentially a charter airline, performing special VIP missions or carrying ultra-high priority cargo. With more than a year of operations under its belt, it was evolving into a regularly scheduled airline operating on a timetable to faraway places, and at those faraway places, such as Natal, Karachi, or Prestwick, TWA began to have permanent staff, including station managers, maintenance chiefs, and mechanics. During this time, Blackie's flights were limited to a maintenance check here or a runway certification there, but that was about to change in late February when Otis Bryan was promoted to

Vice President-War Projects. That involved considerable government contact and also responsibility for ICD along with supervising several training operations and the TWA completion center in Kansas City. One of Bryan's first moves after learning of his promotion was to send a note to his closest associates, Blackie included, thanking them for their hard work and expressing the wish that they would in turn express that appreciation to their subordinates.

On 1 March 1943, Cliff Mutchler moved up to manager of ICD and Blackie became assistant manager while Swede Golien moved from chief pilot to take over Blackie's just-vacated position. Mutchler, not being a pilot, was on station performing some of Bryan's jobs when Otis was frequently away flying VIPs or conducting surveys on new routes or bases. With Mutchler essentially deskbound, Blackie had more time to fly, including one or two long foreign trips each month.

Mutchler never having significant dealings with flight operations, engineering, or maintenance simply increased the extent to which he relied on Blackie. Although TWA had a public relations arm in downtown Washington, Blackie assumed much of that responsibility for ICD, especially relations with the aviation press, and served as the ICD spokesmen whenever *American Aviation* and other magazines called. Blackie was also the point man on relations with other airlines which during wartime were much more cooperative than after the war. In the latter regard, Blackie occasionally found himself functioning as a personnel consultant when executives at other airlines put out feelers for job opportunities. At the other extreme, Blackie frequently followed up on typical wartime requests, some from Mutchler's desk, to slip in a package to a loved one overseas or personally deliver letters to foreign stations, etc. Knowing that every last item put on a C-54 was closely inspected, Blackie became the very frequent bearer of bad tidings to those making such requests. In other cases, Blackie was the one to whom fell the duty of making decisions on what items in pilots' expense reports were legitimate or not. That created the occasional difficult moment for him because in quite a few cases, he had been a co-pilot under the man submitting the reimbursement requests.

In effect, Mutchler and Blackburn became Mister Inside and Mister Outside with Mutchler minding the store while Blackie increased his activities outside of ICD, particularly his involvement with ATC and the other carriers with ATC contracts. Blackie was the one who represented TWA in ATC Contract Carrier Committee meetings. It was through that committee that the airlines helped ease some of ATC's growing pains, especially through mid-1943 before ATC gained critical mass.

In June 1942, the Air Transport Command field organization was restructured into wings which included the North Atlantic, Caribbean, North Atlantic, South Atlantic, Pacific, India-China, and Africa-Middle East Wings (AMEW). It was in the latter two wings, being stretched thin and far down on the Air Force food chain, where friction between the airlines and the military was most pronounced in 1943. Blackie, along with his counterparts from Pan Am, American, and (later in the process) Eastern, was charged with smoothing out operating procedures. The airlines' requests were quite reasonable and among them were the need for AMEW to provide fuel services, mechanics where possible, aircraft cleaning, office space, improved lodging and messing, inventorying parts, equipping repair facilities with appropriate tools and machinery, and maintaining adequate fire/rescue equipment among other basic services. For their part, the airlines accepted responsibility for conducting as much maintenance as possible at their headquarters fields (Miami for Pan Am and Eastern, New York for American) and at Accra so as to minimize demands at fields further to the east. The airlines also pledged to staff locations within AMEW with sufficient supervisory personnel to ensure the greatest efficiency possible.

Blackie and his counterparts in the Carrier Committee were not able to resolve all of their issues with ATC immediately in the summer of 1943. That took time, but Blackie's involvement in the process was excellent preparation for his future and certainly indicative of the confidence Otis Bryan and Jack Frye had in him. As it turned out, problems in the AMEW and India-China Wings became less of an issue for TWA and more of a problem for Pan Am because ATC increasingly used TWA and American for its North Atlantic and European service.

Although Swede Golien was an old hand, Blackie had seen more of the ICD operation at that point and corrected the rumors and grumbles that had been floating around in the previous months. Some of the problems Blackie mentioned were minor procedural matters that were easily corrected. Winter flying suits were always left in bags within the planes. Now and then, some items would be missing. The problem was solved by removing the bags from the planes, checking to make sure all items were there, and then placing them in planes before flights. Oddly enough, there were no approach plates for LaGuardia in pilots' flight kits. LaGuardia and a number of other fields were legitimate alternates. Also some planes were flying with outdated *Air Navigation Radio Aids* brochures. Other problems required better communication – such as ground crews getting one fuel load order from a captain and a different figure from the flight engineer. In some cases at Meeks Field (the new airfield on Iceland), ground crews took it upon themselves to load an extra 1,200 pounds of cargo, mistakenly assuming the fuel load was below capacity. Some station managers felt they got better company information from arriving flight crews than they did from the company. Blackie promptly launched an informal monthly news bulletin that also included useful items such as tax and selective service developments, i.e., useful items not normally covered in *Skyliner*, the monthly company newspaper.

Blackie did observe a few cases of what he considered improper aircraft handling procedures. Nobody was surprised when the few captains Blackie had observed were promptly given thorough line checks. ICD's chief engineer also got instructions to bear down on a few of his errant charges. Blackburn's reputation never came close to that of a martinet, but he was a stickler for proper procedures. In one area he relaxed – but only slightly. That was in the matter of proper uniforming. ATC had given contractors specific directions as to wearing uniforms which were essentially Army uniforms with the only difference being the understated look when it came to insignia. Like any Army pilot, ICD pilots, first officers, and navigators wore regular Army caps, usually with the stiffening material removed, i.e., the "fifty mission crush cap." There was a reason for removing the stiffener because it interfered with wearing a radio headset. Still, some ICD officers went overboard and wore the caps practically on the backs of their heads. Strictly speaking, Natal boots, or "Brazilian boots" as some people called them, were not regulation footwear. One reason Blackie was not overly critical of other pilots wearing Natal boots was his desire not to be seen as a hypocrite. His footwear was hardly the brown oxfords the regulations required. Blackie had acquired a pair of Gaucho boots, essentially cowboy boots with accordion-like flexible tops, which were good at keeping mosquitos away in the tropics while offering some protection at altitude where fleece-lined flying boots were not readily at hand. Blackie was willing to look the other way unless the offender was wearing his pants legs tucked into the boots or his pants were simply too short. Because most of the other contract carriers such as United or American were wearing the dark green shirts instead of the summer shirt, Blackie saw no reason to make an example of ICD crews.

There was naturally some grumbling about accommodations, but this was an area where Blackie had little control, other than to answer that things were improving which was true. He observed that the new barracks in Iceland were far superior to what he had experienced as an aviation cadet. Some ICD men were still housed in hotels and with private families in Prestwick, but new accommodations were being built. Scottish Aviation, Ltd. had the food/hotel concession at Prestwick and frequently had its hands very full, considering it was not only ATC using the airfield. It was the entry point for every bomber group flying across the North Atlantic from the States, so it naturally got crowded. The crowding was not limited to Prestwick; Natal and Accra could occasionally have as many as 200 airplanes parked around their fields.

Blackie always believed that improved communications and personal relations could solve most problems, and he put that conclusion to work by approving job relations training for all captains. He also suggested that Swede set up an informal monthly meeting of captains, or at least the ones who were available, to get things out into the open and nip any problems in the bud.

At ATC bases, ICD people could eat at the officers or enlisted messes. Granted, there were places, such as the Azores, for which the only mess facilities were British which usually meant unimaginative food. The Azores, as part of Portugal, had its compensations; the tasty Portuguese version of champagne was readily available as were many high-quality, attractively priced lace goods.

Blackie had to settle for a C-87 Liberator on his next South Atlantic crossing which was compensated for by his flying into northwest Africa. This time, he had more time to spend in Natal which he had just breezed through on his last Latin trip. He arrived in Natal on 20 March and was not scheduled to leave until 24 March, allowing him a few days to get acquainted with Brazil. Pamamirim Airfield was virtually undistinguishable from any air base in the American South – hot, humid, constant traffic from both aircraft and Jeeps, clouds of dust, rows of parked aircraft. The officers club and PX were not as good as most in the States, but there was ice, ice cream, and the mess hall food was rated as good. Many of the transient bomber crews on their way to China or North Africa got a tent for the night, but ATC crews had the use of the Staff House, a grand hacienda on the edge of Natal, a medium-sized city. With its thick walls and shade trees, it was cooler than the barracks back at Pamamirim to which a shuttle bus ran several times a day for meals at the officers' mess hall. Blackie toured downtown Natal where most pilots bought a pair of Natal boots, cowboy boots of sorts only lower and with a rounded toe. Made from buffalo, the boots were billed as mosquito boots if they were not simply called Natal boots. Tens of thousands of them were sold and they could be seen on Army airfieds everywhere from East Anglia to the Marianas. What he did find near Natal were the white sand beaches and dunes at Genipabu, just north of Natal. Blackie, always a strong swimmer, enjoyed the pristine beaches and surf long before they became trendy tourist attractions. The next few days saw him repeating his first South Atlantic crossing by flying to Ascension and then on to Accra, Rufisque, Dakar, and Marrakesh.

Chapter Seven
The Flying Manager

Increasingly, more of the ICD staff was pulling temporary duty at various ATC bases around the world – six weeks duty abroad was the standard and it applied to aircrews and ground crews as well. Eventually the overseas tours were extended well beyond six weeks at a time. ICD's explosive growth caused major changes in the support structure, and resulted in some operations being decentralized. Frank Busch left Washington to take over the training facility in Kansas City where Hugh Davis from Eagle Nest again was in charge of Link simulator training. Busch assumed he would be back in Kansas City for the long haul and therefore bought a small farm, raising cattle and growing corn. It would prove to be a faulty assumption. TWA acquired a Lockheed Hudson and modified it with the addition of several astrodomes for navigation training in Kansas City. During that first quarter, ICD flew 210 international flights (compared to 483 for the first full year). The creation of both Eagle Nest and ICD created some union issues which were eventually resolved by John Collings, TWA VP Operations, and Dave Behncke of the Air Line Pilots Association. The pilot lists from the TWA domestic and new hires going to domestic and ICD were not so amicably merged. However, they included ways to resolve conflicts for men qualifying on the same day (previous airline seniority date followed by age instead of the previous reliance on the alphabet). All told, TWA pilots numbered 728, including 101 on active duty in the military. Hiring would be largely static in 1943 at ICD, but would pick up again in early 1944 when the C-54 fleet grew larger. The rewards of becoming an ICD pilot, including a few who originally, came from outside of TWA caused considerable weeping, wailing, and gnashing of teeth among TWA domestic pilots. ICD pilots were paid more and would get first crack at TWA's desirable, post-war international routes. Domestic pilots had to concede the fact that they were not flying gas-leaking C-87s, getting shot at, flying through some of the most atrocious weather in the world, encountering some of the world's highest mountains, and being gone for several weeks at a time unless they were on vacation.

Saturday, 9 January 1943 was a busy day for the Army Air Force all over the world, except in England. The only activity there that day was a patrol along the French coast by Spitfire Mark Vs from the Eighth Air Force's Fourth Fighter Group which until the previous September had been better known as the American Eagle Squadrons. Blackie's Eleventh Bomb Squadron had been re-assigned from the Seventh Bomb Group to the 341st which that day was busy bombing Japanese targets in Burma. In North Africa, medium bombers were active against ground targets in Tunisia and shipping off the coast. New bomb groups were starting to arrive in Algeria and making their way eastward closer to the front. Although most of the bombers arrived via the South Atlantic route, others reached Africa from England on a perilous route that Blackie would fly in the near future. In the Southwest Pacific, medium and heavy bomb groups struck Japanese airfields on New Guinea and shipping off its coast. For most TWA observers, that war news was paled by the first flight of an airplane that would come as close to being a Blackie trademark as any other: the Lockheed Constellation.

The plane had been built under great secrecy in Lockheed's experimental engineering shop and pushed out the door in early November for various ground tests. Because the Connie project had been conscripted by the Air Force, the first "L-49" Constellation appeared in standard Army camouflage: olive drab and gray. It was decided that an outsider should be the first pilot to fly the Connie based on his being the only man experienced operating the powerful Wright R-3350 engines. He was Eddie Allen, Boeing's chief test pilot, and he had gotten the big engine experience on the Boeing B-29 which had first flown the previous September. There were a lot of eggs in the basket, or cockpit, for the inaugural flight. In addition to Allen, there were Lockheed's chief engineering test pilot, Milo Burcham, the plane's principal designer, Kelly Johnson, and two Lockheed engineers, Dick Stanton and Rudy Thoren. The plane was unique if not revolutionary. True, it was not the only pressurized airliner – the Boeing 307 was also pressurized, but the Connie had hydraulically assisted controls, a dedicated station for the flight engineer, huge Fowler flaps, and a wing design that closely resembled that of the P-38 Lightning. The shape, like a gracefully streamlined, leaping porpoise was capped off by the distinctive tri-tail. TWA was not going to build new hangars for the plane's maintenance, so to avoid one or two very tall tail surfaces, the engineers came up with three that would comfortably fit within existing hangers. The big engines came with big propellers – with such an arc that the plane required extra tall landing gears which is why the nose of the Connie swept down to house the tall nose gear. The fuselage just oozed speed, but its shark-like shape was also narrow in the front and tail sections which prompted its few critics to comment on how many more passengers the plane could have carried with a more conventional fuselage.

After some fast taxi trials, Eddie Allen took her up for fifty-eight minutes and landed the Connie at Muroc where it was flown five more times that day. There were practically no squawks in the test log so Eddie Allen turned the project back to Lockheed's pilots. There were two tragic postscripts to the maiden flight; within six weeks, Eddie Allen was dead. He was the victim of a B-29 in-flight engine fire which was hardly unusual on the first Superfortresses. Milo Burcham never lived to see 1944; he was killed in the crash of the second prototype of the P-80 Shooting Star jet fighter late in 1943. The engine fire problems drastically slowed down development of the Connie, particularly as the Army grounded the plane for several months. It would take a while, but in less than two years, Blackie would get his hands on a Connie, and it would become the plane with which most people always associated him.

For ICD, 1943 opened with more Blackburn-planned VIP flights, beginning on 13 January when Otis Bryan landed at Bathurst in Gambia in a C-54 and was told to await immediately pending orders. The paperwork arrived and so did the passengers, President Roosevelt, Harry Hopkins, and other VIPs. Bryan flew them to the Casablanca Conference in French Morocco the next day and back to Bathurst via Monrovia, in Liberia, on the twenty-fifth so Roosevelt could inspect the rubber plantations that were so valuable to the Allied war effort. Because the president was the most very important of all VIPs, Bryan had a back-up C-54 piloted by Don Terry just in case a spare was needed. Pan Am flew the Roosevelt contingent from Bathurst back to Natal on one of their Clippers which was deemed safer for the president than crossing the ocean in a C-54. The conference itself was basically held for Roosevelt and Churchill to hammer out policy for 1943. General Arnold wanted to invade France in 1943, but Churchill's recommendation prevailed. The Allies were not strong enough yet in Churchill's view; he favored a thrust into Italy via Sicily since it was clear that Rommel's days were coming to a close in Africa (which they did in May). Churchill's view was accepted, but he was thunderstruck at Roosevelt's proclamation that the Allies would accept nothing but unconditional surrender from the Axis. The wily Churchill was thinking far ahead; his quarrel was with the Nazi leadership, not the average German citizen or infantrymen. Churchill could already see the advantage of the German Army as an ally in a war with Russia if the latter tried to grab more land beyond recovering its own lost territory. Churchill feared the unconditional surrender policy would simply stiffen German resistance, making present matters more difficult. The leaders agreed that the Pacific War would be the second priority after Germany and that support for Russia in terms of planes, tanks, and supplies should be ramped up. The conference roster read like a who's who of the American and British military. The French, in the persons of Generals De Gaulle and Giraud (who detested each other) were invited, but not as full participants. Larry Trimble flew De Gaulle down to the conference from the UK and took the standard precaution of flying well to the west to avoid the *Luftwaffe's* prowling JU-88 interceptors. As the flight was made at night, Trimble ordered all cabin lights turned off. De Gaulle left his light on and his window curtain up despite a personal request from Trimble to lower the curtain and turn the light off. Trimble sent his first officer back, and De Gaulle refused to douse the light, saying "Young man, I have faced the Germans before and am not afraid to face them again." When the first officer came back to the flight deck with that Gallic news, Trimble simply switched off the main cabin electrical circuit. Trimble had a delightful VIP passenger on the flight back to England: Winston Churchill, who came up to the flight deck and asked if he could fly the C-54. Trimble agreed, not knowing that Churchill had qualified as a pilot years before. In fact, Churchill was the one who brought aviation to the Royal Navy and for a brief period in the interwar period sometimes flew with one of the RAF's auxiliary units, 615 "County of Surrey" Squadron. Churchill was occasionally uniformed as a marshal in the RAF, pilot wings included, at wartime conferences, although Casablanca was not one of those occasions.

In mid-January, Blackie dispatched an accident investigation crew to Paramararibo, Dutch Guiana, to determine the cause of a C-54 loss which had occurred on the fifteenth. Swede Golien led the search while Blackie pulled the maintenance log on the plane, serial number 42-32939, and questioned all mechanics who had worked on it. Because the crew of nine and twenty-six passengers had all been killed, it became the second worst flying accident in U.S. history to that date. Even Blackie's old charter client, the FBI, was brought in to investigate. Local fishermen had seen the plane flying, watched it make a 180, and then witnessed a massive explosion and the plane's falling to earth. There had been no radio report of trouble from the crew and the weather was perfect at the time of the accident. The sixteen-page report that ICD submitted to ATC strongly suggested sabotage since everything appeared normal and the maintenance check turned nothing up, although the fuel, oil, and hydraulic lines, and the exhaust systems on the first C-54s did have some problems. The accident was yet another personal loss for Blackie because the captain, Ben Dally, was a very good friend who had been in Blackie's TWA co-pilot class in April 1935. Dally was a graduate of Missouri's Westminster College (where Winston Churchill would give his "Iron Curtain" speech in just a few years) and was occasionally known as "Three Cs," a reference to his TWA career which progressed from courier to co-pilot to captain. Like Blackie, Dally had gotten his wings and commission at Kelly Field.

Dally had recently transitioned to the C-54, had over fifty hours on it, had total flying time of almost 8,000 hours which was only slightly more than Blackie's 7,592 hours. Nothing specific could be identified as the source of the explosion from the surviving wreckage, so ICD's report listed the only three possible, potential causes of the accident: structural failure related to the explosion, sabotage, or, remotely, enemy submarine fire. Dally's Skymaster was carrying some high priority cargo: secret documents intended for British General Harold Alexander and a small fortune in American money. Dally's C-54 had a history of problems with the number four engine and had generated enough other squawks to force a thorough, second pre-flight before takeoff. There were IFF problems and part of a re-fueling hose extension had broken off and was floating around in one of the fuel tanks.

Because President Roosevelt was traveling to the Casablanca Conference at the same time, scuttlebutt had it that the Germans had tried to shoot down any aircraft that might be carrying the president. There were rumors, unconfirmed, that a German submarine had been shooting at Dally's Skymaster which was one of three Skymasters acting as decoys while Roosevelt was actually crossing the South Atlantic in a Pan Am flying boat. Although Dally's name was certainly not widely recognized, he had a well-known person on board: Major Eric Knight, author of *Lassie Come Home*. Knight, originally from Yorkshire, had fought in the Canadian Light Infantry in the First World War before becoming an American citizen. He also was the author of *This Above All* and a co-writer of *Why We Fight*, the stirring, eight-part wartime film directed by Frank Capra.

Robert D. Smith, a young co-pilot from Kansas City, was scheduled to return to his assignment in Natal on the Dally flight, but was caught in a dilemma which had been initiated by the rumor mill suggesting that the Army was about to completely take over ICD. Even stronger rumors were flying about Air Transport Command taking over one or more of Pan Am's wings. Smith had a tentative offer from Eastern Airlines which seemed not to be an ATC takeover target, but was reluctant to leave TWA, the airline he had essentially watched grow up as he himself grew up in Kansas City. Smith came to Blackie and requested a transfer to TWA Domestic. Blackie, seeing that the young man was really torn between the safety he saw at Eastern and his loyalty to the hometown airline, phoned John Colllings and tried to work out a transfer, but Collings was not hiring. The Eastern offer could only be made firm if Smith obtained a letter from TWA saying that the parting was amicable. Blackie could have refused, especially considering how difficult it was to find good co-pilots, but signed the release anyway. Smith was eternally grateful, more for not being on the Dally flight than for the ticket to a new job at Eastern.

As with the Zimmerman service just two months before, Blackie and Swede Golien took one of the C-73s along with a number of other ICD pilots and left Washington shortly after midnight to attend Dally's memorial service which was held at the West Presbyterian Church in St. Louis on 21 February. They flew back the same day. Dally left behind a widow and infant daughter. In 1948 the remains of the crew and passengers were returned to Missouri where they were all interred in one grave in the Jefferson Barracks National Cemetery in Saint Louis. Dally's widow, years later, confided in her young daughter how much she appreciated Blackie's sympathy and help during the family's nightmare. It was a progression: Greeson, Redding, Zimmerman, Dally – and when the war was over, the list would be much longer.

On 15 January, the same day as the Dally loss, another of Blackie's cadet buddies, Harvey Regan, was in serious trouble. Blackie heard about it several days later; it was a flying accident so typical of the many stateside crack-ups that claimed thousands of lives during the war years and illustrates the type of danger Blackie faced every time he took off. During America's direct participation in World War II, nearly 14,000 aircrewmen were killed in flying accidents within the continental United States, and of that number 2,796 were killed in B-24 Liberator crashes. It was not Regan's first brush with death. Just four years before in 1939, he and another acquaintance of Blackie's, Roland Sansbury, had survived a bad accident while flying an OA-9 (Army version of the twin-engine Grumman Goose amphibian) out of Washington's Bolling Field. Sansbury had just returned from China as an instructor with the Chinese Air Force and had been reinstated in the Reserve along with Regan who was the engineering officer of the Twenty-fourth Staff Squadron at Bolling. They lost power in their left engine and made a normal water landing on the Potomac River, but hit a submerged object which cart-wheeled the Goose. Both pilots got out with no injuries, but Regan was not so lucky on his 15 January 1943 mishap when he was riding in the jump seat on a 504th Squadron B-24 when it crashed on takeoff at Salina Army Air Field in Kansas. Regan, then a lieutenant colonel, was on his way to West Palm Beach in Florida as part of the 308th Bomb Group's overseas movement to China when the accident occurred. The pilot, Captain Frank Beales, was criticized but not officially blamed for the crash which killed five of the ten men on board when the Liberator stalled after takeoff. Regan got out with third degree burns to his face and hands, putting him out of commission for several months. It appeared that some of the Liberator's load, namely a lot of personal equipment, duffel bags, and B-4 bags had shifted during takeoff and fouled the control cables of the normally tail-heavy Liberator. The accident probably saved Regan's life; he likely would have been killed in China since the 308th suffered the heaviest losses of any bomb group in China. Fate did not favor Beales either who was killed on 31 December 1944 in Puerto Rico while landing a B-29 Superfortress with engine problems just before shipping out to the Pacific with the 502nd Heavy Bomb Group.

Dally's loss and Harvey Regan's narrow escape was followed three days later by another accident involving an ICD C-87, although it had been flown by Army ATC personnel, not TWA, which exempted Blackie from forming the accident investigation board. The C-87 disappeared on the night of 17/18 January over the South Atlantic; there had been no distress calls made on the radio which promptly invited comparison with Nick Wasil's loss. On 10 February, a life raft containing a decomposed body and the personal effects of six other men was found, and a week later two more rafts, each with one body, were discovered on the shore near Natal and Recife. Nothing could be pinpointed as the cause of the accident, but much was made about a lack of cooperation from the Navy which ATC felt could have been of greater service since the Liberator's route was well known and it was assumed that the men would have still been alive as late as 2 February.

With two C-87s lost, TWA and Pan Am ATC crews formed an ad hoc committee in Natal to lobby for some changes on C-87s that would make the Liberators safer. Pilots had been complaining about the C-87s from the day they first flew them, but their complaints fell on deaf ears because maintenance departments said they did not have the parts, nor did they have authorization from the Army to make in-the-field modifications. American Airlines was also flying the C-87, had the materials to make modifications, and blackmailed the Army into approving the changes by grounding all of its C-87s until the approval was issued. When American offered to share its solutions with any C-87 operator,

TWA and Pan Am put together a high level committee of captains headed by Larry Trimble from TWA and Pan Am's Captain Gordon to clearly elucidate what fixes were necessary. The original copy of their report was sent to Blackie with a copy to Otis Bryan. The recommended changes were a lot more than could be completed on an overnight stop, i.e., every known squawk was written up. The fuel system fixes included improving the transfer plumbing system, re-designed gas caps that would not siphon and allow fumes to circulate within the wing, greater capacity electric fuel pumps that were really up to the task if an engine-driven pump failed, relocating cabin gas valves to a safer location, installation of proper fuel flow meters and quantity indicators that were accurate and rugged, reworked interconnecting gas lines between the leakproof tanks so fuel flowed properly, ruggedization of exposed fuel lines in the cargo section or some sort of protection from shifting cargo, new vents in the wings to prevent build-up of gas fumes, and relocation of radio transmitters as far away from the fuel system as possible. Even with that, the C-87s had inherent problems for which there were absolutely no fixes, including its inborn, wayward lateral instability. When loaded, the plane was a poor climber which had pilots looking nervously at any hills, even low ones, beyond the end of the runway. The C-87's nose wheel was identical to that on the B-24; it rarely collapsed on the bomber version in a hard landing because it was designed with the idea in mind that the Liberator would land minus its four-ton bomb load which had been dropped on some enemy target. The C-87 Express, with the same anemic nose gear, took off loaded and landed with the same load, thus increasing the possibility of a badly stubbed toe if not landed on the main gear first.

The recommendations continued: installation of decent firewalls behind the engines, upgraded fire extinguisher capacity in the engine compartments, installation of an auxiliary cockpit lighting system to at least provide a back-up for the current fluorescent system which forced many captains to make night approaches using a flashlight to illuminate a darkened instrument panel, and replacement of the landing lights which could not stand up to the Liberator's typical approach speed. And finally, one which Blackie could appreciate considering his long afternoon over Albuquerque with Bob Redding: improving the emergency hand pump for gear extension. Consolidated Aircraft did incorporate many of these fixes in later models of the Liberator, but they did not come fast enough for those flying the C-87s. The best fix came on 17 June 1943 when TWA returned its last C-87 to the Army.

The loss rates of ICD and ATC flying C-87 Liberators and Skymasters never approached that of Eighth Air Force operational heavy bomber crews based in England in 1943 which were pretty discouraging – only about a third of the crews could expect to complete a tour of twenty-five missions. TWA suffered only ten accidents from late 1942 through early 1945. TWA's flying was of course free from the hazards of facing talented ME-109 or FW-190 pilots and legions of *Luftwaffe* 88 mm flak gunners. Still, flying the freight and passengers was not a ticket to an old age home.

At about the time the Dally accident report was submitted, Blackie's inner fighter pilot persona surfaced; he bought a used, fire-engine red Indian Motorcycle with a long saddle to satisfy his need for speed. Indians did not have the prestige of their Harley-Davidson competitors, but they had plenty of roar and fun when Blackie took Martha and Betty or Bob out for a spin. Martha soon disapproved but Blackie countered by saying the machine was the perfect response to wartime gas rationing. So, the Indian was the means by which Blackie commuted to work at Washington National. To complete the picture, Blackie, when riding the Indian, always wore a pair of cowboy boots he had acquired in Albuquerque. Although Martha preferred not to be seen on the contrivance, she too had a matching a pair of cowboy boots. What the Blackburn family did not have, but the Milo Campbell family did have, was a pair of Indians. Mrs. Milo Campbell was an enthusiastic motorcyclist in her own right, but riding on Blackie's Indian, not driving one, was about as far as Martha's interest in two-wheeled vehicles went.

There was another informal change in Blackie's uniforming in early March. His application for membership in the RAF Ferry Command's Trans Oceanic Pilot's Association was approved, and he was thus qualified to wear that organization's wings, although not on his ATC uniform. Milo Campbell, who had joined earlier, occasionally did wear his TOPA wings on his ATC uniform causing some pilots in ICD to wonder just what kind of fearless flyer Campbell really was. As a result of his North Atlantic flights and frequent stops in Iceland, Blackie became a member in another exclusive wartime society, The Brotherhood of the F.B.I. This had absolutely nothing to do with Blackie's flights with Jack Zimmerman and the Federal Bureau of Investigation charters. Instead, Blackie's new "FBI" was the Forgotten Bastards of Iceland, a fraternity of Army and Navy men whose misfortune it was to be stationed, even on a temporary basis, in Iceland. His certificate was endorsed by the "Great Cod" and the "Grand Eider."

Blackie scheduled himself for another South Atlantic crossing on 18 March although the trip would take a sharp turn compared to his first flight to Karachi and would last three weeks. There were not that many passengers, but those who did climb aboard Blackie's C-87 were fairly high level: it was an inspection team from ATC. The flight would also show him first-hand what a nomadic life style could await a pilot flying for ATC once he left the American mainland. The start was similar – to Miami, then to Trinidad, with a layover at Atkinson Field in British Guiana. On the next day, he flew the C-87 down to Natal, but lingered there while the inspectors conducted their survey. The inspection over the course of the trip was not limited to TWA, but also included operations of Pan Am and Eastern. Finally, on the twenty-fifth Blackie flew the group to Ascension Island and then on to Accra the next day. Here the flight departed from his previous landing in Accra since his plane sat on the ground for five days while the inspectors inspected and some minor maintenance issues were addressed.

It was not unusual for ATC maintenance crews to pitch in and help work on the Army bombers that passed through ATC bases – and vice versa. In early 1943, ICD had not yet reached manpower levels to correct all problems on an overnight stop, but that was the goal, and it was made much more difficult under brutal operating conditions ranging from the constant freezing winds of Greenland and Iceland to the sweltering temperatures and brutal sun in Africa. Much of the work was done by flashlight and reflectors out in the open. In Accra, by September 1942, a Stratoliner arrived from Brazil daily around midnight. If a pilot or flight engineer complained about an engine not turning up right, mechanics had little choice but to run a compression test right on the ramp to identify a problem cylinder. If pounding on the top of the valve stem did not solve a sticky valve problem, mechanics had to drop the valve for reaming which was a delicate procedure even in the shop back at Washington National. If that failed, there was the time-consuming task of replacing the cylinder itself – and getting the job done so the Stratoliner could take off for Natal in the morning. As soon as possible, TWA started maintaining a small stock of replacement cylinders on the line at each of its foreign bases whereas earlier in the game the mechanics had to wait for a replacement to be flown in which could take days.

When Blackie finally got out of Accra on 31 March, his heading after takeoff was not westbound; rather, it was northbound. He was taking his ATC inspectors to Oran, Algeria, along what became known as the North African ATC route. His first stop on the new route was in Dakar which was reached after nine hours of flying, including one hour on instruments. Dakar, in what was French West Africa and now is Senegal, had been a major base of the Vichy French which had fought a combined Free French and Royal Navy naval force to a standstill in September 1940. With the successful Allied landings in Algeria in November 1942, the Vichy forces in Dakar changed their tune and opened their port and airfield to the Allies. Unfortunately, Blackie had only an overnight stop there and had no time for wandering through the old quarter's shops or relaxing on Cape Vert's beaches. He arrived in Oran on 2 April after having spent the previous night in Marrakesh, another city for which he had insufficient time to explore old city walls, mosques, palaces, and stunning views of the nearby Atlas Mountains. There would be another time for Blackie's introduction to Marrakesh, one of Winston Churchill's favorite places on earth. Churchill, who first visited Marrakesh in 1935, was so taken by the city and its Hotel La Mamounia, that he insisted on hosting President Roosevelt there during a side trip from the Casablanca conference. Oran, in Morroco, Blackie's final destination, had been Vichy French as Dakar and Marrakesh had, but had been occupied as part of Operation Torch in November 1942. And as with Dakar and Marrakesh, Blackie had no time to spend as a tourist in Oran which he left on the third to fly back to

Natal, not via Ascension but as a direct hop from Dakar. Arriving in Natal, Blackie turned the C-87 over to another captain and spent several days shuttling between Natal and Recife in a C-47. Before turning over the C-87, Blackie and his passengers posed for a picture beside their C-87 which was none other than "Old Nosey," the same C-87 that had served as a South Atlantic pathfinder in 1942. In a juxta-position so typical of some World War II aviation photography as best popularized by the photo of another C-87 in Africa surrounded by a camel caravan, there was a monkey running from wingtip to wingtip on top of Old Nosey when the photo was taken. Blackie finally took off for Washington as the pilot of a C-54 on 12 April to some momentous news.

Martha said she thought she was pregnant which was very quickly confirmed. With a baby due in mid-November and Blackie frequently being overseas for two or three weeks at a time, the Blackburns decided that Martha and the kids would take up temporary residence with the Bondurants back in San Antonio where the youngest Blackburn child, Bonnie, had been born in 1941.

Trans World Airlines

Blackie, second from left, was the pilot of the C-87 Liberator Express "Old Nosey" on the Air Transport Command Inspection trip to Northwest African bases in April 1943.

On 10 April 1943, Larry Fritz re-opened the North Atlantic route which had been closed due to weather issues. Blackie scheduled himself for what amounted to his monthly trip and flew a C-54 to Presque Isle on 12 May and then to Goose Bay the following day. Departing on an ATC trip overseas was not as simple as showing up today at the gate with a passport, boarding pass, and some carry-on. After a passenger presented his orders, there might be an impromptu physical exam if one had not been performed recently and inoculations, if expired, were brought up to date. There were intelligence questions and lectures, personal questions including the state of financial affairs and wills, equipment checks, baggage weigh-in, and safety lectures relating to emergency procedures and equipment. ICD pilots received very little formal training in survival or conduct if captured other than a few sentences in the War Department survival booklet with the introduction by Eddie Rickenbacker.

On the fourteenth Blackie stopped twice in Iceland: once at Reykjavik as he had previously and also at Meeks Field, which would become ATC's main terminal for Iceland. He pushed it on to Prestwick where he landed after five hours of flying and would not return until 25 May. Although Blackie was able to see more of Scotland and explore his ancestry including trips to Glasgow, the village of Blackburn, and Edinburgh, his ten-day sojourn was also filled with interfacing with the management of TWA's operation in Prestwick which had changed substantially since his first visit there in the summer of '42. Prestwick had become a TWA station not unlike Philadelphia or Amarillo with ground crews, a dispatcher, station manager, and nearly everything else that would soon represent regular schedules with the obvious exception of ticket sales. ICD was growing rapidly. When Blackie first flew the Atlantic in July '42, ICD had by then completed seventy-three ocean flights. And just for the month of May 1943, ocean flights totaled 172. As the assistant division manager, it was Blackie's business to become intimately familiar with all of TWA's far flung operations. Those operations included ground and administrative crews that numbered forty-two at Natal, thirty-five at Accra, and twenty-nine at Prestwick at the end of 1942.

On the twenty-fifth, Blackie touched down in Marrakesh after a twelve hour flight through airspace that literally bordered on the hostile. Flying direct a direct course from England was out of the question because such a route would take Blackie much too close to France and over neutral Spain. The *Luftwaffe*'s Focke-Wulf Condors and JU-88s operated well out into the Bay of Biscay, so the only safe way to Gibraltar, Morocco, or Algeria was to avoid airspace east of the fifteenth meridian west which was assumed to be about the limit of a JU-88's operating radius. To put a somewhat finer point on that geography, Prestwick is about five degrees west (London is zero) and the tenth degree west meridian cuts right down the west coast of Ireland and just off the coast of Portugal. The last German aircraft to crash land in Ireland was a Focke-Wulf Condor in December 1943, and it had been patrolling off the west coast of Ireland.

The German unit creating the danger was V/KG40; that is the Fifth Group of Kampfgeschwader 40 (*Luftwaffe*'s Fortieth Bomber Wing). The name of the unit implies that it was a bomber wing, and it did in fact start operations with a heavy bomber, the FW-200 Condor which was used for long range maritime patrol, particularly in support of German submarines sailing to and from ports on the French coast. When the Royal Air Force's sub killers, the Short Sunderland, Consolidated Liberator, and Consolidated Catalina, started sinking German

U-boats, the JU-88 was pressed into service to minimize U-boat vulnerability to Allied air attack. The JU-88 was a formidable opponent. During the Battle of Britain, Spitfire pilots had their hands full keeping up while closing in for the kill when a JU-88 stuffed its nose down, fire-walled the throttles, and ran for it. Not only was the maritime version speedy, it was very heavily armed. In addition to combating the Allied sub killers, the *Luftwaffe* discovered another profitable activity: picking off Allied aircraft flying between Southwestern England and North Africa. There was considerable medium and heavy bomber traffic flying from England on its way to North Africa and other bombers on their way to England after having crossed the South Atlantic. Just a few days before Blackie safely flew to Marrakesh from Prestwick, The Royal Air Force lost a Bristol Blenheim light bomber to KG 40. And within days of Blackie landing in Marrakesh, the RAF lost a Liberator bomber to another JU-88. June 1943 proved to be a big month for KG 40 which destroyed fourteen Allied aircraft ranging from one C-47 to several B-17s and B-24s along with more RAF heavies.

Probably the most famous casualty attributed to KG 40 was the film star, Leslie Howard, who was killed on 1 June 1943 when the C-47 in which he was flying was shot down by a JU-88. He was best known for his portrayal of Ashley Wilkes in *Gone with the Wind*, but had become a major director, actor, and producer of British propaganda films. Of much more value to the Germans was the loss of two other fellow passengers in particular. They were both major aces and veterans of the Battle of Britain: Wing Commander Howard "Billy" Burton and Squadron Leader Osgood Hanbury. When Blackie made his first flight from Britain to North Africa, KG 40 was at its peak operating effectiveness. By late summer it, KG 40 the hunter, had very definitely become the hunted, particularly as the Royal Air Force committed more Bristol Beaufighters and DeHavilland Mosquitos to the battles above the Bay of Biscay. By the time of the Normandy invasion, KG 40 had been effectively annihilated; its few surviving aircrews transferred to other units, particularly night fighter squadrons.

Although staying safely to the west of the JU-88's operating radius sounds simple, ICD Skymasters were not equipped with modern navigation systems that would make child's play of such a simple restriction. There were no radio ranges at sea, thus the only dependable means of navigation was astral which was not a problem unless visibility was poor which it happened to be for about a quarter of Blackie's first night flight to Marrakesh. Bad visibility and a stronger than forecast easterly-blowing wind could spell disaster particularly in light of the fact that, some of KG 40's aircraft were radar-equipped. Fortunately, on the very few occasions when ICD pilots identified JU-88s in the distance, there were also significant cloud formations nearby which made good hiding places for westerly escapes. Compounding the danger was the fact that many of KG 40's attacks were carried out, not by a single JU-88, but by a flight consisting of a half dozen or more of them. Unrestricted visibility was wonderful for shooting the sun and stars, but a lack of clouds denied ICD pilots of good places to hide had they been fortunate enough to see the Jerries first instead of vice versa. Fortunately this was an uneventful first passage to North Africa, and Blackie had a full day to explore Marrakesh and its Place Djemaa El-Fna, the old market square where he saw everything from snake charmers to countless food stalls offering fresh-squeezed orange juice, tasty kebabs, and mint tea. On 29 May, Blackie was back in Washington after having returned from Marrakesh via Dakar and Belem. The summer and autumn of 1943 proved to be very lackluster for Blackie in terms of flying; he only flew several foreign flights between August and the end of the year, all to North Africa via the North Atlantic route and the western side of the Bay of Biscay. Fortune smiled on those passages to North Africa; he never saw a German in the air. The summer and early autumn months represented the high water mark of *Kampfgeschwader* 40. That period was very quiet for Blackie whose only noteworthy flight was on 30 January when he flew back to Washington from what was a new transatlantic waypoint for him: Bermuda.

On 15 July 1943, Blackie's day passed quietly at ICD headquarters in Washington. Across world battlefronts, much was going on with the exception of the Eighth Air Force in Britain from which no missions were flown that day. Most of the action was in Sicily where American medium bombers supported Allied infantry by striking many tactical targets in central and eastern Sicily. American ground forces captured key port cities at Empedocle and Agrigento. On mainland Italy, Naples and the airfield complex around Foggia on the Adriatic Coast came in for more heavy bombing by both American and British heavies. The Allies lost only one ship that day to Nazi submarines – a British freighter east of Madagascar. The Germans paid dearly for that one victory, losing one U-boat in the Caribbean Sea and four more in the Atlantic. America was truly living up to the "Arsenal of Democracy" title on that Thursday the fifteenth of July; six new destroyer escorts were launched and keels were laid for three more and one submarine too. Army and Navy pilots in the South Pacific turned back a bombing raid by knocking down twenty of twenty-five Japanese bombers attacking bases in the Central Solomon and eliminating most of the fighter escort. In Russia, the Germans were failing to achieve their objectives as the Battle of Kursk was coming to a climax. The Russians lost more men, tanks, and aircraft, but their lines held. Perhaps the biggest battle of all on 15 July, and one that certainly would impact Blackie's future in terms of what he would fly when and where, was initiated by Jack Frye in a meeting of airline CEOs.

The opening salvo of this new struggle was fired at the conclusion of the five-day conference which was held in Washington. The joint resolution issued on 15 July, although not unanimously agreed to, involved adopting five principles which would govern international, commercial aviation following the war. The announcement to a large extent pre-empted the Civil Aeronautics Board which had, to date, not taken a leadership position in a vision of post-war, international aviation. The principles emphasized free and open competition subject to reasonable regulation, private ownership and management of the airlines, government fostering a sound, worldwide transportation system, worldwide freedom of flight, and establishment of civil and commercial outlets worldwide. All but one of the CEOs agreed to the full declaration which condemned any policy of monopoly. The holdout was Pan Am's CEO Juan Trippe whose vision of the future saw Pan American Airways as the only American flag carrier, i.e., a state-sanctioned total monopoly of travel to or from American shores.

The second barrage of the coming post-war battle for airways beyond the seas followed within days. It, like the statement of international air commerce principles, would have a direct impact on Blackie's post-war fortunes and extend even into the late 1950s. That second salvo was Jack Frye's announcement that TWA had formed a study project on setting up airlines in foreign lands and rehabilitating those foreign national carriers which had been decimated during the war. Frye knew that there would be many surplus transport aircraft after the war and the several hundred thousand Army and Navy pilots were not all going to be offered jobs back home once the shooting stopped. Significantly, the scope and importance of this study was revealed in the form of an interview in *American Aviation* (the forerunner to *Aviation Week and Space Technology*).

These two initiatives became paramount in TWA internal and external communications. Every TWA employee received a little pamphlet explaining the importance of future global markets and whenever a TWA manager spoke in public, there usually was mention of the importance of free international competition. Just a month later, Otis Bryan was the guest speaker at the Boston Club, an organization of women leaders and professionals in that city. Bryan's talk was about ICD and how its aircrews were helping defeat the Nazi U-boat forces in the Atlantic. He cited several examples where C-54 skippers were able to radio sightings which resulted in aircraft or destroyers getting confirmed submarine kills. The latter portion of his talk strayed and concentrated on the five principles of post-war aviation, with emphasis on the evils of monopoly.

Stirring the pot even more and causing a great deal of heartburn in Kansas City was the appearance of the article, "Colombus of the Airways," which appeared in the then widely read weekly, *The Saturday Evening Post*. The article's subject was Juan Trippe, and the gist of it was the antitheses of everything the other major airlines believed in: open international markets.

That TWA was starting to think about itself as Trans WORLD became even more obvious within several months when a major purchase was announced. TWA bought a nearly forty percent stake in in Taca, the Latin American freight airline, for $2,300,000. The principal owner of Taca was an old friend of Howard Hughes, Jack Frye, and Harold Blackburn. He was Lowell Yerex, the man who had taken Blackie aloft nearly twenty-five years before. In less than a year, TWA bought a twenty percent share of Hawaiian Airlines, a move that Pan Am could not have welcomed. Just a month previous to the Taca purchase, TWA had filed for nearly 100 new route extensions in the United States to, in effect, create a much more far-reaching feeder network for its basic east-west route. At nearly the same time, TWA also filed for some international routes with the CAB: to Honolulu, Paris, and London with flights originating from New York, Boston, Chicago, Detroit, and Washington. To better operate those hoped-for routes, TWA streamlined its field operations down to three regions: Eastern, Central, and Western.

TWA was branching out in other ways as well; its predominantly male work force had been reduced by the draft and men volunteering to the extent that the company started the Women Apprentice Mechanics (WAMs) which was the TWA equivalent of the WACs and WAFs. The WAMs were learning a new trade just as Rosie the Riveter was going to work in aircraft factories. Stung by the Army Air Force's insistence that there be no women hostesses on ATC flights, several TWA hostesses (all previously registered nurses), led by Gladys Entrekin, joined the Army Nurse Corps (ANC). They were assigned to Bowman Field in Kentucky to launch the Air Evacuation Group, a service that would facilitate the return of the wounded from all theaters of war. When Entrekin joined the ANC she had most recently been chief hostess of the western region, but Blackie always remembered her as one of his regular hostesses on the Newark to Chicago run. The C-54A could carry up to 28 litter patients and were always accompanied by a nurse and frequently a medical assistant. Those wounded soldiers coming back from Europe were flown into Mitchell Field on Long Island or to Washington National.

Most of the Army's air forces were very busy on Thursday 5 August. In the South and Southwest Pacific, the Fifth and Thirteenth Air Forces bombed ground and naval targets in New Guinea and in the Solomon Islands with support from the U.S. Navy and Royal Australian Air Force. The Ninth Air Force and Northwest Africa Tactical Air Force bombers were active over Sicily, the toe of Italy, and Sardinia. Only the Eighth Air Force was quiet with the sole activity being the transfer of the 351st Fighter Group from Goxhill to Metfield, its station for the next year. There was a reason for the lack of activity at Eighth Air Force bomber bases – the force was staying on the ground, readying itself for the major attacks on Schweinfurt and Regensburg in less than two weeks. Blackie, to his disgust was also staying on the ground; he hadn't flown anything since late May when he flew a C-54 back from Brazil. As Blackie sat at his desk on that Thursday afternoon, there was a sharp aerial encounter several miles off the north coast of Iceland. A Focke-Wulf FW-200 Condor from *Kampfgeschwader* 40 on a shuttle mission between France and Norway had been intercepted by a pair of P-38 Lightnings from the Fiftieth Fighter Squadron of the Fourteenth Fighter Group. In the exchange of fire, one of the P-38s was slightly damaged, but the Condor was fatally wounded so the pilot, Gunter Karte, ditched at sea near Grimsey Island. The crew was rescued and only one of the crew, a gunner, was slightly wounded. Blackie was naturally totally oblivious to this combat and would remain so for fifteen years until Gunter Karte, who became one of Blackie's students, told him all about it.

On 7 August, Blackie got a letter from Lt. Col. James Peyton of the Army Air Forces Headquarters Flight Control Command about the possibility of going back on active duty. Civilian pilots flying for airlines were exempt from the draft due to their occupations having reserved status. For a reservist flying for an airline to go on active duty absolutely required permission of the employer. Jack Frye was a patriotic man and quickly released many TWA pilots to go on active duty, but there were a few he really did not want to lose, and Blackie was one of them. Blackie had let TWA management know that he had a burning desire "to get into the scrap," but never really expounded on just where or as what. He obviously was thinking of his Army training and would preferred to have been sent to an operational bomber command as squadron commander, group exec, or group commander but those possibilities never surfaced. So Blackie had to set his sights lower and consider Peyton's offer for Blackie going on active duty as a regional flying safety officer. Jack Frye and Otis Bryan were unmoved. Blackie had to disappoint Peyton, hope that matters might improve in the future, and continue doing his best for ICD.

As second in command of the Intercontinental Division, it was incumbent upon Blackie to read what was going on in the minds of both the managers and ground crews of all of the ICD offshore operations, not only for current operating efficiency but also for the future. That future for most people at TWA in mid-1943 was somewhat uncertain, but the future course of the war was beginning to come into focus based on some hard facts. North Africa was no longer in Nazi hands, the Russians had blocked repeated attempts by the W*ermacht* and *Luftwaffe* to drive eastward, the *Kriegsmarine* and its many U-boats were rapidly becoming a smaller force, and the Japanese were conclusively defeated at Midway and Guadalcanal. More importantly, the Allies, especially considering America's industrial and organizational might, were only beginning to reach full strength. TWA had yet to hit its full stride as evidenced by the first half 1943 financials; net earnings after taxes were $732,576 compared to the same period of 1942 when net earnings came in at $422,416 – and in 1943 TWA was operating with a smaller airplane fleet.

The deadly serious business of flying the Atlantic with high priority cargo and passengers was occasionally punctuated with a few light moments. One of those instances came at the expense of Colonel Larry Fritz, the volcanic former director of flight operations who was about to ferry a B-24D bomber from Prestwick back to Washington. Just before takeoff, he approved the addition of 3,000 pounds of mail on top of other previously loaded baggage. Seconds later he OK'd the addition two ICD ground crewmen scheduled for rotation back to America. All of the mail went into the back of the waist gunners' compartment as did one of the two passengers. The other passenger had no choice but to ride in the tail gunner's turret. When cleared to taxi, Fritz rammed his cockpit window closed so hard that it jumped its track and fell to the pavement, thus producing such a stream of cursing from the cockpit that a muleskinner would blush upon hearing it. Fritz closed the throttles and missed his four P.M. departure. Finally a new window was found and installed after which the impatient Fritz started the engines, but as soon as he advanced the throttles to start taxiing, the Liberator's tail fell to the ground giving the passenger in the tail gunner's turret quite a jolt. More blue invective followed from the cockpit as the realization sank in; Fritz had not run another weight and balance on the normally tail heavy Liberator after taking on nearly two tons of weight that was located far aft of the center of gravity.

Some of the other light moments were provided by the VIP passengers. The actor Frederick March was on his way to Belem as the head of a USO troupe. Like many actors, March wanted to be where the action was – so he got himself seated in the co-pilot's seat. The captain got up to walk back to the lavatory, leaving March with his hands and feet on the controls. Upon returning and hearing March's comment, the captain said nothing, but smiled at March who, thinking he had been flying the big ship, said, "Wait until I tell my wife about this." The captain was much too considerate to tell March that the Skymaster had been on autopilot. In war and peace, TWA pilots had to tolerate show business idols and their every little whim. The orchestra leader, Andre Kostelanetz, and his opera star wife, Lily Pons, were scheduled to depart Prestwick after a USO show tour. The Skymaster was ready to go and number four had just been started when the purser opened the passenger door and yelled to the ground crew that there was a major problem on board. Miss Pons could not find the flowers that some adoring fans had given her. The flowers had been on board before the plane had been cleaned and Lily was not leaving without them. The wilted bouquet was located in the field's garbage dump, the worse for wear posies were returned to Miss Lily, the ship was buttoned up for a second time, and it departed only a few minutes late. Another TWA C-54 crew, which had taken off from Presque Isle for the North Atlantic crossing, found itself facing bad weather at Dakar and decided to land at a small RAF auxiliary field in the middle of the desert where the temperature was approaching 120 degrees. At such a remote field, there were no guards for the planes and knowing how items could be stolen from unsecured planes, the crew carried all of its personal gear into the operations shack while waiting for a clearance. The resident RAF types' jaws dropped at the sight of the TWA crew which had armfuls of snowshoes, tents, and winter, fur-lined parkas. As the crew had flown the North Atlantic via Iceland, they had been supplied with winter survival gear as a standard operating procedure. Because they had to sign for the gear and turn it in when returning, they weren't taking any chances – not even in the Sahara desert.

On one occasion Blackie assigned his friend Joe Carr to a flight that wound up having a few humorous moments but in reality was more of a cloak and dagger mission. Carr's assignment was to fly to Prestwick and await further orders.

Upon landing and receiving his orders the following morning, Carr flew to St. Mawgans in Cornwall where he picked up a group of high-ranking officers each of whom had a brief case chained to his wrist. Carr then got new orders: fly his passengers to Marrakesh and await further orders. When he got to Marrakesh, Carr and his crew wanted to call it a day and rest overnight, but new orders dictated an immediate flight to Algiers. The props had barely clattered to a halt when new orders were given to Carr. He was to take off immediately and land at a small strip west of Tunis. The C-54 arrived in Tunis where new orders were waiting, but fortunately the crew was able to get some sleep after having been flying for nearly twenty-four hours from Prestwick. Still, more orders were received the next morning; the crew was restricted to the cockpit and crew area up front. Once aboard they were not to have any access to the passenger cabin. A small convoy pulled up and transferred its passengers into the waiting Skymaster which took off for its next destination: Casablanca where it was to refuel for the flight back to Prestwick. After crunching the numbers in his preflight checks, Carr determined that there was no way he could safely become airborne at Casablanca. He was overweight and he would need a longer runway. There was a blistering exchange with the control tower and base operations officer, after which Carr got his orders to de-fuel enough to get him out of Casablanca and to Marrakesh where runways were longer. Several hours out of Marrakesh, Carr heard a vigorous knocking on the cockpit door which opened to reveal General George Patton. The general was an enthusiastic private pilot and had heard about what fine artificial horizons and directional gyros were on C-54s and would Captain Carr mind if the general came up front and learned all about why the instrument panels on Patton's puddlejumpers were obsolete. Patton spent a few hours in the cockpit, taking the C-54 through a few gentle turns, and then asked Carr and his crew never to mention his presence on the plane. The secrecy of Carr's mission was accentuated upon arrival at Prestwick where the follow me Jeep led the C-54 to a remote corner of the airfield where the passengers deplaned in great privacy. In the not so distant future, it became obvious why there was such a hush-hush aura associated with Carr's flight. General Eisenhower did not want the Germans to know that Patton was in England to further prepare forces then in training for D-Day. The Allied planners had even gone so far as to leave an actor with a striking resemblance to Patton back in the Mediterranean theater to complete the charade.

Blackie scheduled himself on as many foreign trips as possible, but, considering his administrative duties, had to assign many of the most appealing flights to others. One of the most interesting was contained in orders from ATC for five Army pilots, fifteen Northwest Airlines pilots, and ten TWA pilots to ferry thirty Curtiss C-46 Commandos full of spare parts along with some Curtiss tech reps to India. Once there the pilots would train C-47 crews on the C-46. All of this was prompted by the much higher performance and load carrying capacity of the C-46 which was needed to keep the flying supply line going over the Hump to

China. Unfortunately, the C-46, while it eventually became a strong performer, was full of bugs when it was first delivered, although many of them had been worked out by early 1943. Blackie split the TWA pilots into two groups headed by Joe Carr and Chick Fredericks. Carr's group was almost exclusively made up of former Eagle Nest captains, including Jim Wheeler and Clint Dombrowski. The TWA pilots were to be checked out on the giant C-46 at LaGuardia by a Curtiss check pilot, but no such check pilot could be found. Blackie got on the phone with the Curtiss plant in Buffalo to have the instructor sent down promptly and was thunderstruck at being told that supplying an instructor was not part of the company's contract with the Army. Fortunately, there were C-46 manuals available, so, as was the case with the Hudsons at Albuquerque, Joe Carr and Chick Fredericks checked themselves out and then taught the remaining eight pilots.

After several false starts caused by maintenance issues, the group assembled and took off from the Ferry Command base in Wilmington, Delaware, for the 15,000 mile journey to Gaya, a field between Delhi and Calcutta. Four of the C-46s failed to get as far as Florida due to hydraulic failures. Chick Fredericks' C-46 arrived at Homestead in Florida on schedule and landed with a timely slogan emblazoned on the nose – "Never fear when Fredericks is near." Joe Carr flatly refused to continue the mission until a satisfactory hydraulic fix could be found. The base commander threatened Joe with a court martial, but had to back down when he realized that Joe, although uniformed, was actually a civilian. Joe was aware of how the Eastern pilots had solved the problem, but the Homestead commander was adamant – until Blackie and Otis Bryan got through to ATC which leaned on the colonel at Homestead. The C-46s were all flown to Eastern's base in Miami where mechanics quickly fixed the problems by replacing the existing pressure relief valves with C-54 units and reduced hard bends in the existing hydraulic lines. There were a few problems on the way to India, including some brief bewilderment by an inexperienced navigator fresh from training and one C-46 with an engine-driven fuel pump failure, but all planes finally reached Gaya where their crews remained for several weeks transitioning C-47 pilots to the ground loop-prone C-46. While in India, several of the pilots actually flew the Hump and met a number of TWA pilots on active duty with the Army Air Force who were flying over the high Himalayas on a regular basis.

A number of Blackie's former captains had gone back into the Army Air Force on active duty from the Reserves unit in Kansas City and were members of ATC's special duties flight which flew Very Very Important People. Lex Klotz was one of those pilots, and he had been the captain on several of Blackie's early flights as a co-pilot. Klotz became difficult when his name was pronounced other than "Kloe-tz." Klotz flew the 1940 Republican presidential candidate turned Roosevelt personal emissary, Wendell Wilkie, on a goodwill tour that included the Middle East, Russia, and China. Klotz was enthused about the trip, but less so by the means – it was a C-87 Liberator named "Gulliver." TWA got its share of heavyweight flights, although Blackie did not insert himself into the crew list on many of them. One of TWA's most significant VIP flights was flown by Swede Golien whose passenger was former Ambassador to Russia Joseph Davies. The primary purpose of the journey was for Davis to lay the groundwork for a top level conference that would involve Roosevelt, Churchill, and Joseph Stalin at Tehran. The plane was an ATC-supplied C-87 which had all of the Liberator's fuel system vices. Golien picked up the plane at Wright Patterson Field in Ohio, but the plane sprang leaks so bad that fuel was dripping into the main cabin from the fuel tanks in the wings. When one or more of the Liberator's twelve fuel tanks in the wings leaked, fuel flowed on the front or back side of the wing spar and eventually dripped out into the cabin. In this case, the leaks were so serious that the special cabin insulation and furniture were soaked. The leaks were repaired, but Golien was so alarmed he complained about it to Blackie whose further complaints to ATC also found their way to the White House. Combat crews flying the Liberator, considering its hazardous fuel system, frequently said that getting shot at was just adding insult to injury. A hurriedly spruced-up ATC C-54 was promptly substituted for the Liberator and Golien was off on 5 May 1943 via the South Atlantic Route to Khartoum and Cairo. Upon reaching Tehran, a Russian radio operator boarded the C-54 for the flight on to Moscow where Davis and the crew remained for ten days. The entire delegation, including the TWA crew, were treated like royalty, wined (vodka'd that is), dined, and put up in the best hotels. Golien was toasted at another reception where he met Stalin himself. The Russians leaned over backwards to ensure safe navigation to Alaska and even sent a Lend-Lease C-47 ahead of the eastbound C-54 to report any potential problems. The C-54, named the "Gude Ship" based on Davies' Swedish ancestry was, of course, a great hit with Swede Golien. Davies immediately reported to the White House at the conclusion of the flight which took almost 140 hours of flying and later wrote a book, *Mission to Moscow*, about his experiences on the journey. Davies sent an autographed copy of the book to each TWA man in the crew. The odyssey was TWA's first complete circumnavigation of the earth. Golien had been gone so long (six weeks) that Blackie, ignoring the old saw about the pot calling the kettle black, publicly stated that he was planning on strapping Golien into a safety harness with a lock and chaining him to his chair at Hanger Two.

Blackie all but quit flying in November when Martha was expected to have her fourth baby. Blackie flew two flight tests on C-54s midmonth, and then headed for San Antonio where the baby daughter, Beverly, was born on the eighteenth. Tragically, the child was in very frail health and died on 1 December. The family was devastated and Martha in particular was crushed. Blackie remained in San Antonio for a while and did not fly again until late January.

During that month, Jack Frye changed his Washington residence to prepare for what he perceived as another world war – this one for obtaining new international airline routes that were sure to be set up after the shooting war was over.

Blackie had no idea at the time that he would be in the front lines of this new war, nor that he and Martha would be enjoying one of the benefits of gearing up for the new conflict. The Fryes' previous Washington residences had been suites in the Mayflower and Wardman Park hotels. Classy as they may have been, the hotels were not in the league of what Jack and Helen Frye saw as a residence of the magnitude required for entertaining Washington's officials and power brokers. Helen Frye had been searching for months and finally found what TWA leased – the Doubleday mansion just across the Potomac River in Virginia. The Fryes set about remodeling it for use as "TWA's White House" and certainly had their hands full considering the size of the property: thirty rooms in an imposing brick structure set on seventy acres, but it would prove worth the leasing costs. The mansion also provided the otherwise very family-focused Hal and Martha Blackburn with an occasional evening out in the company of Washington's elite when the Fryes entertained.

While ATC determined who filled ICD's seats and what filled the cargo holds, Blackie, and later Swede Golien, made the crew assignments – most of the time. One exception was when the passenger was the most important VIP of all: President Roosevelt. Swede Golien's trip with Ambassador Davies had essentially been a scouting trip for Roosevelt's journey to the Cairo and Tehran conferences of late 1943. For those flights, ATC decided on the pilot who came as no surprise: Otis Bryan. What differed on the 1943 flights with Roosevelt was Bryan's being called to temporary duty in the rank of lieutenant colonel in contrast to his 1942 flight with the president when he was a civilian. Pan Am did not fly Roosevelt across the Atlantic on one of its Clippers this time; instead, the new battleship USS *Iowa* did and Otis Bryan, in a special C-54, flew the land portions. There were some new twists in both Roosevelt's and Churchill's transportation. Churchill was trading in his Liberator, "Commando," which had steadfastly refused to blow up on its prominent passenger. His new conveyance was a custom-converted AVRO York which included a specially vented chamber where the prime minister could chain smoke to his heart's content without terrifying his pilots. In an ironic historical twist, Commando did eventually blow up in early 1945 following a takeoff from the Azores, and it was feared that a leaky oil tank, not a lit cigar, precipitated a fire that spread to the fuel tanks which subsequently exploded.

Roosevelt's new C-54 had been modified with the addition of a special hydraulic lift which moved the wheelchair-seated president from the tarmac into the aft passenger cabin. There was another health-oriented procedure. Roosevelt was not the vigorous man he had been just a few short years before, and his personal physician dictated the maximum altitude at which the president could fly: 8,000 feet. Bryan had some concerns which he eventually dismissed by flying the route in advance and determining that there would be no problems. In fact, the only problem encountered on the Cairo and Tehran conference flights was one landing in which the flaps could not be lowered. This occurred during a diversion to Malta and then Sicily in which Bryan lost hydraulic pressure to the flaps on his approach to landing at Luqa Airfield on Malta mid-morning on 8 December. Bryan explained to the jovial Roosevelt that the landing would be a mite hotter than usual, by about 20 mph, to which the president replied, "Well, go ahead, take all precautions, and we will all have a good time." The landing was totally uneventful. As it turned out, the president's cardio-pulmonary conditions were not worsened by the trip, and he frequently commented on how much he enjoyed the scenery.

The three conferences to which Bryan flew the president to date were apparently not consistently contentious and probably did not significantly add to the president's severe hypertension. In fact, the First Cairo and Tehran Conference were not held as one conference for the simple reason of avoiding tension. The First Cairo Conference which ran from 22-26 November 1943 and in which Roosevelt and Churchill were joined by Generalissimo and Madame Chiang Kai-shek, focused on Asia-Pacific matters and called for the unconditional surrender of Japan. It also called for the return of all lands occupied by the Japanese to be returned to China, the establishment of a unified, free Korea, and renounced any territorial claims over any other lands taken from the defeated Japanese. Premier Joseph Stalin did not attend the Cairo Conference because he was not at war with the Japanese and would not be for another two years. It was thought that his presence would create major issues with Chiang Kai-shek, so Stalin remained out of the picture until the four-day Tehran meeting which kicked off on 28 November. The Second Cairo Conference was held immediately after the Tehran Conference and included Roosevelt, Churchill, and Prime Minister Inonu of Turkey. The purpose was to discuss the possibility of Turkey's coming into the war on the Allied side with considerable American support. Stalin would not have been welcome at the latter meeting because the Russian leader was adamant about minimizing British and American involvement in the Balkans.

Otis Bryan was not privy to all that went on at any of the conferences, but a man with whom Blackie would become well acquainted was, and that was the president's son, Elliott Roosevelt, who was acting in the capacity of a military aide to his father. The younger Roosevelt was witness to one prophetic instance in which sparks did fly, and then he leaked the details to the press. Stalin, after a number of vodka toasts, was gloating about how he would execute the top Nazi leadership along with up to 100,000 of Germany's military leaders in order to diminish the possibility of a future war with Germany. Churchill was outraged and stormed out of the room while Roosevelt tried to defuse the situation by making light of it. The die was cast, however, and it gave the Allied leaders a good picture of what they would be dealing with at Yalta and Potsdam. The principal purpose of the Tehran Conference, however, was achieved: more precise agreement on the when and where of the Allied cross-channel invasion in the late spring of 1944. In the aftermath of that crusade, Blackie would again meet Elliott Roosevelt. And, as it had turned out, the leadership of TWA and the Hughes Empire had already more than just casually crossed paths with Elliott Roosevelt.

After several months of flying little more than a few maintenance checks, Blackie took off on 11 March 1944 on another North African trip – this time to a new destination for him, Casablanca, which he reached on the fourteenth via Newfoundland and the Azores. He had no time for exploring the city and was on his way back via Marrakesh where he was delayed for two days, flying maintenance tests on his C-54 before taking off for Prestwick where the Atlantic weather kept him on the ground for two more days. When the weather abated, he flew the Atlantic and reached Stephenville, Newfoundland, just as one of TWA's C-54s was reported overdue and then missing. The weather was not good, and the tower lost radio contact with the captain, H.M. Thompson, as he reported he was starting his initial let down for Stephenville on Newfoundland's west coast, about 450 miles from Presque Isle, Maine, and about 200 miles from the ATC airfield at Gander which is on Newfoundland's northeast coast. Blackie spent two hours searching early on the twenty-third in a C-54 with Colonel Maxwell, the base commander and his exec, Major Robertson. The C-54's captain had struck the ground while in a normal descent and was lucky to get off so lightly; the only injury was suffered by navigator R.K. Smith who had a broken leg and some facial lacerations. The pilot actually had seen the airport lights out in the distance and was starting to descend, but then encountered cloud. He quickly decided to level off and make his descent when closer in, but moments later the left wing hit the mountain, shearing off the wing and its two engines. The impact turned the Skymaster 180 degrees and rolled it inverted in the snow. The crew had tried to use their hand-cranked generator-radio, the "hurdy-gurdy," but had not been heard due to the peaks in the immediate area. Blackie flew two search/rescue missions in a ski-equipped Noorduyn UC-64 Norseman, a Canadian bush plane built in Montreal specifically for operations in Arctic conditions. Ever confident, Blackie had quickly checked himself out in the plane even though he had never landed on skis before. The wreckage was located miles to the south in the Anguille Mountains, but after several hours of flying, Blackie decided that a Piper Cub would be a better rescue ship, considering the terrain's not offering enough open space for a normal Norseman landing. The Piper Cub, with one of the base's pilots flying, returned to the site carrying the base physician who treated Smith who had been carried to the landing site on the cockpit-cabin door. By the time the Piper Cub was airborne with Smith, it started to snow. On the edge of darkness, an Army Arctic Rescue team arrived complete with dog sleds and two Saint Bernards. The doctor, the Skymaster crew, rescue team, and the rest of the crew had no choice but to camp out in the plane's now inverted fuselage. Finally, after three days of snow, the men got fed up and hiked to railroad tracks that the rescue team had crossed on the way in. They hitched a ride back to civilization after flagging down an approaching train. Stephenville, located on Newfoundland's western shore, was a major ATC base that traced its origins back to 1940, but was better known for its very long runway. Transient crew quarters were basic but the officers; mess was well regarded for its steaming hot lobsters and cold beer. Blackie remained at Stephenville writing the accident report, and then on the twenty-sixth returned to Washington where tragic news awaited him.

The grim news relayed the loss of another C-54, and again Blackie had to sign off on the accident report and console the family of another friend whose life had just been claimed. The day before, Captain C.S. Garber had been flying between Lagens Field in the Azores and Casablanca following a routine crew change when his Skymaster was shot down. Prior to departure, the IFF set had been checked thoroughly and was operating properly. Garber was on top of clouds at 8,000 and actually above an Allied convoy that he was unaware of. There were two carriers escorting the convoy – one American and the other from the Royal Navy. American naval fighters approached Garber's C-54 but quickly identified it as a Skymaster and promptly broke away. Very soon thereafter, Garber was intercepted by a flight of Royal Navy fighters which later reported shooting down a German FW-200 Condor. The entire crew was lost, including Royce Welliver the flight engineer from western Nebraska who had flown with Blackie so many times at Eagle Nest and had had such an important hand in writing the B-24 training syllabus. The Royal Navy subsequently accepted responsibility for the accidental shoot down.

After its brief grounding with engine issues, development of the Constellation had proceeded to the point of making a splashy introduction and what better way to do that than set a new non-stop cross country record with it. Howard Hughes and Jack Frye, as co-captains, took off from Burbank at 3:26 A.M. on 17 April with Hughes as the pilot, bound for Washington National. The takeoff was later than planned because Hughes was reportedly busy, unsuccessfully trying to get his latest girlfriend, starlet Ava Gardner, to come along as a passenger. Hughes leveled off at 15,000 and his passengers watched the lights twinkle below as darkness gave way to the morning light. Everybody expected the landmarks below to slip past quietly, and they were not disappointed. Kingman, Arizona, moved into view – the same Kingman where not much more than a decade before commercial planes had to stop for fuel because they did not have the range to fly non-stop to Winslow. At 5 A.M. the eastern sky lightened and soon the panorama below became the dull tan-gray of the desert. Soon, the Connie flew beyond the San Francisco Peaks and continental divide, but started picking up some weather over northern New Mexico which prompted a climb that reached 18,000 over Kansas where Hughes encountered some light icing. Soon, Kansas City and St. Louis passed below the Connie's wings as some of the passengers were in the midst of a serious game of poker. Frye kept his altitude; and, just east of Cincinnati, the down clouds broke – but only temporarily. The descent was delayed due to storms but the Connie's speed was boosted by a nice tailwind. As they neared Washington, the crew and passengers had inkling that the flight might actually be completed in less than seven hours. Nearly everybody stole quick glances at their wristwatches – that is everybody except Howard Hughes. His watch had stopped hours earlier; he had forgotten to wind it before takeoff.

Because Hughes was always heavy on the power and TWA wanted no hiccups in the Connie's introductory tour, TWA mechanics promptly replaced all 144 spark plugs in the four Wright R-3350s before the demonstration flights began. Uncharacteristically, Hughes at one point used something less than full power during one of the flights on which most of the CAB higher-ups were on board. He cut power to two of the engines which prompted TWA V. P. Jack Nichols to rush forward to the cockpit to see what was going on because the CAB people and other government bigwigs were petrified, thinking there had been a catastrophic failure and that they were about to die. Just as Nichols reached the cockpit, Hughes was reaching for the throttle to shut down a third engine. Nichols managed to convince Hughes that any mishap would spell the end of TWA; sanity prevailed and Hughes brought the Connie in uneventfully on all four. Some of the Army brass objected to the demonstration flights because as they understood it, insurance had not been taken out for carrying VIPs on introductory rides. The deep pocketed Hughes told the brass that he would self-insure the plane which was good enough for the generals who were among the first passengers.

Blackie was not there at Washington National to meet the Connie with Jack Frye in the left seat after its six hours and fifty minute flight. Blackie was in Marrakesh flying several Skymaster maintenance check flights. The Connie did set the new record, beating the old record of seven hours, twenty eight minutes which had been set on 19 January 1937 by none other than Howard Hughes in a Hughes H-1 racing plane of his own design. TWA's public affairs department had pulled out all stops and lined up a blue-ribbon crowd for the landing after which Jack Frye claimed, "She flew like a dream." First to greet the pilots was Jesse Jones, Secretary of Commerce. Others followed, including CAB chief, Charles I. Stanton, and head of the ATC, General Harold George. There were a number of Lockheed personnel on the plane, many of whom had been on the first flight, and the Army officer who was in charge of the Air Force acceptance flight at Lockheed Burbank facility: Major Clarence Shoop. Blackie's old cadet buddy had moved from production testing at Douglas to Lockheed where he was in charge of the acceptance flight test program, particularly on P-38 Lightnings and Hudson light bombers. Leader of the TWA test pilots was Blackie's old mentor and until recently an ICD pilot, Larry Chiappino. There followed more VIP inspections of the plane, which utterly dwarfed a B-24 bomber that landed after it, and VIP demonstration flights as well. The Connie was in bare metal finish, but the interior was all Army: 100 seats which could quickly be converted to hospital plane configuration. General Hap Arnold was among the VIPs as were two United States senators of whom the first was Alben Barkley. Arnold was livid about the appearance of the plane which, technically speaking, was property of the Army Air Force. The plane's exterior was emblazoned with the TWA logo, the company's red stripes, and a bold "The Transcontinental Line." Nor was Arnold happy about the plane's being flown for several days of VIP demonstration flights before eventually being turned over to the Army Air Force.

The second senator was soon headed for a bigger role, much bigger, and his decisions would eventually have a massive impact on both TWA and Blackie's flying fortunes; he was Senator Harry Truman from Missouri. Among the other senators inspecting the plane was Ralph O. Brewster of Maine, a man who, in contrast to Senator Truman, would prove not to be a TWA ally. Two ladies also inspected the giant Connie: Mrs. Woodrow Wilson, the former first lady, and then the most prominent aviatrix of her day (and close personal friend of Blackie), Jacqueline Cochran. TWA laid on a major party in downtown Washington at the Willard Hotel as the culmination of the Connie's record setting flight.

The plane was also on tour at LaGuardia and at Caldwell, New Jersey, where the engine manufacturer, Wright Aeronautical, had its own field. Later it was flown to the Air Material Command headquarters at Wright Field in Dayton, Ohio, where the plane was turned over to the Army Air Force as the first C-69. Unfortunately a pall was cast over the triumphant tour when Larry Chiappino collapsed and died suddenly on the very day he had taken a giant of American aviation aloft for a Connie ride. That day, 26 May, was also Chiappino's birthday. The guest was none other than Orville Wright who wryly noted that the Connie's wingspan of 123 feet was longer than his first powered flight in 1903. Chiappino died at only forty years of age; it became known that heart attacks at an early age were part of his family background. Blackie, at this point, was becoming hardened to hearing about the deaths of friends and acquaintances, but Chiappino's passing was among the most difficult. At this point in ICD's history, Blackie was in a superior position to many of the pilots with whom he had flown the line, so he assiduously tried to avoid playing favorites. Although Chiappino had graduated out of ICD by the time of his death, he had in fact been one of Blackie's best friends.

Hardly an issue of the company newspaper, *Skyliner*, arrived without news of a former TWA pilot who had gone back on active duty and won the Distinguished Flying Cross. In other cases, an article would appear about the exploits of formerly non-flight crew employees who had been accepted into Aviation Cadets, gotten their silver or gold wings, and had been decorated for heroism while on combat operations. Pride and frustration joined when Blackie met Eagle Nest graduates when he was in Britain, and he encountered one or two almost every time he flew to the UK or North Africa which is not surprising considering how many pilots the school graduated. The young men, most of them just completing tours with the Eighth or Fifteenth Air Forces' pioneer bomber groups were full of gratitude for being taught so well, but that simply made Blackie all the more anxious to get in uniform and into a bomber outfit.

Probably the most frustrating to Blackie was news of his former ATC public affairs officer friend, Bob Hotz, who, although not a rated pilot, had nonetheless flown combat missions with the Fourteenth Air Force in China. Hotz actually had civilian flying experience before the war and had logged a few hours in one of ICD's Link trainers, but when he flew in China it was as a top turret gunner in a B-25 Mitchell. He only flew several missions, but they were eventful and one of the sorties proved the value of a co-pilot. On that mission, the pilot took off without a co-pilot, assuming that the sortie was not especially risky. As fortune would have it, the B-25 was badly shot up by flak and the pilot wounded. Hotz jumped into the empty co-pilot's seat and brought the B-25 back to home base where the pilot was by then sufficiently revived to pull off a safe landing. Hotz was later shot down behind enemy lines, and although wounded, had to walk his way back to friendly lines. News of that type amply fueled the "get back into uniform" fires that were constantly burning in Blackie's mind. Thus Blackie again petitioned Otis Bryan for a release from employment with TWA in order to go on active duty. Blackie had just heard from his sources embedded in high places that the Air Force was about to ask the airlines to release more Army Reserve-affiliated pilots for active duty. In his letter to Bryan, Blackie quickly acknowledged the facts that his career, salary, and other advantages would be sacrificed if released but again made it clear that he wanted to be in uniform in time of war. This time, Jack Frye was very specific in just why he was denying permission, and pointed out the obvious. The Army would never give Blackie the rank commensurate with his ability and, furthermore, it was highly unlikely the Army had anything for him that was as important as what he was already doing. Frye reminded Blackie that a great future would be opening up for TWA as the war wound down and that by staying with TWA, Blackie would have a greater share in that future. The letter made the point that Blackie could take his share of the credit for building up TWA to what it had become and that he had an obligation to keep building the company into an even larger organization that could welcome back those pilots who had gone on active duty. Blackie took the bitter pill and never made another attempt to get into combat by leaving TWA for the duration. Blackie described his situation as being a pilot in "the swivel chair squadron." His opportunity for self-realization as a military pilot in time of war, however, would come from an unexpected source and it would not be that long in coming.

Although he could derive much satisfaction from seeing ICD performance figures and knowing that each one of the division's C-54s was in the air every day for almost thirteen hours, there was some displeasure in seeing turnover in the administrative staff and the women assigned as his secretary in particular. His first secretary, Marjorie Perkins, moved back to Kansas City with her young son after receiving word that her husband had been killed in action. His next secretary, Nancy Hanks had married a Navy lieutenant commander, left on maternity leave in mid-1944, and never came back, preferring to be a full-time mother. Miss Jane Rossiter, his secretary through the end of the war was such an attractive young woman that Blackie teased her with threats of putting "No Trespassing" signs around her desk.

On 13 May there was another C-54 accident at Stephenville when a Skymaster nose wheel collapsed on takeoff. Although the aircraft was destroyed by fire, everybody got out, including eight Army officers and twelve nurses. On that day Blackie had taken off from the very same field on his way to Lagens Field on Terceira Island in the Azores but did not witness the accident. It was one of those months which Martha Blackburn would have just as soon forgotten, although Blackie would have just as soon have forgotten the previous month, April, when the only thing he flew was his desk. Blackie was gone for eleven days and returned from Casablanca to Stephenville on the sixteenth, but did not continue on to Washington. Instead, he took two more Newfoundland to Prestwick and return flights before finally getting back to Washington on the twenty-first. Blackie's flying took a back seat to his management duties for the next two months; he flew less than ten days and logged two roundtrips: one to Casablanca and the other to Prestwick.

It was at about this time that Martha and the kids returned to Washington, and it was to a different address. Blackie had rented a new place – predictably a two-story brick home on a well-shaded street. The new location, at 3343 South Stafford Street, put Blackie just a couple of miles from Washington National. Martha missed some of the social whirl in Washington which, in spite of the war, had its share of receptions and celebrations. Just a few weeks before Martha and the kids left Texas, Blackie was invited to a cocktail party at the Chinese Embassy along with both TWA brass (Frye, Bryan, and Mutchler) and the Army Air Force topside (Generals Arnold and George). The party's purpose was the presentation of Chinese military decorations to ICD's Captain Shelton and his crew for their recent roundtrip to China with Madame Chiang Kai-Shek as the principal passenger. Blackie was already acquainted with the man presenting the medals, Chinese Air Force General Mow who had been on Blackie's flight with T.V. Soong to Karachi in late 1942. The Madame proved to be a hardy traveler – she was hardly fazed when her C-54 lost two engines over the Caribbean before reaching Florida. Blackie had a little recognition of his own at the same time: TWA was paying for his new $10,000 accident insurance policy which was restricted to death or accident in any aircraft mishap – whether the plane was owned by TWA and flown by Blackie or a non-TWA pilot. That policy and personal commendations from both Jack Frye and Otis Bryan clearly showed that Blackie was a very highly valued member of the TWA management community.

June 1944 was a bad month for ICD. On the twentieth, just a day before Blackie took off for Marrakesh via Bermuda and the Azores, an ICD Skymaster flown by Captain Robert "Rollie" Inman disappeared at night on a flight from

Newfoundland to Presque Isle. Just a year before, a couple of VIPs were visiting Prestwick and Inman was selected to show them, the King and Queen of England, the inner workings of a Douglas C-54. The circumstances surrounding Inman's disappearance were hauntingly similar to the loss of Bob Redding and his crew at the Philmont Ranch in April 1942. There were heavy storms about, radio range reception was disrupted, and the wind was much stronger than forecast. Inman had checked in with the air traffic control system as he came over the coast and later reported that he was on the airway between Blissville, New Brunswick, and Bangor, Maine, prior to that airway's making a hard left turn at Millinocket, north of Bangor. Another C-54 pilot who was flying the same airway just an hour behind Inman, reported such strong southeast winds that he was forty miles north of his course before radio reception finally improved enough to allow him to accurately fix his position. Rotten weather and low ceilings kept searchers on the ground for two days, but based on an analysis back at Presque Isle, the hunt moved seventy miles north of the regular airway. The new search zone was in the vicinity of Mount Katahdin, Maine's highest mountain at just over a mile high. Other peaks and ridges in the immediate area exceed 4,000 feet above sea level. The wreckage was found three days after the accident by a C-47, and the recovery party reached the remote site on Fort Mountain, elevation 3,900 feet, four days later. Fort Mountain is little more than four miles northwest of Mt. Katahdin. The crew had been killed instantly, and as with the Liberator crash at Trail Peak, just a couple hundred more feet of altitude would have saved the men. Unlike the Liberator accident at Philmont, there was an explosion and fire when the C-54 hit the mountain.

Then on 26 July when Blackie was starting a flight from La Guardia to Prestwick, ICD lost another C-54 – one full of wounded soldiers. This missing C-54 was en route between Iceland and Newfoundland when it simply disappeared. Blackie always doubted that it was sabotage or mechanical failure, always maintaining that the captain, Robert Funkhouser, had strayed too close to a Nazi sub or possibly a convoy whose trigger-happy Navy escorts opened up on the Skymaster assuming that they were shooting at a Focke-Wulf Condor. Blackie took off for Iceland deliberately trying to backtrack on Funkhouser's exact course, flying only a few hundred feet above the waves and trying to find some clue as to exactly what happened. Just past Greenland, he spotted a submarine which was promptly reported to the Icelandic Submarine Patrol. Blackie continued on to Iceland and landed at Prestwick the following day. Neither Blackie nor any other authority ever discovered whether the downing was caused by a German U-boat or naval gunners, uncertain of aircraft identification, deciding to shoot first and ask questions later. Nor did he ever learn if the submarine he reported was ever attacked, although at that stage of the war it probably was. The engineer on the Funkhouser flight was Rayburn Darst, the man whom Blackie had publicly commended for devotion to duty during ICD's early days in 1942. There was a very bittersweet post-script to Darst's death. He had been a member of Otis Bryan's crew that took President Roosevelt to the Cairo Conference the previous year. The entire crew, not just Otis Bryan, had been placed on sixty-days of active duty at the time of the conference. About a year after the flight, each member of the crew was awarded the Air Medal for meritorious aerial achievement, with General Harold George presenting the decorations. On the next day, Colonel Harold Harris, ATC chief of staff, met with Darst's widow, a former TWA employee, and her ten-day old son, Darryl. Harris pinned the medal onto little Darryl's flannel nightshirt while the infant snoozed peacefully in his mother's arms.

Although late spring and summer 1944 were costly in terms of flying accidents to ICD, TWA did receive some basically good news – in late April the Army started returning the Stratoliners which would go back into TWA's domestic service, although Blackie never flew one again. Actually, getting the Strats back was not 100 per cent good news because the weary ships needed a lot of work. About a year before, one of Blackie's close friends, Johnny Guy, attended a joint meeting with Pan Am and the CAA about the upcoming certification of the Stratoliners for civil use again. One of the Strat's known weakness was brought up: cracks in the tubing which was part of the wing spars. The prewar quick fix was just simply drilling holes at the ends of the cracks, painting them, and subjecting them to 100-hour inspections. That was acceptable two years previously, but now TWA's five Strats had been flown hard and collectively logged over 21,000 hours. According to the CAA, watchful waiting wasn't good enough. It was found that changing out wings from a B-17 G model was a satisfactory fix – for starters. The planes were flown to Seattle for more than just a superficial clean-up. What they got were new landing gears from the B-17G, 1,200 horsepower engines instead of the old 1,100 hp powerplants, new tails, B-29 type electrical systems, new propellers, and a two-stage supercharger system. Whereas the old Strats' claim to fame was pressurization, the made-over Stratoliners had none. They would be relegated to hauling passengers at the same altitudes as the DC-3s; Connies would fly the rarified heights in the postwar era. Although the end of the war was almost in sight, it took until April 1945 before all five Stratoliners were back in service.

It was during late spring that the Civil Aeronautics Board formally advised the nation's airlines that it would accept applications for the awarding of international route certificates. Much earlier, the CAB had ruled against TWA in a preliminary recommendation concerning the choicest routes. Jack Frye was hardly overwhelmed. He was a whiskey-sipping, poker player who frequently entertained another whiskey-sipping poker player: Vice President Harry Truman. Although the CAB announced the regions in which certificates would be awarded, the North Atlantic was of the greatest interest to Jack Frye who knew that Pan Am had already pioneered the Pacific and Latin America. The route to Europe was where nearly all of the postwar traffic would be headed, a fact not lost on Pam Am either. Pan Am painted a dismal projection of anticipated North

Atlantic traffic which, if realistic, would have suggested that there would not be enough passengers to even think about serving them with competing airlines. One very authoritative observer put the matter to rest; it was ATC's General Harold George who saw the numbers of airline passengers crossing the Atlantic as "staggering." TWA was seriously outgunned in terms of lobbying firepower in Washington by Pan Am's presence which emanated from its imposing office building on "F" Street. In the year leading up to the CAB announcement, Pan Am had lobbied hard for something revolutionary: the creation of one single American airline to fly all foreign routes. At the time, there were only two other American airlines operating from American shores to foreign destinations: Panagra and American Export. The latter was owned by a steam ship company, American Export Lines, while the former was a joint venture between Pan Am and Grace Lines, another ocean liner company.

Formation of American Export was a bloody nose for Juan Trippe who felt that President Roosevelt had personally exerted his influence in favor of American Export. On the surface, one might have expected a favorable FDR decision for Trippe because both men were from old New York money and were Ivy League graduates (Harvard and Yale, respectively), but then Trippe was not a democrat. Trippe had offended another man who would have no direct role in awarding post-war international routes – a man whose was nonetheless very important in all aviation matters, military or civil. It was General Henry Arnold who had at one time tried to recruit Trippe into the Army Air Force as a brigadier general in charge of military air transportation. Trippe was interested in the role, but not the military cloak that went with it; he would be an underling to Arnold if not some major general and was not used to taking orders from anybody.

The counter-proposals and intense lobbying from Pan Am, Congressional committee work, investigations, wrangling, debates, and hearings would continue for another full year before the post-war routes to Europe would be awarded. Before that year ended, Harry Truman, Jack Frye's whiskey-sipping, poker-playing friend from greater Kansas City was firmly ensconced in the White House. When the final North Atlantic rulings eventually were handed down, Blackie was at the right place at the right time.

As a result of the CAB opening the door to domestic airlines for overseas routes, Jack Frye promptly filed for an around the world route by extending the existing ATC routes to Europe, Africa, the Middle East, India, and beyond by establishing service to Mandalay, Hanoi, Shanghai, Tokyo, and Anchorage. In August, Frye gave a major speech to the Kansas City Chamber of Commerce, as hosted by Mayor John Gage, in which he outlined the future of the airline which was so important to that city. He foresaw the day when TWA would have over 45,000 employees worldwide, and that the end of the war would spell dramatic growth not only for TWA domestic but especially its new international operations which should function in an environment free of monopoly. The Army Air

Force fully endorsed the speech and, indeed, the State Department was preparing for the eventual negotiations that would lead to landing rights in foreign lands. Frye's talk also reminded the audience and media of TWA's leadership roles in aircraft (DC-2, Stratoliner, the Connie) and its scientific achievements. In the latter category, TWA was far ahead of its rivals in thunderstorm research, advancing the state of the art in radios, very high altitude flying, and more recently pressure pattern flying.

One of Blackie's job responsibilities was generating statistics related to monthly performance of the division, and the numbers partially explained the rash of accidents. Activity just prior to and immediately after the D-Day invasion had soared compared to previous periods, and with that much higher utilization, there were bound to be more accidents. Departures from Washington National had averaged around 100 per month in the very early spring of 1944 but jumped to 158, 161, 171 in May, June, and July respectively. Captains' flying hours averaged about sixty-eight per month early in 1944, but during the months noted above they climbed to sixty-eight, ninety-five, and ninety-seven. For a man whose job description said absolutely nothing about flying planes, Blackie averaged fifty-three hours per month during the three-month high traffic period. He was not flying at all in June until the twenty-first when he took off on a four-day roundtrip to Marrakesh via Bermuda and Lagens Field, Azores.

His passengers on outbound flights before D-Day were the predictable military brass and senior diplomats. Many of them came back on subsequent return trips along with field grade officers with high priority assignments after their combat tours expired. The return traffic changed after D-Day when Blackie was among the first ATC pilots to bring home soldiers wounded in France. As the Allies drove eastward, the names on his return flight manifests occasionally had a few unanticipated, very Teutonic spellings: those belonging to captured German generals. On several occasions later in the war, Blackie returned from Prestwick not with human cargo, but with some captured, high-tech German war booty instead: Nazi V-1 Doodlebug flying bombs.

In the meantime, the pattern of flying one foreign roundtrip a month continued until August when he touched down at Prestwick for what became a longer than average stay. That atypical sojourn would provide Blackie with the most exciting flying in his life.

Chapter Eight
Watching the Jerries

On 12 August 1944, Blackie set out on another transatlantic crossing that began routinely enough with its origin at Washington National and subsequent legs to LaGuardia and Stephenville, Newfoundland. On the fourteenth, blessed by favorable weather, he flew the Atlantic, arriving in Prestwick after an eleven-hour, forty-minute flight. Early the next morning, his old friend from aviation cadet days, the now Colonel Clarence A. Shoop, arrived at Prestwick and began Blackie's checkout on the Martin B-26 Marauder. They flew the fast, medium bomber to Shoop's base, Mount Farm, where the Eighth Air Force's Seventh Photographic Reconnaissance Group was stationed. The paint on that B-26 was barely dry – it had arrived at Mount Farm only six weeks before, having been accepted from Martin's plant in Baltimore on 27 May. It was then flown to Dow Field in Bangor, Maine, before taking the North Atlantic route over to Mount Farm and the Seventh Group.

Colonel James G. Hall, who was the group's first operational commander, had flown the unit's first mission in early March 1943 in an F-4 Lightning, was decorated with the Distinguished Flying Cross, and had returned home to become the Army Air Force's Chief of Reconnaissance. The DFC was not Hall's first decoration. He had received a Distinguished Service Cross and a *Croix de Guerre* during World War I when he served as a pilot in France. Hall was also five years older than Blackie which served to simply add more fuel to Blackie's burning desire to get onto active duty and directly into the fray. Although commissioned in the U.S. Army, Hall flew with the French as did quite a few Americans whose final flight training was completed in French aviation schools. Hall distinguished himself in aerial combat, shooting down at least one German machine. In the interwar years, his seat on the New York Stock Exchange provided him with the wealth needed to own a Lockheed Altair, participate in air races, and own quarter horses. Seven months after Pearl Harbor, Hall, as an Army Air Force reservist, was in the right place at the right time and became the commander of the Thirteenth Photographic Reconnaissance Squadron which was activated at Peterson Field, Colorado Springs. From that beginning, three additional squadrons followed the Thirteenth to eventually become the Seventh Photographic Reconnaissance Group. The group was first assigned a base near the Bedfordshire village of Podington (about sixty miles north of London in the western half of Eighth Air Force "bomber country"). Hall objected to the location and got his wish when the group was moved to an airfield closer to the RAF's reconnaissance operations in Oxfordshire.

The Mount Farm base was near the town of Dorchester-on-Thames and roughly fifteen miles west of "Pine Tree," code name for the former girls' school that was headquarters for the Eighth Air Force's headquarters at High Wycombe. The final approach to Mount Farm, due to the Marauder's exceptionally high wing loading, was the fastest Blackie had ever flown – at nearly 140 mph. The hot landing set Blackie to thinking about one of his Kelly classmates. Roland Sansbury, one of Kelly class of February 31's old China hands, who was now chief production test pilot at Glenn L. Martin Company in Baltimore where the Marauder was manufactured – and where 140 mph touchdowns were the rule, not the exception. Blackie's quarters for the night were not in a damp Nissen hut. As a VIP guest he stayed in a spare bedroom in Shoop's accommodations – the former manor house dating back to Mount Farm's days as a working farm.

The Seventh Group's primary mission was photographing targets, both potential and actual, for the Eighth Air Force Bomber Command and obtaining photos following the strikes for bomb damage assessment. The list of targets far exceeded the obvious Nazi aircraft plants, airfields, shipyards, and refineries. Nearly anything of potential interest was photographed – from Norway to the Spanish border and from Hitler's Western Wall to the Eastern Front. Canals, docks, bridges, prisoner-of-war camps, radio towers, flak batteries, highways, power generating plants, prisons, dams, warehouses, and railroads were among the priority targets.

Seventh Photographic Reconnaissance Group

Blackie with his great friend, Colonel Clarence Shoop, at Mount Farm, the Seventh Photographic Reconnaissance Group's base near Oxford, England.

Just looking at real estate from 30,000 or 300 feet could turn up something useful – including the V-1 and V-2 sites. The Seventh's sorties were usually flown at altitudes of 30,000 to 40,000 feet, although the group also flew very low altitude "dicing" missions to monitor German activities such as the establishment of new radar sites, launching platforms for rockets, etc. Another important mission was photo-mapping which required very precise flying on specific headings and altitudes. Other recon groups flew tactical reconnaissance, especially after the invasion, in direct support of ground forces. Specialized units flew night missions using flares for illumination while some squadrons flew weather reconnaissance sorties. The latter missions were usually flown in target areas just before a major strike or routinely far out in the Atlantic to help forecast weather over Britain and the continent. Occasionally, the Seventh Group flew meteorological flights, but the vast majority of those missions were assigned to other squadrons in the wing.

Mount Farm ("Station 234" in Eighth Air Force parlance and code-named "Home Plate") was near Oxford or about forty miles northwest of London. The field was home to the group's four squadrons of which three (the Thirteenth, Twenty-second, and Twenty-seventh) were equipped with a more recent version of the Lightning, the F-5. The remaining squadron in Shoop's group was the Fourteenth which flew the British Spitfire PR (Photo Reconnaissance) Mark XI in what amounted to a reverse Lend-Lease arrangement. The Spitfire PR Mk. XI was essentially a Mark IX with guns and ammunition bays removed to make room for additional fuel tanks. The Mark IX was the RAF's standard day fighter from late 1942 well into 1944. There were five other noticeable differences: the Mark XI was fitted with a retractable tail wheel, the engine cowling was deepened to accommodate the larger oil tank needed for longer range missions, and, in the last production units, the rudder's top was pointed instead of rounded. On the PR Spit XI, there was no bullet-proof windscreen with its two mounting bars running up from the fuselage to the canopy bow. Instead, the forward part of the canopy was *all glass*; only the usual canopy bow remained, thus contributing to improved visibility. In the cockpit, the gunsight was removed and the camera control unit was installed. Both the Spitfires and Lightnings were painted in "PR Blue" to blend in with the blue sky as much as possible. From late 1943 until the end of the war, the group was given twenty-one Spit XIs by the RAF of which seven were lost to enemy action. Due to their speed and maneuverability at high altitude, Spitfires were assigned the continent's most distant and sensitive targets. To extend their flights as far as possible, Seventh Group Spitfire pilots frequently flew the eighty miles to Bradwell Bay, a night fighter base on the Norfolk coast, to top off their tanks before continuing their missions across the Channel.

Mount Farm was located in the heart of Allied "reconnaissance country" and had started as a satellite of the Royal Air Force base at Benson, just a few miles away. Benson had been the early headquarters for the RAF photo reconnaissance function. It had grown substantially, and long before Blackie was aware of Mount Farm the base had smaller, satellite fields in northern Scotland and southern Cornwall to cover Scandinavia and southwestern France respectively. When Blackie arrived at Mount Farm in 1944, Benson was home for four squadrons of RAF PR Mosquitos and Spitfires. At both bases, off-loaded film was developed and very briefly analyzed, but the prints and negatives were then immediately sent to Danesfield House near Medmenham in Buckinghamshire (about thirty miles west-northwest of London and roughly ten miles east of Mount Farm). Medmenham was no small operation; at its peak RAF personnel were joined by American and other Allied photo-interpreters to create a Central Intelligence Unit staff of over 1,700 expert photo-interpreters responsible for second and third stage analysis of tens of millions of photographs that came in during World War II. Of those millions of images, five million came from the Seventh alone.

Medmenham was not your garden variety Royal Air Force base. It was very different, and quite a few of the personnel were really civilian specialists masquerading in uniform for the duration. There were many women on the staff (as WAAFs – Women's Auxiliary Air Force). The average photo-interpreter at Medmenham was much older than the typical twenty-one year old fighter pilot. There were actually a very few who had been professional photo-interpreters before the war (principally map makers and specialists involved in locating mineral and petroleum deposits). Joining them were mathematicians, artists, sculptors, writers, architects, musicians, industrial engineers, archeologists, and many more specialized occupational types. Winston Churchill's daughter, Sarah, started at Medmenham in the section that analyzed shipping photographs. Dirk Bogarde, who became a British cinema star several years later, also served at Medmenham as a photo-interpreter.

Photo reconnaissance had its detractors in the first days of the war, but came into its own very quickly, especially during the early stages of the Battle of Britain as RAF recce pilots watched the build-up of the *Luftwaffe* in France and the concentration of landing barges in French ports. Photographs taken by Mount Farm pilots went to one of several sections at Medmenham. There were separate departments for airfields, aircraft, shipping, rail transportation, German army bases, camouflage, industrial sites, and even one section for dummy or decoy sites. There was another department where the artists, sculptors, illustrators, and carpenters worked in Danesfield House which had previously been the magnificent country home of a British industrial tycoon. They toiled in the basement, not on the top floor where they would have expected to have plenty of natural light. Their charge was making models of radar installations, invasion beaches, port facilities, rocket sites, and many other high value targets. Photographers and cinema crews photographed the targets from different approaches in different levels of lighting, thus simulating what a bomber pilot could expect while attacking a target from low altitude in the moonlight or what commando raiders would see as they emerged from a submarine near an enemy shore installation in the dark of night.

Much could be gleaned from the photos. Because the specialists in the naval section knew how long it took to build a submarine and how long its shakedown trials took, Winston Churchill was provided with very accurate forecasts of how many Nazi U-boats would be joining the wolf packs on a monthly basis well into the future. The airfield section could determine precisely which *Luftwaffe* fields were used by the new German Me-262 jet fighters even in the absence of photographing the jets on the ground. The evidence was found in jet exhaust burn marks on the runways. Medmenham's industrial engineers, looking at bomb damage assessment photos, could predict when the Germans would have a plant up and running again, thus providing valuable inputs to RAF and Eighth Air Force intelligence officers wondering when a target would be ripe for another visit. The British Admiralty could be advised of just how long it would take for a German surface raider, battleship, or supply convoy to reach a certain destination. A ship's speed could be derived by mathematical analysis of the vessel's bow wave as captured by a photograph.

To accomplish its work, Medmenham used technology much more advanced than magnifying glasses (which were widely used at Medmenham nonetheless). Photo interpreters utilized stereoscopes to see targets three-dimensionally. That was accomplished by viewing two overlapping photographs of the same target under the stereoscope, resulting in the target appearing much more realistic and, therefore potentially offering more information. Shortly after the war broke out, Medmenham had acquired some advanced Swiss technology from a private British company – that was the Wild ("Vilt") Stereo Plotter, a sophisticated, very heavy optical machine that could dramatically magnify photos and transform them into highly detailed maps.

Shoop had only taken command of the Seventh on 5 June after transferring from the P-38 equipped Fifty-fifth Fighter Group. He was originally attached to the Fifty-fifth for thirty days to gain operational experience in the P-38 which was to prove useful, considering his regular assignment at Lockheed Aircraft in Burbank. After a Lockheed production test pilot signed off on a new Lightning, Shoop or one of his officers wrung it out; if all went well, the ship was accepted into the Army Air Force inventory and then promptly ferried, usually by pilots from the Sixth Ferrying Group based at nearby Long Beach, to its first location. As his logbook confirms, Shoop had flown the Army acceptance test flights on over 1,000 Lightnings or what amounts to ten percent of all Lightnings produced, and he had done experimental test flying on a number of later model Lightnings' improvements such as the dive recovery flaps.

Shoop had also been charged with preparing a report on the effectiveness of the P-38 Lightning in the European theater following his brief combat tour with the Fifty-fifth and interviewing personnel in other P-38 groups. In early June 1944 however, he had been ordered to leave the Fifty-fifth's base at

Wormingford, about forty miles northeast of London, and report to Eighth Air Force Headquarters. There he was told that he had been appointed to command the Seventh Photographic Reconnaissance Group with effect from 5 June 1944. Shoop's visit to the Fifty-fifth concluded just in time – the group converted to P-51 Mustangs several weeks after Shoop moved to the Seventh.

Shoop was graying at his temples when he joined the Fifty-fifth Fighter Group, i.e., he was an old man among the twenty-two year-olds. But he joined the unit with much more fighter time than the average new pilot in the group and had also gone through the Third Air Force Gunnery School at Sarasota, Florida, in December 1943. There, in addition to the P-38, he also flew P-39s, P-40s, P-47s, and P-51s. Shoop was thrown into harm's way at once and flew his first escort mission on 7 May as the wingman of Major Val Bollwerk, skipper of the group's 338th Squadron. The *Luftwaffe* had not challenged the group on its most recent missions, thus the Fifth-fifth was in the midst of a dry spell – that would end on Shoop's first combat sortie, a bomber escort mission to Germany. In fact, the Eighth Air Force was flying a maximum effort that day: mission number 342 in which 922 heavy bombers escorted by 764 fighters attacked Berlin, Munster, and Osnabruck, Germany.

At 27,000 feet after entering Nazi airspace, the group encountered some flak which was not accurate enough to be worried about; what concerned the pilots were contrails at six o'clock high. Suddenly the squadron was bounced from above. Shoop had fire-walled the throttles to keep up with Bollwerk who was trying to attack the topmost of two Focke-Wulf FW-190s on the tail of another Lightning. Shoop went after the second 190, but both Germans broke and started to dive. Shoop had knocked a few pieces off his 190, but then abandoned the descent, realizing that the dive was rapidly reaching the point where compressibility was going to become a problem. He pulled out, saw one more 190 pass in front of him, and fired at that German's wingman, who also dove away. As so often happened, Shoop then found himself completely alone – not another plane, German or American, could be seen anywhere in the sky. At 12,000 feet after turning and chasing the Germans, Shoop then set course for Wormingford and eventually did join up with Bollwerk. The flight was then ordered to return to Antwerp after reaching the coast to escort a damaged B-24 but was unable to find it. The flak gunners did find Bollwerk's flight which quickly evaded the black bursts.

Spirits were soaring at the debriefing. It had been the first scrap the group had gotten into in quite a while, and three victories were chalked up on the group scoreboard. There was a cost; one pilot missing and another wounded. The bombers suffered remarkably few losses on that Sunday, but five fighters from other groups, mostly Lightnings, did not return. The *Luftwaffe* was becoming selective about how and when it would attack the bombers. May 1944 was not a big month for the Fifty-fifth Fighter Group. It destroyed thirteen German aircraft and probably destroyed another six, but paid for it with the loss of nine of its own pilots, including Major Bollwerk. By the time of his last mission, Bollwerk had been credited with destroying three enemy aircraft, one probable, and one damaged, but was shot down and captured forty miles north of Paris on 21 May.

Shoop flew another dozen escort missions over Germany, France, Belgium, and Holland with the Fifty-fifth, but was never in a position to tangle with German pilots at close range again. However, he did see the broad panorama of air war as marked by all kinds of smoke: black puffs of exploding flak, smoke from German air-to-air rockets being fired into the vast formations of American heavy bombers, smoke rising from the objectives as target markers struck, great columns of smoke rising from targets as they were hit by tons and tons of falling bombs, and greasy plumes of black smoke pouring from stricken Liberators and Flying Fortresses as they were knocked out of the sky.

Shoop's transition to operational reconnaissance flying can only be described as approaching the sudden. After being in charge of the Seventh Group for only several hours, he was up at four A.M. and flew the group's first mission over the invasion front on 6 June. He and the other Lightning, flown by Deputy Group Commander Lt. Col. Norris Hartwell, encountered poor weather, no German interceptors, but did attract the ire of many German light anti-aircraft gunners.

Returning to the United States in late June, Shoop delivered his findings to the Pentagon, and also found time to marry a movie star. His new bride was Julie Bishop who had appeared on the silver screen with such leading men as John Wayne and Humphrey Bogart. The best man at his wedding was Jack Frye. Indeed, the wedding was performed at Frye's suburban Washington residence. Shoop's time with his beautiful new bride was limited, and he was back in command of the Seventh, just one week before meeting Blackie at Prestwick.

There was a major party on the night of 16 August at Mount Farm's officers' club and Blackie was in the thick of it. The group had that day received a commendation from the Eighth's commander, General Jimmy Doolittle, for its extraordinary achievements in the period preceding D-Day and throughout the month following the invasion. The group had also just received a formal memorandum congratulating the Seventh for its performance from the commander of the 325th Reconnaissance Wing, the Seventh's parent organization. That letter had been written by President Franklin Delano Roosevelt's son, Col. Elliott Roosevelt who had been placed in command of all Eighth Air Force reconnaissance units leading up to D-Day.

The more immediate reason for the party was to celebrate what had transpired on that day – completion of the group's 3,000th sortie. August was proving to be a very busy month for the Seventh which had averaged twenty-four sorties a day so far that month, including the low of three missions on 6 August to a high of fifty-five sorties on 14 August. The 3,000 milestone had been achieved in the afternoon when a flight of four Lightnings photographed the

German airfields at Norderney and Langeoog in northern Germany followed by Soesterberg, Deelen, and Harskamp in Holland. Bomb damage assessment was the reason for photographing Soesterburg and Deelen because each had been struck the day before by nearly 100 Royal Air Force Lancasters on daylight raids. General surveillance was the basis for photographing the other three targets.

Deelen, located roughly forty miles southwest of Amsterdam, and Soesterberg, twenty-five miles southwest of Amsterdam, were important day and night fighter bases for the *Luftwaffe*. Both fields dated back to the World War I era when they were established by the Royal Netherlands Air Force. Deelen had served as home to several squadrons of Focke-Wulf 190 day fighters from *Jagdegescwader* 54 (Fighter Wing 54), a *Zerstorergeschwader* (a destroyer or attack unit) flying the twin-engined Messerschmidt 110, and a group of Junkers 88 night fighters. In mid-August 1944 it was home to several squadrons of Junkers 88 medium bombers and was protected by four heavy flak emplacements and eleven lighter flak guns. Based on its potential for causing problems during Operation Market Garden, the airborne invasion in Holland that would follow in September, Deelen would be bombed so heavily in the days ahead that the *Luftwaffe* soon abandoned it completely.

Luftwaffe bomber units flew from Soesterburg during the Battle of Britain. In addition to squadrons from two separate fighter wings, Soesterburg also served as a base for one German paratroop division and a searchlight group that supported nearby heavy flak batteries. As Allied air power grew, Soesterberg evolved into a defensive fighter base – not surprising since it was on a direct line from the many bomber bases in East Anglia to Berlin and other major targets in central Germany. Therefore, Soesterberg received all of the attention it deserved in terms of frequent Allied bombings. Soesterberg was defended by three heavy flak emplacements and ten lighter flak cannons. Within four weeks of being photographed by the Seventh Group on 16 August, the *Luftwaffe* wrote off Soesterberg after having downgraded it to forward field basis only later in August. Eventually, Soesterberg became an important base once again for the Royal Netherlands Air Force, but it took six full years to restore and modernize the field which reopened in 1951.

Although Harskamp was listed by Seventh Group operations as a grass-surfaced airfield, it served only light, liaison planes such as the short take-off and landing Fiesler Storch. Harskamp had a history going back to World War I as a gunnery range, and it was taken over by the German Army in 1940 which continued to use it as a gunnery range. Its importance was magnified by the fact that its location put it in the sights of planners of the upcoming Operation Market Garden. Harskamp also served as a decoy or dummy airfield, designed to draw attention away from nearby *Luftwaffe* airfields. There was, however, nothing false about the several antiaircraft gun emplacements located on the edges of Harskamp.

Because those Dutch fields were of great interest to Market Garden planners, they were photographed several times a month by the group in July and August. In fact, the three Dutch targets had been photographed only the day before by Lt. Gerald Adams from the 27th Squadron, but the results, due to some down cloud, were not satisfactory, thus putting the three targets back on the board for the sixteenth.

Coming within range of light and medium flak gunners was a good way of courting death. Of the Eighth Air Force's top twelve fighter aces, six were shot down. Within that half dozen, there was only one fatality, and that was a loss to an enemy fighter in aerial combat. The other five were shot down by flak, surviving the experience but proving the points that German anti-aircraft gunners were the best in the business and that strafing ground targets was an exceptionally dangerous undertaking.

Langeoog and Norderney comprise part of the Frisian Islands just off the north coast of Germany. Langeoog was not a large base by *Luftwaffe* standards, but served periodically as a fighter field earlier in the war and was still in periodic use as a forward base in 1944. Occasionally, the *Luftwaffe* temporarily based fighters on the nearby islands of Borkum and Wangeroog as well.

By the time of the 16 August mission, Langeoog was primarily used by naval mine detection aircraft because the RAF's mine-laying activities were having a punishing impact on German coastal shipping. The *Luftwaffe* also based an *Ergaenzungmeinensuchstaffel* (training squadron for mine detection aircraft) on Langeoog. Norderney was a much larger facility whose naval aviation heritage as a *Seefliegerhorst* (seaplane base) dated back to the World War I era. During the interwar years, Lufthansa flew vacationers to and from the island, but during the first months of the war, Norderney had been home to six squadrons of minelaying aircraft that seeded the Thames Estuary and British harbors.

Blackburn Family Collection

Summer 1944 – photo Lightnings over Oxfordshire, England.

Those squadrons moved west in the spring of 1940 after the Nazis conquered Holland, Belgium, and France, but Norderney remained as an important air-sea rescue base. There was plenty of business for those squadrons as the Frisian Islands were under the path of Allied bomber streams frequently headed for such important targets as Bremerhaven, Cuxhaven, Kiel, Lubeck, and Wilhelmshaven. It was not only *Luftwaffe* fighter pilots who had run afoul of bomber gunners and escorting Allied fighters that *Luftwaffe* air-sea rescue crews plucked from the North Sea's chilly waters; many RAF and USAAF crews were also saved. There was a darker side to Nazi use of Norderney – it also was a slave labor camp whose prisoners were used to build northern sections of Hitler's Western Wall.

Nazi bases in the East Frisian Islands were visited by the Seventh Group only infrequently, but still had to be watched because the *Luftwaffe* occasionally based interceptors there. Any increase or decrease in air-sea rescue or interceptor aircraft spotted at Norderney was worth knowing about because it might say something about how the Germans were allocating assets elsewhere in northern Germany.

The Lightnings on the 3,000th sortie milestone returned to Mount Farm from their Frisian Islands and Dutch objectives in the late afternoon in two separate echelon formations, diving down low over the field for a 360 degree overhead approach with a left-hand pull up into a normal downwind leg for individual landings at thirty-second intervals. Most of the Seventh Group's sixteen sorties that day were flown over northern Germany and Holland because the weather was rotten elsewhere in Germany. Although there was plenty of down cloud over central and southern Germany on the sixteenth, the Eighth Air Force Bomber Command flew mission number 554 to bomb airfields in Germany and Belgium. By that point in 1944, the bomber streams did not need perfect weather because the lead ships were bombing using radar. When the lead bombardier toggled his load, the rest of the ships in the formation immediately dropped their eggs even if the target was obscured by smoke or clouds. Not everybody made it back to England on 16 August. Eleven B-17 Flying Fortresses, five B-24 Liberators, and five fighters were shot down. The Ninth Air Force was active over northern and western France on that Wednesday as nearly 130 B-26s and A-20s struck rail bridges and ammunition dumps.

Blackie was in high spirits at the party that evening because he had been one of the four pilots completing the memorable mission! The sortie had also come very close to being Blackie's very last flight on earth. His participation in the three-hour, forty-five-minute sortie was not publicized at the time. Blackie's close friend, Colonel Shoop, had taken a big risk and authorized the unauthorizable: Blackie's quick, one hour and fifteen minute check-out in one of the Thirteenth Squadron's Lightnings that morning, which included three landings, before the flight took off several hours later. That particular Lightning, serial number 43-28616, was not, unlike Blackie, to lead a charmed life. It was later transferred to the Tenth Photographic Reconnaissance Group which had been based at nearby Chalgrove. On 22 November 1944 it was lost in action after the Tenth, a tactical recce group, had moved to a forward base in France.

Had Blackie, technically a civilian, been killed, gone missing, or been taken prisoner, Shoop would have faced a certain court martial. At that point in the war, a P-38 group commander could expect a new, replacement pilot to have logged nearly 400 hours, including about 220 hours from aviation cadets and the rest in transition and operational training on Lightnings. Even with that, a new pilot would complete about twenty hours in local orientation before going on his first combat mission. Although Seventh Group pilots cracked-up a few Lightnings, the ship was a relatively safe plane to fly. With an accident rate of 145 incidents per 100,000 hours of flying, the Lightning was statistically safer than the P-39 Airacobra and P-40 Warhawk series, but not quite in the same league as the P-51 Mustang or P-47 Thunderbolt. Actually, the safest aircraft of the fighter designs was the F-6, the photo version of the Mustang at only sixty-two accidents per 100,000 hours of flying. Blackie had logged thousands of hours in multi-engine aircraft before checking out in the F-5, so the experience proved to be exhilarating instead of intimidating. Still, the familiarization contained a few hurdles even for a man with a mind as quick as Blackie's. He had to pass the blindfold test– reaching for the correct lever, handle, or instrument without being able to see it as it was read from a checklist. Other than the camera controls, there were a couple of other items Blackie had not expected in the Lightning. First was the Sperry autopilot which was of great value on very long range flights and especially when the mission was mapping. Second was another instrument that helped to ensure a proper track over the ground – an optical drift sight.

That flight of four pilots taking off at one P.M. in the early afternoon was an extraordinary group by any measure. Colonel Shoop was arguably the most experienced Lightning pilot in the Army Air Force. Flying the number three position in that flight was Major Carl Chapman, an iron-jawed, eagle-eyed former pilot in the 22nd Squadron. Now he was the group operations officer with thirty trips over the continent under his belt. The youngest pilot of the four was Lt. John Anderson from the Thirteenth Squadron; he had flown his first mission on 21 April. By mid-August he had flown nearly twenty missions of which his 10 June flight to Breskens in Holland was by far the most memorable. His Lightning had been nearly shot down by anti-aircraft fire as he was approaching his objective, but although he had lost his electrical and hydraulic systems, he successfully photographed the target and brought his valuable film home.

Loss of the electrical system put nearly everything on his instrument panel to sleep, and with hydraulics shot, thus taking out his brakes and flaps, Anderson faced a hot, memorable landing back at Mount Farm. Fortunately, he was able to manually crank down his landing gear, but still used every foot of the field's longest runway, 6-24, which was just short of a mile long. His Lightning had over 100 holes in it; he was awarded the Distinguished Flying Cross for his accomplishment. When Chapman's tour expired, Anderson became the new group operations officer. Unlike Anderson, Blackie had never flown in a hostile sky before, but he was the most experienced pilot of the four. He was closing in on a total time of 9,000 flying hours with nearly 1,000 hours of that time on instruments. At that point in the war, Chapman and Anderson would have a long way to go before reaching 1,000 total hours.

Preparation for the memorable sortie had actually begun before Blackie had even eaten breakfast at the Officers' Mess hall with Shoop. Many of the pilots about to take off in the morning avoided the officers' mess and grabbed something in their squadron ready rooms, i.e., some tea, and toast with jelly or jam and peanut butter. The teleprinter in Group Operations had clattered away well before breakfast, assigning targets for the day. Due to the weather over most of Germany and France, only thirteen Lightnings took off from Mount Farm on the sixteenth. Normally, a pilot knew he would be flying on a certain day since his name would be on an alert list. The pilot would get a phone call where he was waiting at his squadron's ready room, and be told to report to Group Operations where he would receive a detailed briefing from operations, weather, and intelligence officers. Unlike bomber and fighter pilots, recon pilots were pretty much free to determine their own course to the objective. The briefing was followed by a Jeep ride back to his squadron's Nissen hut to pick up gear and then a short ride out to the aircraft's hardstand for the preflight. Runways and taxiways at Mount Farm were paved, but the hardstands were not exactly paved. Rather they were formed from Marsden matting, better known as "PSP" (pierced steel planking, made of perforated steel plates).

Because he was a civilian, Blackie's participation in a combat mission required a slight deviation from the usual procedure. Nobody questioned the group commander when he introduced "Lieutenant Blackburn" as a flying school classmate. There may have been a few furled brows and rolling eyes because Blackie's appearance was certainly not as youthful as any of the many other lieutenants at Mount Farm. Some may have wondered if Blackburn had been a discipline case or some sort of a problem pilot because, after all, he was still just a lieutenant and his cadet classmate was a colonel. Most of the pilots were fairly recent graduates of advanced flying schools where they had learned to fly the AT-6 trainer which usually led to operational training in fighters. That was very much in contrast to how the RAF recruited its PR pilots; it avoided fighter pilots and preferred selecting "mature" flyers with superior navigation skills, i.e., from RAF Bomber or Coastal Commands. Blackie was certainly mature enough – he was old enough to be the father of many of the younger pilots.

"Shoopy," as he was called by very close friends, had escorted Blackie to the Mount Farm base stores very early that morning for a serious updating in flying clothing and equipment. Blackie, the former apparel salesman, was in for a noteworthy makeover. Flying a fighter called for significantly different clothing and equipment from that worn by C-54 pilots. Blackie signed for a pair of fleece-lined A-6 flying shoes, a pair of summer flying gloves, an A-4 gabardine summer flying suit, an A-14 oxygen mask (an item completely new to Blackie who had not done much very high altitude flying in over two years), and a pair of B-8 type goggles. Capping it all off in size 38 was an alpaca-lined, B-10 summer flying jacket which filled the gap when the more popular leather A-2 jacket was not available. He did not need a leather flying helmet or radio headset since he had brought his own along on the flight down to Mount Farm.

Lunch was on base, that is to say, the officers' mess, and the fare was predictable. Nourishment would seem somewhat Spartan by standards of what most pilots had experienced for lunch back when they were aviation cadets in the land of plenty. In those earlier days they could expect something along the lines of chicken fricassee, mashed potatoes, gravy, and string beans or carrots preceded by a nice garden salad and rolls – all washed down with milk, iced tea, fruit juices, or coffee. Pies or cake and ice cream were standard dessert fare back home. But this was wartime Britain; lunch at Mount Farm's mess halls was frequently based on spam or other canned delights served with mashed potatoes and Brussels sprouts ("bubble and squeak"). There was, however, special dispensation for pilots scheduled for missions: beans and any other flatulence-producing foods were not part of the program.

The pre-mission briefing was a new experience for Blackie (with maps in hand), although the weather briefing and assigned radio channels for voice communications were quite routine. According to the weather officer, the surface winds would be from the south at a brisk fifteen mph. The pilots would encounter unlimited visibility aloft with some scattered cumulus clouds with tops at 10,000 feet, a thin cirrus layer at 28,000 feet, and slant visibility greater than twenty miles in the target area. The contrail level was predicted to be at 25,000 feet. The briefing turned out to be somewhat inaccurate; conditions were actually somewhat better than forecast, over the northern part of the mission. The intelligence officer's briefing included a few items that could induce some gnawing fears: location of *Luftwaffe* fighter bases near the intended route, the latest reports on flak battery positions, and escape/evasion techniques in the event that the flight turned out not to be of the roundtrip variety.

The briefing took longer than average because it was a four-ship mission and Blackie was on a steep learning curve. Radio channels were reviewed – a simple procedure in those days because there were only four channels (unlike today's VHF navigation/communications transceivers with hundreds of frequencies). There were separate channels for Mount Farm control, another for air-to-air, a primary channel for air-sea rescue which also offered a direction finding vectoring service for pilots needing some navigation help on the return, and a back-up emergency channel in case radio traffic was heavy on the primary emergency frequency. Blackie was also given a quick orientation on operating the oxygen system which was very different from that on the bomber and transport aircraft he had been flying. Prior to his orientation at Mount Farm, Blackie's only previous high altitude flying had been at 20,000 feet in B-24s on "graduation" flights at the CCTS in Albuquerque over two years before.

The IFF unit ("Identification-Friend-Foe," an early type of transponder) was explained to him although he was already familiar with it. Colonel Shoop explained the mission profile in depth, not only for Blackie's benefit but also because a four-ship mission to the same objective, such as an enemy airfield, was somewhat atypical, although it was a very regular procedure to assign as many as eight or nine Lightnings in formation to a mapping mission. Shoop identified the targets and the sequence in which they would be photographed and from which intended headings, speed, and altitude. Shoop and Blackie would photograph Langeoog while Chapman and Anderson would cover nearby Norderney. After each pilot made several runs over his target, the flight would proceed down to Holland where Shoop and Blackie would photograph Harskamp and

Soesterberg, leaving Deelen to Chapman and Anderson. It was important for each man to understand the full scenario because once within range of enemy radio reception, radio silence was observed for as long as possible. When they had to communicate it was by code: Red One for Shoop, Red Two for Blackie, Red Three for Chapman and Red Four for Anderson.

The check-out on a Lightning in the morning before the mission posed no problems for Blackie – his bedtime reading the night before had been the Army Air Force pilot's manual for the P-38 Lightning. Shoop signed Blackie off on the Lightning after just one hour and fifteen minutes of flying, including three landings. The F-5 was hardly as complex as a C-54 and the engine start-up sequence was practically identical to that of other aircraft he had been flying for years. Operating a modern, inline, liquid-cooled engine with its plumbing system and double radiator flaps was certainly new, considering his vast and exclusive experience on great round, air-cooled engines that belched fire, coughed smoke, and dripped oil. The check-out's only significantly new aspect was learning how to operate the Fairchild aerial cameras (the Lightning could carry four cameras that provided vertical mapping photos and both vertical and oblique reconnaissance shots). The cockpit layout was cramped as might be expected for a fighter, but with one exception the controls were where Blackie expected them to be. The throttle-mixture-prop pitch quadrant was on the left as were the fuel tank selector switches. The P-38's reflector gun sight was replaced on the F-5 by a camera control box, a unit that established the rate at which photos were taken and also contained camera on-off switches along with annunciator lights. The camera control box was not located where the gunsight had been; instead it was at the bottom of the instrument panel between Blackie's knees. Thus, forward visibility was much better in an F-5 Lightning than the pure fighter version.

While the flaps and landing gear selector controls were on opposing sides of the cockpit, the "stick" was a bit of a departure from the C-54 and Liberator. It consisted of a control yoke located atop a control column which was hardly unorthodox except in the Lightning the control column was not located between the pilot's legs as with the C-54 and the other major American fighters. Rather, it emerged from the cockpit floor just outside of the pilot's right knee – exactly as did the co-pilot's control column on a DC-3.

There were several other novel aspects to Blackie's orientation in the Lightning. First, there was no torque on takeoff because the props were contra-rotating, making the Lightning particularly smooth-handling for aerobatics. Most striking was the acceleration on takeoff – the Lightning roared down the runway like a scalded cat, and Blackie had never flown anything that was capable of nearly 400 mph in level flight. It was not the first time Blackie had taken off in a single-seat fighter; he had a few hours in the Boeing P-12 which was still in Air National Guard and Reserve squadron use even into the late 1930s. But the resemblance ended with the seat; the P-12 was a biplane fighter with the fixed undercarriage, wing struts, bracing wires, open cockpit, and a 550 hp radial engine that marked it as a relic from a past generation of fighters.

Lockheed engineers had provided Lightning pilots with an innovative means of getting aboard their mounts – a method that was not so far removed from stirrups in fact. Blackie was used to getting into cockpits by the usual means: handgrips, toe slots, wing walks, ladders, platforms, and even through retracted bomb bay doors. The trailing edge of a Lightning wing was nearly six from the ground or just about eye level for a man of Blackie's height. Pulling a handle at the aft of the central gondola released a spring-loaded drop-down assembly that can best be described as a mini-ladder complete with three stirrups.

Following the briefing, two Jeeps took the four pilots to their planes' respective squadron operations rooms, which were located in Nissen huts at three different squadron dispersal areas, to pick up parachutes, emergency kits, flare pistols, and Mae West life preservers. Unlike other fighters, the group's Lightnings had no prominent letters on their aft fuselages to identify squadron and individual planes – not surprising since a Lightning tail boom barely had room for the Army Air Force national insignia. Instead, the Seventh Group's squadrons relied on different colored prop spinners and rudders to differentiate squadron affiliations. Later in the war, all Seventh Group Spitfires and Lightnings had deep blue spinners and a broad, horizontal red stripe on the nose, but that standardization did not occur until much later in 1944. Blackie's ship, from the 22nd Squadron, already had the dark blue spinner, upon which the group would later standardize, and white rudders. The other squadrons' aircraft were painted in a mixture of red, yellow, blue, and green combinations for their spinners and rudders. All four ships were still sporting their "invasion stripes," broad black and white bands around the aft tail booms and outer wing panels which were designed to alert Allies on the ground during the invasion and weeks immediately after that these planes were "friendlies."

Shoop was the first off his Jeep; he would be flying one of the Twenty-seventh Squadron's older Lightnings, a C model, serial number 42-67132 with red rudders. The same Jeep dropped Blackie off at a Twenty-second Squadron hardstand where he started pre-flighting 43-29006, a very new E model. The other Jeep took Chapman and Anderson to the Thirteenth Squadron where their Lightnings awaited. Blackie's walk-around inspection revealed nothing out of place, so he signed the Form One which was given to him by the crew chief whose own inspection confirmed the ship as ready to fly. The crew chief had also tested the cameras to make sure they were working and that the protective covers had been placed over the lenses to make sure no specks of mud or other debris picked up prior to becoming airborne would degrade the photos (the covers fluttered to earth after takeoff). The crew chief verbally indicated that each camera was OK to Blackie. There followed a near universal tradition among Eighth Air Force aircrew: a quick pause to answer nature's call on the grass just beyond the tail unit. Then Blackie walked to the rear of the central gondola and pulled the release handle to drop the little ladder. Getting aboard was an awkward undertaking, considering the extra weight of the Mae West life preserver, maps and oxygen mask in hand, the dangling radio plug, and emergency kit.

When finally aboard, he started going through his final checks – a procedure Blackie, as a stickler for cockpit discipline, took very seriously. After re-adjusting the seat height and rudder pedals, he connected the shoulder harness, secured the canopy, making sure the locking pins were in position, and rolled both windows shut (failure to do so would burble the airflow over the elevators, making for a less than optimal takeoff). Checks for correct positions of rudder and aileron trim, fuel selectors (on reserve tanks), crossfeed and booster pump switches, and drop tank selector and arming switches followed. He had already plugged in his microphone, moved the flight controls to check for proper of movement, and adjusted all trim tabs. Then, '006 was ready to be started--an easier process than a B-24 or C-54 in one respect: only two engines instead of four. On the other hand, there was no co-pilot to share the duties, so the workload was about the same.

In less time than it takes to tell, Blackie cracked the two throttles, confirmed mixture controls as being in idle cutoff, opened the carburetor air filters, and made sure the prop circuit breakers were turned on. Quickly continuing, he checked the clock to be certain it matched his wristwatch as set during the briefing and then corrected the altimeter slightly to make its indicated altitude equal to Mount Farm's elevation. More eyeballing followed: camera switches off for now, flaps up, and radio off. A few cycles of the auxiliary hydraulic hand pump was next. All of this took Blackie's mind off the fact that his flight path over the next few hours put him within reach of thousands of people who dearly wanted to kill him. Blackie, like most American pilots, but unlike many other Allied pilots whose families and friends had been bombed, killed, imprisoned, wounded, or tortured by the Nazis, harbored no real, burning hatred of Germans. His was a very simple motivation; the Germans were his country's enemy, and as an American pilot, his duty was to carry out the mission assigned him. Blackie had seen with his own eyes the darker side of the war, having been through London in the summer of 1942 when signs of the Blitz were still very obvious. Unbeknownst to Blackie on that August afternoon, he would, in time, become very well acquainted with many *Luftwaffe* pilots, including some who had bombed London. In spite of their wartime activities, Blackie would eventually have a very high opinion of his former opponents' flying training and their piloting abilities.

More eyeballing followed to check positions of prop controls, flap handle, radiator shutters, IFF controls, and that oxygen pressure was at least 400 psi. Then Blackie was ready to start the left engine. For pilots who needed a little help remembering all of the pre-takeoff procedures, there was a checklist card in the cockpit map pocket. One item that he did not have to worry about was a turbocharger control. On Lightnings, the automatic turbocharger controls were linked to the throttles. By then, the ground crew had already inserted the trolley cart plug for external power and one of them was standing by with an item that was not exactly encouraging – a fire extinguisher. Blackie pulled and turned the prime handle to the left and gave it three shots of prime. Then, in quick sequence, he switched the ignition, generator, and fuel pump switches to on. Once he had fuel flow, he turned off the fuel pump and engaged the starter. The three-bladed propeller commenced turning with a metallic whirring sound, started coughing and barking with light bluish-white smoke pouring from the exhausts, and within a few seconds started settling down to a throaty purr. Convinced the engine had caught, Blackie quickly shoved the mixture control into auto rich. Once he saw oil pressure starting to rise in the left engine, he repeated the procedure for the right engine. Once that was started, he moved the battery switch into the on position and turned the radio, IFF, and compass switches on and then, with a well-practiced sweep, confirmed that all temperature, fuel, and hydraulic pressures were in the green.

Shoop taxied past Blackie's hardstand followed at some distance by Chapman and Anderson. Blackie looked out of the canopy and suddenly gave a double-thumbs-up motion with both arms whereupon the chocks were pulled and the ground crewman with the fire extinguisher retreated. The battery cart cable had already been pulled. Blackie then released the brakes, advanced the throttles slightly, taxied out onto the perimeter track, and took his position behind Shoop and ahead of Chapman. After taxiing for several hundred yards to a point just short of Runway 17, Shoop turned thirty degrees to the left and stopped, as did the other three. All four pilots went through their checks for cockpit integrity and proper trim tab positions before making final checks of their fuel systems: reserve tanks selected, fuel pumps back on, drop tank switches on and arming switches in the safe position, prop controls forward and their selector switches in automatic, generators on, oil and coolant shutters in automatic, and flaps up. The run-up was done at 2300 rpm to check the props, magnetos, ammeter, and voltmeter. After the run-up Blackie idled the engines, at 1200 rpm.

Shoop taxied out onto the runway, lined up on the centerline, and after a brief pause pushed the throttles forward and then released his brakes. It was one in the afternoon. Rather than initiate a climbing left turn after reaching 400 feet of altitude, he kept on flying straight. Thirty seconds after Shoop started his takeoff run, it was Blackie's turn. He taxied out and turned to perfectly align the Lightning with the runway centerline and promptly stopped. With throttles advanced and then with brakes released, the Allison engines were howling. As the next few seconds passed, Blackie's eyes were focused outside the cockpit and down the runway with only an occasional stolen glance at the airspeed indicator. As soon as he saw the airspeed pointer moving through eighty mph, he firmly and steadily pulled back on the yoke with his right hand while his left hand was still gripped on the throttles. When the Lightning swept through 100 mph, the rumbling and bumping stopped as Blackie became airborne and was off to war.

Chapman and Anderson followed at brief intervals. Blackie and the other two pilots climbed and turned inside. Shoop, who had completed a 180 degree climbing turn, was by then approaching the downwind leg. By turning inside each other, the men all joined on Shoop while he was approaching midfield. They quickly slid into a finger-four formation which was a standard Eighth Air Force defensive formation for fighters – it was something Shoop had learned at the Fifty-fifth Fighter Group. The formation appears much like positions of fingertips on an extended hand (thumb excluded), maximizes visibility in which the pilots can watch each other's tails, and was the best way of avoiding being bounced by German fighters. Blackie was flying on the left side of the formation as Shoop's number two and Anderson was on the far right, flying number four to Chapman who was in position three.

Once formed up, Shoop immediately took up a northeasterly heading for the Frisian Islands of which Norderney was roughly 400 air miles from Mount Farm. Unless they were jumped by Jerry fighters, the flight would cruise at 250 mph indicated. Shortly after takeoff, each pilot had turned on the electric fuel pump, selected drop tanks, and after suffering no hiccups or deadly silences, returned that fuel pump switch position to off. At that point, the formation was relatively compact. When they neared the Dutch coast, the formation spread well apart with nearly a half mile from one side to the other. Blackie suddenly tensed as they neared the Dutch coast, but then relaxed. He had been startled by the unexpected whine in his radio, but then he remembered the briefing – it was German radar painting their approach and interfering with the flight's radio reception. By that point, they were at altitude – 25,000 feet and climbing to an eventual 30,000. That altitude did not put them completely out of the flak cannons' reach, but it considerably reduced the danger. It was unlikely that an ME-109 or an FW-190 just taking off could intercept them – the Americans would be long gone by the time any Jerry got to altitude. If Germans were already at altitude and starting to close, the only avenue of escape was to shed the drop tanks and dive away, preferably into a nearby, all-concealing fluffy-puffy cumulus cloud formation.

The airspace they were entering had been the site of savage air battles for several years – first with the RAF at night and then with the Eighth Air Force by day. Although there were very few enemy fighters on Langeoog, and then on an intermittent basis only, there was a major *Luftwaffe* fighter base and wing headquarters at Jever. That airfield was only about twenty-five miles to the south or literally just seven or eight minutes away. There were also other major *Luftwaffe* fields south and northeast of Jever. That was too far for Allied radar controllers to warn the formation of danger, so the best defense was maintaining a constant vigil, i.e., swiveling one's neck constantly and glancing frequently at the little rearview mirrors on the outside of the canopy bow.

When the flight reached a point several miles north of the Dutch island of Terschelling (one of the West Frisian Islands, just off Holland's northwest coast), Shoop turned right ten degrees so as to maintain a seaward side approach to the objectives, thus minimizing any possible flak. Blackie never made the turn and was observed continuing straight ahead, i.e., gradually leaving the formation. Shoop knew at once that something was very wrong, yelled "Blackie" on the radio, nudged his throttles, and rolled left to catch up with Blackie. What Shoop saw was a man not really with it; Blackie seemed dazed and was clearly in that stage of anoxia where unconsciousness is about to take over. Tucking in close to Blackie, Shoop motioned and held up his oxygen hose. Blackie still had enough presence of mind to see and understand, checked his oxygen flow indicator, and then checked the hose connection. That was it. The bayonet clip attaching the oxygen mask hose to the oxygen regulator hose had worked loose. As Blackie came back into focus, he and Shoop started looking for the other two Lightnings. They were higher and off to the right, leaving occasional contrails that Blackie always remembered as a pair of intermittent, white quotation marks in the sky. Within a few minutes Blackie was completely back on top of things, and the two were able to rejoin the formation and carry on with the mission. But it had been a very, very close call indeed. Occasionally, a pilot who had been overcome by anoxia got lucky – his plane would settle into a gentle dive and the pilot would start to revive at an altitude somewhere below 10 to 12,000 feet which would still allow him to pull the nose up and resume normal flight. In the more typical scenario, the plane with an anoxic pilot (or a dead one already having succumbed to hypoxia) would go out of control, culminating in a smoking crater in the ground or, as nearly happened to Blackie, a great splash at sea with only a gradually disappearing oil slick to mark the pilot's final resting place. Although Shoop had planned on flying over the targets at 30,000 feet, the Lightnings were leaving contrails at 28,500, so the flight dropped down to 28,000 to eliminate them. From the easternmost border of the East Frisian Islands, Langeoog was very easily identified as the third major island in the group and the one with the most pronounced southern-jutting arm. Shoop led Blackie in a textbook approach to a target by overflying the island, occasionally dipping his right wing, commencing a turn that brought him back to the objective at a thirty or forty degree intercept angle and then rolling into a final turn that would take his track directly over the *Luftwaffe* field 28,000 feet below. The Nazi base was near the western end of the island, just in the crook of the arm that descended southward. Shoop completed his first run over Langeoog and then two more, each from an approach angle roughly thirty-five degrees different from the first run, and then it was Blackie's turn. Great care had to be taken during his three runs because his Lightning had to maintain a constant altitude, airspeed, and heading. The intervalometer, which also contained the camera on-off switch, had been set to establish the exact interval at which exposures would be taken, usually providing a sixty percent overlap between frames which was preferred by the interpreters as Medmenham. If the mission profile changed while in-flight, i.e., a significant change in altitude and/or speed, a pilot could adjust for the change with the intervalometer and still provide the Medmenham staff with the preferred overlap.

What was going through Blackie's mind as he rolled in for a steady, straight flight over the German aerodrome below? Nobody can say for sure now, but it was probably a combination of Martha and the kids at first along with tremendous relief that there were no German flak gunners shooting at them and that there were no Jerry interceptors closing in on him. But after that, Blackie, the disciplined check pilot, most likely was completely dominated by sweeping his vertical speed indicator, airspeed indicator directional gyro, and turn and bank indicator until he figured it was time to reach forward and turn the camera off.

With Blackie off target, the Lightnings rendezvoused and set course directly for Holland via Emden in Germany where the anticipated flak never materialized. Contrails had again started to form behind them, so the formation had to drop down to 24,000 feet. When Shoop estimated the flight was about fifty miles due east of Amsterdam, the formation split up; Chapman and Anderson would cover the larger base at Deelen, and Shoop and Blackie would photograph Soesterberg and Harskamp. Although the visibility was not particularly good, Blackie and Shoop each made three runs over the easily identified Soesterberg without attracting any flak and then turned east for the three-minute flight to Harskamp, an objective which proved to be a problem. The grass-surfaced Harskamp dummy airfield was difficult to pinpoint, and to confirm the location the flight circled over the suspected target, gradually reducing altitude. Shoop was beginning to become concerned about remaining fuel. Just as their time over Harskamp reached thirty minutes and their altitude had dropped to 12,000 feet, the sky erupted in little black puffs with red centers. Even wearing a helmet with radio pads, and in spite of the roar of two very loud engines within feet of him, Blackie could still hear the "crump" of exploding 88 mm shells. With the flak rapidly getting closer, Shoop quickly radioed Blackie and called it a day. Throttles were advanced smartly for a climbing turn onto a course direct for Mount Farm. Harskamp would have to wait for another day. At this point Blackie's Lightning was much more responsive than it had been on the way to Langeoog – at 12,000 he was in denser air where the controls grabbed better and his ship was much lighter with less drag because he had pickled the drop tanks before reaching Harskamp. Only in cases involving a shallow mission where there was little or no chance of running into German fighters would a Seventh Group pilot return to Mount Farm with his wing tanks still attached.

Seventh Photographic Reconnaissance Group

Blackie just back from his very eventful, first combat mission – a three-hour, forty-five minute sortie over Luftwaffe bases in Holland and northern Germany.

After landing back at Mount Farm, photographs were taken of Blackie and Shoop together and individually. Then came the debriefing and a serious choice: a cup of coffee or two slugs of medicinal, American rye whiskey. The interrogation took only a few minutes and was based on completing a straightforward form. The questionnaire required a brief description of the target, reason(s) for not photographing it if the mission had failed, mission number, squadron number, aircraft number, time in/out, tanks dropped or not, and remarks on flak and enemy aircraft. About an hour after the flight returned, Blackie was able to see some of the photographs. Harskamp, with no photos at all, was the only disappointment. The prints of Langeoog, Norderney, Deelen, and Soesterberg were all excellent. Bomb damage assessment of the two major Dutch airfields resulted in a mixed picture. The bombing of Deelen was nearly complete; it never recovered from the plastering by the RAF's ninety-five Lancasters. Soesterberg got off so lightly on that 15 August (just one Dornier 217 destroyed and very little physical damage to the field and facilities) that it had to be promptly revisited.

The next day, 18 August, was another day of wonder for Blackburn, because he flew the group's British-built Airspeed Oxford (a light twin) as co-pilot with Lt. R.S. Quiggins to LeMolay, an airstrip inland from the Normandy coast. Starting in early August, the group flew courier missions to LeMolay (Station A-9) which is located nine miles northwest of Bayeux. The base had been completed as a forward landing field in early July and was also used by the top Allied brass, including Eisenhower and Montgomery since their field headquarters were located only minutes away. The field was still very much in range of German fighter-bombers, although the front had quite recently moved about thirty miles to the east following the breakout from nearby Caen. While the group's primary mission was to support the strategic objectives of the Eighth Air Force, many of its photographs were of great interest to Army ground commanders in France, hence the courier flights. In early August, the Seventh began temporarily flying tactical photo-reconnaissance missions for one of its neighbors, the Tenth Photographic Reconnaissance Group which was located at nearby Chalgrove. The Tenth, a Ninth Air Force unit, was moving en masse to its new base in France, and the Seventh had been ordered to carry the tactical load until the Tenth could become fully operational again.

Blackie's visit to Mount Farm still held one more surprise. On the eighteenth, he accomplished something that nearly every Allied pilot (and many *Luftwaffe* pilots too) only dreamed about. He checked out in the British Spitfire. Granted it was not the PR Mark XI which the Fourteenth Squadron regularly flew – it was a Mark V which was no longer in frontline service, having been retired in 1942 in favor of the Mark IX which was quite capable of besting the Germans' Focke-Wulf 190. But it was a Spitfire nonetheless. Blackie logged an hour and ten minutes of airwork with the elliptical-winged fighter, one of three Mk. V Spitfires which the Seventh used to transition new pilots to the Spitfire Mark XI, and shot three landings with it. That particular Spitfire Mark Vb came to the Seventh with a pedigree. It had been seconded to the Seventh by one of the RAF's former American Eagle Squadrons in 1943. That was after the three squadrons of Yankee volunteers left RAF Fighter Command, became the Eighth Air Force's Fourth Fighter Group, and started flying Republic P-47 Thunderbolts.

Blackie emerged from the experience thrilled at having flown a legend, but was not impressed with the poor ground handling characteristics related to its knock-kneed landing gear and great mullet-shaped engine cowling. The terribly cramped cockpit was spartan by Lightning standards, and one item immediately caught Blackie's attention. It was a small crowbar attached to the inside of the cockpit entry door (useful in case of a jammed canopy or door when faced with sudden need for a bailout or hacking one's way out of a wreckage). Surprisingly, the Spit's Merlin engine was not as smooth-running as the Lightning's Allison powerplants; the Spitfire's stronger vibrations were instantly noticeable compared to the Lightning. The British fighter was delicate, light, and almost feminine compared to the Lightning. Most marks of the smaller British thoroughbred had a wingspan of thirty-six feet compared to the Lightning's fifty-two feet. The Lightning was almost as large as one of its PR rivals, the DeHavilland Mosquito. The first Spitfires weighed less than three tons, fully loaded. When empty, Blackie's Lightning was pushing seven tons and over eight tons when fully loaded with fuel. Not surprisingly, the Lightning's wing loading was over double that of the Spitfire. Flying the "Spit" would, however, remain a treasured memory for Blackie and one shared with very few Allied pilots. Blackburn returned to Washington as a passenger on a C-54 flight from Prestwick the day after chasing clouds with his Spitfire and did not fly again for two weeks during which time he was occupied with administrative details at ICD headquarters.

It is doubtful that Martha Blackburn would have agreed that Shoop's putting her husband's name on the mission board for flights over Occupied Europe was a wonderful opportunity for Blackie. German fighter pilots and legions of 88 mm cannon flak gunners were only part of dangers faced by Seventh Group pilots. During the two years that it was in combat, sixty-two of the group's pilots were killed, including twenty who died in flying accidents. Twenty-three Seventh Group pilots became unwilling guests in prisoner of war camps while another seven who were shot down managed to evade the Germans (six got back to England and one was interned in Sweden). Two other lucky pilots survived bailing out over the North Sea and were rescued; hypothermia, exposure, and exhaustion frequently claimed those who got so far as getting into their life rafts.

By late summer of 1944, the *Luftwaffe* was hardly the force it had been during the Battle of Britain in 1940, but in spite of fuel shortages, the German fighter arm was still a formidable opponent. While the *Luftwaffe* did not pay much attention to tactical bombers such as the B-26 Marauder or B-25 Mitchell operating over France and eastern Germany from mid-1944 on, it unleashed most of its fury on the heavy strategic bombers and the little reconnaissance planes that preceded and followed the bomber streams over the Nazi heartland. The Seventh Group faced such opposition, including increasingly frequent interceptions by ME-262 jet fighters that in the autumn of 1944 fighter escorts were arranged. The first Mustang escorts were very briefly provided by Eighth Air Force Fighter Command and then by the group's own dedicated flight of P-51D fighters.

Photo-recon pilots were highly regarded in the pilot community as well they should have been. Although PR pilots flew fighter planes, they received neither the glory nor the medals to the extent that the fighter jocks did. Like their bomber brethren who had to fly a straight and steady course during the actual bomb run regardless of flak and enemy fighters, the PR pilots had to fly a steady speed, altitude, and heading while their cameras were clicking away. And unlike fighter and bomber pilots, Lightning and Spitfire PR flyers had no guns in their ships. All they had were their wits and speed to avoid being shot down. "Alone, unarmed, and unafraid" was the popular motto ascribed to the PR boys. When questioned about it, they usually said the parts about being alone and unarmed surely were the truth.

Seventh Group pilots were only too aware of what could happen on a deep penetration into Germany. On 12 April, the Allies' leading photo-recon ace, the RAF's Wing Commander Adrian Warburton, DSO, DFC, disappeared flying a Seventh Group Lightning. Warburton had been assigned to liaison duties with the Americans as Col. Elliott Roosevelt's assistant commander at the 325th Reconnaissance Wing. Like Blackie, Warburton was flying a "guest mission" with the Seventh. On that mid-April day, Warburton and Major Carl Chapman, the Seventh Group operations officer who had flown with Blackie to the airfields in northern Germany and Holland on 16 August, comprised a two-ship sortie for bomb damage assessment at Regensburg and Schweinfurt, both well into southern Germany. Warburton had attained iconic status based on his photo-recon exploits in the Mediterranean and had even shot down enough enemy aircraft to qualify as an ace (some of the RAF's recce ships were armed). The pair flew together to a point 100 miles north of Munich and then split up; Warburton was to photograph Regensburg while Chapman covered Schweinfurt.

Then the pair was to rendezvous and continue on to Sardinia, returning the following day in what amounted to a shuttle mission. Warburton never made the rendezvous and his body was discovered fifty-eight years later deep in a southern Bavarian wheat field along with the shattered, burned-out remains of his Lightning, the victim of German anti-aircraft gunners.

The Allied recon pilot community was also abuzz about the disappearance of another prominent reconnaissance flyer – the famous French author Antoine de Saint-Exupéry who had disappeared on a mission from Corsica to southern France on 31 July 1944. Too young for active flying in World War I, he had learned to fly in the post-war French Air Force and after active duty, had remained in the Reserves. Until the French capitulated in 1940, Saint-Exupéry had flown reconnaissance missions against the advancing Germans, but later fled from France to America where he could concentrate on his literary career. By 1943, he was back in uniform and he was flying a photo Lightning on operations with the Free French in North Africa. Unfortunately, a fall on some marble steps in the darkened hall of a guest's home resulted in back injuries that kept him sidelined for over a year until he started flying reconnaissance missions from

Sardinia and later Corsica. He was on his ninth sortie after returning to operations in the summer of 1944 that he disappeared on a mapping mission east of Lyon. In stark contrast to Blackburn, Saint-Exupéry's health was not good; his bones alone, broken in aircraft crashes and accidents over the years, ached painfully at the 25,000 to 30,000 foot altitude levels where recon flyers in unpressurized aircraft spent their operational hours.

The remains of Saint-Exupéry's F-5B Lightning, later confirmed as his, were discovered on the sea floor just south of Marseille in year 2000. An identity bracelet bearing his name had been found in the same area by a fisherman nearly two years before. While inconclusive, an official French inquiry, including analysis of the Lightning's remaining parts and structures, captured German records and transcripts of *Luftwaffe* signals, suggested that St. Exupéry had not been hit by flak or shot down by a German fighter. Could hypoxia, which came so close to killing Blackie, been responsible? Saint-Exupéry was about one and one-half years older than Blackie when he disappeared – *un vieux* (an old man) making his contribution to the war effort by flying in a young man's game.

Any Lightning pilot faced the specter of mechanical failure, including oxygen system malfunctions of the type that had very nearly taken Blackie's life. The Spitfire PR Mark XI performed well, even at very high altitudes, especially with its Rolls-Royce Merlin engine and was faster than the Lightning. Although not as fast as the new German Messerscmidt 262 jet, the "Spit" could easily outrun, outturn, and outclimb any intercepting Germans flying the latest Messerschmidt 109 or Focke-Wulf 190 fighters, but the Lockheed Lightning, which Blackie flew on all of his missions, was another matter. To escape a closing ME-109 or FW-190, Lightning pilots had to quickly get rid of their drop tanks and dive which could result in the compressibility that was only beginning to be understood during the early war years. Occasionally, Lightnings shed their tails in high-speed pullouts – something that also happened during the stresses encountered in severe thunderstorms. Knowing that he had two engines on a Lightning was comforting to Blackie, but their location was not as reassuring. The turbochargers in the engines were only a few feet from his location in the central gondola, and when and if one blew up (as they occasionally did) the pilot was invariably hit by whirling, nearly red-hot shrapnel. The usual problem was with the turbocharger regulator which was prone to failure at high altitude, thus leaving the pilot with the not so cheering prospect of crossing hundreds of miles of enemy territory on one engine (assuming he had not been disabled by shrapnel if the event resulted in an explosion). Engine failure, in the near future, would be an unwelcome companion when Blackie and Shoop next flew together. Indeed the Lightning's engines also had a reputation for broken rods and thrown pistons at high altitudes in the European theater, although some of that was traced to fuel issues to which the Merlin-powered Spitfire seemed immune.

Blackie had been warned about a takeoff hazard <u>not</u> associated with the single-engined P-39 Airacobra, P-40 Warhawk, P-47 Thunderbolt, or P-51 Mustang. To emerge alive from a P-38 engine failure on takeoff required instant reflexes, because unless checked immediately the resulting asymmetrical thrust invariably, and instantly, rolled the P-38 into the ground on the side of the failed engine. Make that double trouble if a pilot forgot or did not have time to get rid of the two drop tanks – each filled with 165 gallons of 100 octane gas. Blackie also had to be concerned about air traffic, albeit the threat of a mid-air was not as great as that associated with flying bombers. Still, southern England was very high density airspace, and four of the group's twenty flying accident fatalities were the result of mid-air collisions.

Blackburn took off from Washington National again on 9 September, stopping at LaGuardia and Stephenville before arriving at Prestwick late on the tenth. His Seventh Group hosts, this time including Col. Elliott Roosevelt, continued Blackie's checkout in the B-26 Marauder during the nearly two-hour flight down to Mount Farm. Shoop was lucky to be joining Blackie that day. A week after the pair's mission to northern Germany and Holland in August, Shoop took on a mission to photograph bridges along the Loire just beyond the Allied lines in France. Low altitude photography of bridges was a very dangerous undertaking, and the group had lost quite a few pilots on those missions. Figuring that it would be best to make it a multi-ship flight, Shoop selected the redoubtable Carl Chapman and the group's resident bridge expert, Major Hubert Childress, skipper of the Twenty-seventh Squadron, as his wingmen. Childress was very familiar with Loire River bridges, having won a Silver Star in June for low altitude photography of bridges between Nantes and Tours. He was fortunate to have made it back alive – flak took out his left engine and the right tail boom was badly shot up.

The trio crossed over the invasion beaches in Normandy and proceeded into France where the ceiling was lowering. The primary objectives were covered in ragged, low clouds, rapidly fading light, and rain, so the Lightnings swung northeast to cover a last-minute secondary objective: the city of Paris which had reportedly been liberated that day. The weather improved considerably as they sped toward the City of Light, a place over which Shoop had some reservations based on how much flak it had put up when he was on temporary duty with the Fifty-Fifth Fighter Group. Nor had he forgotten about all of the flak that was directed at him while over Paris on the return leg of his June shuttle mission to Italy. And this time they were going in relatively low at 6,000 feet because dramatic photography had been requested from Eighth Air Force headquarters in conjunction with the Allied liberation of the city. Just before reaching *L'Arc de Triomphe* in an echelon right formation, what seemed like every German flak cannon in the "liberated" city opened up on the Lightnings. Shoop rammed the throttles fully forward and went into a climbing turn, hanging right on the edge of a high speed stall. The black puffs had him completely bracketed – fore and aft and on both sides. He had heard that if flak is bursting immediately in front of him, he could count on buying the farm in the next few seconds. Moments later, but what seemed like hours, he was out of the flak, pulling the throttles back, and heading northwest to rendezvous with Chapman and Childress. The latter pair had immediately broken in different directions and headed for the deck, thereby creating more difficult targets for the German gunners. At the debriefing, Shoop quietly resolved that the next time he took pictures of a "newly freed" city, it was going to be from a Jeep with a Kodak box camera and with an army of American infantrymen surrounding him.

There were a few light moments amidst the tension of flying combat missions. Shoop had been able to convince one of the base service officers to get a copy of *Northern Pursuit*, a movie starring Errol Flynn (Hollywood's leading Lothario at the time), Helmut Dantine, and Shoop's wife, Julie Bishop. The film, shown in Mount Farm's theater, was about how Flynn's character, a Royal Canadian Mountie, foils a German plot to drop a pilot into the frozen Canadian north to fly a previously landed and cached bomber to destroy key targets along the American-Canadian border, thus severely crippling the shipment of war supplies. Although the story line was based around Hudson's Bay's northern shores, the film was shot in Sun Valley, Idaho. Shoop had to tolerate a lot of good-natured ribbing about just why he was getting shot at over Occupied Europe while Errol Flynn was chasing his wife, Julie, back in the States.

Roosevelt, now sharing the flight down to Mount Farm, would play a major part in Blackie's next combat adventure which would extend all the way to North Africa. Roosevelt was a rather controversial figure in Army Air Force circles. After graduating from prep school and deciding not to attend college, his ten-year, scandal-swaddled business adventures in the 1930s were spent primarily in broadcasting, ranching, and aviation (regional airlines, journalism, promoting, and lobbying). His personal life was just as checkered as his business career: by the time Blackburn met him, Roosevelt had just shed the second of his five wives.

Roosevelt had been a student pilot in the 1930s, but had never progressed to the point of earning a private pilot certificate. He had been commissioned directly into the Army Air Force in 1940 not as a second lieutenant, but as a captain, which provided newspapers with plenty of fodder about favoritism in high places. In 1941 Roosevelt conducted valuable aerial surveys in Labrador, Baffin Island, and Greenland that would lead to new American bases on the North Atlantic Ferry route. His accomplishments on that assignment led to success on subsequent assignments in intelligence and reconnaissance in North Africa that also were of great value in the subsequent invasion of North Africa in 1942. Before that invasion, Roosevelt, who had completed navigator training in late 1941, was given command of the Third Photographic Reconnaissance Group which was forming up at Peterson Field in Colorado Springs. He could not pass a military pilot's physical exam (his very poor eyesight alone was disqualifying) and was frequently absent because he attended President Roosevelt's wartime conferences at Casablanca, Cairo, and Teheran as FDR's military aide.

Nevertheless, Elliot Roosevelt's bravery under fire and his strong leadership of the group in North Africa, particularly during the invasions of Sicily and Italy, resulted in his becoming commander of the Ninetieth Reconnaissance Wing. The wing was located at San Severo, Italy, when Roosevelt was tapped for the top reconnaissance job on the upcoming invasion of France. Although his peacetime managerial talents were dubious, his managerial talent blossomed in the military environment which was responsible for his rapid advancement (he made general before the war ended). Roosevelt was not popular with the British who saw him as a control freak who wanted to break up the joint RAF-USAAF responsibility for photo interpretation and set up separate commands. Splitting the two groups would take up the better part of two months – something that would be completely unacceptable during the run up to D-Day. General Spaatz, the Eighth Air Force commander and Blackie's former group C.O., decided to retain the existing structure, and when Roosevelt was posted back to the Pentagon in November, few people in the RAF were sorry to see him depart. Roosevelt had also wanted to get the latest DeHavilland PR Mosquitos for the Seventh Group based largely on the plane's excellent range; the RAF emphatically said "no" and kept the latest PR Mossies for itself. When Roosevelt flew on operations, he flew as an observer or navigator and mission commander while his aircraft was flown by a rated pilot. Roosevelt had very good connections. While Shoop had the luxury of flying a B-26 Marauder for personal transportation when not on operations, Roosevelt's personal plane was nothing less than a B-17 Flying Fortress.

Blackie was aloft early on Sunday 13 September in an F-5 Lightning, Army serial number 42-68106, flying the group's sortie number 3307. It was a four hour, forty-five minute mission flown over deep southeastern France, although it began by crossing into France right over the Normandy invasion beaches which had by then become a great distribution center, moving men and material inland. Southern France had been invaded only a month before and much had to be known about how the German defenses were being established in the upper Rhine, an area that the Seventh had visited only infrequently. Allied intelligence had determined that the Germans would establish a winter line in the Vosges Mountains to thwart the American Seventh's Army's advance. Allied planners also needed to know what resources were available to the Germans just east of the Maginot Line, France's failed defense which was still in German hands. The distance was such that a shuttle mission to Italy was justified. It was a two-ship sortie, and again Blackie's senior partner was Colonel Shoop who had decided to fly the mission himself. Shoop was also interested in a little "inter-theater cooperation" and wanted more contact with photo-recon operations in Italy. There was plenty of precedent for cooperation – the Seventh had a pilot exchange program with the Fifth Photographic Reconnaissance Group while the latter had been based at Bari, at the top of the Italian heel on the Adriatic Coast. In the back of Shoop's mind was the possibility of extending the flight down to

Egypt and getting together with a favorite, former commanding officer: Major General Benjamin Giles, the theater commander in North Africa. Giles had been instrumental in advancing the career of young Lieut. Shoop in the 1930s.

The weather was not especially conducive to good photography on that thirteenth day of September. At takeoff, there was a stiff breeze of nearly twenty mph from the south along with considerable clouds aloft. Sky conditions were scattered cumulus with bases and tops at 3,000 and 5,000, respectively, and scattered altocumulus between 15,000 and 18,000. Visibility was unrestricted at altitude, but downward and slant visibility was poor. Conditions would worsen into central and southern France.

Fortunately, the *Luftwaffe* did not intercept their flight – most of the closest *Luftwaffe* fighter assets were being used to thwart the Eighth Air Force's missions 628-630 in which 1,015 bombers struck industrial targets in Stuttgart, Darmstadt, Wiesbaden, and Ludwigshafen in southwestern Germany that day. Fifteen American bombers and eight fighters were shot down on those three missions. Over France, it was a relatively quiet day in which several bomb, fighter, and reconnaissance groups moved to France from their bases in England. Still, Ninth Air Force fighter-bombers were particularly busy supporting Allied ground troops in the Nancy-Metz areas – off Blackie's left wing on the flight down to Italy.

Whereas the *Luftwaffe* and flak gunners did not challenge Blackie and Shoop, Mother Nature and her concealing clouds certainly did. The Lightnings did their best to photograph the French targets, but the pair still had a lot of remaining film in their magazines after reaching Italian airspace where the weather steadily improved as they flew south. Knowing that there was plenty of film left, the pair photographed Rome and then turned southeast to complete their mission early in the afternoon. Cruising over Rome was a wonderful new experience for Blackie who had enjoyed his school courses in geography, history, and Latin. He was quite taken with the Coliseum, St. Peter's, and Rome's other major landmarks, but was disappointed in the River Tiber which, in early autumn, struck him as really not much more than a flow-challenged drainage ditch compared to the other sights. Little did he know then about how familiar Rome would become to him not so far into the future.

Beyond Blackie's right wingtip was Naples and the Tyrrhenian Sea. Just north of Naples was the town of Fondi; then lost in the distance. Only eight months before, a very young fighter pilot from the 525th Squadron of the Eighty-Sixth Fighter Group had been shot down over Fondi while on an armed recon mission. The young lieutenant was flying an A-36, the dive bomber version of the P-51 Mustang, on his thirtieth mission. Because he was the newest man in his squadron, he had the unenviable assignment of being the last pilot to dive on the target, a German truck convoy. By the time he released his two 500-pound bombs, the German gunners had tracked him perfectly. Lieutenant Robert Drew was literally blown out of his plane when it was hit on the climb-out at 3,000 feet after bomb release at 1,000 feet, but he managed to evade the Germans after his parachute landing on a rooftop in the middle of town. Blackie had no knowledge of this of course, nor was he aware that Drew's quarters back at his base outside Naples in Pomigliano had had a guest – the noted war correspondent, Ernie Pyle. Intrigued by why Lieutenant Robert Drew was writing a letter addressed to Lieutenant Robert Drew, Pyle inquired what the coincidence was all about. The young Drew was writing to his father, who had the same name and rank. The elder Lieutenant Robert Drew was a service pilot in the Air Transport Command, most recently ferrying planes from their American factories, to Gander, and then across the North Atlantic to Prestwick. The elder Drew had served in the Navy during World War I, had learned to fly during the interwar years, and had earlier been delivering Army planes to India, North Africa, and the Pacific theaters. Pyle was the younger Drew's first contact with a journalist. It would not be the last. Drew would go on to become an award-winning, internationally respected documentary film maker. His subjects were varied, but all had an interesting story to tell, and in less than twenty years, Drew would be telling the Captain Harold F. Blackburn story.

Blackie's flight landed in clear conditions under high, scattered clouds at San Severo where the Fifth Reconnaissance Group and the Thirty-first Fighter Group were based. Northern Italy still had *Luftwaffe* fighter fields and was supposedly the home of the *Aeronautica Nazionale Republica*, that portion of the *Regia Aeronautica* that did not change sides in 1943 and was still loyal to Benito Mussolini who had been deposed as the Italian leader the previous year. The Thirty-first was a top-scoring fighter group, but most of its recent victories had occurred on bomber escort missions to southern Germany, Austria, Poland, Czechoslovakia, Yugoslavia, and, until it surrendered to the Allies in August, Romania. Although reconnaissance showed that the *Luftwaffe* still had a presence in northern Italy, the Thirty first had not bagged any German fighters over southeastern France or northern Italy in several months. That did not, however, stop Blackie and Shoop from constantly turning heads and scanning the skies for Germans trying to bounce them. In all of that neck craning and sweating, Blackie realized why those white silk scarves were so popular in the fighter community – they helped prevent the chafing that came with constant neck turning.

There was, however, a very active air war over northern Italy as Blackburn and Shoop entered Italian airspace. Out to the east, the Twelfth Air Force (responsible for tactical operations) was bombing bridges between Milan and Verona. Fighter-bombers were hitting trains and bridges from Florence well up into northern Italy. Little of this could be seen because there was a heavy undercast below the Lightnings in the far Northern provinces.

San Severo had been home to Roosevelt's original Third Photographic Reconnaissance Group before it moved to Corsica with the Twelfth Air Force to support the invasion of Southern France only a month before. The base was one of two dozen Army Air Force and Royal Air Force fields in the Allied fifty-mile wide complex in and around Foggia on Italy's east coast. San Severo is located just inland from the Adriatic Sea in the province of Apulia – roughly 170 miles north of the bottom of the heel in the Italian boot. Blackie's was not the first shuttle mission flown by a Seventh Group Lightning; there had been an average of one shuttle per month in 1944. Previous shuttles had taken Seventh Group pilots to Sardinia, Poltava in Russia, and Bari, about eighty miles south of San Severo in Italy. Colonel Shoop had flown a shuttle on 13 June over Paris and Lyon, landing in Italy outbound. He photographed more targets on the way back and attracted plenty of flak near Paris in the process. Just a few miles away and several minutes after Shoop overflew Paris, an F-5B Lightning from the group's Thirteenth Squadron was hit by flak. The pilot Lt. Adam Tymowicz was lucky; he bailed out, and although wounded, reached the French Resistance which sheltered and returned him to Allied control exactly two months later.

San Severo was extremely busy on that thirteenth day of September. The Thirty-first had escorted heavy bombers to Odertal, Germany, to hit a refinery, but encountered no German fighters. One pair of the group's Mustangs had left the formation near the target due to oxygen problems and was returning on the deck, shooting down two German fighters along the way. Transient quarters were in short supply that night; thirty Mustangs from an Eighth Air Force fighter group had touched down that afternoon on the second leg of another shuttle mission from Russia and would not be departing until the following morning. At the debrief after landing at San Severo, Shoop learned that orders had come through to the Fifth Photographic Reconnaissance Group to photograph four of the German airfields on Crete. After a little wheeling and dealing with the group commander and operations officer, Shoop had his orders to photograph those four fields and then continue the mission to Cairo with a return to San Severo on the fifteenth.

When taking off from San Severo, the beautiful city of Dubrovnik was on the other side of the Adriatic Sea, and as the two Lightnings headed south, out to the east were the mountains, seaside cliffs, and deep bays of Croatia, Montenegro, and Albania. After passing over the mountainous core of Greece, the pair flew over places sacred to Greek history, including Patra, an important center of Mycenean culture, and Sparti off to the east with its legendary Spartan warriors. Flying over Kalamata, famous for its olives and figs, the two pilots emerged over the Laconian Gulf with the Ionian Sea and Messenian Gulf to the west and the Aegean Sea to the east. Ahead and to the left were the many Greek Islands. Two hundred miles to the southeast was Crete and its four target airfields. East of Crete was Rhodes, still occupied by the Nazis. Crete had been in German hands for over three years and had been taken by a large-scale aerial assault primarily by parachute troops. The Germans had paid a disproportionally huge price for Crete which they saw as a strategic threat to their eastern Mediterranean and Balkan plans in 1941. With North Africa, Sicily, and most of Italy in

Allied hands by autumn 1944, Crete had become less of a problem, but was still worth periodic monitoring and bombing.

When briefed by the group intelligence officer, Shoop and Blackie learned that the Germans had seven airfields on Crete, but that two of them, Kissamu and Reythymnon, were little more than emergency landing fields. Timpaki, a major field on the south coast, had been practically bombed out of existence and was only infrequently photographed. Three of the mission's objectives, Maleme, Suda Bay, and Heraklion were all along the north coast, and Kastelli was about fifteen miles southeast of Heraklion in central Crete. Shoop decided to photograph Maleme first and head east along the north coast to cover Suda Bay and Heraklion before turning southeast to overfly Kastelli on the way out to Cairo. Maleme, Heraklion, and Kastelli had been bases for *Luftwaffe* fighter, transport, and bomber units, but in September 1944 the only units stationed on the island, with one exception, were there on a temporary basis. The exception was at Suda Bay which technically was not an airfield; it was an established seaplane base that had been used by the RAF's four-engine Sunderland flying boats before the island fell to the Germans in 1941. Suda Bay's full squadron of Arado 196 single-engine float planes were active right up until the last day of the war flying reconnaissance and air-sea rescue missions. Other than temporarily based fighters, the airfield defenses ranged from Suda Bay's eight heavy and twenty-four light flak guns to nearly twice that firepower at both Maleme and Kastelli.

As Blackburn continued his odyssey on 14 September, it was a relatively quiet day at San Severo. The Fifteenth Air Force flew no missions that day, so the Thirty-first was standing down except for a flight of eight of its Mustangs which escorted one B-17 with VIP passengers to Bucharest, Romania. The Twelfth Air Force was busy, however, pounding German defensive positions in the mountainous Gothic Line and hitting transportation and communications targets in the Po Valley to the north

As long as Crete could be supplied from Greece, the *Luftwaffe* on Crete was a problem for Allied shipping in the eastern Mediterranean. The Germans were run out of Greece by the end of the year, but the garrisons on Crete and other eastern Mediterranean islands such as Rhodes, held out until VE Day. Although the *Luftwaffe* still had a presence in Greece, particularly around Athens, most of the Allied bombing activity in the early autumn of 1944 was directed at blocking German exit routes from Greece and southern Yugoslavia.

Shoop planned the mission to be flown at 25,000 feet, but arriving over Maleme airfield on Crete's northwestern coast, he dropped the flight down to 15,000, hoping for a better look at the target below which was largely obscured by clouds. Maleme had been visited little more than a week before by a flight of RAF Wellington bombers which could not claim any significant damage although one secondary explosion and several fires were recorded as the flight swung off target. Between breaks in the clouds over Maleme, Blackie and Shoop could occasionally see *Luftwaffe* aircraft below but soon gave up and headed due east for Suda Bay where conditions were no better. Unseen below the clouds hanging over Heraklion as the Lightnings flew east were the ruins of Knossos Palace which several thousand years before was at the core of the once highly advanced Minoan culture. The German occupiers below were the latest in a long series of calamities plaguing Crete, including earthquakes, tidal waves, crop failures, invasions by Greeks, and occupation by another northern force – the Vikings. Remaining fuel was becoming a concern, so the Lightnings quickly turned southeast for Kastelli with Blackie settling into a position slightly below and roughly 400 yards to the right of Shoop. The clouds also obscured Kastelli which had been bombed, with reportedly good results, in late July by a full squadron of Wellingtons. There were almost two hours of flying left before reaching Cairo, and it was the type of flying to which Blackie had become very accustomed: over vast expanses of ocean with no land whatsoever in sight ahead. That was very much in contrast that was to what he had seen on the first stage of this flight.

But after leaving Crete, the monotony of flying two more hours in a featureless environment set in. Still, vigilance was required both inside the planes and outside. Blackie's sweep took in the quantity, pressures, and temperatures of the Lightning's vital, bodily fluids along with frequent glances at all of the air data instruments. He also had to keep an eye on his friend off to the left and above. On one of his regular visual sweeps outside the cockpit, Blackie was startled to see that Shoop was gone! No, there he was behind and further out to the left now, falling back, trailing some thin smoke from his left engine. The radio crackled to life announcing the problem just as Blackie instantly pulled his throttles back to rejoin Shoop who had been startled by a loud "bang" and smoke pouring from his left turbocharger. Shoop quickly pushed his right rudder pedal and retrimmed the ship, realizing that the supercharger had not exploded but that he wasn't getting full power from the engine either. By re-adjusting the mixture, throttle, and propeller pitch controls and changing fuel tanks, he was able to maintain some power in the engine. With his tachometer, manifold pressure, and temperature indicators settling down, Shoop cautiously proceeded into what was going to be a slower, somewhat longer than planned for flight. Fuel was a concern, but not a major issue, and if the left engine had completely packed it up, Shoop still had crossfeed controls to keep the right engine turning.

As with Blackie and his oxygen problem on the way to Langeoog, Shoop got lucky because the supercharger had not actually exploded, thus no turbine parts hit him nor had the engine caught fire. It was one of those moments when it was best to be flying the twin-engined Lightning. A Spitfire driver in a similar situation, in spite of his plane's exceptional glide ratio, would be facing an outcome that would at the very best only result in sitting in a very small life raft being drenched by very big breaking waves a very long way from anywhere.

Shoop and Blackie had several layover days in Cairo after arriving because Shoop's plane was in the shop for substantial work at Payne Field, a major Air

Transport Command base. His F-5 Lightning, a B Model was not the oldest ship in the group, nor was it the youngest either. It is possible that Shoop may have actually flown the acceptance flight on it himself; it came off the Burbank production line and was accepted on 20 December, 1943, while Shoop was still assigned to Lockheed. That Lightning was picked up by a ferry pilot from the Sixth Ferrying Group in Long Beach and flown to Dallas, Texas, on the twenty-eighth via Tucson. The plane was at the Love Field Modification Center in Dallas being transformed from a stock P-38G without the usual four .50 machine guns, one 20 mm cannon, and reflector gun sight into an F-5B-1 with camera controls, mounts, and camera ports along with updated drop tank hard points. On 16 January it was flown to Newark, New Jersey, where it languished until 28 January when it was placed on a ship, finally arriving at Mount Farm on 23 February 1944. There followed fourteen months of being "rode hard." The ship never fell to German guns; instead, it was destroyed by its owners – scrapped back in the United States on 24 June 1946. Fate was not kind to the recce Lightnings which were bested in speed, range, ceiling, and, especially, operating costs by the F-6D, the Mustang conversion for photographic reconnaissance. The F-6D continued to be used as late as the Korean War for tactical reconnaissance – which had been its recce origin in both Europe and the Pacific late in World War II. Blackie's Lightning never even made it back to the United States; it was scrapped in England on 21 April 1945.

Just after landing in Cairo, Shoop called his former commander from California, General Benjamin Giles, just expecting a quiet dinner and a chance to reminisce about all of the good hunting and fishing in California. Giles had something else in mind: cocktails and some live theater where they saw "No, No Nanette" followed by dinner with their Egyptian hosts around midnight. The food was a bit exotic by the transient pilots' standards as was breakfast which consisted primarily of mangoes. After breakfast Shoop was told that his Lightning was going to be down for at least two days, if not more, which suited him fine because General Giles wanted the two pilots to see more of Egypt while they were in Cairo. Using the Jeep placed at their disposal, the two pilots spent the morning with a guide touring the pyramids which had actually been visible from General Giles' penthouse atop an apartment building located on the shores of the Nile. Now enjoying the scenery, superb flying weather, balmy nights, and exotic surroundings, Shoop started having doubts about his fighting the weather-cursed war in the European theater of operations.

Giles put one of his officers in charge of the afternoon tour, a trip to the bazaar in which Blackie developed a particular interest in Middle Eastern art – an interest that future travel would allow him to further develop over the following decades. Shoop bought a gold-inlaid cigarette box for his new bride. Walking along Tahrir Square, Blackie noticed a shop with the unlikely name of "Ding Dong Bazaar." Intrigued he went into the shop and was greeted by a gnome-like, energetic little Coptic Christian fellow named Bassili Tawadros, proprietor of the establishment which sold all sorts of Egyptiana, Middle Eastern clothing, and reproductions of various antiquities. Blackie purchased nothing significant that day, but he would be back many times over the years.

On Saturday the sixteenth, Blackie flew a C-47 up to Palestine and Tel Aviv's Lydda Airfield which, like Payne Field, was an Air Transport Command base. The objective was to visit Biblical sites beginning with Jerusalem; their guide was told that the pilots only had a half-day to see it all. The whirlwind tour included the Mount of Olives, the Garden of Gethsemane Church of the Crucifixion, the Wailing Wall, Stations of the Cross, and marketplace. The pilots had a few mixed impressions about Jerusalem – being in a place where reverence was certainly called for, but at the same time there was some disappointment in the rampant commercialism. Thus, the continuation to Bethlehem with its Church of the Nativity was a breath of fresh air because it was not burdened with so many peddlers. Shoop paused briefly near the Church to buy a Bible for Julie. Unfortunately, the trip was a short one because they were on a tight schedule and had nearly 270 miles to fly back to Cairo where Blackie and Shoop enjoyed a nice dinner with General Giles and a movie courtesy of Army Special Services. Sitting at the dinner table that night, Blackie had no idea what the future might hold for Giles and Shoop or that both of their paths would lead to TWA. In Giles' case the association would be sooner rather than later. The work on Shoop's left engine still was not done on the morning of the seventeenth, so the two pilots flew a C-47 up to Alexandria with General Giles who was on his first day off in a month and enjoyed Princess Beach on the coast. The Egyptian food started tasting better after a day of riding donkeys, sailing, and swimming in the Mediterranean.

Blackburn and Shoop began the return leg of the shuttle the next day with a six and one-half hour flight back to San Severo. This was by a different route: via Benghazi, Libya, where they stopped for fuel. There was an ulterior motive (wanderlust) in each of the pilots, and that was to see the remains of battlefields and cities, both ancient and modern, along the way. The flight started with several spectacular monuments only minutes to the southwest from Paine Field – the great Pyramids and the Sphinx which they had seen from the ground just days before. The pair photographed those wonders from antiquity at little more than 1,000 feet of altitude, including direct overhead and oblique angles before taking up a northwesterly westerly heading.

Shoop had originally planned on flying back via Crete to photograph the four airfields, but had been told by the Payne Field weather officer that all he could expect was cloud cover comparable to that experienced on the fourteenth. Below their track to Benghazi was the trail of ancient civilizations. Alexander the Great had marched in the shifting sands under their wings. Great civilizations had come and gone thousands of feet below and many centuries ago, including the Byzantines, Carthaginians, Phoenicians, Romans, Muslims, and even the Vandals. Flying out of Payne Field to the coast they soon saw El Alamein where Rommel, the Desert Fox, had met his Waterloo in the autumn of '42. Marsa Matruh and Sidi Barrani in Egypt had been the scenes of battles between the *Africa Korps* and the British Eighth Army. In Libya's coastal plain, the Lightnings flew over two cities where the battles had see-sawed back and forth not that long ago: Derna and Tobruk. Little more than a year ago, on 1 August 1943, Benghazi's airfields had been home to the five heavy bomber groups comprising Operation Tidal Wave – the very low altitude bombing of the petroleum center in Ploesti, Romania. After Tidal wave, those bomb groups had moved with the war – several to Italy and the rest back to England. Benghazi was now a stop on the Air Transport Command's North African route structure, no longer a complex of Liberator bomber bases. The date of the raid or the name "Benghazi," even decades later, dredged up a bittersweet memory for Blackie. Alphabetically, the first man in Blackie's cadet class at Kelly was Addison Baker, a young man from Akron, Ohio. Baker, six years younger than Blackie, was commander of the Ninety-third Bomb Group, the second group in Operation Tidal Wave, and it was Baker who realized that the first group had committed a navigation error that would take all five groups to Bucharest, not the objective: the Ploesti petroleum complex. Baker swung his formation and those following onto the right course. Although his B-24 was hit by flak and on fire, Baker led the formation to the target, successfully bombed it, and only when past the target broke formation to attempt a crash landing. The Liberator exploded on impact with no survivors. Baker and his copilot, John Jerstad, a Luke Field graduate in Cadet Class 42-B, were awarded the Medal of Honor.

Seventh Photographic Reconnaissance Group

Using his Lightning's oblique camera, Blackie photographed Colonel Clarence Shoop over the Pyramids of Giza on the flight back to San Severo, Italy.

The balance of the flight to San Severo proved uneventful, but the visibility was marginal during most of the flight, making it difficult to see Malta, in 1941-1942 the site of air battles second only in ferocity to the Battle of Britain in 1940. Also off his left wing was Sicily which had been taken from the Germans and Italians only the year before. Blackburn and Shoop landed at San Severo with twenty miles of visibility under scattered cumulus clouds after the Thirty-first Fighter Group returned from an uneventful mission that day – escorting several heavy bomb groups to destroy bridges near Budapest. It had been a long day in the cockpit for Blackburn and Shoop – six and one-half hours of being cooped up and strapped into a noisy, cramped environment without much to do. Blackie could, and indeed had to, scan the sky and the instrument panel, inspect the engine cowlings for any leaking oil; he looked at a loose rivet twisting in the 250 mph wind, wonder about the black dot at the nine o'clock high position – some Jerry on a bounce, another Allied fighter on the way to Sicily, or a tiny speck of dried oil on the canopy glass?

On the next day, a Tuesday, the pair left sunny San Severo very early on a nearly five-hour return flight over France to Mount Farm. San Severo was relatively quiet the morning Blackie left, other than Thirty-first's 308th Fighter Squadron taking off to escort a PR Mosquito from the South African Air Force on a ten-target mission over southern Germany. The Fifteenth Air Force would bomb railroad bridges in Yugoslavia later that morning, although the escorts were not the Thirty-first's Mustangs, but rather P-38 Lightnings of the First and Fourteenth Fighter Groups flying from other Foggia complex airfields. Several hours after Blackie was airborne, San Severo's traffic pattern started heating up – another Eighth Air Force fighter group was landing after flying in from a Russian shuttle. The weather that day was yet again not favorable, a factor thwarting the 796 bombers from the Eighth which hit marshaling yards in western Germany. Many of the bombers struck alternates or other targets of opportunity due to the weather which also forced Blackie to fly on instruments for the second half of the return flight to Mount Farm.

The nineteenth was also the third day of Operation Market Garden which was designed to cross the lower Rhine by a major airborne assault through Holland, thus eliminating the problems associated with attacking the northern end of the German Siegfried Line head-on. Market Garden was a costly failure of which the abysmal weather, through which Blackie was flying on instruments, was an important factor. Another one of Blackie's Kelly classmates was impacted by the poor mid-September weather – Lt. Col. Lloyd C. Waldorf had just taken command of the 440th Troop Carrier Group the day before. Waldorf had flown on the seventeenth, towing gliders into Holland, when his group commander, Col. Francis X. Krebs, had been shot down and was posted as missing. Fortunately, Krebs had been taken prisoner but managed to escape nearly six weeks later. The Ninth Air Force was just barely able to pull off medium bomber raids on marshaling yards where German reinforcements were being sent by rail to the front. Also occurring on 19 September, the opening day of the Battle of Huertgen Forest was being fought on the eastern side of the Belgian-German border. The day did not go well for the American First Army which would suffer over 33,000 casualties during the three month campaign.

Mount Farm, with a very low ceiling on the afternoon of the 19th, was progressively getting socked in as were most bases in the middle and south of England. It was just a poor day to be flying. Although the high cirrus layer at 25,000 to 30,000 that day was not an issue, the problem was the layered stratocumulus varying from 600 to 1,000 feet all the way up to 13,000 feet.

In 1943, at Colonel Hall's insistence, a VHF direction finding (DF) system was installed at Mount Farm to reduce reliance on the DF system at nearby Benson. The range of Mount Farm's VHF direction finding system (call sign "Skate Rink") was nearly 100 miles, so once a pilot was close to England's shores, he could at least be vectored to Mount Farm. The group's Spitfires and Lightnings were, unlike many bombers and transport aircraft, bereft of any instrumentation facilitating an instrument approach such as that afforded by a non-directional beacon or the new instrument landing system that was just starting to come into use back in the United States. Once a pilot was over Mount Farm, he was pretty much on his own to get down safely although the tower could provide what amounted to very rudimentary ground-controlled approach headings which were derived from radio direction-finding equipment on the ground. The pilot was given headings to fly that would result in his being on the final approach course. Flares would be fired off once he was close, and if he could not see them he would hopefully be vectored elsewhere. If elsewhere was no better than Mount Farm, the tower advised the pilot to head for open country and bail out once at a proper altitude. Getting down safely was not a problem for an experienced instrument instructor pilot with an airline transport rating like Blackie. It was a problem for most younger pilots who had neither a "green card" (a military instrument rating) or hours of instrument experience beyond the basics acquired in cadet training.

Even in good weather, many airfields in England were hard to find because the landscape below was largely free of distinguishing landmarks. Mount Farm was one of the exceptions.

Prominent Mount Farm landmarks included the spires and domes at Oxford University, particularly the tower at Magdelen College, which was little more than seven miles to the northwest. The field was just northeast of a large bend in the Thames River, and the main road from London to Oxford also was nearby. The adjacent land was essentially flat to barely rolling, but the dominant feature, just south of the field, was Wittenham Clumps, a small hill with a wooded top (rumored to be a fort of the ancient Celts).

Thus the return portion of Blackie's shuttle mission, sortie number 3381, was a weather-induced failure – not a single photo was taken over eastern France on the way back from San Severo. Of the 297 sorties the group flew in September 1944, thirty-one were thwarted by bad weather over the targets. In that September, eleven of the group's pilots were killed or captured while Lt. John Ross of the 22nd Squadron was plucked from the channel after ditching his F-5. Just in the week that Blackie had been on his shuttle mission, the group flew seventy-five sorties of which ten were failures due to down cloud, but worse was the loss of two Lightnings and one Spitfire – with all three pilots killed in action.

September 19, 1944, was supposed to have been a red-letter day at Mount Farm. In the morning sixteen sorties were still flown – mostly to southern Germany where the weather was somewhat better. Shortly after Blackie's flight landed on a drizzle-slickened Runway 9, there was a rumbling, droning sound aloft in the gloaming over Mount Farm. Soon a B-17 in natural aluminum finish emerged from the low-hanging clouds and landed. Out stepped Colonel Roosevelt who had been hoping for a bright sunny day for the major awards parade that was to be held on the field late that afternoon. By the time he got out of the Flying Fortress, it was raining hard; so the ceremonies were conducted in one of the large maintenance hangers. Roosevelt awarded a number of Air Medals and a few Distinguished Flying Crosses of which one of the DFCs went to Colonel Shoop. Then Roosevelt called Shoop forward again and pinned the ribbon of the

Presidential Unit Citation above Shoop's left pocket. All personnel in the group were then entitled to wear the ribbon which was awarded for the outstanding performance in the many missions flown leading up to D-Day. Blackie, wearing a truly damp raincoat, observed the ceremony from the front row. Lt. Grover Parker, who had flown a Twenty-seventh Squadron Lightning to cover Operation Market Garden could not attend the ceremonies; he was hiding in a Dutch farmhouse, having been shot down by flak that morning.

Blackie's shuttle mission was the first of only two shuttle missions flown to Italy in September 1944. A week later, Captain Frank Carney flew a Twenty-seventh Squadron Lightning to photograph marshaling yards and airfields at Debrecan and Szolnok in Hungary. On the next day, he returned to Mount Farm from Italy, photographing targets in eastern France.

Blackie's landing back at Mount Farm concluded an adventure of which he would always derive a great deal of private satisfaction over the following decades. It soothed some of the burning frustration he felt when his persistent attempts to get back on active duty and into the fight were consistently thwarted. Time has erased the runways and buildings of Station 234, Mount Farm. Where so long ago the four squadrons' blue-hued Spitfires and Lightnings with white stars and bars on their wings and fuselages rose up from the ground, now corn and barley reach skyward every summer.

Had he been on active duty, the six combat flights with the Seventh would have qualified Blackie for an Air Medal, a standard Eighth Air Force recognition for satisfactory completion of five sorties. Looked at another way, his missions amounted to twenty-five percent of what was an average tour for operational photo-recon pilots in the late summer of 1944. One of the group's young stars, Lt. John Blyth, flew fifty-one missions primarily in Spitfires, more than any other pilot in the group. Others were rotated home after roughly twenty-five missions. The Air Medal was never awarded, but even in its absence Blackie knew that he had done his combat flying bit for the war effort. On 20 September, he and Shoop flew the group's Marauder on the two-hour flight up to Prestwick to begin the journey home.

One month later Shoop was back in England, but only briefly. That same Marauder, Army serial number 42-197481, was again flown up to Prestwick, in late October by Col. Shoop, but he would not be bringing it back to Mount Farm as he had on 20 September; his good friend Carl Chapman would. Shoop too was about to return home – this time to a new assignment: evaluating the new Lockheed P-80 jet fighter as a photo-reconnaissance platform. Shoop's and Blackie's paths would cross again. Although Shoop made it home safe and sound, his Marauder never did – it was scrapped in the UK on 21 September 1945.

Chapter Nine
Far Horizons

As if being punished for having broken the rules and experiencing one of life's greatest exhilarations (according to Winston Churchill anyway), that is being shot at without effect, Blackie's flying after his exciting sojourns with the Seventh PRG took a major nosedive. From late September through the end of the year, he flew with several new pilots on a nine-hour training exercise on December third and then a six-day roundtrip to North Africa via the Azores mid-month, and that was it for the rest of 1944. His employer was making sweeping changes for what was expected to be a post-war explosion in both domestic and international travel. At about the same time Blackie returned from his combat flying, TWA announced several significant organizational changes which were to prepare for both the growth in passenger traffic and the re-employment of many TWA employees returning from duty once the war was over. In a few cases, TWA had already welcomed back some returning veterans – those who had recuperated from wounds, but were no longer able to serve on active duty. At a higher level, former board chairman T.B. Wilson returned to TWA after his latest job in the Army Air Force (establishing General Stilwell's transportation systems in the China-Burma-India Theater of war) had been completed. The Army had originally seen Wilson's tour as lasting six months – it had turned into two and one-half years.

The principal change was greater decentralization of the operations department into four separate regions. Although this was a major change within TWA, it had no impact on Blackie since his future had already been dictated by the nature of ICD's flying. Although June and July 1944 had been ICD's biggest months with almost a million and a quarter miles flown, the monthly average following that peak was still well over a million miles per month. ICD was becoming a major employer within TWA with almost 1,600 employees in Washington and TWA bases around the world in the late summer of 1944. In September another major change was announced – one that impacted the Blackburn family quite directly. As part of its gearing up for peace, Cliff Mutchler would be returning to his old job in Kansas City effective on 1 January. And Blackie was promoted to become the new manager of ICD. That was good news for the Blackburn family in terms of prestige, responsibility, and income although one aspect of the job was unappetizing, that being fewer prospects than ever for Blackie's flying time.

The change in management structure at ICD was practically a wholesale one with many of Blackie's closest friends moving up one notch. Swede Golien became the division's assistant manager and Frank Busch took Golien's old job as supervisor of flight operations. Joe Carr gave up the training manager's job to become chief check pilot while Milo Campbell went into supervising administrative operations. Fred Richardson and Dutch Holloway became chief and assistant pilot respectively.

News of Blackie's ascent to the top of the ICD ladder was not well received in every last quarter of the company. There were a few people, particularly those in ground jobs, who took Blackie's rigid insistence on strict cockpit discipline as a clear indicator that he was a martinet. To illustrate that image, there was a classic example when he reamed out an experienced pilot, Robert Sherwood, who lost all four while over the Atlantic and was far from any possible salvation. Sherwood's C-54 sputtered its way down to a warmer, lower altitude where he was able to get all four engines turning again. Passengers and crew made certain that news of Sherwood's expertise in saving their lives reached Blackie who then called Sherwood into his office to explain how simply injecting some alcohol into the carburetors would have nipped the problem in the bud. "Now you're a damned hero," Blackie grumbled. Blackie was well known for never asking a pilot to do something he would not do – or had not already done. It was common knowledge that ICD had received an urgent request from ATC in late 1942 to ferry a C-87 Liberator from Washington to Miami on very short notice. The order came at a moment when ICD was stretched very lean indeed, and Blackie could not locate a single available crew – not even a copilot. So Blackie soloed the C-87 to Miami himself even though it was a flagrant violation of crewing requirements. The C-87 checklist is substantially longer than that on Blackie's old Curtiss Condor; the engine instruments are over on the co-pilot's side, and some of the controls, including prop pitch, are also on the right side, thus turning Blackie into the very busy one-armed paper hanger.

As far as Blackie was concerned, however, flying without a co-pilot was a much lesser evil than telling the transportation department at ATC that TWA's Intercontinental Division was not up to the job. If anybody else had pulled that off, Otis Bryan would have fired him at once. The C-87 caper did reinforce what a few of Blackie's detractors felt was a fault – that he was a "company man" in an organization where many pilot allegiances favored fellow pilots, not corporate management.

Johnny Guy, chief engineer at ICD, was among those who initially did not welcome the announcement that Hal Blackburn would become the new ICD chief. Guy was no rookie; he was a fifteen-year man who had been an automobile mechanic in Los Angeles when one of his customers, Tommy Tomlinson, recruited him into Maddux Airlines as a mechanic. When Maddux went into what became TWA, Guy was an overhaul inspector in Kansas City. His career grew as TWA prospered, and he mirrored the lack of confidence about Blackie's managerial abilities among the non-flying employees. Years later he said, "We were wrong. I don't think in all my years at TWA that I ever saw an organization function as a group better than ICD when Blackburn was running it." Blackie had plenty of training for both the flying operations and ground-based departments. Very early in life he had led temperamental musicians, negotiated with hard-nosed theater and dance hall owners, met payrolls, and had managed Eagle Nest's many employees while simultaneously dealing with demanding Army Air Force senior officers.

Johnny Guy also related what happened three months later after Blackie formally became the ICD manager. A meeting of all department heads was called, and Blackie announced that anybody could call anybody else a son of a bitch – so long as both parties were in the room at the same time. Blackie announced, "I don't want anybody coming to me with stories about somebody else." Then he added, "There's only one real problem in the world and if you can solve it, you've got it made. There are no technical problems or political problems – just human relations problems." Blackie always taught his kids the value of supporting the underdog, and very early in his leadership of ICD a prime example of that approach surfaced. A new telephone operator had been hired and became the victim of some sloppy paperwork in the personnel department. Over a month after coming to ICD, she had not received a single paycheck. Co-workers were loaning her money, and her immediate supervisor submitted a request for a paycheck advance which eventually made its way to Blackie's desk. He not only signed the form at once, but corrected the problem in payroll on his way down to the switchboard where he personally apologized to the young lady.

Much of Blackie's approach to maintaining good human relations was built on a foundation of good communications, which, of course, had always been Blackburn's strength. He started a program of periodically hosting an informal dinner meeting with various departments, and of those departments he paid special attention to those within the engineering and maintenance function. The dinners were more than just a nice salad and some rubber chicken; they were give and take sessions in which each party brought forth ideas and grievances right out into the open. Blackie's predecessors, Otis Bryan and Cliff Mutchler, had proven to be good managers, but they had spent their careers to date as a pilot/manager and business administrator, respectively. On the other hand, Blackie had been an automobile mechanic, heavy road equipment mechanic, and airplane mechanic – and therefore could relate quite easily to what was going on in the hangars and on the tarmac. The mechanics also heard about Blackie's problems which were related to ATC's expectations of ICD and the availability of aircraft allocated to ICD. Particularly gratifying for Blackie was a follow-up thank you note he received after hosting the twenty-seven members of the Engine Build-Up Department; it was personally signed by all twenty-seven members of that department.

Even in the interim period before formally taking the reins in January, Blackie's public appearances increased significantly. In mid-September he was the guest speaker at the Explorer's Club in downtown Washington where his talk emphasized new developments in aviation. One focal point was his demonstration of a new still that converted seawater into drinkable water. The unit was carried on all ICD planes and was being rushed into service with the U.S. Navy. That forum also gave Blackie a platform to discuss recent developments in TWA's weather research flying, including pressure pattern flying, and a new series of TWA-developed polar projection maps then under evaluation by the military. Another September appearance was a natural for Blackie; it was at one of the monthly dances sponsored by either an ICD department or the TWA Club and held after hours in such local venues such as the Nightingale Jazz Club in Alexandria. September 1944 was special because it was the first appearance of ICD's own dance band which opened at the Arlington Fire Hall. Blackie led a few numbers, but gracefully begged off leading the group in future performances.

When the ICD band debuted, Jack Frye had an introduction of his own. It was his plan for flying the North Atlantic in a post-war world. He announced, in late October, 1944, that TWA was ready to begin service as early as January 1945. That was not to be for several reasons not the least of which was the fact that the Constellation continued to have engine problems at high altitude. And the Commerce Department would take much longer than three months to complete its work leading up to the decision of just who would be flying into London, Paris, Rome, and other European and Asian capitals.

In some TWA quarters, ICD was regarded as a young upstart with no history and no traditions. In late summer, TWA's corporate personnel department looked into the matter and conducted some research into just how well experienced ICD's leadership was, making its report in mid-September. At the time there were twenty-nine men with supervisory responsibilities. The most experienced man in terms of service with the company was Otis Bryan who had been with TWA for fifteen years. Cliff Mutchler was within a few weeks of getting his fifteen year service pin (in a company that made much of those lapel pins). Blackie was still a little more than a half year away from his ten-year pin, but there was no shortage of ICD pilots with much lower seniority numbers than Blackie. Many of those men included pilots who were already captains when Blackie joined them as a rookie co-pilot. Those old hands included Milo Campbell, Felix Preeg, Jim Eischeid, Howard Hall, Mel Bowen, Earl Fleet, and Don Terry among others.

Blackie routinely informed the Airline Pilots Association (ALPA) of his advancement as a pilot-manager after not doing so when he became director of flight operations at ICD in late 1942. The union had been tracking Blackie's progression into management anyway. In early June, ALPA's membership director wrote to Blackie advising that his status would be determined at ALPA's next executive council meeting, but until then Blackie was maintained as an active member of the union. On 12 October 1944, the union moved him to its inactive list which prompted Blackie to ask to have a portion of his previously paid annual dues returned. A portion was returned, but not the amount Blackie had in mind. ALPA replied saying that it determined refunds based on which quarter the decision was made, not on an accumulated days basis, thus positioning ALPA in Blackie's mind as a petty bean counter in financial matters. The relationship would not improve in the years ahead. When Blackie received notification that he was becoming an inactive member, ALPA included an explanation of just how the central executive committee interpreted its by-laws and that vice presidents, directors of operations, division managers and their assistants, and related management titles were also becoming inactive members. There was some wiggle room in the committee's position in that "under extraordinary emergency situations" a pilot-manager could remain as an active member subject to approval by the committee. The implication was that a division manager's work put him far outside of the emergency situation definition.

All told with 1,600 employees at the end of 1944, ICD was a significant portion of the TWA entity whose total employment then was slightly over 10,000. There was much more to ICD than one might think which posed a greater management challenge for Blackie. Beyond the C-54s, flight crews, engineering and maintenance departments there was an accounting department with a staff of nearly fifty people, including six supervisors. The personnel department had a sub-department not exactly typical of a Fortune 500 Company; that was the medical department which was staffed by two M.D.s and supported by a squad of R.N.s. Their primary goal was to maintain the health of all employees, with particular emphasis on the flight crews which might return from the tropics one day and later head for Polar regions. The department administered countless physical examinations before and after flights in addition to having given nearly 50,000 inoculations by the time Blackie became manager in January 1945. The medical department chief, Dr. R. Bretney Miller, was also responsible for the medical exams flight crews required for keeping their licenses current. There were two ills that Miller could not cure: fatigue and the common cold, the latter being a typical complication of flying in Arctic regions. Miller did not initially conquer an ailment common to the tropics, malaria, but with education of flight crews and ICD staff stationed abroad, the last case of malaria among TWA personnel was reported in October 1943. Amoebic and bacillary dysentery also showed declines as Miller's preventative campaigns were expanded.

There were a few differences between ICD's headquarters operations and those back at the company's facilities in Kansas City. The most obvious was the number of posters on ocean ditching procedures, lifesaving, and the safest water landing techniques. The dispatch office in Washington had one function completely lacking in Kansas City: checking passports. One of the notable differences in Washington was the lack of facilities for meteorology, cargo loading, and ground radio. Those functions were provided by the Army Air Force which was also responsible for final boarding of passengers. The training facilities, especially the Link Trainer room, were essentially the same, as was maintenance and engineering, although ICD had a unique offshoot of the latter department called industrial engineering. Its function was completely decentralized and involved working with all ICD departments to simplify jobs and extract as much efficiency as possible. Maintenance and engineering took up most of TWA's two hangars at Washington National and ran three shifts, twenty-four hours a day. As often happened in wartime, necessity bred a few inventions. TWA maintenance personnel designed and built a machine that would soon become an industry standard. The new tire changer completed in seconds what had originally taken many minutes.

Blackie's management duties kept him largely ground-bound during the holiday period. Blackie's family was beginning to grow up. Betty, musically inclined like her dad, was almost twelve, Bob would turn ten several weeks after Christmas, and the baby, Bonnie, was now a little girl nearly four years old. The Blackburns never sent off-the-shelf Christmas cards, but created their own photo cards for Christmas. 1944 reflected one reality of the war years, but was still uplifting. The family was shown on four bicycles with the text reading, "No Rationing on Holiday Greetings! The Hal Blackburns." Although Blackie was not flying during the holidays, ICD was, and he was on hand at Washington National on 30 December with a photographer and several ATC officers to welcome ICD's 6000th flight – a returning C-54 captained by Russell Dick. Non-pilots were about to get into flying with Blackie's approval. About two dozen ICD'ers formed what they called the National Flying Club and started pooling funds for the purchase of a light plane. Blackie approved the use of ICD's training department for both ground and flight instruction. Although his appointment as manager was not effective until New Year's Day and with Cliff Mutchler already having moved back to Kansas City, Blackie decided that the ICD community deserved some sort of recognition for a job well done in 1944. So he put together a major holiday open house for all ICD employees and their families. Blackie's immediate bosses, Jack Frye and Otis Bryan, could not attend, but T.B. Wilson did and in early January sent Blackie a warm note indicating how much he enjoyed the gathering.

Blackie did not fly at all from mid-December for almost two months at a time when the war news was dominated by the Battle of the Bulge in Europe and the re-taking of the Philippines in the Pacific. Gladys Entrekin, among the first hostesses when Blackie was a co-pilot, was now chief evacuation nurse in Europe and busy overseeing transportation of the severely wounded back to the United States. TWA suffered a few fatalities of its own during the holiday period. On 1 December 1944, the DC-3 flying Flight 18 from San Francisco to Burbank crashed in bad visibility. The pilots, John Snowden and Thomas Bamberger were killed along with six others. The impact removed the nose and wings while the rest of the fuselage came to rest in a muddy field just short of the airport. Thanks to the soft mud, the hostess and remaining fourteen passengers survived. Little more than four weeks later, ICD lost a C-54 at Lagens Field in the Azores. The runway was short and hemmed in by ridges on both sides and a sheer drop-off at the approach end which even on a good day was subject to a lot of chop. The captain, Russell Jones, was following another C-54 which aborted two attempted landings due extreme turbulence. That C-54 finally made it down on his third approach, but Jones struck the retaining wall at the approach end of the runway, damaging the left wing and removing the landing gear on that side. There was an explosion and fire in which the pilot and first officer, W.S. Turner, were injured (thrown forward right through the windshield); the navigator and flight engineer were killed. Fortunately, the passengers all survived. Blackie was very familiar with the C-54 accident since he read the report in great detail and was familiar with the Burbank accident from word of mouth. Fortunately, the Turner accident was the last ICD accident of the war. There was another ATC accident on 12 December which Blackie heard about through the grapevine, but did not see the report because it involved an Army ATC crew. He was, however, very familiar with the surroundings – Gander in Newfoundland where ICD had twenty employees stationed. The crash involved an older first lieutenant service pilot and his B-24L which disappeared after takeoff. The lieutenant, Robert Drew, who was based at Romulus, Michigan, with the Third Ferrying Group, had picked up a new B-24 Liberator at Presque Isle, Maine, and had stopped at Gander before crossing the Atlantic. There were snow showers in the area, but not of the severity to prevent safe operation after takeoff. When he failed to arrive in the Azores, a local search around Gander was initiated the following day. Drew's bomber was found completely wrecked only several miles away near the top of a forested mountain in circumstances disconcertingly similar to the Redding crash at the Philmont Ranch in early 1942. A thorough technical investigation could not identify a cause of Drew's accident other than controlled flight into rising terrain, i.e., suspected pilot error. Blackie promptly forgot about it – until an important somebody reminded him of it almost twenty years later.

Counterbalancing the bad news was some of the more humorous variety. Colonel Elliott Roosevelt, whom Blackie had met at Mount Farm in September 1944, got his orders to return home later in the autumn. Roosevelt threw his weight around and managed to get two of his four large English mastiff dogs on board the flight home. Together the two dogs weighed 185 pounds. The older of the two was a giant male named Blaze whose mate, Blitz, had been shot stone dead by a Mount Farm sentry who thought the massive canine was an intruder. Word got out about the canine passengers, and the media had a field day with why a pair of pooches had priority over wounded American soldiers on a flight back to the States. There followed an executive order that no dogs would be carried on any American flights, military or civil. At about the time the doggy stories were still on the front pages, the Battle of the Bulge was reaching its climax. Winter weather on the continent was atrocious, and normal movement over the snow-covered terrain in many cases was nigh impossible. It was decided to use dogsleds to evacuate the wounded from the battlefield. The nearest dogs and sleds were in Greenland at Bluie West One, and a special order was sent down that an exception could be made to the previous no-dog rule. An Army ATC C-54 would take the dog teams to Prestwick where strict rules on quarantining incoming dogs were temporarily lifted, but unfortunately the westerly winds were such that extra fuel was loaded for the canine flight. The weight and balance analysis showed that one team of men and dogs along with their sled would have to be left behind. It just so happened that there was an eastbound ICD flight scheduled for departure at about the same time – and it did have room for the sled and team. However, the captain was essentially told that he would be spending the rest of the war as a prisoner at Fort Leavenworth if any word even got out about the pooch passengers and that he was to observe strict radio silence in the matter. Unfortunately, a ground radio operator at Bluie West One sent a coded message to the Prestwick base commander reading, "Expect woof and attendants on Mess House Dog." The Prestwick C.O.'s imagination went into overdrive; he was convinced that the C-54 was a VIP flight of the highest magnitude and that a figure none other than President Franklin Roosevelt would shortly be deplaning at his field. The commander was well prepared when the ICD C-54 taxied up – there was an honor guard of white-helmeted /white-gloved men, and a special ramp was wheeled into place to accommodate the President's wheel chair. But when the cabin door was opened the Prestwick commander was greeted with nothing more than a cacophony of barking, howling, and baying from the hungry huskies.

The Roosevelt part of the story did not end with the arrival of Blaze back in the states, although TWA was off the hook because the rest of the story became a monkey on the back of American Airlines. American was complicit in flying the dog terror, Blaze, across the country to Elliott Roosevelt's new wife, Faye Emerson, a starlet in Hollywood who would go on to a modest TV career and before that an even shorter career as Elliott Roosevelt's latest wife. The problem for American and the Roosevelts was that Blaze bumped quite a few people off their flights – servicemen home on compassionate leave to visit dying parents, other servicemen trying to reach new assignments on time, etc. Anybody with a B, C, or D priority got bumped. The vicious dog enjoyed an "A" rating.

President Roosevelt was very much in the news at about the same time, and with it came more good publicity for TWA. It was the Yalta Conference to which Roosevelt traveled by sea as far as Malta on the cruiser *USS Quincy*. The convoy, consisting of two cruisers and a screen of destroyers, was met on Malta by Prime Minister Churchill. From there Otis Bryan flew the President aboard a specially modified C-54 named "Sacred Cow" to the conference in the Crimean resort town of Yalta. Although TWA and the other airline ATC contractors carried nearly all of the topmost VIPs during the war, the Army's special VIP flight flew the VIPs more frequently. The one glaring exception was TWA flying Roosevelt, although flying with TWA was not his decision to make; it was a choice made by the Secret Service. Flight details were left to IDC and Bryan, including whether or not to fly with a fighter escort. On Roosevelt's previous trip to Teheran, a P-38 Lightning escort was provided. Bryan chose not to fly with fighter cover on the Yalta trip because he thought those fighter jocks could create more hazards than the Germans. He also thought they probably would not make a difference if an enemy attacked in large numbers, especially during the most vulnerable phases of flight: takeoff and landing.

Blackie closed out the year 1944 on 16 December with a return to Washington after a flight down from Stephenville. As the year 1945 began, Blackie was nearing his tenth year with TWA which was reflected in his pilot seniority number. As of New Year's Day, TWA's pilot employment stood at 754 of which Captain Harold F. Blackburn stood at position number seventy-five on the seniority list. His health as confirmed by his frequent flight physicals was absolutely robust – something that could not be said for Franklin D. Roosevelt whose health was deteriorating very publicly and precipitously. The president soldiered on, trying to accomplish his objectives for a post-war world, in the Allied leaders' week-long Yalta conference that opened on 4 February. Premier Stalin had vetoed Roosevelt's request that the meeting be held somewhere in the Mediterranean based on medical reasons, although Roosevelt was the only one of the Big Three who appeared frail. On that day, Blackie flew from New York back to Washington after having flown a C-54 from Washington to Bermuda the day before.

On the surface, the immediate outcome of the Yalta Conference seemed very positive. Eastern and Central European countries were to have open elections while Berlin and Germany would be divided into American, British, French, and Russian sectors. Russia would join the Pacific War ninety days after the unconditional surrender of Germany, and Russia would join the United Nations, including veto power in the UN's Security Council. The British bulldog, Winston Churchill, had well-known misgivings about how free the elections would be in central and eastern Europe, and time would tell he was right. At the Teheran Conference, the Russians were still fighting on their own soil, but in February 1945 they were approaching Berlin and adopted a policy of possession being nine-tenths of the law when it came to freeing land once controlled by Nazi Germany.

Twenty-six February 1945 marked an important milestone for ICD: three years of flying for ATC. Blackie was not a plank owner at ICD, having been assigned there four months after its formation. However, he was very much a part of building ICD which could look back on three years of important support of the war effort. That included carrying over 112,000 passengers and 20,000,000 pounds of vital supplies including blood plasma, medicines, special munitions, V-mail, and other high priority cargo. Particularly satisfying was one other statistic: TWA had returned over 10,000 wounded soldiers back to the United States. The routes ICD flew had changed just before the 1944-45 winter flying season, and TWA's pressure pattern research was a major reason for the change. For all practical purposes, the South Atlantic ATC route was history during the last year of the war. TWA's pioneering work for the North Atlantic Command of ATC, headed by Col. Larry Fritz, resulted in ATC aircraft taking advantage of the winter weather rather than being stopped cold by it. Thus, TWA was using the North Atlantic route to reach Casablanca, Prestwick, and in late 1944, Paris.

During the week following the conclusion of the Yalta Conference, Blackie reached an important milestone in his flying career – he checked out on the plane that would essentially become his trademark for the next sixteen years: the Lockheed Constellation. While Blackie was learning to fly the Connie, Jack Frye was trying to seal the deal on just where the new plane would be flying. He followed up his original filing with the CAB for North Atlantic routes by submitting paperwork for the North Pacific. That application involved routes from the Lower Forty-Eight to Alaska and then down to Tokyo, Seoul, Shanghai, Hong Kong, Calcutta, and westward until they met eastbound flights originating in India.

Blackie's first flights with a Connie consisted of ten hours aloft outside of Washington although only about four hours of that time were required for instruction and the flight test which qualified Blackie for his Constellation type rating. The rest was devoted to operational flight testing of the C-69. Unfortunately for TWA, neither Lockheed nor the Army had supplied much in the way of flight data on the Connie, including cruise charts. Swede Golien took off on one mid-February day looking for ice, and in spite of being vectored to areas where traffic controllers had been advised there was ice, he found none. Although the Army had briefly tested the Connie in Alaska, the plane's performance in icing conditions was still essentially unknown. To generate the data, some of the first qualified Connie pilots had to put in very long hours obtaining the information. Several months later Blackie took charge of a Connie that had just returned to Washington from more icing tests. That Connie had a sick engine, but Blackie flew it, serial number 43-10314, from Washington to Burbank anyway and came back the next day with a newer Connie. The Connie with the engine problem was not to enjoy a long service life; it was ditched in the Mediterranean in 1946, sold to Air France which used it for parts, and then was scrapped in 1951. Fixing that inoperative engine was something for the manufacturer to take care of, and performance data was needed on flight characteristics with one engine out anyway. That was Blackie's second trip to Burbank in three months; he had flown there in early March to turn over some of the first performance data that TWA had generated.

What Blackie, Golien, and the other pilots in the evaluation program found were a number of other problems of which some were simply annoying such as poor cockpit sealing which was reminiscent of the early Douglas transports. Although Blackie only experienced it once, there were occasional failures of the hydraulically boosted flight controls, thus requiring both pilot and co-pilot to strain on the controls to get an appreciable response. The fuel tanks leaked because they had not been designed with sealants that could cope with changing temperatures and flexing wings. There were no immediate fixes for the engine problems that were also common to the B-29, but until they could be addressed, TWA never operated the Wright R-3350s in high blower. A week after he got back from Burbank in early March, Blackie passed his instrument check ride in the Connie, but did not fly again for a month. That particular C-69 Connie that Blackie first flew enjoyed a relatively long and productive life; in 1947 it was sold back to Lockheed which upgraded it into the civil 049 Constellation model. Then it was sold to Capitol Airlines which operated it before it was sold several more times, eventually being scrapped in 1965.

When Blackie did get back into the air on 12 April, it was after the nation had been shocked by President Roosevelt's sudden death only several days before. It was not a Connie he was flying, but a C-54 instead. The destination was brand new for Blackie, and within a year it would become a city that became an important part of TWA's post-war operations: Paris. Blackie's return flight from Paris via Prestwick and Meeks Field on Iceland proved the value of pressure pattern flying and proper fuel load planning along with en route decision-making under weather constraints. Blackie was given Gander or Moncton as alternate for Stephenville. As it turned out Gander was going below minimums, and air traffic control gave Blackie Presque Isle as the new first alternate. Over the radio, Blackie heard a flight one hour ahead of him being vectored to Bangor, Maine, as a refueling stop. Estimated winds from Stephenville to Presque Isle were estimated at seventy knots right on the nose. Coming up on Gander and knowing that visibility at Presque Isle was going down rapidly, he requested clearance to Bangor, planning to re-file in flight for LaGuardia if it appeared he had enough fuel to do so. Blackie had taken off from Iceland with 3,105 gallons of fuel which would allow him to reach Stephenville even with the substantial headwind and still have a comfortable reserve to reach an alternate.

The C-54 was full of wounded soldiers, and Blackie was reluctant to prolong their trip home to Washington by landing at Bangor. Clearance was received over Bangor to proceed to LaGuardia with Newark as the alternate, and when Blackie touched down at LaGuardia, he still had 300 gallons of useable fuel remaining. That was made possible by not flying the conventional great circle route for the shortest distance between Iceland and Stephenville. The flight, for which the portion out of Iceland was flown well south of the conventional one, had lasted sixteen hours and twenty minutes – made possible by playing the winds of a large high pressure system to his advantage and the fact that the passenger/cargo load was relatively light. The rest stop at LaGuardia took an hour and a quarter; the landing at Washington was made one hour, twenty-five minutes later. Blackie challenged TWA's chief pilot at the time, Fred Richardson, to review the Army's requirements for landing at an East Coast facility first before subsequently taking off for Washington since it was clear that he could have made it to Washington without the stop at LaGuardia.

That flight was completed during a time of great change for TWA. Although the European war was all but over, the month of April was a bloody one in the Pacific with Okinawa having been invaded on 1 April. The Army Air Force was starting to release C-47s to the airlines, and TWA's Kansas City base was beginning to modify them for civilian use. The Stratoliners were going back into scheduled service as well, and the Air Force was beginning to release another valuable resource to the airlines: pilots. TWA became the first airline to employ a former military pilot when Lt. James Ardery from the Troop Carrier Command reported to Kansas City for his transition training. TWA was also expanding its feeder network with newly approved routes to such second tier markets as Albany, New York, and Willliamsport, Pennsylvania.

In May, the war in Europe ended and Blackie's flying took off again. Early in the month he was busy checking crews out on the Connie with several four-hour roundtrips to Charleston, South Carolina, including both night and instrument time. His most important flying in May and June qualified as "special missions," i.e., flying military leaders back from Europe. On this first of two trips to Paris, Blackie's passenger manifest included General Omar Bradley, most recently commander of the U.S. First Army. Bradley went on to become director of the Veterans Administration in the peace that followed the war. Another of Blackie's passengers was his old Seventh Bombardment Group commander, General Joseph McNarney, who had been Allied air commander in the Mediterranean theater. McNarney would shortly return to Europe where he became commander of all U.S. forces in post-war Europe. Another returning general was Dwight Eisenhower although Blackie was not the pilot. Eisenhower's return to America ended at Kansas City where the Abilene native had a quiet reunion with his eighty-three year old mother and four brothers inside TWA's executive offices and away from the crowds. TWA publicity had made much of the fact that it was busy getting former aircrew who had gone on active duty back into their original roles as TWA pilots, co-pilots, navigators, and engineers. The reality was somewhere short of that. On his post-VE Day trips, Blackie continued to encounter pilot colleagues who felt that TWA was not doing enough to get them released from active duty as quickly as possible. Lt. Colonel Roger Kruse was one of the most vocal when Blackie saw him in Paris. Blackie got the same story from Major Roy Noland at Lagens Field in the Azores. Blackie was inclined to agree with the two ATC officers who had both been TWA pilots. Noland had been in the Kelly Field cadet class just one year behind Blackie, and eventually went on to a thirty-four year with TWA. Kruse and Blackie were old friends; both were Kelly grads and in the same incoming TWA co-pilot class in April 1935. From those two meetings, Blackie wrote a no-holds barred letter to TWA VP John Collings, who had taken Larry Fritz's place as the operations chief, on how to expedite the release of former TWA pilots from the military. It was, in effect, a form letter wherein all Collings had to do was address the letter to the ex-TWA pilot's commanding officer.

Counterbalancing Blackie's flying was the ground-side part of his job, some of which was ceremonial. In May he opened the "TWA Club" which was actually a pavilion on some leased land between Mount Vernon and Alexandria. There were picnic facilities and even a diamond for ICD's softball league. Much more important for Blackie, the club also had tennis courts which he promptly took advantage of. Within a few weeks, more finishing touches had been put on the club's facilities, and a formal open house took place with Mrs. and Mrs. Jack Frye and Mr. and Mrs. Harold Blackburn as the honored guests.

Although not part of his job description in early 1945, Blackie strengthened his position as the de facto media publicity agent for ICD in spite of TWA's powerful PR operation in downtown Washington. That is not surprising considering the fact that most of the editors were pilots by trade and training, and the G Street agents were not. As a result, ICD frequently enjoyed sustained, favorable press in the nation's leading aviation magazines including *Flying*, *Aviation*, *Air Facts*, and many others. Another duty was inspiring all departments in ICD to pony up and buy more war bonds in the mighty Seventh War Bond Drive which was strongly supported throughout TWA. By the end of the first half in 1945, ICD stood at eighty-eight percent of its quota. The only weak departments were Accounting and Procurement which stood at sixty-nine and fifty-three per cent respectively. Blackie's department, "Managers Office" led by example at 187 percent.

Blackie and Swede Golien both happened to be in Prestwick after VE Day and attended a small banquet at a newly opened hotel in the village. The high point was the awarding of five year pins to nearly a dozen employees. Those employees came to the banquet in shifts because planes had to be turned around at all hours of the day. One guest was in the hotel illegally; it was "Constellation Leo," an English mastiff dog that was smuggled into the banquet hall in spite of a sign prohibiting dogs on the premises. Blackie's own ten year pin was awarded at about the same time, although its lateness due to clerical error was apologized for personally in a letter to Blackie from Jack Frye.

Blackie was at ICD headquarters on 5 July when the CAB finally awarded the new international routes. TWA had gotten some leaked information over the preceding few months that its chances of gaining access to European cities on the North Atlantic route were slim to nil. Jack Frye seemed not to pay much attention to those discouraging rumors. After all, Harry Truman was in the driver's seat, and when the routes were formally announced, Jack Frye and everybody in TWA were beaming. TWA would serve Paris, Lisbon, Rome, Switzerland, Athens, and several other stops as far as Ceylon. American, which had absorbed American Export Airlines, won most of the United Kingdom, Holland, northern Germany, Scandinavia, the Baltic States, and Russia. Pan American got London, Marseille, southern Germany, Austria, Eastern Europe, and what remained of the Middle East.

With TWA's triumph came some hard realities. The company did not have any airplanes to serve the new routes. The Connie, technically, was an Army airplane, not that many had been completed, and those that were flying did so under certain restrictions. Having been certified for the new routes did nothing to clear some practical hurdles, including obtaining foreign landing rights and setting up completely new organizations where English was not necessarily spoken very well. Gearing up for flying in European airspace would not happen overnight. TWA already had its Route Planning Division, but although it had provided management with the research and recommendations needed to complete foreign route applications, its focus was primarily on domestic routes. Thus that division was gearing up for the work it would take to establish TWA in new countries where it previously had no ICD presence. Much negotiation remained in the future. In the case of Switzerland, the CAB award said nothing about which city would be served by TWA, thus forcing TWA to consult with Swiss officials to decide if Berne, Geneva, or Zurich would be the launch city.

Juan Trippe's plan for one American flag carrier in the international arena had been defeated, but time would tell that he was not finished with the fight by a long shot. Undermining TWA's joy at the North Atlantic decision was Pan Am's application to the CAB for the one advantage that TWA had in spades: a domestic feeder network that would terminate in the jump-off point for international flights. The feeder concept was taken a step further by Jack Frye who was starting to think about how all of the people in foreign lands would get to TWA's gateway cities in those countries. For that reason he was very conducive to the idea of obtaining partial ownership in foreign airlines or at least launching or rehabilitating air carriers in countries whose transportation systems were still in ruins from the war.

Two days after the CAB announcement, Blackie was involved in another ceremonial event – one that was very representative of how busy ICD was in spite of the European war having been concluded. It was welcoming home a crew which had just brought its C-54 back to Washington to conclude ICD's 7,777th ocean flight (which conveniently took place on the seventh day of the seventh month, although not at the seventh hour and seventh minute of that day). In August 1945, ICD reached its peak employment of 1,943 employees with 708 in flight operations, including 219 pilots of which nine were in supervisory positions. Maintenance was the largest department with an employment of 893 people of whom 70 were stationed abroad in just six locations (Stephenville, the Azores, Prestwick, Paris, Iceland, and Casablanca) since the South Atlantic Route was no longer in regular use.

Statistics kept by ATC showed that among its several civilian contractors, TWA was consistently turning in the best performance by key measures such as pilot hours flown per month (eighty-five) and aircraft utilization. It was a constant battle to maintain the figures reached in August and September of 1945, and Blackie was caught right in the middle. TWA's fleet reached its maximum in August 1945, twenty-seven C-54s of which about a quarter were new C-54Es. The problem was downtime for scheduled and unscheduled maintenance complicated by later than promised delivery of the new E-model Skymasters. A further complication was ICD being ordered to transfer older C-54s and C-54As out of Washington into the Army Air Force Training Command. Occasionally, one or two flights a day out of Washington National had to be cancelled simply because the Army had not delivered the new planes on time. The only authority to which Blackie could appeal was the now Brigadier General Larry Fritz, then in charge of ATC's North Atlantic Wing. Another complaint related to the fact that ICD was still operating older C-54s, and spending more time maintaining and updating them, when the other ATC contractors were getting higher allotments of the newest model. Blackie's plea for a more even-handed approach essentially got ICD nowhere. Fritz indicated that the problems were beyond ATC North Atlantic Wing's control, but that the matter would be taken up with ATC headquarters. The salutations in the exchange of letters between Fritz and Blackburn on the matter of aircraft transfers in and out of ICD along with maintenance scheduling reflected the friction; it was no longer "Larry" and "Hal" or "Blackie," but rather Mister Blackburn and General Fritz instead.

As if to highlight the maintenance issues and the resulting downtime, Blackie assigned a special mission to Captain Bob Power for a flight to Cairo and Karachi. All went well on the flight across the Atlantic via the Azores and Casablanca. The segment to Cairo was along the North African coast where the crew could look down at scenery where a little over two years before General Erwin Rommel and his Afrika Korps had been defeated. Cairo to Karachi was a night flight in which all electrical power was lost immediately after takeoff. Flashlights facilitated an emergency landing back in Cairo where the two flight engineers changed out an electrical inverter before the flight to Abadan, Iran, (just north of Kuwait) resumed. The plane, carrying a delegation of Chinese VIPs and some pilots just assigned to fly the Hump in China, was refueled in Abadan and took off at 3:30 A.M. several hours behind schedule. Very shortly after takeoff, one of the passengers came up to the cockpit to inquire why rain water was streaming back across the windows when nary a cloud could be seen in the sky. The "rain water" was gasoline, and it was streaming because the refueling crew back in Abadan had not properly secured the cap on the fuselage tank. Within moments of that discovery, the number three engine started to vibrate and its oil pressure light came on. It was promptly feathered. As soon as that crisis passed, number four started losing power. Nothing, not even a full rich mixture, could bring it back to anything except running at idle. It too was feathered, leaving the pilots with the rudder fully trimmed for the imbalance which could only be corrected by both of them straining on the opposite rudder pedal.

The engine failures occurred at a rather inconvenient moment inasmuch as the plane was then over the Persian Gulf. By radio, Abadan suggested that Captain Power try for Bushire, also in Iran, but on the Persian Gulf's eastern shore. The field at Bushire was rough with just one dirt runway, but the C-54 was landed safely. C-47s from Abadan flew in the next day with a new engine and an engine change crew of five mechanics. The C-47s flew most of the passengers and crew out, but the flight engineers remained to help change the number three engine. It was summer and just too hot to work on the plane in the afternoon, hence most work was done at night and in the morning. The fix on number four was to reattach the two halves of the exhaust collector ring that had worked apart permitting hot exhaust gases to directly enter the carburetor air scoop. The engine change crew and two flight engineers were treated like kings by the local British consul until the plane was flown out several days later. While this story had a happy ending for those changing the engine, the resulting downtime was not so pleasing for those like Blackie whose job it was to maintain as much uptime as possible. Counterbalancing the discouraging maintenance issues, Blackie got some very good news at home from Martha. She was pregnant and expecting a child in April 1946 which was a cheering prospect after the Blackburns lost their infant daughter Beverly in December 1943.

Blackie's one foreign trip in August was what had become his typical overseas run – to Paris and back via the Azores and Stephenville. On the twenty-seventh, he attended a meeting at ATC's North Atlantic Wing administrative headquarters in Manchester, New Hampshire. Although the wing's principal operations were at Presque Isle, Maine, General Larry Fritz had moved his office to Grenier Army Airfield near Manchester where contractor airlines met with him several times a year. When the meeting on the twenty-seventh was completed, Fritz took the TWA contingent out to dinner and then went bowling. When Fritz walked into the bowling alley, the manager offered to clear a lane just for the general and his party, but Fritz said he would wait his turn like anybody else. It appeared not only to Blackie, but to others as well that Fritz, whom the Kansas City maintenance staff had called "Mister Bang" due to his firecracker temper, was starting to mellow. Another treat awaited Blackie the next morning when he learned that a B-17F Flying Fortress had to be ferried down to La Guardia, and the man first scheduled to fly it was an old friend, Colonel T. L. Boyd. Blackie's checkout on the "Queen of the Skies" was short and sweet; so was the subsequent flight itself – just ninety minutes. Blackie left behind no reactions to flying the plane, but would certainly have enjoyed the experience, at least in contrast to the other big bomber in which he already had plenty of experience, the B-24 Liberator.

Although V-J Day would not take place until 2 September, TWA was already busy doing its part to wind down the war by getting returning veterans back to their homes as soon as possible. The Army turned over seventeen C-47s to TWA Domestic Division for that purpose and ICD started flying a limited number of Connies between Washington National and Hamilton Field in California. At the same time, the government announced that it was phasing out the passenger priority system and that ninety-five percent of the seats would not be subject to priority requirements by mid-September and that the system would go completely out of effect by the end of October.

The biggest news to effect Blackie personally came out in early September. On the first of that month, he turned over management of ICD to Frank Busch and became manager of negotiations and route surveys, foreign routes. At about the same time, Jack Frye announced that TWA was buying thirty-six Connies for a purchase price of $300,000,000 with the first twelve arriving by 1 October. Blackie, however, would not get his hands on one of them for nearly three months because he was traveling outside the country. The beefing up of the route planning function had been announced earlier, but in September the field work began with two separate parties taking off for Europe. Swede Golien took one detachment to Europe in a modified (civilianized) C-54. Another group consisting of board Chairman T.B. Wilson, Blackie, George Spater (TWA's legal counsel), and several others headed for Ireland based on the anticipation that a Connie would not be flying non-stop from Gander to Paris. Ironically, the best transportation to Ireland at the time was via Pan Am and its seaplane base at Foynes, on the River Shannon where Blackie arrived on 7 September. His journeys and meetings with foreign officials kept him out of the country for two months during which time he went as far west as Cairo and Athens. Other important stops included Geneva, Naples, and Rome. The flight between Naples and Rome was not courtesy of Swede Golien. In Naples, Blackie encountered a former Army acquaintance, Lt. Col. J. Johnston, who was flying a B-25 Mitchell bomber up to the Eternal City, thus Blackie enjoyed a busman's holiday and flew the lightly loaded Mitchell himself in less than 40 minutes. The survey party also visited Benghazi, Tripoli, Algiers, Casablanca, and Lisbon on the way back to Paris. Some of these cities, such as Cairo, Paris, and Casablanca were well known to Blackie. Others, such as Benghazi, had been ATC bases during the war, but all of the airports in those cities were or would shortly be run by a different cast of characters which took time to cultivate and convince before any agreements could be signed.

By V-J Day, TWA had a payroll exceeding 11,000 people, although Blackie's attention was focused more on the employees at ICD which would be undergoing a significant transformation in the immediate months ahead. Thus he only glossed over the hiring of John Furbay, Ph.D., who was becoming the manager of TWA's new Air Education Program. Furbay was very widely traveled, had written articles and books on world travel, was an educator, and a lecturer. He was hired in conjunction with TWA's expansion into international markets and was still around fifteen years later in a role which would become very important to Blackie once his active flying career was over. On 5 September, President Truman signed an agreement with the government of Ethiopia to establish an airline in that country much as President Roosevelt had done with Saudi Arabia in early 1945. Blackie was aware of those two foreign developments from reading TWA's employee newspaper. The news also noted a Saudi delegation visiting TWA's operations in Kansas City following the San Francisco Conference which was held in mid-1945 to establish the United Nations. He paid little attention to those developments at the time, not knowing that two years later he would very much be headed to the Middle East on special assignment.

The foreign route survey was completed by 1 December when there were more announcements which were to have a major impact on Blackie. The leadership of the new International Division was introduced. General T.B. Wilson, the chairman of the board was named managing director, Otis Bryan became chief of operations, Blackie became manager of the North Atlantic Division, and Swede Golien was appointed chief pilot. Blackie's new job was essentially a narrow version of his previous job with the principal difference being that he was managing a much smaller fleet of aircraft on a more focused route structure. At the time, TWA had well under a dozen Connies in its own fleet, not the nearly three dozen C-54s and a handful of Connies Blackie was in charge of just several months before.

TWA was moving forward in its domestic activities, and had leased North American Aviation's B-25 modification center at Fairfax Airport back in Kansas City to modify the C-47s and C-54s that were coming into service on the line. TWA opened a new ticket sales office in Cairo – located in the lobby of the Shepeard Hotel which was the choice of most international travelers in Cairo. TWA's domestic ticket offices had already received over 500 requests for tickets to Europe even before it announced its inaugural flights, primarily from businessmen seeking to establish or renew commercial ties broken by the war.

Beyond the horizon, far beyond what Blackie's 20-20 eyes could see, storm clouds were beginning to build. Jack Frye was already talking to Lockheed about an improved Connie – one that would have better range. Other problems would surface with the Connies, too, but they had not been flying long enough for some of the real gremlins to appear. Jack Frye and Howard Hughes were slowly, but surely drifting apart. Economists were still rosy about the future and the economic boom that would come to aviation, but then their forecasts never came with probabilities attached. ALPA was growing stronger with an increased membership in which many pilots were not happy with perceived injustices as to the way they were treated by management. TWA was becoming a large organization and one that critics might have described as bloated with its 11,000 employees. And then there was the big financial nut to crack; where was the 30,000,000 bucks coming from to pay for the new Connies? Among the other financial hurdles were the costs of building the infrastructure required for foreign routes now that ATC was not footing the bills.

Those thunderheads were still well over the horizon, and in the meantime Blackie and his close associates would be riding the crest of a wave for months to come. In late November, Blackie, with his co-pilot Joe Carr, took a Connie on a shakedown flight up to Gander from Washington National and back to

LaGuardia. It was in preparation for the first of two flights to Paris that would cement his reputation as one of TWA's finest pilots. The first of those journeys was to take off on 3 December 1945 at Washington National with a blue-ribbon list of passengers on what was the prototype commercial flight to Paris with Joe Carr again as the co-pilot.

The year 1945 was coming to an end, and although ICD would continue as a going concern well into 1946, it could look back on a job well done, certainly when compared with the other civilian contractors serving ATC. From its beginning through 1945, ICD flew 9,376 over-ocean flights of which the vast majority was over the Atlantic, topping its other competitors American (8,668), United (7,311), and Pan Am (6,155). The various divisions of Air Transport Command itself flew more than the others – 19,215 in the Pacific alone and 26,578 in the Caribbean and North Atlantic. The Naval Air Transport Service was not a major player in the Atlantic (1,227) but was a significant force in the Pacific (18,505).

Blackie's demonstration flight to Paris was not the first the first trans-Atlantic flight of a Connie, since ATC had flown a limited number of Connies to Europe beginning in the summer. Those early flights were usually attended by engine problems and as such were not anticipated with any joy by maintenance departments. The same problems plagued domestic flights, and both suffered from B-29s still having top priority on replacement engines. Still, the Connie was a formidable airplane. General Larry Fritz took a Connie non-stop from Newark to Paris on 4 August in fourteen hours and twelve minutes. The plane, however, was still beset with problems. Pan Am was involved in the same program of moving returning servicemen from one coast to the other as was TWA, and both contractors flew most of their flights in C-47s. Occasionally, Pam Am would be assigned a C-69 Constellation for those flights, but the airline only loaded them with twenty-five passengers. That was done deliberately in case there was an accident in which case it was anticipated that the press would assume it was a C-47 that had gone down. Pan Am was gun shy of the Constellation because it had lost a Connie on 18 September when one caught fire in midair; fortunately the crew was able to pancake it onto a Kansas crop field. Pan Am simply assumed that it was another induction system fire. Connies were briefly grounded and inspected, but were soon back in the air. It was not until seven months later when Pan Am suffered another in-flight fire that the real culprit was discovered. The cabin supercharger drive shaft had failed and severed the oil pump, thus starting a fire. There certainly was potential for heat and failure considering the drive shaft operated on a fourteen to one ratio from the engine rpm, itself turning at 2,000 rpm, i.e., the supercharger shaft was turning at 28,000 rpm. The fix was a simple one: switch to a graphite-based lubricant system for the bearings supporting the drive shaft. The problems did little to boost the Air Force's confidence in the Connie. The military had originally ordered 260 C-69s, then the order was cut to seventy-two, and when the war ended, all orders were cancelled, leaving Lockheed with some hard decisions. At that time, only fifteen Connies had been built as military transports, and Lockheed was forced to decide whether to lay off thousands of workers and go back to square one to redesign the plane or buy back the military C-69 Connies and upgrade them into civilian passenger planes before resuming production for TWA and Pan Am. Lockheed chose the latter course and produced two models: the 049 Constellation for TWA and a somewhat modified, longer-range 149 model for Pan Am.

Just as TWA was completing final training for its demonstration flight to Paris, terrible news was received from the Army about a plane crash in North Africa. The Army had provided TWA with a B-25 Mitchell and crew to finish some follow-up surveys not completed by the T.B. Wilson and Swede Golien survey groups. The Mitchell had crashed with all aboard killed in a takeoff accident at Asmara, Eritrea, on a survey flight to Addis Ababa in Ethiopia. On board were Milo Campbell, TWA vice president Jack Nichols, and several other TWA personnel in addition to the two Army pilots. Campbell was more than a mentor to Blackie, he was a very good friend; it was Milo who had popped in and kept an eye on the Blackburn household when Blackie was away for extended periods. And Milo had only recently helped Bob Blackburn repair his bicycle while Blackie was in Paris. At the same time, Blackie was under the gun to get the program going in high gear. Although TWA had been given the choicest European routes, both American and Pan Am had already launched flights to their European destinations, initiating service 23 and 27 October, respectively. One reason they could launch service faster was the aircraft type: C-54 Skymasters. The Skymasters may have been first, but in the beginning they were configured with only a limited number of passenger seats and the typical airspeed was dramatically slower than that of a Connie. The principal reason for TWA's delay, however, was that negotiations with the French for landing rights at Orly Field in Paris took much longer than anticipated. And TWA was flying what was essentially still a very new airplane on the flight, one whose airframe and engines were hardly gremlin-proof.

Amidst a great deal of publicity and top management presence, Blackie and his crew were given a major send off at Washington National for the Paris demonstration flight on 3 December 1945 – but not before several speeches were made and many flash bulbs popped. Jack Frye spoke first and was followed by Robert Brennan, Minister of Ireland, and Henri Bonnet, Ambassador of France, and Welch Pogue, Chairman of the CAB. Bonnet's wife christened the new Connie, "Paris Sky Chief." Blackie and his crew of nine were on their way to breaking five international records while carrying a cabin full of twenty-three VIPs and some very important cargo as well. The VIPs were high-level members of the House and Senate, the CAB, Post Office Department, and representatives from France and Ireland. Among the non-flying TWA personnel were Otis Bryan, Cliff Mutchler, and several advertising and publicity managers. The cargo was a precious shipment of what was then a miracle of science, penicillin, which would very shortly be put to good use in Irish and French hospitals.

The "Paris Sky Chief" was off at 2:39 p.m. and cruised at 11,000 feet averaging just over 300 mph for the 1,330 mile flight to Gander which was reached after four hours and twenty-six minutes aloft. After a brief (forty-eight minutes) stop for refueling, Blackie took off on the 2,005 mile journey to Shannon and blessed by favorable weather at 17,000 feet, the Connie landed after a six-hour flight, but remained only for one hour. On 4 December, Blackie had only 565 miles to go, reaching Orly after almost exactly two hours of flying. The French reception proved as grand as the send-off, although among the passengers only TWA's International Division PR Manager, Eugene Warner, could speak French. Blackie's crew was put up in style – at the Ritz. The TWA team was on the ground for two days before commencing the return flight. During the visit, Blackie visited a hospital patient; it was Captain Henri Launey, still gaunt but on the mend after a three-week battle with blood poisoning. Without the penicillin, Launey never would have survived. While in Paris, the TWA delegation was feted almost non-stop at receptions, teas, and cocktail parties by Air France, the French Foreign Minister, Minister of Air Transport, and the American Ambassador to France. The uninitiated were busy signing short snorters, although that held little attraction for Blackie who explained what the procedure was, although he was, of course, strictly observing the time limitations between bottle and throttle.

The flight home on 7 December began leisurely and included circling over two invasion sites in France. The first was Cherbourg and the second was one with which Blackie was already familiar, Normandy, having flown over it in a Lockheed Lighting little more than a year before. Blackie reached Shannon after two hours and forty minutes, a period in which the weather was beginning to look not so rosy. The flight stayed overnight at Shannon, being wined and dined again, before leaving shortly after midnight on 9 December for a seven and one-half hour flight to Gander. In spite of significant headwinds over much of the North Atlantic, Blackie was able to put his pressure pattern flying skills to good use and actually enjoyed limited tailwinds over portions of the flight. No such advantage could be gained on the final leg to Washington which took almost six hours to complete. Even so, the flight set five new records: Gander to Shannon, Washington to Shannon, Washington to Paris, Shannon to Gander, and Shannon to Washington. Blackie would be setting speed records even fifteen years later, but none of the records lasted very long (some only by a matter of hours and minutes, but that's another story). After all of the heady stuff establishing new records, there came a significant anti-climax on the day after Blackie got back and that was his only other flight in December: having his Constellation instrument flying skills evaluated during an hour and a quarter check ride out of Washington National. The airplane used for the check ride was the same Connie (registration number NC86505) Blackie flew on the Paris trip. He never flew it again. That Connie did not enjoy a long life; it was destroyed in a landing accident on Saturday 28 December 1946 at Shannon. The year-end holiday season was a bit of a trial in the Blackburn household; all three kids were in varying stages of the chicken pox.

Blackie was also scheduled to pilot the TWA introductory passenger flight to Paris in February, i.e, the flight inaugurating regular revenue service to Paris. The service began with a twice weekly schedule before it soon became daily. Before that first flight came together, Blackie spent the better part January shuttling between Washington, Wilmington, Delaware, and Newark, New Jersey, for the purpose of checking out crews on the Constellation. At this point, TWA had almost 100 airplanes of which half were DC-3s, and there were another five DC-3s used for training. From ATC, TWA inherited eight DC-4s which were being civilianized just as the five Stratoliners had been converted. The first ten Connies had been received by year end, leaving TWA with only one other plane in its inventory. That was a B-17 bought from Boeing at a fraction of its original cost and converted into a TWA executive transport for which the first use was in route surveys.

Flying the first revenue flight to Paris really represented the easiest part of the program, i.e., the shining tip of an iceberg whose mass was the supporting infrastructure of people and facilities that had to be in place first. In that regard, Blackie depended on many resources, especially the Industrial Relations Department and its sub-divisions. During the autumn and early winter representatives from Medical, Food Service, and Properties and Construction had been sent to Europe to lay the groundwork. Sales offices were established in Dublin, London, Paris, Geneva, Rome, Athens, Cairo, and Jerusalem. Some of these cities, particularly Paris, had still not recovered from the war and such basics as heat and light were not back to normal. Passenger relations representatives had to be hired. There were language hurdles to overcome – not surprising when putting together a 10,000-mile foreign route structure. By the time, Blackie was ready to takeoff on the historic Paris flight, the International Division had 900 employees, of which many were located in New York City. Blackie himself was moving from ICD's hangar offices at Washington National to a smaller facility near Gravelly Point in Arlington, although it was very close to the north side of the airport. ICD for that matter would be moving also, and would soon find itself based at New Castle (just southwest of Wilmington, Delaware) which had been a major ATC/Ferry Command base during the war. Even at that late stage, Blackie was still looking for good, senior pilots and had pinned his hopes on recruiting one TWA captain in particular, his old Eleventh Bombardment Squadron comrade, Bud Gorman. It was not to be; Gorman was quite content to stay on the sunny West Coast and politely declined Blackie's offer.

If there was one month in the history of TWA to date when the company enjoyed a super abundance of good publicity, it was February 1946 – a time that would eventually prove to be the calm before the storm. There were four epic

Connie flights in February, a month in which Blackie's flight to Paris was almost lost in the sea of positive ink that ran for TWA. Jack Frye took another demonstration flight on 3 February from Burbank to LaGuardia non-stop in seven hours and thirty-eight minutes. He had actually flight planned it for eight hours, and blessed by a tailwind, arrived at LaGuardia with 610 gallons of fuel on board after having started with 5,050. One reason that TWA enjoyed so much favorable press was the passenger list: forty-four members of the nation's aviation news media. The route was unconventional; it was a great circle route and off the standard airways and certainly aided by the tailwind which at times allowed the Connie to attain a groundspeed of 375 mph. Only days before, there had been another "publicity" flight that broke records although that westbound flight did have a stop in Kansas City.

Blackie's 5 February flight to Paris took off at 1:21 P.M. and arrived in Paris after sixteen hours and twenty-one minutes of flying via Gander and Shannon. The wind gods were howling, and certainly not in Blackie's favor eastbound. His "Star of Paris" Connie was christened by Helen Frye before takeoff from LaGuardia where the ceremony's principal speakers were Jack Frye and Otis Bryan. Frye made a point that this flight, unlike Blackie's December 1945 flight to Paris, was not filled with VIPs. Rather the thirty-six passengers had all bought tickets on a first-come, first-served basis. Among the passengers were a former GI heading to Brussels to marry his Belgian fiancé, a diamond expert, several fashion designers, a few French engineers, and a large contingent of French businessmen. The flight over did have some VIP cargo – more penicillin, this time bound for onward shipment to Rome. The return flight, flown four days later, returned via Shannon and Boston, took almost exactly eighteen hours. If there was any one flight in Blackie's career that TWA historians and editors associated with him, it was this 5 February flight to Paris. Blackie's Connie on this flight (CAB registration NC86511) had a career that lasted almost as long as Blackie's. It was involved in a landing accident at Long Beach, California, in late 1950, but was destroyed in a crash at Chicago's Midway Airport in early December 1961, just three weeks before Blackie's mandatory retirement at age sixty.

News of Blackie's flight to Paris was eclipsed several days later when TWA, on 10 and 11 February launched its service to Italy from New York and Chicago via Gander, Shannon, and Paris. Although these were revenue flights, there were some VIPs on board each of the two separate flights: one Catholic bishop and four archbishops on their way to Rome where Pope Pius XII would elevate them to cardinals, including the popular Archbishop Francis Spellman of New York.

Blackie took a flight to Paris in early March and another to Rome in early June, but other than those two trips, he spent very little time in the cockpit for the next year and a half. During that period his flying was limited to an occasional maintenance check flight, evaluating new instrument systems, and checking out new pilots on Connies and Skymasters. If that period was quiet for him personally, it was anything but quiet for TWA. Jack Frye had been expanding TWA's overseas interests and had just bought twenty-eight percent of another war-flattened carrier, Philippine Airlines. Otis Bryan was at about the same time very busy restoring Italy's air transport system which actually dated back to Blackie's participation in the route surveys and his spending considerable time in Italy during that two-month international tour. The Italians were very appreciative and awarded Blackie a pair of Alitalia Airline pilot wings and a decoration by making him a Commander of the Crown of Italy Order. The awards were later presented at the Italian Embassy in Washington in early June. At about the same time, Swede Golien was dispatched to Ethiopia to plan the establishment of an airline in that country, and T.B. Wilson was aloft in a Connie to South Africa on another route survey.

April 1946 brought a passing and an arrival into Blackie's life. The arrival was that of a baby daughter on the eleventh. Her name was selected after Blackie's considering how many "Bs" there already were (Betty, Bob, and Bonnie) and posing the logical question of "To be or not to be." Thus, the new Blackburn child was christened, BeBe, better known simply as "Be." The passing was that of TWA's Intercontinental Division itself. On 15 April 1946, Mel Bowen, one of Blackie's old DC-3 mentors, supervised the departure of the last ICD flight – the second of the two daily departures for Paris. It was ICD's 3,840th departure from America.

Blackburn Family Collection

The Blackburn family in 1946. From left: Blackie, Betty, Bonny, Martha, Be, and Bob.

Very shortly after Blackie got back from his early June roundtrip to Rome, he was interviewed by the *Washington Star* (whose assets were acquired by the *Washington Post* in 1981) for one of their ongoing profiles of local business and political leaders. When asked why a man whose job description suggests a schedule with both feet permanently on the ground was still flying, Blackie quickly explained why he occasionally scheduled himself for an international flight.

> I fly for two principal reasons. First, I have about 7,740 miles of air routes, half a dozen major stations, and about 1,200 employees in my region. By flying those routes and stopping at each station, I get a better idea of the problems and needs along the system and get to know our personnel better. Secondly, it's the best relaxation I know of. I come back feeling fine with more ideas for the improvement of our service.

The bulk of the article was a breezy biography of Blackie's days as a student at Colorado A&M, band leader, Army flyer, TWA pilot/manager, but also mentioned how proud Blackie was of his family. The text also mentioned that he was frequently putting in quite a few twelve to fourteen hour work days. Reading between the lines one could easily see a man who was increasingly frustrated by the "pile of letters, reports, and other business papers on his desk." No wonder he occasionally "picked up his telephone and called Flight Scheduling."

On 17 June, Blackie was back in the news, this time with the *Washington Post*, DC's dominant newspaper. The two-column article had little to do with Blackie's flying, but was all about his becoming the newspaper's Father of the Year for 1946. The title was an annual contest of the paper and was based on editors receiving and reviewing the many letters received from children on just why their dads deserved the title. Bob Blackburn, at age eleven, had composed the letter in which he described his dad as "aces high." Blackie was described as the handsome pilot with a thin moustache and notable crows' feet around his eyes when he laughed. Young Bob mentioned all of the Army planes his dad had flown during the war, the more than two million miles he had logged, and that his father had flown all over the world. He also mentioned how good his dad was at playing the piano and that they all had lots of fun when the head of the household took the family fishing. There were some benefits to the title, including a broadcast interview with the entire family on radio station WINX as conducted by the popular host and singer, Jerry Strong. Then the family was off to dinner at the Statler Hotel and an evening at the Capitol Theater where they were seated in the presidential box. Unfortunately, Blackie smelled a rat. While the sentiments of the winning letter were certainly those of young Bob Blackburn, some of the syntax and penmanship just did not add up. Blackie rode along with the honor and pleasant day session interview at the radio station, dinner, and the movies, but privately suspected that the letter had been penned by a couple of publicity writers from TWA's headquarters on G Street. And he was right.

Almost a month to the day after Blackie got back from the Rome flight mentioned in the article, his phone rang with news of an accident. It was not just an ordinary accident report since this one involved fatalities and the complete destruction of a Constellation. The date was 11 July and the accident location was Reading, Pennsylvania. Blackie had been part of the committee that wrote the report on the first ever TWA Constellation accident which had occurred at Washington National on a rainy, twenty-ninth day of March when a Connie landed hot and skidded off a wet runway. The much more serious 11 July Connie accident at Reading Airport almost amounted to a nail in the TWA heart because one immediate impact was the CAB's grounding of all Lockheed Constellations. Blackie hurried to Reading that afternoon and became part of the accident investigation team.

At 11:21 A.M. the "Star of Lisbon" had departed TWA's training base at Reading to practice instrument approaches and fifteen minutes later was about four miles west of the field when the five-man crew detected what they took to be burning insulation. The flight engineer was dispatched aft to investigate the problem only to find that the entire cabin was filled with thick smoke. The heat and smoke became so dense that the crew abandoned its fire extinguishers and had to retreat to the cockpit where the instruments were already obscured by smoke. The crew hatch was also opened, but that only sucked more smoke into the cockpit. The aircraft commander tried flying with his head out of the side window, but after getting lined up with the active runway, he decided to crash land immediately since the heat and smoke were becoming unbearable. The plane was quickly put in landing configuration and the throttles closed, but the Connie hit some low-hanging power lines and rocks on top of a low hill, turned 160 degrees, and careened 1,000 feet through a hay field. It finally came to rest as a blazing, road-side wreck minus much of its left wing 150 feet beyond the field after hitting trees and telephone poles. Four of the six crewmen died in the crash; the student pilot being checked out in the plane, Captain Art Nielsen, expired shortly after being found near the plane. Only the instructor, twenty-seven year old Captain Richard Brown survived, but his flying days were over; his left arm was so badly mangled that it had to be amputated. Dick Brown was well known to Blackie who had checked him out as a Connie captain only six months before.

There had been yet another TWA-related crash just days before the Reading tragedy. TWA's principal owner, Howard Hughes, had nearly been killed in the crash of his F-11 prototype recon aircraft in California. The twin-engine aircraft had contra-rotating propellers of which one failed over the ocean; Hughes tried to put it down on a long fairway at Beverly Hills Country Club instead of bailing out, but lost control and never made it. After taking out several houses and trees he was rushed to the hospital where he was a patient for well over a month.

Hughes, already eccentric, was becoming even more so considering the pain killers he was taking and would continue to rely upon. There were more months of recuperation in which he was unable to involve himself in TWA's affairs to the great relief of TWA's topside.

Within twenty-four hours of the Reading crash, Blackie was managing a small international airline that had no Connies – the CAB had promptly grounded the Constellation fleet, an action that impacted both Pan Am and American Overseas. The investigation focused on the Connie's electrical system which utilized a conductor to pass electrical current between the plane's pressurized and unpressurized locations. The CAB believed that one of the electrical through studs, as the conductor assemblies were called, had loosened itself, started to arc, and one of its components had penetrated the adjacent hydraulic line, thus producing a serious fire. The CAB found fault with the conductor assembly's design, metallurgy, and placement so close to tubing carrying pressurized, highly inflammable hydraulic fluid. Also at fault was an inspection process that essentially overlooked such a likely source of trouble. It took over two months to certify a smoke detection system, modify the Connies, update inspection procedures, and bring them back into service which cost TWA a bundle. Blackie did have a few DC-4s to use for European flights, but that meant TWA's domestic operations had to rely completely on DC-3s which were a bit passé at that point considering American and United were flying DC-4s. Blackie would not fly another Connie until the end of October – 30 October in fact, a date on which he ferried a Connie with no passengers between Washington and New York. The flight created a number of Blackburn critics because only nine days before, on 21 October, TWA's ALPA members had gone out on strike.

Adding to Blackie's problems had been another Constellation accident on Saturday 12 October when the "Star of Geneva" was ferried down to TWA's new maintenance base at New Castle which had previously been ICD's operations base. The weather was windy and rainy when the captain chose to land there instead of diverting to Philadelphia which had been suggested by the Wilmington tower. The captain attempted to land with a sixteen knot tailwind, touched down long, momentarily gunned the engines as if to initiate a go-around, promptly changed his mind, and then overshot the runway. Although there were no fatalities, the Connie crunched two cars on an adjacent highway and turned into a blazing wreck, thus reducing Blackie's fleet of available Connies at a time when they were desperately needed to extend European routes. At least there had been no fatalities at New Castle; the same could not be said for the next accident which took place not that far into the future.

It is not the purpose here to thoroughly analyze reasons for the strike other than to briefly note that there were several, of which probably the most fundamental was the fact that Jack Frye and the ALPA president, Dave Behncke, really could not stand the sight of each other. Whereas Frye was normally friendly and easy-going, Behncke was cranky and zealous to the point that, even a few of his own staff was uncomfortable around him. Behncke despised Frye for starting a TWA Company Union, even though it was short-lived, and Frye was outraged that ALPA had the audacity to threaten an industry-wide strike back in 1933. The Constellation was the catalyst that provoked Behncke to call a strike because he believed that speed and four engines created a "hazardous" situation which called for hazardous duty pay. The average pilot bought into being paid more for flying a complex airplane but backed off on the dangerous duty angle. It was not the first time that Behncke had brought up the matter of equating four engines for more pay; he had approached the matter before the war when the Stratoliners were introduced but quickly backed off once America became part of the global conflict in 1941. He had also come to a conclusion that industry-wide bargaining probably ran counter to the long term interests of ALPA members, i.e., it made more sense to take on one airline at a time and build upon concessions gained one after another. Another issue was co-pilot pay which had been very low before the war, but now that they had an expanded role in flying these sophisticated, new aircraft, they wanted compensation that was commensurate with their new responsibilities.

Blackie was not the only pilot who flew during the strike. Frank Busch also ferried a Connie without passengers, and like Blackie, he was in some quarters branded as a scab. The strike was not without a few ironies of which the most obvious was Jack Frye's having almost no involvement with the day to day negotiation process itself. TWA was represented by the VP operations John Collings and the company's legal eagle, George Spater, who really did not believe the pilots would strike in the first place. After twenty-five days, the two sides were at an impasse, but did agree to binding arbitration by a three-man panel. That group was composed of the then chief pilot, Bob Buck, representing the union, George Spater representing the company, and the National Mediation Board's selection as the neutral member of the panel. The neutral was Judge Frank Swacker, an elderly man who was learned in labor law but reportedly not up to speed on modern aviation. Blackie had given Bob Buck his captain's check ride just before the war and was one of the witnesses called during the arbitration panel's hearings. The personnel responsibilities and paperwork aspects of being chief pilot were wearing thin for Buck and he soon left the job to resume flying weather research for the Army Air Force in a P-61 Black Widow. He later returned to the line as a captain rather than be caught up in more management-labor disputes which did indeed break many friendships within the TWA pilot community. The arbitration panel had a lot on its plate which was full of questions beyond just what a captain's salary should be, i.e., maximum number of hours to be flown in a month, co-pilot pay, etc.

Blackie's views on the strike were widely known if for no other reason than Blackie's forthright nature. Nearly all of his peers and young pilots who joined TWA right after the war had gone from two or more years of college straight into the military and from that service directly to TWA. Blackie was a major exception because between college and the Air Corps he had been an owner/manager of two businesses. He understood what excessive costs could do to a company, especially one on the brink of bankruptcy. He cautioned other pilots about not killing the goose that laid the golden egg and suggested that by all pushing together and striving for excellence, they would be rewarded in the long run without having to strike.

After three weeks of turmoil, the pilots went back to work and let the arbitration panel settle the issues once and for all. The transcript for the hearings reads more like a court case, one in which Blackie appeared as a "witness" for company management and was questioned by management's lawyer and then cross examined by the union's representative. Many of the questions were related to pay, moving expenses, hours flown, and other economic issues. On some questions, particularly those relating to the domestic side of TWA's operations, Blackie was a disappointing subject from the union's viewpoint, because he claimed to be somewhat ill-informed since he had been on the international side for more than four years. He proved to be even more disappointing when it came to the general subject of responsibility and authority of the first pilot relative to co-pilots, first officers, and second officers. The union was trying to build a case for increasing the pay for all crewmembers based on a chain of command argument, i.e., the first officer is second in command, the second officer is third, etc. Blackie sharply countered by stating his personal belief that the captain is the only authority with whom all responsibility rests. And he said that was company policy as well. In that belief, Blackie aligned himself with the old-line captains of the early 1930s who were practically a law unto themselves. However, Blackie made a distinction which reveals why TWA regarded him so highly as a check pilot and instructor. The Blackburn philosophy was based on treating first and second officers as "captains in training" and that it was the first pilot's responsibility to do all he could to further that education. In the many times Blackie had been quoted in print to that date and in the many articles that would be written in the decades ahead, nothing ever came as close to outlining Blackie's credo as well as his testimony in those labor hearings.

The panel released its rulings in late January when it announced captains on international routes would be paid $925 to $1,100 per month depending on seniority. Co-pilots would make $290 to $520 per month also depending on seniority and emergency duties such as celestial navigation. The domestic pilots' pay scales were somewhat lower, but all were subject to flying a maximum of 900 hours per year except during extenuating circumstances when they would be paid an extra $13.50 per hour overtime. This did not apply to Blackie, who was well situated near the top of the management pyramid and already making considerably more money than any international captain. At about the same time the pilots returned to the cockpit, the International Division announced the appointments of several new managers, and one turned out to be an old friend. It was General Ben Giles who had hosted Blackie and Col. Clarence Shoop in Cairo almost two years before while Shoop's F-5 Lightning was down for maintenance. Giles, whose twin brother Barney also had been an Air Force general, became TWA's new North Africa and Middle East regional manager, a job he accepted on the basis that it would not be a long term assignment. He would stay only long enough to set up the infrastructure that would take TWA from the Middle East farther east – to Bombay.

By late November, TWA had gotten its domestic schedules back to seventy percent of normal, and while international was at about the same state of recovery, fares to Europe were reduced in an attempt to boost sales. Connies were used on North Atlantic flights while DC-4s were used to reach North Africa via the Azores. Service by then had been expanded to Lisbon, Madrid, Tunis, and Tripoli. Although Blackie was part of its planning, he did not fly up to Gander at the end of November for the dedication of new facilities there. Earlier, Gander had been a military base, but now that it had become an international civilian passenger terminal, the airlines and the Newfoundland government spruced up the facilities to welcome international travelers. Whereas during the war, hundreds of aircraft with Army Air Force stars and bars or RAF/RCAF roundels on their wings and fuselages passed through Gander, in 1947 the logos of TWA, BOAC, American Overseas, Pan Am, and Trans Canada joined the traffic at Gander. For passengers there was a new terminal building and a hotel while airline crews enjoyed new, but more modest, hotel accommodations.

Just as the tumultuous year was to end, Blackie got some very bad news in a telegram which informed him that the Constellation "Star of Cairo" had crashed at Shannon in Ireland on Saturday 28 December in bad weather with both crew and passenger fatalities. That Connie, which Blackie had flown to France before, was inbound from Paris when the Shannon tower reported a 400-foot ceiling with one mile visibility after clearing the plane to initiate the range approach. Captain Herbert Tansey hit the ground with his left wing when coming out of the procedure turn for Runway 14 which was out of view below a low hill. Over half of the passengers were killed as were three of the seven-man crew. The investigation focused on poor visibility from the cockpit due to unheated windshield panels, but more importantly on faulty altimeter inputs since the primary and alternative static source lines had been crossed, a situation that a maintenance check failed to identify and one that led Tansey to believe he was safely coming out of the turn at 500 feet. The final report also suggested that the Shannon meteorology office did not relay completely accurate conditions to Tansey who ultimately suffered a leg amputation as a result of his injuries. His story was a sad one, involving not only the loss of his career at TWA but eventually being killed in another flying accident. He later became a flight instructor and suffered a heart attack while aloft with a student who subsequently lost control and crashed, killing them both.

Most news coming out of the TWA public affairs office in late January 1947 was not good, but there were a few exceptions, largely technical in nature, that did impact Blackie. One outcome of the Shannon crash of late December was the decision to minimize the extent to which there was any misinformation about ground proximity resulting from altimeter errors. The fix was not exactly the latest technology, the radio altimeter, but new or not TWA was retrofitting the entire Connie fleet with them and thereby expected to put an end to controlled flight into rising terrain accidents. Blackie was familiar with the new instrument landing system (ILS) from his experience during the war, but as so often has been the case, it would take a while before new military technology could work its way into the civil aviation universe. Blackie flew into New Castle frequently in early 1947 to test and demonstrate the system which consisted of a localizer beam on the ground and a left-right indicator in the cockpit to depict where the plane was on the approach: left, right, or dead center. The other part of the system was a glide slope which provided the same information, but in the vertical plane, i.e., high, low, or what was the goal, right in the middle. The ground controlled approach (GCA) was set up at New Castle at the same time and consisted of radar painting the approach of an airplane. The pilot was given the headings, speed, and rate of descent that would bring him down onto a short final approach where he could take over visually to make the landing. The system had already been demonstrated at Kansas City, but between city administrations and government agencies, the two new systems would require years to install at all of the nation's major, controlled airports.

The cumulative effects of the Connie grounding, the strike, getting back on line with the domestic division, and building the overseas network caused TWA to run deeply, very deeply into the red. At the end of 1946, a year when sales were the highest ever, TWA lost over $14,000,000, of which nearly $3,500,000 could be directly related to the strike and its aftermath alone. One untoward side effect of the strong financial downdraft and Hughes' recuperation from his F-11 accident was Noah Dietrich, Hughes right hand man, again becoming involved with managing TWA. Only this time the involvement was more substantial than ever before. Jack Frye and TWA's executive staff looked upon Dietrich as more than just a snake in the grass, but rather as a poisonous snake in the grass chief among whose goals was fatally biting Jack Frye. Hughes was successful in gaining a bigger role for Dietrich in managing TWA, and one of the first, most obvious results, announced on 13 February 1947, was that thirteen new directors would be joining the board at its next meeting in April. Many were from Houston, thus increasing the number of "Hughes-Dietrich" loyalists. As TWA was approaching the end of 1946, the board consisted of twelve members of which only a minority of five were TWA insiders: T.B. Wilson, the chairman, Jack Frye, Paul Richter, Lee Tallman, and John Collings. At that point, there was only one man from Hughes Tool Company: Charles W. Perrelle who was vice president in charge of production at Hughes Tool Company. T.B. Wilson resigned from the board on 6 February 1947. However, the possibility of Frye's resignation was in the air long before the board expansion announcement was made in mid-February. Thus it might be concluded that the outsiders on the TWA board even in late 1946 had lost confidence in Frye in spite of Frye's denial of any upcoming changes of top management in his New Year's message to employees.

This seismic shift would impact Blackie within several months when the dust settled, but even the handwriting on the wall for the past several weeks strongly suggested that Jack Frye's days as TWA's skipper were numbered. That did matter to Blackie whose loyalty to Jack Frye was undivided. Frye was the man who replied to his employment inquiry letters in the early 1930s, the one who did offer him a job in 1934 and again in 1935, and the one who promoted him over the years from pilot into senior management positions.

The executive suite intrigue, starting in the early autumn of 1946, spilled out beyond TWA and eventually reached Merrill Lynch and the Equitable Life Assurance Society. In late 1946, it became clear to Jack Frye and his financial chief, Lee Tallman, that TWA needed a massive infusion of capital to pay for the new Connies and building the worldwide infrastructure that put the "world" into "Trans World Airline" (singular) as it was then billing itself. Hughes was aware that Jack Frye had been talking to banks and, based on TWA having become an international operator, the Export-Import Bank too. Frye had floated the idea, on 5 December 1946, that a new stock offering involving one to three million new shares of TWA common stock might solve TWA's financial problems even though TWA's stock price had fallen dramatically as its woes mounted. Naturally, such a proposition was dead on arrival at Hughes headquarters where any dilution of Howard Hughes' stake in TWA was intolerable. The board meeting on 23 December, in which voting on the new stock offering appeared on the agenda was cancelled due to a lack of a quorum because nobody representing the Hughes Tool Company attended the meeting. By that time, the rumor mill was starting to crank up with hints of an executive level shake-up, but if Frye was concerned about how he was perceived at the offices of TWA's principal owner, Howard Hughes, he could derive some satisfaction that his hard work was appreciated in other quarters – the United States government. On 21 December, Secretary of War Robert Patterson awarded the Medal of Merit to Jack Frye. It was the highest decoration a civilian could be awarded during wartime. The year 1946 ended on a positive note for potential TWA customers; the airline was starting to install its new reservation system, based on new, high speed teletype circuits, which could promptly result in telling an inquirer if a seat was available. And, as might be expected, TWA announced a major new sales contest which, it was hoped, would get TWA's total load factor up to seventy-five percent or higher and thereby bring in more hard cash. The program came close, attaining ninety percent of its goals.

What better way, Hughes had earlier reasoned, to quickly amass plenty of cash in a hurry than to give that assignment to his own financial wizard, Noah Dietrich? Very quickly, Dietrich was tasked with lining up $40,000,000, which he initiated by calling his contacts at Merrill Lynch in early October 1946. The investment house declined the opportunity while suggesting that Dietrich contact Equitable Life which soon purchased $40,000,000 worth of TWA bonds. The deal was thus struck, but was based on the strict proviso that TWA absolutely not use the funds to pay off old debts. As it turned out, TWA was indeed retiring the debt and paying for aviation gasoline with the proceeds from the bond sales, and it looked like forty million was still not enough money to keep the airline afloat. Hughes had earlier offered to loan TWA $10,000,000 from the financially prospering Hughes Tool Company, but was reluctant to do so until Jack Frye was let go, fearing that a pro-Frye board would not approve the loan. Dietrich apprised Equitable's CEO Thomas Parkinson of that fact while explaining that without the Hughes loan, TWA would promptly go down the drain and take all of Equitable's money with it. Even as late as the first week of January, Parkinson had gone on record stating that Equitable was not pushing for changes in TWA's executive leadership.

The only solution, from Noah Dietrich's account, was to have Parkinson, as the airline's principal creditor, advise all members of the TWA board that each would personally be held financially liable for any default. His other two essential, first steps in a turn-around were the sacking of Jack Frye and getting fresh blood into the board of directors. Once those objectives were accomplished, according to Dietrich's autobiography, which cannot be considered a truly objective record, Hughes could loan TWA enough money to keep it out of the red. Naturally, Parkinson ran into a buzz saw when he took the matter of resignation up with Frye whose influential Democrat friends in Washington rallied to his side (chiefly Postmaster Hannegan and Attorney General Thomas Clark). Powerful as those men were, they had enemies, and among them were Jesse Jones, the recently deposed Secretary of Commerce, and the Republican leader in the House, Charlie Halleck, who threatened Congressional hearings if the Democrats persisted. Parkinson of Equitable was a patient man and preferred to wait for TWA's next quarterly report before any further action. However, the adversities of 1946 (pilots' walkout, Connie grounding, escalating costs of building the international business) combined with what was turning into a sluggish economy for 1947 were too much to be ignored. The end results were Jack Frye's dismissal shortly after the Hughes loan was announced on 3 February and a board of directors that was certainly not universally pro-Frye as it had been before.

Details of Frye's firing are lost to history, but signs point to Hughes himself as the axe-man who met with Frye the day before the resignation was publicly announced. Both Hughes and Dietrich had stated their belief that Frye was a true airline visionary, but in the end Hughes lost confidence in Frye as a financial manager. Frye, certainly antagonistic toward Dietrich, was never bitter about his departure as far as Hughes was concerned, but to a large extent blamed Lee Tallman, TWA's financial chief, for not keeping him sufficiently well informed about how bleak the financial picture really was. Tallman's days were numbered too; when TWA's new leadership was announced later in the spring of 1947, his name was not on the list.

There was a sequel to the Hughes loan which he originally offered as a straight loan not tied to any stock purchase. Dietrich eventually did convince Hughes to exercise his rights to convert the loan to TWA stock within a stipulated period and was able persuade the board to approve the exchange at a cost of $10 per share (below market price at the time). Dietrich, who had rejoined the board by then, excused himself from the voting which resulted in Hughes, with interest, gaining $10,390,000 worth of TWA stock. About twenty years later Hughes sold that block of stock for over $83,000,000.

Jack Frye officially resigned on 21 February 1947 several days after Blackie happened to encounter him at the TWA's downtown offices. Blackie asked if the rumors were true and Frye confirmed them saying it was impossible for him to stay under the circumstances laid down by Hughes and Dietrich. Blackie was already beginning to consider Hughes a meddling micro-manager; that impression would only deepen as the years rolled by. Very shortly after Jack Frye returned to his ranch in Arizona, Blackie was summoned to Paul Richter's house, a somewhat unusual invitation since Blackie and Richter, although friendly, were certainly not close friends. Richter, of constitutionally small stature and very intense, was very much a contrast to the tall, beefy, and outgoing Jack Frye. Nonetheless, Blackie, who was not a drinking man, put a significant dent in a bottle of bourbon with Richter during their discussion of Frye's demise and what that would do to TWA. Like many other TWA employees, Blackie assumed that Richter, as executive vice president, would become the next CEO and asked Richter that question directly. Blackie and nearly every TWA employee assumed that the very able Paul Richter would take on the leadership of TWA with Frye gone, but that was not to be. Although certainly capable of the job, Richter was not favored by Hughes. Furthermore, his loyalty to Frye and sense of what a future would be like under the eccentric Hughes made staying at TWA an unrealistic choice. Richter, knowing the certainty of conflict with Hughes and Dietrich, left TWA in April to become president of TACA, TWA's affiliate in Latin America. TACA was not a small operation; it flew to 126 cities in sixteen Latin American countries and had a terminal in Miami. Richter later moved to Oakland, California, where he owned a very profitable Coca Cola distributorship. He died at just age fifty-three in 1949. With his departure, TWA was certainly no longer the "airline run by flyers."

The early spring of 1947 brought little in the way of encouragement for Blackie. He had not taken a flight to France in nearly a year; and, what little flying he logged was quite routine – taking an occasional flight down to Washington or ferrying a Connie to New Castle for maintenance. The essential character of TWA was still changing. When T.B. Wilson resigned, Otis Bryan was placed fully in charge of the International Division, but the division's structure changed in early April when a new management team was put in place. Warren Pierson, a member of the old board who survived the recent housecleaning, took over T.B. Wilson's job as managing director of international. Pierson's background did not include aviation; he was Harvard Law School trained, was CEO of American Cable and Radio Company, and had also been president of the Export-Import Bank. Hughes installed another old board member, LaMotte Cohu, as CEO of TWA, but both parties knew that Cohu's job was really an interim presidency. Cohu's credentials were quite strong; he was CEO of Northrup Aircraft, Inc., was a graduate of Princeton '17, and had gone into naval aviation before a career on Wall Street and the aviation industry. At its April board and stockholders' meeting, Paul Richter's old position, Executive Vice President, was eliminated and replaced with an executive committee consisting of Cohu, Pierson, and A.V. Leslie, a Hughes Tool Company official. The new fourteen-member board was approved, but only one was a TWA insider, John A. Collings. By far, the majority were from Houston and very much in Hughes' personal orbit. One noteworthy addition was a Kansas City banker: Arthur E. Eisenhower, brother of General Dwight D. Eisenhower.

Predictably, Cohu did not get along with Howard Hughes. Otis Bryan found it increasingly difficult to work with both Cohu and Pierson, and although he had been encouraged to stay on by both Frye and Hughes, he later left to become CEO of Philippine Airlines. Bryan's departure was upsetting for Blackie who for seven years had followed in Bryan's footsteps: Eagle Nest, ICD, and International. In that track, Blackie's close friend, Frank Busch, had followed in Blackie's footsteps right up the same ladder. Not long after the pilot's strike, Busch opted out of the Washington atmosphere and returned to Kansas City as assistant flight operations manager under John Collings.

There were few diversions and distractions for Blackie in the spring of 1947. In mid-March, he went up to the New York offices of WOR to appear on the Mutual Broadcasting System's national radio show "Boss on the Spot." On the show, managers were asked questions over the telephone by their employees without the bosses having any previous knowledge of the questions headed their way. The other three panelists came from Madison Avenue, Hollywood, and a national consumer research firm. The tape of that segment is forever gone, but it can be assumed that Blackie, with his mind for detail and being very fast on his feet, was his usual articulate self.

On 10 March, Blackie lost another friend in what probably was TWA's most freak, yet terrifying accident to date. Fortunately, there was the loss of only one life in a sudden decompression accident. That type of accident, as would sadly be demonstrated in the future, could result in the deaths all passengers and crew when some structure in the fuselage ruptured under the stress of pressurization. Navigator George Hart, with whom Blackie had occasionally flown, was using his octant in the astrodome of the Connie "Star of Hollywood" while it was on a flight from New York to Geneva. At a point 500 miles out from Gander and at 19,000 feet, Hart's glass bubble blew off at 7:30 P.M., and he was instantly sucked out of the plane and into the night sky over the cold North Atlantic. The only other injury was to a hostess who suffered a minor facial cut when she was hit by some flying debris. The Connie promptly returned to Gander where a replacement Constellation eventually took off the next day to continue the flight. Inasmuch as there was no reported evidence of Martin striking the tail, one could only hope that he perished instantly of fright. Unfortunately, from 19,000 feet he might hit the water without having passed out from anoxia because with with his quickly reaching terminal velocity (120 mph) it took little more than a minute to fall 10,000 feet and then less than sixty more seconds to hit the water.

Trans World Airlines

Of all the aircraft types flown by Harold F. Blackburn, the one most frequently associated with him is the Lockheed Constellation.

The Connie fleet, which had suffered a few lesser decompression accidents before, was not grounded, but was ordered to fly lower and/or unpressurized until modifications could be made. The modifications involved strengthening the astrodome attachment and providing a harness for navigators which was certainly appreciated by the roughly sixty navigators TWA employed on transatlantic flights at the time. Navigators at that point in TWA's history were essential to the success of transatlantic flight which depended on pressure pattern flying and constantly figuring the effects of wind. Drift was easily estimated by focusing on the ocean waves below with the driftmeter. For those rare occasions when there was dead flat calm below, Connies carried a smoke flare which could be dropped for drift tracking.

In the early summer of 1947, Blackie's flying was not providing a great deal of satisfaction, and consisted of giving a few check rides, demonstrating the ILS at New Castle, or taking a Connie or Skymaster up for hour-long test flights after engine changes. Complicating Connie availability was the loss of another one on 11 May. On that day a crew took off from New Castle on a routine training flight in which Captain Robert Weeks was checking out a new Connie pilot, Patrick McKiernan. The Connie was believed to be practicing emergency procedures at low altitude when it plunged into the Delaware Bay just off from Cape May, New Jersey, killing all four aboard. Blackie and the other investigators were stumped – there had been no indication from the crew that they were in difficulty. The accident board concluded the report by noting the cause was undetermined.

It was time for a change in Blackie's career trajectory. He missed flying terribly, and although the salary and prestige of management were attractive, flying a desk was not where Blackie wanted to spend the next fifteen years. An international captain's vacancy, based in New York, was announced in mid-June and Blackie put in his bid, having submitted his resignation at the same time. Bryan, and even Warren Pierson, were disappointed, but accepted Blackie's request. With Blackie's seniority and experience at the time, he would easily have gotten the captain's job, but Otis Bryan pleaded with him to temporarily withdraw the bid and take on an interim job. As it turned out the interim job was a big one and related to Swede Golien's mission to help Ethiopia develop a modern airline. Blackie's official resignation as assistant manager of the international division took effect on 1 July. Then he was to go on a brief special assignment to Saudi Arabia which was to be followed by becoming acting manager of Ethiopian Airlines. Before leaving he and Otis Bryan worked out a specific agreement concerning the special assignment's duration: six months. Blackie estimated he could lay out routes, establish schedules, and train enough pilots, enabling his successor to take over at the end of six months. And that period would finally end his days in management which he had come to believe was the best way to earn stomach ulcers. Blackie also limited his foreign tour to just six months because he was unwilling to bring Martha and the baby over and put Betty, Bob, and Bonny in Swiss boarding schools because there were no English language schools in Addis Ababa. He also did want not any long term involvement with other management duties: attending official receptions, conducting VIP tours, interfacing with government ministers and bureaucrats, making speeches, etc. In deciding to head for Saudi Arabia and Ethiopia Blackie was experiencing a yearning typically felt by many aviators returning from the war, or at least by those pilots who had had "a good war." The excitement and sense of accomplishment from surviving successful combat missions had, to a certain extent, become addictive for many veterans. That longing to re-live some of the recent past's adventures contrasted sharply with Blackie's day to day life as a business manager: working long hours to put other pilots in cockpits, seeing that their planes arrived on time and were properly serviced at destinations where he was not going, and completing that with a bottom line that was black, not red.

Wanderlust was part of Blackie's inner being, and heading so far east was a good way to experience new lands and cultures just as he had been able to do in the early days of the Intercontinental Division back in 1942 and 1943. The cultures were, for all practical purposes not that far removed from Biblical times, although King Saud of Saudi Arabia had traded in his camel for a Cadillac. Still, when the king wanted to visit some of his sheiks in central Saudi Arabia, his entourage rode with him in a convoy of hundreds of cars needed to carry his wives, their servants, his harem, and yet more servants. These trips could last for two weeks and always involved setting up tents in the desert each evening. President Roosevelt had given the king a DC-3 with which Saud was so thrilled that he bought five more and signed agreements with the U.S. State Department and TWA to establish an airline in Saudi Arabia. TWA would convert the five C-47s to civilian DC-3 configuration with airline seats, fly the routes, and train Saudi pilots. It was a major enterprise eventually involving many more aircraft and over 700 TWA employees who at first lived in a military style barracks in Jeddah, the gateway city.

Blackie began his Middle East adventure by flying west – on 27 July to California where TWA was converting one of the C-47s at Glendale. Joining Blackie was Mark Outhwaite, an Eagle Nest and ICD veteran who was rated as both a flight engineer and pilot. Over the next three days, Blackie test flew one of the DC-3s, civil registration number NC67826, and spent several hours in a Link trainer to refresh his instrument skills. Since he had a little time on his hands while the C-47 was being modified, he spent a few hours with his former neighbors in Roscoe where he had been the flying grocer and service station manager twelve years before. He also enjoyed a reunion with his Seventh Photographic Group buddy Clarence Shoop who was living in Beverly Hills. His movie star wife, Julie Bishop, had co-starred in two films that year, *Last of the Redmen*, an adaptation of *Last of the Mohicans* by James Fennimore Cooper, and a film noir about newspapermen and politicians, *High Tide*. The three got together at the Beverly Hills Hotel, a familiar haunt for Shoop because that is where his boss lived. The boss was none other than Howard Hughes! Blackie was aware that Shoopy had gone from early test flying of the Lockheed P-80 Shooting Star jet fighter to command Muroc Army Air Field in early 1946. Muroc later gained famed as Edwards Air Force Base, the test flying center. However, life in Hollywood was not possible on a colonel's salary, so Shoop had gotten in on the ground floor of Hughes Aircraft as chief test pilot. That flying did not include any new Hughes racers or eight-engine transports, but did soon result in test flying helicopters and much later test firing missile systems. As it would transpire, Shoop had a very successful career with Hughes and was one of very few people whom Hughes trusted and got along with. Shoopy, who eventually joined TWA's board of directors, was still flying fighters, though not with the Air Force. He was flying on weekends with the California Air National Guard in which he would eventually rise to the highest level.

Then on the thirty-first, Blackie flew the DC-3 back East to New Castle. That particular airplane, the second of Saudi Arabia's DC-3s, led a long and useful life, being withdrawn from Saudi service in 1972, more than a decade after Blackie retired from airline flying. He then flew it up to LaGuardia and departed for Gander on 5 August for a long, long journey, finally reaching Jeddah on the sixteenth after flying via the Azores, Lisbon, Rome, and Cairo. Several of those legs were exceptionally long for a DC-3: nine hours between Gander and the Azores along with eight hours between Rome and Cairo. Blackie's tour with Saudi Arabian was brief because much of the foundation had been laid by the original TWA team of Mark Outhwaite, pilot Joe Grant, and radio operator Zed Lockhart each of whom had arrived in Saudi Arabia earlier in 1946. When Blackie arrived in Jeddah, TWA's contingent consisted of only twenty pilots, but that number would grow considerably in the years ahead.

Setting up an airline for the Saudis in 1946 was essentially one of beginning from scratch, quite literally. An airport existed at Dharhan – sort of. There was a small hangar, one tent, a small range station, and a runway which was lined with fifty-gallon drums. Water was brought in by plane in the same size drums and all food had to be thoroughly washed before cooking. Outhwaite, Wells, and Lockhart lived at the king's nearby summer palace; if they remained overnight elsewhere they slept under a wing using tents, cots, and mosquito nets. Poor visibility in thunderstorms was never an issue, but the massive dust storms that rolled in from the northwest were. Pilots had little choice but to land where they could find some level stretch of desert and hurriedly cover the engines and intakes. The TWA'ers were trying to set up a scheduled airline, but they had their hands full when the king frequently would demand one of the DC-3s for his private use which he frequently did, especially for his hunting trips.

Blackie's sojourn with Saudi was brief, lasting one month before heading to Ethiopia, but long enough to qualify him for a pair of the Saudi's pilot wings. Life was not that Spartan in Jeddah where TWA's assignment was of a long term nature. Had Blackie lingered there he would have enjoyed swimming and snorkeling, fishing, and swimming in the Red Sea. The barracks gave way to homes in Jeddah, but still the transition was substantial for most TWA'ers on temporary duty. The coldest day of the year might see a thermometer registering seventy degrees. Rain gauges were non-existent because the annual rainfall was somewhere in the magnitude of one-half inch per year. With TWA's help Saudi, the airline, grew and became a prosperous company. It was substantially helped along by the fact that hundreds of thousands of Muslims flew Saudi to make pilgrimages to Mecca which is located close to Jeddah. Saudi's first flights were out of Jeddah to Riyadh and Dhahran and later in 1947 to Cairo and Damascus. Although his time in Jeddah was brief, Blackie did meet the State Department's top man in Saudi Arabia, King Saud's chief advisor Najeeb Halaby who had been a U.S. Navy test pilot during the war. Much later, Halaby became CEO of two organizations which Blackie did not hold in the highest esteem: Pan American Airways and the Federal Aviation Administration.

Blackie's role with Ethiopian was both as acting general manager and instructor pilot flying from Addis Ababa, primarily to regional capitals in Ethiopia, but also to locations quite familiar from ICD days: Aden and Asmara. Before returning to the United States he also took flights and trainees as far as Nairobi in Kenya and Luxor in Egypt. He soon established Ethiopian schedules to Jerusalem and Cairo where passengers could connect with regular TWA flights. Of all destinations Blackie reached during and after the war, Ethiopia was among the most fascinating and professionally satisfying. It was a land of the twentieth century in many respects yet many ancient customs were in common use. The three major religions, and offshoots of each, were practiced in the land as were old customs such as adults feeding each other by hand. Indeed, feeding by fingers was and remains commonplace which explains why finding knives, forks, and spoons can be a challenge in all but the major cities. Thus Blackie's dinner might consist of a spicy stew and *incrja,* a type of flat bread in which pieces are broken off to serve as a means of scooping up the stew.

The lanky and outgoing Swede Golien preceded Blackie to Ehtiopia on a survey mission and became a favorite of Hailie Selassie, the diminutive emperor. The emperor's airline at that time consisted of little more than a dozen DC-3s that would soon serve provincial capitals where as many as eighty different languages and dialects could be heard. Runways were varied; some were paved and others were just hard-packed dirt fields. Among the most memorable was one in the heart of Ethiopian coffee-growing country – Gore with a runway that had a three percent grade. Other destinations included Dire Dawa, a city in the north that was spared of bombing during the Italian invasion in 1936, and Gondar, a city in the north that had once been the country's capital where ancient fortresses and castles still remained. Blackie's students were very young pilots some of whom had begun life in little more than huts. It would always be a source of great satisfaction when Blackie looked back, many years later, to know that some of his pupils had become international jet captains. One reason Blackie enjoyed the flying so much was related to Ethiopia's varied terrain which included mountains and high, grassy plateaus where most of the airports were located – nearly all of them at 5,000 feet above sea level. Because the country has a comparatively high elevation, average daytime temperatures were cooler than one might expect for that part of Africa. When Blackie left Ethiopian, his friend

Swede Golien returned and became the new acting manager of Ethiopian Airlines.

In spite of the good flying and sense of accomplishment, Blackie was faced with many problems at Ethiopian of which the most pressing was financing. One of his first assignments after arrival had been to prepare a business plan which included a projected profit and loss statement for the balance of 1947 and the full year of 1948. Even with the financial support of the government and an anticipated flying of 100,000 miles per month, 1947 would see a loss of nearly $800,000, but 1948 would incur a loss of only $75,000. That was all predicated on Blackie having the sole authority to lay out the routes and set schedules. Blackie also recommended new passenger rates and called attention to the fact that Ethiopian was not charging nearly enough for airmail. The government agreed to offer all assistance in building an airline infrastructure with emphasis on modernizing regional airports and, especially, setting up a radio network. Although the flying weather was generally very good, local weather could be bad, not to mention the monsoon season, and in late 1947, pilots were taking off with scant knowledge of what the weather was like at their destinations. Blackie estimated that the government would have to outlay $750,000 to bring Ethiopian into the twentieth century in terms of engine/airframe parts, maintenance tools, test equipment, and a communications system.

When Blackie arrived, Ethiopian was on the doorstep of a financial crisis, not unlike the one TWA itself was facing in early 1947. The Shell agent in Cairo, from which Ethiopian got all of its aviation fuel, was owed $100,000 and had put Ethiopian on a cash in advance basis. Personnel problems were another headache; Blackie's secretary was on leave in Nairobi and within a month of his arrival, two of his American mechanics quit. Ethiopian's board was composed primarily of government ministers in a country where there was no equivalent of the CAB, thus part of Blackie's job was that of education. He reported to the Minister of Justice who was essentially just another frustrant; that man was in chronic ill health and not available for meetings, sometimes for weeks on end. Blackie had hoped to have the minister available to fly on a regularly scheduled Ethiopian flight to Cairo on 17 September when TWA's board chairman, Warren Pierson, was on a Middle East fact-finding tour. The meeting never came off because the minister was too sick to travel so Blackie took another flight instead – to Nairobi. Blackie had set down a foundation from which Swede Golien and subsequent American acting managers could build upon, and build they did. Ethiopian eventually became one of the largest of the African airlines with more destinations in Africa than any other carrier. Still, Blackie's recommendations awaiting Swede Golien ran to thirty-four action items.

After he returned from his Ethiopian sojourn, Blackie always looked back on it as a time of rejuvenation in spite of the many managerial headaches – one that took him back to his early days with TWA when the company was blazing new trails with the DC-3. The only truly untoward memory of Ethiopia was how cruel justice could be in some Third World countries. A prime example of that justice, or perhaps more appropriately injustice, was suffered by a member of the ground crew in Addis Ababa who was towing a DC-3 when a wing clipped the corner of a building. The damage was minimal, only a dented wingtip, but the poor man paid dearly for his mistake; one of his hands was cut off by the authorities.

If the air was hot and dry on the Horn of Africa, at least it was fresh in comparison with that of Kansas City. Howard Hughes by then was less engaged in financial matters and more involved in becoming a major nuisance to TWA's technical and operations managers. He introduced a "new" radar for terrain avoidance, one that was quickly retrofitted to the Connie fleet. Hughes personally demonstrated the system on a Connie by flying in close proximity to mountains to showcase the innovation, and promptly gained substantial national publicity from his flights. The unit included a warning light and alarm to warn a pilot of ground proximity, but the technology was not new – it had been used by the Army Air Force during the war as an aircraft system which warned of an approaching enemy plane. Hughes, in the autumn of 1947, had other distractions including the first (and only) flight of his H-4 troop carrying, eight-engine seaplane, the wooden "Spruce Goose." Although he was content to let Dietrich, Cohu, and Pierson deal with financial matters, Hughes became a nightmare for TWA's transportation department. He had his own personal Connie which he kept at Burbank, but was forever demanding that TWA bump passengers off regular flights so his Hollywood friends could get a seat at short notice. At the time, TWA was the only airline that had a transcontinental four-engine sleeper which was very popular with the Hollywood idols who made TWA, the airline once run by pilots into one "flown by the stars." While Blackie was in Ethiopia, TWA started to produce a movie to promote more air travel, but the film turned out to be little more than a travelogue focusing on some of the more scenic waypoints, such as the Grand Canyon, along TWA's routes. Hughes at the time was starting to think about getting back into the movie business as a studio owner and not just as producer which was the extent of his involvement with films since the mid-1930s. His purchase of RKO in 1947 would have a direct impact on Blackie in just five years. RKO would suffer the same type of micromanagement that Hughes' had rained down on TWA and eventually be sold after years of declining revenue.

While Blackie was away, there were major structural changes at TWA in which Cohu was streamlining the company. He was turning it into a straight line organization from the system of distributed management that had prevailed in recent years. Today's business jargon would call this "right sizing" instead of the downsizing it really was. The resizing resulted in the reduction of several thousand TWA jobs including some at high levels. Otis Bryan was essentially demoted and made VP-Operations of International and moved from TWA's growing eastern headquarters in New York to New Castle. A new executive committee was named consisting of Cohu, Pierson (who was in effect the CEO of International), Albert Leslie, who replaced John Lockhart as treasurer, and Noah Dietrich. Concurrently the domestic operations had added new cities in the TWA network and moved all DC-3 flights to and from the New York area back to Newark, thus turning LaGuardia into a Connie-only base. In order to boost the top line, passenger sales, cargo, and on-time contests were launched.

As the year 1947 ended, TWA had 120 aircraft in its fleet, was rapidly adding cities to its domestic network, and had signed an agreement with Northwest Airlines by which TWA passengers could make connections in Detroit or Chicago with Northwest, and then fly to Tokyo, Seoul, and other Asian capitals to eventually link up with TWA in India. TWA International was ahead of the other airlines in total payloads flown to Europe, and doing it with fewer flights as well. The "right sizing," cost controls, and sales contests were having a positive impact; TWA had a couple of promising top line quarters in 1947. Yet there was a growing burr under the corporate saddle in the form of competitors with new DC-6s powered by Pratt and Whitney R-2800 engines which gave the DC-6 an edge in top speed. Not unlike the first Connies, the DC-6 had some gremlins and one of them, a tendency for inflight fires caused by poorly located fuel vents, grounded all Sixes in service for four months while modifcations were completed. At about the same time, TWA was buying a new Connie, the 749, with the latest version of the Wright 3350 engine, and although the range was outstanding (New York to Paris non-stop), it was still not quite as fast as the DC-6. All of the new 749s, twelve of them in early 1948, were assigned to International which released its older Connies to Domestic.

Just before Blackie went back on the line, TWA lost another Constellation, although this one was attributed to pilot error and not the Wright engines or any other of the Connie's systems. On Tuesday, 18 November the Connie "Star of Madrid" was shooting instrument approaches at New Castle with a CAB examiner on board. The tower approved an ILS approach to Runway 1 over which the crew flew a low pass before pulling up into a normal downwind leg. All appeared normal until short final when the Connie seemed too low; moments later the nose gear hit an embankment 120 feet short of the runway, ballooned up, and then fell to the runway, losing its right wing in the process. The fuselage rolled inverted just as fuel sprayed and ignited from the ruptured fuel tanks and exploded. The three pilots and CAB examiner all perished in the ensuing fire. The pilots, Francis Winkler, Virgil Kennedy, and Emery Christensen were all acquaintances of Blackie's as was the inspector, Herbert Dowsett.

Chapter Ten
Across The Seas

Blackie had some accumulated vacation time when he arrived back home from Ethiopia and much of that time was spent looking for a new home and relocating Martha and the kids from Washington. The Blackburns moved back to East Williston to the predictable English Tudor style home on a quiet, tree-lined road, this time at 21 School Street. Blackie's return was confirmed by his good friend Frank Busch who was firmly planted high on the management side and was then working as John Collings' operations assistant in Kansas City. When Blackie first picked up his company mail upon returning to LaGuardia on 6 January, there was a large envelope from Swede Golien who had sent him a certificate proclaiming Mister Hal Blackburn as a *Convaricator Shellback Solarus Empyrus* in celebration of Blackie's first flight to Nairobi from Addis Ababa with an Ethiopian DC-3. That flight, of roughly 720 miles distance, had crossed the equator at a point roughly 100 miles north of Nairobi. The document was far from Blackie's first certificate for crossing the equator and it would not be the last.

Blackie always had mixed feelings about work as a check pilot because he had to sort the good from the bad which was the way the system worked. Giving a down check was essentially career-terminating, but Blackie was never about to let personal sympathy get in the way of professionalism. How a pilot handles real adversity was the ultimate test of a pilot's ability, and fortunately, the number of occasions in which he observed a captain's handling of a truly tough situation were limited. One of the more memorable check flights occurred early on a flight from Rome to Athens where a major thunderstorm was encountered shortly after takeoff. The Connie was struck by lightning several times, resulting in a few perforations in the aircraft's skin. The storm and static were so bad that radio communications were basically lost. They were able to pick up Naples soon after takeoff from Rome, but the other stations along the way, Brindisi, Otranto, and Catanzaro, were all unintelligible due to precipitation static.

As the Connie approached Athens, the indications from the radio range were intermittent and highly erratic. They were so suspect that Captain Talbott and Blackie concluded, in spite of not getting an indication that they had passed over the cone, that they had indeed overflown the station. The navigator celestially confirmed that and Talbott turned to intercept the beam which again proved completely unreliable. Talbott decided that the underpowered range station at Athens was completely unreliable for making an instrument approach, and then promptly diverted to Cairo, a decision which Blackie completely supported. There was no gray area in a check pilot's report; the only grades were satisfactory or unsatisfactory. After giving Talbott a satisfactory report, Blackie followed up on the problem and learned that the power leads into the radio range had somehow just been reversed during recent maintenance, thus creating the navigation problems. It was not Blackie's place to give a check ride to the station manager in Athens who should have been on top of such a problem, but if Blackie had the right of evaluation, the station manager would have gotten a down check. Instead, Blackie reported the problem to operations in Kansas City which did update its procedure for reporting technical problems. Coincidentally, Blackie subsequently conducted a survey on all radio facilities in the Mediterranean area and made recommendations for improvement, although many stations, other than those at major air terminals, were beyond TWA's influence. Nevertheless, having stronger, well maintained stations at such places as Araxos and Heraklion, over which Blackie had flown in a Lightning in 1944, would certainly minimize any flying extra miles, wasting gasoline, and getting passengers in late.

The pattern for Blackie's flying was set for the next seven years, and if it had been more flying that he wanted, he surely got it. His first flights back on the line were those he would be flying regularly throughout the rest of his TWA career: New York, Shannon, Paris, Geneva, Rome and return. The Lisbon, Madrid, Rome, Cairo, and Athens route was the other routine flight. The only rare exceptions were the occasional charter flight to places such as Mexico City or a special cargo flight resulting in nearly eighty hours per month and sometimes more which was based on two roundtrip flights as far as Cairo. And just to maintain his touch, he got in at least two hours of Link trainer time monthly. One of the captains to whom he gave an instrument check in early 1948 was Ernest Pretsch who had made captain just before the war. Pretsch, born of German parents, came across as a somewhat dour man with whom Blackie would share an interesting assignment just eight years into the future.

In early 1948, Blackie's flying was split almost evenly between DC-4s and 049 Connies, until 25 May when he logged over four hours transitioning to the new 749 Constellation. Martha and the kids benefitted from Blackie's new schedule because, in spite of being gone two weeks out of a month, he was also home two weeks every month and not spending any fourteen-hour workdays at the office. Blackie always despised any form of idleness, and the first beneficiary of his industriousness was the Tudor home at 21 School Street. With its neatly manicured lawn and landscaping, the Blackburn residence was, according to several neighbors, the showcase home in its part of East Williston.

Just as his transition to the new Connie was completed, Blackie received some long overdue good news. General Ben Giles had retired from TWA as the North Africa regional manager before Blackie returned from Ethiopia, but had cut through some bureaucratic red tape at the War Department's Adjutant General's Office to finally resolve Blackie's inquiry as to what had happened to his application for an Air Force command pilot rating. The rating required 3,000 hours as a rated pilot and fifteen years' service within an active and/or Reserve component of the military. At this point Blackie had close to seventeen years of active and Reserve duty and well over 9,000 hours as a pilot in command. Blackie finally received the rating in orders dated 22 December 1947. Blackie's relationship with the military continued to be a vexing, disappointing issue. In acknowledging his application for command pilot wings, the War Department informed Blackie that his commission in the Air Force Reserve for an additional six months after the duration of the "present emergency" (Cold War) would expire because he was otherwise ineligible based on the fact that he had not served on active duty during World War II. That was, of course, certainly not of Blackie's doing but rather because TWA had kept him in a reserved occupation. And then there was the matter of being overage in grade. Blackie accepted the wings with gratitude as undoubtedly the oldest, lowest ranked command pilot in the service. In his thank you note to the office of the Secretary of the Air Force, Blackie pre-volunteered his services for active duty in case of another shooting war "before being relegated to the shelf for old soldiers."

Giles also interceded, but had less luck, in his efforts to have Otis Bryan appointed to the grade of lieutenant colonel in the Air Force Reserve. Otis Bryan was further disappointed several days before Blackie got his command pilot wings when it was announced that he was moving back to the New York offices. There he was to take charge of TWA's efforts to rehabilitate key foreign airlines beyond what were already major involvements with Saudi Arabia and Ethiopia. Bryan's responsibilities as the operations manager at International were assumed by John Collings who was the operations VP in charge of Domestic. Collings, at one time a Ford Trimotor test pilot, had been on the TWA scene since the beginning and had been a Kelly classmate of Larry Fritz. News of the change was unsettling for Bryan and disappointing for Blackie since he had followed in Bryan's footsteps for nearly seven years.

Although Blackie was by then free from earning ulcers in the swivel chair world, he followed corporate developments closely, knowing that TWA was still in danger of foundering in the seemingly endless financial storm. There was mixed news of which some was very good. The CAB, after many TWA applications, finally approved air mail rate increases that would significantly help reduce losses. At the time, TWA was losing money at a rate which would equal 1947's full year loss of $8,000,000. The new rates came just in time because in late May, TWA was due to pay $1,100,000 back to Equitable. Shortly thereafter, TWA increased the number of weekly flights to Europe from thirty-five to forty-four. Technical advances included an improved landing gear for Connie that minimized the heavy loads on the struts, and for travelers, a nice new terminal at Orly Field in Paris.

But there was still turmoil back in Kansas City and New York, resulting in more work for the executive committee because it lost a key member. LaMotte Cohu resigned on 6 May, stating that he felt that the work he had been brought in for had been accomplished. Very quickly, he surfaced as CEO at Convair in

California. Within weeks, Bob Loomis, Blackie's friend who had become the top engineering and test pilot in Kansas City resigned. The announcement claimed that the resignation was under doctor's orders and that Loomis was suffering so much from pollen allergies, especially to Kansas sunflowers, that had to move to a less pollen-dense location. Los Angeles fit the bill nicely and so did his new job at Convair. Although Loomis was a good friend, the two departures for Convair were not especially upsetting, but the next one was. Otis Bryan resigned on 17 June to become the chief operating officer at Philippine Airlines.

TWA had no CEO at the helm, but would be run by a committee until a new president could be found. So for the interim, Warren Lee Pierson became the de facto chief at the company and promptly took off for Europe on another fact-finding trip on 21 August for which Blackie was not the pilot. Blackie may have encountered the Pierson group since both were briefly in Shannon on the morning of the twenty-second. Before Pierson took off for Europe, he issued a new organization chart in which TWA's V.P. traffic, Oswald Cocke, joined the board of directors. Of much more interest to Blackie were the names under director of operations/domestic and director of operations/international: Frank Busch and Kemper Jacks respectively. As those men assumed their new responsibilities, Warren Pierson was approving any project that would increase the top line or cut costs which is why the company sponsored so many sales and on-time arrival contests while doing what it could to grow the market for international travel. Pierson beat the bushes giving speeches that were based on some attractive realities. A traveler could get to Paris by first taking an ocean liner to a French port and then travel another day to reach Paris or he could do it on a TWA Connie overnight sleeper flight for about $450 less and still have champagne with his dinner and not worry about rough seas.

While Blackie was enjoying his flying in Ethiopia late in 1947 and returning to transatlantic flying in 1948, the airways over the ocean were much calmer than the atmosphere in several boardrooms and congressional hearing rooms. For a while, it looked as though Blackie and his fellow International Division pilots could be flying for a different company – if they were flying at all. Yet out of the wheeling and dealing, a new successor to LaMotte Cohu would emerge. The year 1947 had been rough on the airlines and 1948 resulted in a major loss for TWA; it would have been worse had the international business not been so strong. Compared to the 1946 and 1947 losses, $13,666,000 and $8,079,761 respectively, 1948 was a good year which tended to point to a strong 1949. Juan Trippe was still pushing his "Chosen Instrument" plan, i.e., one international flag carrier for the United States as controlled by Pan Am. Within this cauldron there also stewed American Overseas (AOA) Airlines which had nearly drowned in 1947's red ink. The loss was of such a magnitude that AOA's owner, American Airlines with its legendary CEO, C.R. Smith, was considering giving up the international business entirely. Unlike TWA, American's domestic business was very profitable. Earlier in 1947, Senator Owen Brewster of Maine, whom everybody considered to be a Pan Am puppet, had reintroduced the Chosen Instrument Bill with a new wrinkle. It was based on pre-dating the existence of the new international entity to October 1945, thus cutting TWA out of the picture since it did not launch its commercial international flights until early 1946. Brewster's membership on the Senate War Investigations Committee was another real thorn in Hughes' side especially when it investigated Hughes for alleged failure to follow through on aircraft development projects such as the Spruce Goose. Hughes and Brewster met privately before the hearings, and it was assumed that they also talked about the Chosen Instrument business and if TWA might be willing to sell its international routes to Pan Am. Hughes even met with Juan Trippe who was talking about a merger between TWA and Pan Am, although there is no evidence to suggest Hughes was doing anything other than sizing up his archrival and keeping him guessing about where Hughes might take his airline.

When Blackie was airborne, on his way to Ethiopia the previous July, the Chosen Instrument bill augured in at the Senate and House committee level. At the same time, there was a quickening of Brewster's pulse at the thought of hearings that would hang Hughes as a war profiteer. Out of Hughes' well publicized and crafty performance which made laughing stocks of Brewster and the committee during the early August 1947 hearings, there came one major and one minor side effect. Brewster lost his next election bid much to the satisfaction of Hughes who contributed generously to Brewster's senatorial opponent. The far more important ramification was Trippe's abandoning his anticipated take over TWA's international routes in favor of AOA. C.R. Smith, CEO of the parent American Airlines, was delighted to shed his airline of the overseas burden. In December 1948, while Blackie was on his way back from Lisbon, Trippe and Smith struck a deal, which they promptly filed with the CAB for approval. For almost $18,000,000 Pan Am would get AOA's overseas routes and twenty airplanes (Connies and DC-4s). Out of that deal, as a side effect, TWA got a very respected, new president: Ralph Damon.

Damon was president at American and was angered by Smith's sweeping the rug out from under him. He had a stellar resume: Army pilot in World War I, cum laude Harvard graduate, and experience with manufacturers (Curtiss-Wright and Republic Aviation) in addition to American Airlines. Damon did not officially become CEO until mid-February 1949, and when he did, there were a series of changes that had an impact on Blackie's flying. Most ferry flights to New Castle for maintenance became a thing of the past; only major overhauls would be done there and everything else would move to LaGuardia. Shortly after Damon's arrival, all major maintenance on Connies was moved to Kansas City, and New Castle was closed as a TWA base. Zurich was added to TWA's schedule making it the second largest airline in the world as measured by weekly, scheduled miles flown. The leader was Aeroflot, the Russian state airline, although the statistics may have been manipulated. There was no doubting

TWA's reach which extended all the way to Bombay. When Blackie first flew into Zurich on 21 January 1949, he and his passengers were very impressed – by the runways only. The passenger terminal then consisted of little more than line shacks; the new airport's long runways had only recently been completed but within a couple years the grand new terminal, one of Europe's finest, would add to Switzerland's tourist appeal as a summer or winter destination.

By mid-year, TWA passenger loads were so high that the company announced plans to buy twenty more Connies. This time, management would not have to beg Hughes Tool for financing because a consortium of five banks, with J.P. Morgan and Chase in New York, became the source. In mid-1949, Blackie's employer was riding a wave, celebrating its twentieth year by crossing the country with a Ford Trimotor shadowed by a Connnie, and showing a first half year profit of $361,907. Ralph Damon signed an intent agreement to lease space at Idlewild Airport on Long Island's south shore for twenty-five years. Signing on behalf of New York was Governor Thomas Dewey whose state pledged $8,000,000 to build TWA's terminals and hangars at the airport which was opened on a limited basis in 1948.

While Blackie was never originally enthralled with being a check pilot, the transition back to the line did aggravate Blackie's touchy relationship with ALPA. From early in the war, Blackie had been placed on inactive status because he was "an official" of the airline, i.e., a member of company management. Blackie was naturally interested in becoming an active member again because, although his first loyalty was to TWA, there were benefits to membership in ALPA. The union had, however, hardly forgiven him for walking across the picket line and ferrying an empty airplane during the strike against TWA. Blackie always believed the union had no basis in fact for its accusation because the plane was empty, he was a member of company management at the time, and his ALPA membership was on inactive status anyway. The union did not see it that way, and its master council in Chicago formally expelled Blackie completely from ALPA while he was away in Ethiopia. Now there was a price to pay since Blackie was no longer in upper management, but did want to rejoin as an active member. The matter festered well into 1949, finally reaching a conclusion on 29 August 1949. Blackie's rejoining ALPA as an active member in good standing had to be preceded by two years of probationary membership under the watchful eye of the ALPA local council, his payable dues would be dated back to the time he started flying the line again 1948, and he was fined $1,000 for his performance as a strike breaker. He complied with ALPA's terms, but the one grand fine remained as a sore point forever. That was a substantial sum in 1949, and even though Blackie was well paid as an international captain, the cost of living on Long Island was a far cry from Albuquerque or Kansas City. And just around the corner was Blackie's being faced with the prospect of college tuition payments. George Spater, TWA's legal counsel felt badly about Blackie's being fined and suggested to the new president, Ralph Damon, that the company reimburse Blackie for the stiff fine paid directly out of Blackie's pocket. Damon agreed, but for whatever reason there never was any reimbursement. Nonetheless, Blackie held Damon in high regard as did TWA employees pretty much across the board. Damon had a good working relationship with Howard Hughes, although his association with Noah Dietrich was not an especially cordial one.

Blackie had many ALPA friends throughout the airline pilot community, including Pan Am, American, and AOA. Therefore he followed TWA and Pan Am sparring over Trippe's Chosen Instrument policy which was still swirling around Washington hearing rooms in 1949. One personal attribute Blackie always admired and commended to his staff when he was at Eagle Nest, ICD, and then International was persistence, a quality Pan Am's Juan Trippe had in spades. Late in 1949 TWA filed another brief at the CAB objecting to Pan Am's planned takeover of AOA. Within days after announcing a pre-tax profit of \$3,931,910 for the first nine months of 1949, TWA sued Pan Am for soliciting charters to Rome which was in violation of the July 1945 ruling on North Atlantic carriers. The latest fracas was over the Holy Year of 1950 as decreed by Pope Pius XII in which thousands of Catholics were expected to make pilgrimages to Rome. Pan Am agreed to cease soliciting until the CAB ruled. As it turned out the decision went in TWA's favor, but another preliminary ruling did not go its way. The CAB's trial examiner, Thomas L. Warren, recommended that the Pan Am-AOA merger be approved. TWA formally objected, citing bias on Warren's part because his recommendation in 1945 was to give Pan Am all of the prime routes to Europe. The legal wrangling continued well into 1950 at a time when TWA was rapidly expanding its domestic fleet and network. The CAB ruling finally did come, and TWA was favorably surprised in late July when the CAB panel ruled three to two in TWA's favor. Then the unthinkable happened; President Truman not only overruled the CAB and permitted the Pan Am takeover of AOA, he also allowed Pan Am to fly into Paris and Rome. Pan Am did not carry the day completely, because TWA was allowed to fly into London and Frankfurt. One can only speculate on the possibility of Truman's decision going the other way had Jack Frye remained at the TWA helm. What had happened was a good example of Pan Am's lobbying machine at work. One of Trippe's Yale classmates was a man named Sam Pryor, a fix-it man if ever there was one. His lobbying convinced Secretary of State George Marshall and Attorney General J. Howard McGrath that Pam Am should be able to conclude the merger and fly to two more European capitals. Congressman Sam Rayburn was a good friend of American's C.R. Smith and of Harry Truman. The three men in turn influenced Truman who did overturn the CAB decision, but also cracked the Pan Am corporate knuckles by denying and domestic routes to the carrier, including the coveted Miami to New York run.

Jack Frye's reaction to this unexpected turn is unknown. He was in a much different orbit in 1950 and was in his third year of leading GAF Corporation where he was installed as CEO in April 1947. GAF was the American arm of the huge German company, I.G. Farben. The United States government seized the company immediately upon America's entering World War II just as it had seized the company in World War I. GAF was not in the airlines business, but at a salary of several times what he made at TWA, Jack Frye was not complaining.

August 1950 brought mixed news for TWA. The good news was the company reporting net income for the first half year of nearly $2,000,000, and the third quarter was looking very strong with record bookings in spite of the Korean War breaking out in June. On 30 August, Blackie was enjoying a few days off after having returned from the Paris-Gander-Boston-New York run on the twenty-ninth. On the thirty-first the Connie "Star of Maryland" crashed forty-eight miles north of Cairo while attempting an emergency landing. Shortly after takeoff the number three engine caught fire and fell off the wing but the fire was still raging. Captain Walton Webb attempted an emergency landing in the desert, knowing that he would never make it back to Cairo. The landing was unsuccessful and the Connie was destroyed with all seven crew and forty-eight passengers killed. Blackie had flown the exact same Cairo to Rome route, most likely in the same Connie, only the week before. He had also been the check pilot flying with Captain Webb two years earlier on a Rome to Cairo flight. At the time of the accident, the forty-five year old Webb was domiciled in Cairo and, like Blackie, had a well-known interest in all of the Mediterranean basin cultures. The cause of the fire was eventually traced after the number three engine was finally found three miles from the crash site; the rear engine row master rod bearing had burned out.

One month later, Blackie was among the first to fly the New York-Shannon-London-Frankfurt route and returned from it very impressed. Unfortunately, TWA lost money on Blackie's first flight to Germany in six years because the passenger cabin was almost empty. Two weeks after returning from his first flight to Frankfurt, Blackie experienced one of the aggravations of the long distance airline pilot: a medical emergency. Rather remarkably, in the millions of miles he flew for TWA, he only experienced one significant medical crisis in a profession where pilots with full careers usually face more than one death, birth, or other major medical event. On 16 November 1950, about halfway between Rome and Cairo at 15,500 feet, the navigator came into the cockpit and said, "The purser is down." Sure enough, the purser was sprawled on the galley floor, bleeding from a cut on his left hand, and was coldly clammy to the touch. Blackie promptly descended to 13,500 and asked the flight engineer to bring the cabin down as far as possible. The crew was trained for such emergencies and promptly bandaged the hand, loosened all restrictive items of clothing, and quickly got the purser into the upper crew bunk. The lower bunk would have been more convenient, but it was carrying some tied-down, extra cargo. After a few minutes of pure oxygen from one of the carry-around bottles and a few whiffs of smelling salts, the purser came around; the passengers were never aware of the mini-crisis. The event prompted another Blackburn question to operations in Kansas City. Why were there no oxygen system outlets in bunk locations?

A week later, Blackie was flying back from a trip to that began in Madrid with continuing legs to Cairo and Athens. He returned to New York on 22 November with visions of fluffy, puffy cumulus floating over Cyprus, of sun-drenched Greek Islands, and the brilliantly lit cliffs of the Adriatic, on his mind. The images were quickly obliterated by what happened beginning on the twenty-fourth. A monster, early winter storm rolled into the New York metro area with winds up to 108 miles per hour and attending, record rain, sleet, and snow. The dikes keeping Flushing Bay at bay around LaGuardia failed and sixteen Connies and Martins out in the open were subjected to flooding waters almost to their wheel tops. After the storm passed, special maintenance crews were flown in from Kansas City to clean up the aircraft landing gears. Fortunately, Blackie was holed up at home, content in knowing that he was not scheduled fly again until his 3 December flight to Paris.

Throughout the early spring of 1951, Blackie was flying two overseas trips per month – usually to Paris or London and return early in the month and then a flight to Italy, Greece, or Egypt later in the month. TWA was flying high financially, making a profit of $6,495,000 for the year 1950. Some of that success was working its way back to TWA employees who were able to take advantage of such benefits as retirement plans, term life insurance, and health care plans. There were some concerns related to the Korean War which at that point was almost a year old. Several of TWA's DC-4s were transferred to Northwest Airlines which was then a contract airline for the Military Air Transport Service. Blackie had flown each one of those Skymasters from 1948 through 1950 before his flying became limited just to Constellations. One of the Northwest-operated C-54s did not live to a ripe old age; it was lost in the ocean just off the British Columbia coast in January 1952 as the result of a botched go around attempt. That C-54, N45342, had lost an engine on the Anchorage-Seattle stage of a flight from Tokyo. Rather than push it to McChord Air Force Base near Tacoma, Washington, the pilot was cleared to land at Sandspit Airport on Moresby Island, British Columbia. That regional airport also served as an emergency field for the RCAF. The plane failed to attain flying speed in the go-around and crashed into the sea where it floated briefly. Unfortunately, the current took it further out to sea; thirty-six people perished due to exposure before rescuers arrived, but seven were rescued. Blackie had first flown that C-54 while checking pilots out of New Castle and then had later flown it to Madrid and Cairo several times.

One could not help but wonder if more of TWA's fleet might also wind up flying for Uncle Sam just as it had in1942. The war years and TWA's role in the Air Transport Command were remembered by TWA aircrews in early May when the airline sold all five of its Stratoliners to the French Aigle Azur Airline on 3 May for $525,000. The Strats would be flown out of Paris to French colonies worldwide, including French Indochina where two years later a pair of the Strats were hit, but not brought down by Communist forces.

Five days later, Blackie took flight 810 to London and returned on the tenth. The flight was routine as possible, with one glaring exception. Flight 810 originated at New York's Idlewild airport (later renamed Kennedy). For a Connie pilot, Idlewild was pure luxury: it had ten miles of runways of which the longest was 9,430 feet. Covering 4,900 acres, it was then the largest airport in the world. It also had 108 employees in the customs department to check on what passengers were bringing back from distant lands. There were the predictable: perfume, hats, watches, Dutch tulip bulbs, optical equipment, leather goods, fur coats, linens, and other items fancied by tourists. There was the occasional odd item: a marble chip from the Parthenon in Athens, a camel saddle here and there, or a stuffed gazelle. Blackie's luggage was occasionally heavier when returning than departing. His flights to Cairo meant he could barter with his old acquaintance Bassili Tawadros at the Ding Dong Bazaar in downtown Cairo. Blackie took items that were difficult to obtain in Egypt, such as silk stockings, and returned with carpeting and interesting art that became decorations in the Blackburn home. Blackie also brought Tawadros fancy undergarments for Mrs. Tawadros, a former Egyptian movie queen. In return Bassili gave Blackie many artifacts, wall hangings, carved wooden jewelry boxes, and scarabs made into jewelry.

On 29 May 1951, Blackie was enjoying some leisure after returning from Shannon to Idlewild on the twenty-fifth. In Germany, Transportation Minister, Dr. Hans Christoph Seebohm, assembled a committee under Hans M. Bongers, to give advice on the future of German aviation and the possibility of forming a German national airline. Bongers was no lightweight; had been in high positions with Lufthansa prior to World War II. That activity was well below TWA's and Blackie's radar screens at the time, but the eventual outcome would impact Blackie and TWA in a very major way.

July of 1951 was a routine month for Blackie, but anything but routine for TWA. He had taken off from Idlewild for Paris via Boston and Shannon on 12 May. After a two-day layover in Paris, he took off for Rome and Cairo; he was back in Paris on the eighteenth and departed Shannon for New York on the nineteenth. Back in Kansas City, there was a major mess to deal with because violent storms had caused the two local rivers, the Missouri and the Kaw, to overflow their banks. Kansas City's municipal airport was not flooded, but the company's massive repair facility on the other side of the river was. Fortunately, most of TWA's aircraft, that is those which were flyable, had taken off and headed for higher country. The repair base was under several feet of water which wrecked major repair schedules and kept planes off the line. It was a matter of weeks before TWA's schedules were returned to normal. That bad news was counterbalanced by a lot of good news over the next six months. TWA president Ralph Damon proclaimed 1951 as the year when traveling by air enjoyed mass acceptance. As if to confirm that announcement, TWA flew its 11,000,000th passenger during this period, a forty-nine year old salesman from Topeka who boarded a Connie in Los Angeles bound for Europe in what was his second only airplane flight. Blackie disagreed with Damon's contention and was still saying the market was still underdeveloped ten years later. On the other hand, Damon was entirely correct about airlines enjoying record passenger loads, making airline tickets more affordable to passengers, and making record profits at the same time. The year 1951 closed well for TWA which enjoyed an after tax profit of $6,816,000 on a gross revenue of $144,912,000.

The year also ended well for the Blackburns. Blackie was doing what he loved best, flying. After Blackie's roundtrip to London in early November, the family took its annual vacation consisting of three weeks back in San Antonio. The family did not take young Bob on the flight to Texas because he was already there as a student at the Texas Military Academy where his grandfather was the president. Betty could not join the family because she was in college at the State University of New York at New Paltz. Although Blackie's family enjoyed the home in East Williston, the East Coast in the early 1950s was a culinary dead zone as far as Blackie was concerned. He enjoyed Chinese food immensely, but that was relatively hard to find on Main Street America in those days. The Asian food marketer La Choy had not yet hit the big time and much the same could be said for the manufacturers of other ethnic foods. As far as Blackie was concerned, another tremendous gap at the local supermarket was Mexican food. Because the Old El Paso brand had not reached far enough back East in those days, the Blackburns always looked forward to trips to Martha's hometown where they could stock up on plenty of Tex-Mex ingredients to bring back East.

One of the benefits of being an international captain was the opportunity to enjoy so many different cuisines. That was particularly true for pilots like Black ie who was anything but a timid diner to begin with. It was at least a partial consolation for being removed from hearth and home about two weeks out of every month. On layovers in Europe, TWA pilots stayed at hotels that were far from being seedy, but by the same token none would earn five stars either. In Rome, it was usually the Mediterraneo which Blackie always enjoyed. In Paris, TWA crews of that era frequently stayed at the Hotel Celtic which was located just off the Champs Elysées where Blackie missed some excitement on 23 February 1952 when the place suffered a fire. Blackie was in London that day, but the fire was not a major one. It did not take that long for TWA pilots to see all of the usual sights in the European cities where they happened to be on layovers. Some went hunting for their ancestral homes and many, like Blackie, took long leisurely walks and enjoyed having meals at outdoor cafes. In all of the cities in which TWA had a presence, there were company sporting teams which played other TWA teams in such games as soccer or tennis, but those activities were usually not joined by transient flight crews. Blackie was still able to play tennis occasionally and even get in some swimming while in most TWA cities. Tennis was available in Ireland at the Limerick Lawn Tennis Club which was located only about seven miles from Shannon, but that airport was not a popular layover point for most TWA pilots who did not care for the climate or the company's uncomfortable accommodations which were located right on the field itself.

On layovers in Cairo while staying at the Heliopolis Hotel, Blackie could swim at the Airline Club which was attended by crews from numerous other airlines. Or he could play tennis at the Ghizara Sporting Club which was located on an island right in the middle of the Nile River. One city became a Blackburn favorite: Lisbon. After retirement, it was an early destination for Blackie and Martha, but early in the game Lisbon always dredged up a slightly painful memory. Lisbon was enjoyable for great seafood and for nearby beaches where Blackie occasionally went swimming. On one occasion while in the surf, he stepped on a sting ray which returned the favor by stinging Blackie on his right heel. The sting of the common sting ray and the pelagic sting ray, both of which inhabit Portugal's coastal waters, is rarely fatal, but always painful. The relief pilot and first officer were the real beneficiaries of the unpleasant incident because Blackie was reluctant to exert much pressure on the rudder pedals during the rest of that flight to Cairo and back to New York. Whereas cities like Madrid, Lisbon, and Zurich had not been touched by the war, others had suffered varying degrees of damage. London had gotten off lightly compared to another TWA city: Frankfurt, which had been flattened by both the USAAF and the RAF. For that reason, Blackie stayed at the Ritter Park Hotel well north of the Frankfurt airport in the small town of Bad Homborg.

In early 1952, there were only a few worries in TWA's boardroom. There had been an oil refinery workers strike which impacted schedules for several weeks. The War Mobilization Board announced the creation of the Civil Reserve Fleet which was subject to call-up by the Air Force if demands on the Military Air Transport Service schedules to Japan could not be met. The Reserve Fleet was to consist of 400 four-engine aircraft made available by the nation's airlines. That figure could result in twelve times the number of aircraft the airlines had made available to the Air Corps in early 1942. Here and there, TWA was losing employees as reservists were called up for active duty, but the losses were far below the levels experienced during World War II. Board chairman Warren Pierson did comment on what was a growing concern among America's airlines by stating confidence in the nation's airlines and the aircraft companies in their drive to meet the European lead in jet transport aircraft. Pierson noted that an appropriation of $100,000,000 had been voted down in Washington at a time when it was estimated the British government had spent three times that figure developing its capability to place jet transports in service. There was some reason for concern; the British deHavilland Comet, with its four jet engines, had first flown in 1949 and had gone into commercial service only two weeks before Pierson's comments appeared in the media. The Comet would go onto a very troubled future, but that was not apparent in May of 1952, a time when Sud Aviation was developing the Caravelle as the French contestant in the jet transport arena.

Blackie's energies were increasingly also being used outside of the cockpit in the early 1950s. Of all of the major airlines, TWA was probably the most respected for its marketing savvy, including public relations which had just been elevated in importance by the promotion of TWA's PR ace, Gordon Gilmore, to a vice presidency. Gilmore was very adept at seeing that TWA-positive ink ran in all media. Among his techniques were directly courting journalists on promotional flights, such as the one to demonstrate TWA's southwest vacation packages, and maintaining a speakers bureau at the New York office. Gilmore placed TWA's most articulate speakers on key podia, and Captain Blackburn was among the most articulate. Although Blackie had spoken frequently to men and women's service groups while based in Washington with ICD, he was getting assignments from New York's growing operations for a wide range of audiences. Typical of these speeches was one given to the meeting of the Foreign Transportation Institute at American University. In that talk, Gilmore collected inputs from several TWA department heads, condensed their thoughts into an outline, and gave it to Blackie who inserted some of his own observations. Blackie was not the only senior pilot speaker, but, with his stage presence, he was emerging as one much admired by Gilmore. In fact, it was the beginning of a productive partnership with Gordon Gilmore and TWA's travel guru, Dr. John Furbay that would extend beyond Blackie's days as a transatlantic captain.

TWA's promotion using movies had previously been quite predictable and focused on wonderful places to visit; that is to say, they were essentially travelogues. Promotion using film media was nothing new at TWA which even then had a motion picture department. Previously, the actual flying of the plane had been very secondary to seeing attractive hostesses, comfortable seats, delicious meals, and smiling faces before, during, and after flights. For 1952, TWA's marketing plans called for a new kind of movie; it would take an average American couple on an actual flight to a foreign land with emphasis on the safety and comfort of the flight itself and the expertise of the men flying the airplane. The thrust of the film was to chip away at the major hindrance to flying: aerophobia or fear of flying. To produce the film, TWA picked who else but RKO, then owned by Howard Hughes. The film need a "star," a senior captain who was tall, slender, authoritative (preferably with a little gray hair and crow's feet to quickly confirm years of experience), wise, and friendly. In other words, the film's casting director needed a Captain Harold F. Blackburn. Fortunately the director didn't have to look for one when the original was available.

Blackie became the narrator and central figure of the twenty-five minute film which was shot in sixteen millimeter Kodachrome film for which RKO's Harry L. Smith was the director. The film opens in Topeka, Kansas, where a travel agent answers questions from a couple who want to spend a month in Europe, but have some reservations about flying so far over the ocean in spite of the fact that it takes less than a full day from New York to Paris compared to ocean voyages. The travel agent tells the couple he will get answers to their questions from the source whereupon the screen dissolves to Blackie who begins by introducing himself and his crew as they arrive at the airport to prepare for the flight. There is a certain amount of Hollywood in the film which uses stage names for Blackie and his crew. Blackie becomes "Bill Barrett" who hails from Lincoln, Nebraska. He announces his age as forty-six which is a bit shy of the reality of fifty-one at that time. There was nothing fictional about his experience: flying the Atlantic since 1942 and a total of 15,000 hours in his logbook. The weather briefing is emphasized, and Blackie answers the couple's questions about flying direct instead of zig-zagging across the ocean. Blackie quickly turns the light on for the couple with diagrams of pressure pattern flying and how longer can mean faster. The safety of landing options in Gander, Greenland, Iceland, and the Azores puts the couple at ease as does the knowledge that there are ships stationed along the way to provide weather and navigational assistance to the eastbound Connie at 19,000 feet. The on-board filming was initiated on 11 May at Idlewild after which Blackie took the Connie to Boston (one hour and four minutes flying time) and then to Paris via Santa Maria in the Azores, logging another fifteen hours and eighteen minutes.

The Kansas couple meets two passengers, a woman who is anxious to see Paris and a silk merchant on his way back to Bombay. The latter two were the only professional actors in the production which otherwise used real people. The actors give first-person testimonials on the comforts of the flight: free champagne, a wonderful steak dinner, and a comfortable, private sleeping compartment. Then they respond to the couple's questions about seeing the great sights of Europe for which older TWA footage is brought in to highlight Big Ben, Parliament, bull fights in Madrid, Notre Dame, Versailles, the Eiffel Tower, the Coliseum in Rome, Capri, Venetian canals, and other enticing tourist destinations.

The actual filming aboard the Connie was limited to a direct sleeper flight, but the film also depicts Gander and Shannon. The weather and radio facilities are highlighted at Shannon which is depicted as a wonderful place to buy Irish whiskey and linens. Much of the film is dominated by Blackie and his crew as they interact to safely fly the great plane across the ocean. Going through the various checklists did not appear in the film, and with good reason. With over 130 items spread over pre-flight and pre-landing checklists on a model 749 Connie, it would have taken too long. Even the DC-3 checklist had over 60 items to be read off and checked. We see and hear Blackie calling "Up gear," and "Cruise power" and can listen to the high-pitched tone in his radio headset as he homes in on the Gander radio range. Shortly after departing Gander, "the watches changed" as Blackie took off his necktie, loosened his shirt, bounded up into the top crew bunk with athleticism typical of a B-17 pilot swinging himself up through the forward hatch of a Flying Fortress. Movie viewers thus learned how much depth there was in the crew when the relief pilot slipped into he left seat for the next few hours. We see the navigator shooting the stars, listen to the radioman talking with weather ships, and watch as Blackie observes the first light of dawn on the eastern sky as they near the continent. As the film nears its end, Blackie is cleared to land at Orly by a controller speaking English with a thick French accent. The film easily establishes the crew's professionalism and safety of flying the Atlantic; it closes with a modest commercial message, "Flying is the way to travel and TWA is the best way to fly."

The film was released through the RKO-Pathé network on a most appropriate day: 20 May which was the twenty-fifth anniversary of Lindbergh's flight to Paris from New York. Blackie introduced the film by observing that Lindbergh's flight had become a popular overseas run which the Constellation had turned into a milk run. He also contrasted the cost of "The Spirit of Saint Louis" with the Connie he was flying: $13,000 for the Spirit and $1,000,000 for Blackie's Connie. Although the film occasionally played as a theater short subject, its primary audiences were associations of travel agents and the many travel clubs that were popular in urban areas during the 1950s. For Blackie, however, there was an unpleasantness related to attaining stardom on the silver screen. For the next several years, while he was back home between flights, he had to attend many of the travel club showings in person to introduce the film and answer questions afterward. Unfortunately, from Blackie's point of view, most of those presentations were double features in which he shared the evening's billing with the likes of United States Lines, United Airlines, and other companies anxious for the travelers' dollars.

If the film missed anything, it was the extraordinary amount of work that went into readying Blackie's Connie for that night flight. Blackie and his crew were living at the top of a food chain which started moving within an hour of the Connie's earlier landing back in New York. The maintenance foreman and his assistant then made a detailed, stem to stern inspection of the entire plane, paying particular attention to any squawks written down by the previous captain. They were looking for the less obvious items that were much harder to find than a balding patch on a tire, loose radio antenna, spark plugs that needed changing, or fresh oil. By the time the plane was ready to fly again, up to sixty specialists had gone over it and tested all systems. The Connie's compasses were checked by towing the plane into a marked position on a compass rose and confirming that the compasses were reading true. Next, the ship was cleaned on the inside and washed on the outside. Several hours before Blackie was to take off, the twenty cooks in the TWA kitchen were starting to prepare meals consisting of wines, cocktails, prime steaks, vegetables, and desserts. One hour before departure, the station agent, ticket agents, and load control agents began their work establishing how much baggage, fuel, mail, and cargo would (or could) go aboard. Thirty minutes before Blackie started the engines, 5,000 pounds of baggage, mail, and cargo were secured on board.

That particular Connie, a 749 model, went on to distinguished service with TWA and a career that outlasted even Blackie's. It sported the name "Star of Kentucky" and carried a registration number of N6020C although its less obvious TWA line number was simply 820. Blackie had flown it several times during the previous year and would occasionally fly it over the next four years. TWA Connie Number 820 soldiered on for the next fifteen years and had the distinction of being the last Constellation to carry passengers on a regularly scheduled flight in the United States – on 6 April 1967 when it was flown to St. Louis from New York with intermediate stops at Philadelphia, Pittsburgh, Louisville, and St. Louis. From St. Louis it was flown to Kansas City where it sat on the tarmac before eventually meeting its fate in Tucson where it was scrapped in 1969.

There was a downside to Blackie's serving as a spokesman for TWA and introducing the film. His off duty days were increasingly monopolized by the public relations and sales promotion departments. Some of his speaking engagements were to local and nearby school student bodies and parent-teacher association meetings (in East Williston, Hempstead, and Roslyn, Woodmere, etc.). Others were out of town and went even as far as Chicago where Blackie spoke to a meeting of physicians and surgeons. Gordon Gilmore and the sales promotion manager, R.L. Robinson, became so anxious to have Blackie's services that they did the unthinkable: contacting flight crew scheduling to have Blackie's assignments altered to suit their own plans. Blackie put his foot down on that and set a limit to the number of speaking gigs per month with instructions to flight planning never to alter his flying schedule in the future. There was a more immediate upside to Blackie's speaking at schools in East Williston, the Blackburn daughters' stock soared among their classmates who thought Blackie was "cool."

Just as Blackie's second career as a public speaker was taking off, TWA was ordering its third generation of the Constellation: the model 1049. The new Connie was eighteen feet longer with a total rating of 11,600 hp. The gross weight topped out at 120,000 pounds or roughly that of two B-24 Liberators. When those new Constellations joined the fleet, TWA had more Connies, seventy-six of them, than any other airline. The new 1049s were equipped with "jet stacks" which were modified exhaust stacks in which the escaping gases actually added ten mph to the Connie's speed. From an airline economist's viewpoint, the jet stacks were an important innovation and could translate into an extra 200 miles in range or as much as 1,500 more pounds of payload. Blackie could also enjoy a more ergonomically correct cockpit; the instrument lighting was changed to red based on what wartime studies had shown about the effects of different color lighting on night vision. To lure more passengers to Europe, TWA launched a new service, Ambassador Flights, which were marked by luxury services including perfume and orchids for women and a greater selection of liquor and wine. Although Blackie was barely a moderate social drinker, he humorously told passengers that the only alcohol he was involved with was that sprayed on the propellers to minimize ice accumulation. One benefit rarely seen on Connies in the United States was the "Speed Pak" which Blackie flew with occasionally in Europe. Depending on the passenger load, extra freight could be carried between European cities in the Speed Pak which was a large pod attached to a Connie's belly.

On 28 June, the CAB made the route blazed by Blackie, New York to Paris, a permanent TWA route – the European routes for all airlines routes announced in 1945 had only been temporary awards. The ruling came at a time when TWA International Division was flying high; passenger traffic was up sixteen percent on international flights and up fourteen percent on domestic routes compared to the same period in 1951. TWA was doing all it could to promote foreign travel ranging from increasing the baggage allowance to sixty-six pounds per person and helping passengers apply for passports. The company also introduced a new marketing program to employees, "It's yours to give, a world of courteous service," with the goal of positioning TWA as the leader in bringing personal service and good will to many foreign lands. The pitch was designed to take advantage of what was perceived as a flaw, imperiousness, in Pan Am's public image. Although Blackie was not flying on 31 July 1952, TWA set a new international passenger mileage record that day: 2,094,741 miles compared to the previous high of 2,030,081 which was set on 2 September 1951. In late 1952, the combination of savvy marketing and new Martin 404s on domestic routes seemed to be working; net profit after taxes for the first half was $1,957,402.

By the end of 1952, Blackie had established a truce of sorts with ALPA and was serving on his ALPA local council much to the chagrin of many who were aware of his past dealings with the union. The local council was glad to have him, and the chairman, an old friend who had just become the chairman, promptly tried to get him to continue serving as the ALPA accident investigation representative and, on top of that, as head of the ALPA Air Safety Council. Blackie had been named to the former while he was on an overseas flight, and promptly said no to the Safety Council job and resigned as crash investigator at the same time. In his letter of resignation, he revealed some past history and present motivation. Blackie admitted to being a controversial figure, certainly as far as ALPA was concerned, but also with TWA upper management which was dismayed on two occasions when he turned down offers of a vice presidency (both dated back to the late 1940s). By then he saw that top management would mean the end of flying for sure and that the only way he could continue flying was quite simply out on the line as a pilot. If he had any other doubts about management, he had no further to look than his close friend Frank Busch who was second in command of operations back in Kansas City. Frank's job was a good one, but it was not putting any entries into his pilot's logbook.

On two other occasions, in late 1952 and again in early 1953, Blackie received firm offers of supervisory positions, and again he turned them down in order to continue flying Connies, spend more time with his family, and avoid the reams of paperwork that seemed to be the curse of desk jobs. He concluded his letter with,

> Now I only want to die in peace. No part of management. Either of my company or my union. I am almost 52 years old – let me fly my last eight years without controversy.

Times and circumstances change, of course, and Blackie would eventually have a change of heart about supervisory positions, but that would be three years into the future. Until then, Blackie would be maximizing both time in the air and with his family as well.

The domestic division of TWA was in the company news more than the international division in the autumn of 1952, primarily due to the national elections. During a two-day period in October, TWA had all of the major candidates aloft as passengers on charter flights: General Eisenhower, Governor Stevenson, and Senators Nixon and Sparkman. Blackie's friend, Joe Carr, was given an interesting assignment, but it took him away from home for over a month and half. It was as pilot of the Connie, "Star of America," that took a management committee, consisting of Chairman Pierson, and Vice Presidents Cocke and Gilmour among others, on a six-week survey tour to explore routes from Bombay to Tokyo. The goal was to establish regular service beyond Bombay to Delhi, Calcutta, Ceylon, Bangkok, and Taipei before reaching Tokyo. Ironically, one leg of the route was accepted little more than a month after the survey returned home in mid-December when service to Colombo, Ceylon was approved. At the same time Joe Carr returned from the survey group, Blackie was off on the first of two trips to Europe, only this crossing was different. Or at least what was in the cargo hold was atypical. The Connie was carrying blankets, clothing, and relief supplies for victims of the devastating floods rampaging in northern Europe that December. In practically no time, TWA sales offices had been flooded with supplies as the result of a plea made by popular television and radio personality, Arthur Godfrey. The year may have ended disastrously for the flood victims, but TWA was enjoying entirely different fortunes. Even before Christmas, the company had made another payment on its loan to Equitable Life, bringing the debt down to $26,500,000 from the original $40,000,000. The year would end solidly in the black with after tax earnings of $7,660,588 on gross revenue of $160,702,790. Prospects for 1953 looked even better with advance bookings for spring and summer travel up 100 percent from 1952. Unfortunately, TWA could not predict or control air mail rates to the same extent, and that would prove to be a problem for both domestic and international in 1953, although that was not immediately apparent when TWA closed its books for the year 1953. Under Ralph Damon's leadership the company had another good year with after tax net earnings of $5,064,392. Costs were going up as quickly as sales which explained the reason for several cost-cutting contests in 1953, a year in which TWA also suffered a strike, albeit not as severe as the pilot's strike of 1946. In July 1953, the company tried to eliminate navigators on its Rome to Zurich schedule which, from the company viewpoint, made all of the sense in the world.

After all, that stage was not over water, but over land where there were adequate navigation facilities. The same Connies flying domestic schedules back in the United States flew without navigators, but nonetheless, the eleven-day walk-out was financially costly for the company. In one sense, it sealed the fate of the navigators' union because TWA started paying much more attention to the latest developments in navigation systems, including inertial navigation. Although TWA's utilization of Doppler radar was still over a decade into the future, it would spell doom for TWA's small band of Magellans.

On 6 January 1953, Blackie took Flight 902 on a ten-hour journey from Paris to Cairo with stops in Rome and Athens. On that day, a new company, *Atkiengesellschaft für Luftverkesbedarf* was formed in Cologne, Germany. Fortunately, it became better known for the time being as *Luftag*, a company with capitalization of 6,000,000 Deutsch marks. The executive board included Hans Bongers, the old Lufthansa hand. The new company's next move would create far greater publicity than that generated by formation of the new company.

With the future looking rosy, particularly based on advanced bookings, TWA upped its order for the model 1049 Connies with delivery of twenty of them scheduled through 1955. Immediately after announcing the new Connie purchase, Chairman Pierson had news on the international front that would impact Blackie in just a few years. The company had renewed its agreement with Ethiopian Airlines, where Swede Golien was still CEO, for an additional three year period. In making that announcement Pierson stated that TWA was also very interested in signing agreements with other countries to develop their airlines.

With TWA becoming profitable, it was also investing heavily in systems from which Blackie would be a prime beneficiary. Along with other pilots he was spending a couple hours per month in a Link trainer, which by the early 1950s was essentially obsolete. TWA invested in new "flight duplicators" built by Curtiss Wright, and those units brought simulation up to the present. The old Link units could simulate a radio range, but the new simulators could maintain a pilot's skill on the instrument landing system, automatic and standard direction finding, the new VHF equipment, and fan markers.

TWA ordered fourteen of the units at $80,000 apiece. But it would not be until late in the year that Blackie would see the inside of one, since the first four were going to Kansas City. But when he got into one, he found bells and whistles including engine sound effects and emergencies perpetrated on him by the controller who sat at a console on the side of the duplicator. Another important development intended for Blackie's flight engineers was an engine analyzer built by Sperry. With this system, which took up comparatively little space at the flight engineer's console, incipient engine problems could be spotted fifty hours before they might actually become apparent to a crew or maintenance inspector. New VHF radios were also soon to arrive, and with them static would practically become a nuisance of the past.

In Germany, the embryonic Luftag was considering the purchase of aircraft for its airline to be and decided not to acquire some clapped out, previously owned DC-4s or 749 Connies. On 16 June 1953, Luftag created a major stir in airline circles by placing an order with Lockheed for four new 1049G model Constellations. In the autumn, Luftag's newly hired pilots, flight engineers, and navigators traveled to California to begin their orientation on the Constellation.

Overseas Run proved to have remarkable staying power and was still being shown in late 1953 when the New York Port authority sponsored it on WCBS-TV in New York. The Port Authority was impressed with the way its airport was presented and wanted the greater recognition that sponsoring the film on television would bring. The film ran on the evenings of 20 and 27 September 1953, but began with a taped interview of Blackie by a man whose original name was James Kenneth McManus. He had grown up in Philadelphia but moved to Baltimore for his education and early career development as a newspaper journalist.

It did not take long for McManus, who had been a very youthful minesweeper skipper in World War II, to jump to the new medium of television. When breaking into television, McManus changed his name and later became very well known as Jim McKay, host of the mainstay sports series, *Wide World of Sports* which ran from 1961 through 1998.

The early 1950s with Ralph Damon at the helm proved to be some of the best years for Blackie's employer which announced the plan to invest $18,000,000 in a massive repair base in Platte County just north of Kansas City. With a net profit of $10,336,132 in 1954, Damon publicized the net earnings as being subsidy-free which was something of which archrival, Pan Am, could not boast. The debt to Equitable was completely paid off, and the new Constellation G models started to arrive in 1955. In addition to the Sperry engine monitors, Blackie would be sitting in back of a Weather Radar, when he first got his hands on the yoke of a G model which was Lockheed's answer to the progression of Douglas transports as evolved from the wartime C-54 (DC-4).

The powerplants were more powerful turbo-compound engines, and for the cabin staff there was a larger, entirely new galley.

Trans World Airlines

Blackie, at left, in his favorite office – the cockpit of a Lockheed Constellation.

The Blackburn family was in transition and although he considered himself a real family man, he and Martha were half the way to becoming empty nesters. On the other hand, they had also become grandparents. Oldest daughter Betty had married Arthur Beard in late summer 1952 and son Bob was a junior at Oberlin College in Ohio where he had married Barbara Wendell in 1953. With the arrival of Karl Andrew Beard on 12 July 1953, Blackie and Martha became grandparents of the first of five grandchildren

In Blackie's logbook, the year 1954 was a good one too. He was averaging seventy-five to ninety-five hours a month and two, sometimes three, trips to Europe each month. His flights as far west as Athens and Cairo took him over some of the world's greatest scenery, that is when it could be seen. Nearly half of his flying time that year was at night and about a quarter of the total time was on instruments. But with ceiling and visibility unlimited, the view from the flight deck was superb. The vistas included many mountains, sunlight-silvered seas, islands jutting up from the oceans, billowing cloud formations that tugged at the imagination during their interplay with the sun, and the progeny of that relationship – shadows that danced on deserts and the seas alike. Nearly all of Europe's great mountains were visible to Blackie as he overflew Germany,

France, Switzerland, and Italy. If they were not directly in his flight path he could see them in the distance – Mount Blanc, the Zugspitze, that trio of the Monch, Jungfrau, and Eiger in Switzerland along with Monte Rosa. The most striking of all was the Matterhorn, past which Blackie flew on 8 July 1954. Unknown to him as the captain for Flight 903, thousands of feet below and 50 miles to the northeast when he was little more than mid-way between Rome and Paris, an old friend was spending a day at leisure in the village of Zermatt. The picturesque town is situated at the base of the 14,692-foot high Matterhorn. Blackie's associate, Earl Korf, was taking it easy that day considering his labors of the day before: climbing the Matterhorn on his fiftieth birthday. Korf had joined TWA as a radio operator after work as a radioman in the merchant marine, but had retrained as a navigator and flown with Blackie on many wartime ICD trips. Adventure was in his blood; he had been shipwrecked near Samoa in the Pacific and had once managed a salmon packing plant in Alaska before joining TWA. He had no inkling, and neither did Blackie, that the two of them would share a unique, multi-continent flying adventure after both eventually retired from TWA little more than a decade later.

Ralph Damon's days as an adventurer were over, other than dealing with Howard Hughes. While Damon did not have the reputation as the visionary that Jack Frye was, he did moderate a panel of industry leaders on 27 January 1955 at the Hayden Planetarium in New York. The subject was what airliners would be like in thirty years. The panel had some heavyweights on board including Werner von Braun, the German rocket scientist, Hall Hibbard, vice president of engineering at Lockheed, and Dr. Fred Whipple, the Harvard astronomer. With the advantage of looking back, it is obvious that the panel's vision of the future was rather optimistic in some regards. There were some heady figures flying about for 1985 when a peak of 500 billion passenger miles were predicted for the nation's airlines which might be flying nuclear-powered aircraft carrying passengers in their wings while flying at 75,000 feet. The group anticipated greater use of helicopters and vertical take/off landing aircraft. Top speeds for transports were estimated to be in the neighborhood of 2,000 mph, pushed that fast by the widespread use of ramjet engines. The reality was that only the Anglo-French Concorde SST eventually came close with its maximum speed of just under 1,400 mph and a top ceiling of 60,000 feet.

That panel, in spite of its optimism, did reflect one reality that was on Blackie's mind, and that was when an American-developed jet transport would be flying with a TWA logo on its fuselage. Enter Howard Hughes who essentially hedged TWA's bets on just when Boeing or Douglas would come forth with a jet transport that could satisfy all airline management requirements. Hughes seriously entertained the idea of buying DeHavilland Comets but was not convinced that it was well designed. He was proven right in that conclusion, but even after the Comet's structural issues were rectified, it would still have been difficult to justify buying one for the same reason that the Concorde was not wildly successful: passenger capacity was limited. TWA was very much aware of what was going on at Boeing which was developing the KC-135 as a jet tanker/transport. Boeing had ignored the civil market for years, other than backing into it with knock-offs from military models such as the B-17 Flying Fortress from which the Stratoliner was developed or the B-29/B-50 Superfortresses which gave birth to the KC-97 Stratotanker and C-97 Stratofreighter.

At the end of the first quarter of 1955, TWA had two important announcements to make – both of which eventually affected Blackie. Chairman Pierson announced TWA's new involvement with Lufthansa for which TWA supposedly would start training crews in June. That was a bit optimistic, but it would not be too much longer after that when Blackie would earn his Lufthansa pilot insignia, bringing his wings collection to seven. Since Blackie's first involvement with airlines in Italy, Saudi Arabia, and Ethiopian, TWA had been active with the state airlines in Greece and Ireland as well. Pierson advised that eleven TWA pilots, all German speakers, would be in the vanguard of a contingent of instructors who would guide German pilots on flights out of Hamburg to Dusseldorf, Frankfurt, Shannon and New York (but not Berlin) at first with eventual expansion to South America and the Orient.

All of the Lufthansa news was subordinated as far as Blackie was concerned to the second announcement that Howard Hughes had decided to stick with piston-engined aircraft for the time being. Hughes had, therefore, authorized the development of the next generation of Constellation, the 1649 model, which was to be delivered in 1957. Ralph Damon was steamrollered by the decision which would add twenty-seven more feet to the Connie's wingspan, take the speed up to nearly 400 mph, and extend the range to 6,500 miles. With the latest version of Wright's 3350 turbo-compound engine, gross weight went up to 156,000 pounds or about the equivalent of three loaded B-24 bombers. It would be well into 1957 before Blackie got his hands on a model 1649 Connie and when he did it was not in TWA livery, but rather in Lufthansa colors. The 1649 could justifiable called the finest piston-engined airliner ever developed, but even Howard Hughes knew that it would enjoy only a short life before the KC-135's commercial variant, the Boeing 707, would come on stream. With the knowledge of what was going on in Seattle, Douglas Aircraft would not be found behind the times and announced its plans for the DC-8 jet liner shortly after the announcement of the Boeing 707 was made. Pan Am did not hesitate as did Howard Hughes; it placed orders for both the DC-8 and Boeing 707, thus forcing TWA's hand which eventually bought the Boeing jet. 1957 was still two years away, and Hughes firmly decided that TWA would fly with an all piston fleet for the time being.

Blackie was puttering around his home in East Williston on 7 April, a day before taking Flight 802 to Shannon. It was his twentieth anniversary with Trans World Airlines, and the occasion was marked by little more than his name appearing in the company newspaper with many, many other employee anniversaries. In 1940 and 1945 Blackie received his service pins with the personal best wishes directly from Jack Frye who included Blackie in that inner circle of employees who were also good friends. Before Frye left TWA, Blackie and Martha had been guests of Jack and Helen Frye at their ranch near Sedona, Arizona. When Blackie joined the company, its total employment was under 1,000 and in 1955 it was pushing 20,000. In a few years, he would say that there were three TWAs: one in Kansas City, a fast growing one in New York City, and then "the rest of us." He was right. TWA was becoming a huge, impersonal company.

On 1 May 1955, Blackie had some back-pedaling to do because he accepted the position of supervisor of flying in the International Division on that day. He had once proclaimed that he nevermore wanted anything to do with the management of his company or union, but here he was accepting a position that was not 100 percent flying in the captain's seat. The reality of son Bob's college tuition and upcoming college tuitions for his daughters at home were chief among the reasons for his change of outlook.

After Blackie brought Flight 840 back from Frankfurt on 11 May 1955, he found a note from Public Relations VP Gordon Gilmore outlining a new opportunity that would take Blackie from the silver screen to the little screen. The Disney Studios had contacted TWA about the possibility of cooperation in filming a feature within the upcoming Mickey Mouse Club television series which went on to become a very popular with elementary and junior high school students of the day. Gilmore explained that the sub-series would be entitled "What I Want to Be" and would explore various exciting careers for youngsters. TWA was selected because it had a reputation of being the "airline of the stars" and because it had a concession at Disneyland which was scheduled to open in mid-July.

Before Blackie could take on the Disney assignment, there was another duty: seeing son Bob off to what would become an eye-opening, life-altering journey to Africa.

All of the gifts and art Blackie had brought back from Ethiopia while Bob was only thirteen-years old had whetted his son's appetite for learning more about Africa, its people, and cultures. In the summer of 1955 twenty-year old Bob traveled to Cairo and the Sudan. He also spent several days tracing many of the air routes Blackie had flown in Ethiopia and then down to Nairobi, Moshi and Arusha. While in Cairo, Bob became the guest of Bassili Tawadros who treated Blackie's son like visiting royalty. Bob still had several more semesters of college left, but the summer of '55 would stick with him and within a decade form the foundation of yet another return to Africa as a Peace Corps administrator.

Disney planned to take a couple of typical junior high school students and turn them into a pilot and a hostess under the guidance of the appropriate real life airline professionals. Blackie was interested and on this occasion did not object to flight scheduling being told that Captain Blackburn would not be available for transatlantic flying for about six weeks starting on 1 June. The segments built around "I Want To Be A TWA Pilot When I grow Up" and "I want To Be A TWA Hostess When I Grow Up" were scheduled for airing in October. The production began for Blackie on 16 June when he flew into Kansas City, a flight that was memorable for him because it was his first in a 1049 "Super G" model Connie. TWA's involvement was expanded early when the series was extended to include ten fifteen-minute segments – one to run each weekday for two weeks. The project had the full support of TWA management including Frank Busch who lined up the various TWA departments participating in the filming. Between those appearing on screen and those supporting behind the scenes, over 600 TWA employees were involved in the project.

Disney assigned some heavyweights to the production. The director, Charlie Haas, had many popular television series and movies to his credit. The 166-page script was written by Stirling Silliphant, a novelist and prominent screen writer. The cameraman, Eddie Coleman, was at the same time also behind the camera for one of the most popular TV shows of that era – *Dragnet*. The unit manager, Gordon McLean, was no stranger to aviation; he had spent World War II in charge of thirty-three training films for the Army Air Force. The professional actors included Alvy Moore, a popular TV character actor in the 1950s who played a reporter shepherding two young people on their journeys to becoming a pilot and hostess. Alvy Moore visits the airport, finding "Bonnie," played by child actress Patricia Morrow. She leads Alvy to the boy, Duncan Richardson, who spends a lot of time with model airplanes, hoping to become an airline pilot. Unknown to Blackie, there would be a coincidence that would someday link Patricia Morrow to Blackie's wartime buddy, Clarence Shoop. Shoop's daughter, Pamela Susan, followed in her mother's footsteps and became a beautiful and talented actress who later appeared with Patricia Morrow in the television soap opera, *Return to Peyton Place*. But in 1955, Morrow was "Bonnie," the little girl who wanted to follow in the footsteps of her assigned role models, real TWA hostesses Pat McReynolds and Dixie Strele.

Although the series was told from a child's viewpoint, the journey of the two children through the various departments sent a subliminal message about how thorough TWA was and that flying was both safe and efficient. Young Duncan Richardson logs time in a simulator, sits in the driver's seat of a tug which tows a massive Connie out of a hangar, and helps mechanics service one of the complex Wright engines. Patricia Morrow goes through TWA hostess school, is thrilled with her new suit, but is devastated to learn that she will have to suffer a haircut to comply with TWA's rules on hair length. Eventually the youngsters complete their training, earn their wings, and are rewarded by flying from Los Angeles to New York. On 26 June, Blackie flew a Connie on a regularly scheduled flight out to California for the one week of shooting at Los Angeles where two special treats awaited him.

He got together with Clarence Shoop who was prospering in both of his careers with Hughes Aircraft and the California Air National Guard. By 1955, Shoop was moving from chief test pilot at Hughes Aircraft to corporate executive in the company which was emerging as a major defense contractor in missiles, electronics, and aircraft subsystems. Shoopy was still in uniform with the Guard in which he had been promoted to brigadier general in January. Blackie outlined the script of the Disney film and indicated that it called for him to come to the aid of an Air Force jet trainer in distress while he was flying Patricia, Duncan, and the other passengers from LAX to Idlewild. Because the script called for Blackie to use his radio to communicate with the Air Force pilot, the latter's squadron leader, and military control facilities, Shoopy and the Disney Studios thought that Blackie would benefit from a jet orientation to facilitate that part in the film where he was interfacing with Air Force authorities. After several phone calls to the right people, on 8 July, Blackie found himself at the Palmdale Airport where he met Major Ed Rackham who, in 1955, had the same job that Shoop once had: Air Force acceptance pilot at Lockheed.

The inner fighter pilot in Blackie was about to be awakened by Rackham who had flown both Mustangs and jets in the Korean War. Blackie logged an hour and a half in a two-place T-33 jet trainer which was essentially a training version of the older subsonic F-80 Shooting Star fighter. Blackie was enthralled with the flight, especially the speed, quiet power compared to four Wright 3350s hanging on his wings, and lack of torque. He described the flight as being, "A piece of cake – in fact a grandfather's rocking chair." With Rackham, Blackie was able to fly the aerobatic routines he had learned at Kelly, but Major Rackham did have one trick up his sleeve, weightlessness, which he demonstrated to Blackie. The major started a climb in which Blackie at first felt something less than 2 Gs, but after the climb reached forty-five degrees, Rackham pushed over and was essentially flying the top of an outside loop. For about fifteen seconds, Blackie was at zero Gs: weightless. He could not fully enjoy it because he was strapped in tight, but it was a novel experience seeing a folded map drifting above his lap.

The series climaxes with the flight to New York in which Duncan pilots the Connie through a perilous storm and Patricia turns a cranky, ill-mannered little boy into a contented passenger. The little pilot and hostess are met in New York by radiant, regular Mousketeers which concludes the series. When he flew back to Idlewild, filming was in the can. In another four months, daughters Bonnie and Be got all kinds of questions from their classmates about their movie star father when the Mickey Mouse Club show debuted on ABC in October. The show's little heroine, Patricia Morrow, became quite fond of Blackie and stayed in touch for many years. As with many child stars, her movie career tanked by the time she became a teenager, although she later salvaged it and became successful in television before deciding on a career change and going to law school.

The boy, Duncan Richardson, was not so fortunate: his acting career augured in permanently by the time he reached age fourteen.

Although it was of little concern to Blackie, the "What I Want To Be" series never took off, due largely to creative differences between Walt Disney himself and the writer, Stirling Silliphant. Disney, whom Blackie met during his week in California, thought the series was pitched too high (more at the high school level) and completely disagreed with Silliphant's idea for the next occupational subject in the show, dairy farming. Considering where the Blackburn family was headed in a couple of years, farming might have made for some productive televiewing, but, alas, the Blackburn daughters' introduction to rural life would be a rather sudden one.

Two items in the company newspaper caught Blackie's attention in mid-September, both of them good as far as Blackie was concerned. His old friend Swede Golien was leaving Ethiopia to become manager-flying, Middle East and would be based in Cairo. Unfortunately, most of Blackie's flying no longer got so far east, and he would have to wait until February 1956 before breaking bread with Swede who by then was very comfortable in his role as a manager just as Frank Busch was. The other news was that Lockheed had just informed Ralph Damon that it had started to cut metal for the model 1649 Constellation. There was a growing need for Connies then because TWA had just announced two additional weekly flights to both London and Frankfurt which partially explained why Blackie was occasionally flying three shorter trips to Europe instead of his older schedules which took him to Athens and Cairo more frequently.

Other than tolerating some good-natured teasing about becoming a Hollywood star after the autumn showing of the Disney segments, Blackie's life returned to normal, and he averaged two trips per month to Europe and an occasional ferry job to Kansas City in the autumn of 1955. In early November, he had a special passenger on a roundtrip to Paris – a passenger who did not ride in the passenger cabin, but rather rode on a jump seat up in the cockpit. The passenger's name was John Bainbridge and he was a staff writer from *The New Yorker* magazine. While he was best known for his *New Yorker* connection, Bainbridge, who was about thirteen years younger than Blackie, had a couple of bestselling books under his belt already. He had taken several articles on Toots Shor, the well-known New York restauranteur, and integrated them into his 1951 book, *The Wonderful World of Toots Shor*. His 1953 biography of Greta Garbo, the elusive actress, had also proven popular.

It is unclear at this point just what set of circumstances brought Bainbridge to the flight deck. It may have been the magazine's initiative in looking for an interesting article that would appeal to its upscale audience. Or TWA may have proposed the article based on an insider's view of a flight to Paris which would do double duty as a market development tool. In either case, the long arm of Gordon Gilmore was wrapped around the article which was intended to depict the life of an experienced airline pilot on a four-day roundtrip to Paris. John Bainbridge was a Midwesterner with a journalism degree from a particularly good journalism school, Northwestern University. He wrote many of the profiles appearing in *The New Yorker*, a magazine which, in spite of its title, has a healthy following in the nation's major metropolitan areas, not just New York City itself. Still, much of the editorial and advertising in the magazine relates to New York where Broadway was enjoying a stellar season in late 1955; among the shows playing were "Pajama Game," "Damn Yankees," "My Fair lady," and "Cat On A Hot Tin Roof." TWA was a major advertiser in the magazine which in late 1955 had a paid circulation of 416,205 primary readers. With pass-along readership, the total audience could easily be estimated as one and a half million readers of whom nearly all were very affluent, i.e., this was prime country for Gilmore's subliminal "Fly with TWA" messages. Among the advertisers with heavy schedules in the magazine at the time were Dobbs hats, the Greenbrier resort, Henredon furniture, Magnavox televisions, Movado, Nieman Marcus stores, Zeiss cameras, and competing airlines.

Bainbridge spent just under thirty hours in the flights to France and back, but it would take another seven months for the article to be published. When it appeared in the 5 May 1956 issue in the "Profiles" column, there was a pen and marker illustration of Blackie which accentuated his eyes. The artist somehow omitted the crow's feet, but the eyes come through loud and clear as those of an experienced captain of the sky. Bainbridge actually joined Blackie from the moment he left his home in East Williston from which Blackie departed with a brief "so long," to Martha who was on the phone at the time. Blackie talks about the safety of his job and comments that the most dangerous part of it is driving the fourteen miles on the Northern State Parkway from his home to the airport. In meeting Blackie, Bainbridge took the pilot to be much younger than fifty-four years to which Blackie replies with a few details such as not being a smoker and devoting a lot of his energies to regular exercise and hard work on his showcase property in East Williston. Before reaching the TWA operations building, Blackie explains how pilots are evaluated, claiming that the usual criteria, the number of hours, seniority, and the quality of their last landings are not as important as what really counts: the quality of the *next* landing. Readers learn that the flight they are about to enjoy with Blackie and Bainbridge, Flight 830, will be the captain's 569th intercontinental flight. Bainbridge follows Blackie around to the various departments following their arrival at two-forty P.M. in preparation for the five P.M. departure. Blackie checks his mail and picks up a wad of Notams (Notices to Airmen which alert pilots to significant changes to airports or facilities along the way) and observes nothing for Gander, Shannon, Orly in Paris, or any of several alternate fields.

The sequence of events that followed were eye-openers for Bainbridge but, of course old hat for Blackie: picking up and distributing passports to his crew of nine, sitting down with dispatchers and weather briefers to complete the flight plan, figuring the fuel load, working up a weight and balance form, and getting an early picture of weather in Paris for the following morning. All transatlantic flights were flown with three pilots, two flight engineers, the radioman, the navigator and a hostess. On tourist class flights there was only one purser, but because 830 was a first class flight there were two pursers. Some of the weather briefing appeared on a large blackboard which was completely unintelligible to Bainbridge – it consisted of, as he described it, an aeronautical shorthand, but the seemingly run-together collection of arrows, symbols, abbreviations, and numbers on the sequence reports and forecasts could be inhaled and digested in seconds by old hands like Blackie. He was now approving the flight analysis which showed a fuel load of 5,050 gallons of gas for the Connie, a model 749 named "Star of New Mexico," they would take to Paris. The forecast weather at Shannon sounded all right – broken at 2,500, ten miles visibility, and a twelve knot breeze out of the north, but the long range forecast for Gander was discouraging. Blackie dismissed Bainbridge's obvious concern by explaining that there were alternates to Gander and not to worry about a low between Gander and Shannon on the way over. Blackie planned the flight for 19,000 instead of the usual 17,000 in order to avoid any significant weather. He would be very well informed on the way over, considering the two weather ships stationed in midatlantic and radio reports from Shannon itself. Soon Blackie was looking at the flight clearance as worked up by his dispatcher; after studying it thoroughly, he signed it and moved on to the flight briefing room where he led the crew in a thorough review of safety and ditching procedures. In addition to his one company medical exam and two CAA medical exams, Blackie and all crews had to participate in an actual ditching exercise conducted by the Coast Guard at nearby Jamaica Bay. Even as Blackie was going through the question and answer safety session with his crew, he could hear mechanics putting the Star of New Mexico's engines through their final tests. Once the crew session was over, each member headed for more individual preparations. The radio operator picked up a small kit containing radio parts (just in case the plane was hit by lightning) and the flight engineers sat down with the mechanics to review the plane's health and any recent issues. Not leaving anything to chance, the engineers took their long wooden dripsticks and physically checked fuel levels in the tanks to make certain the gauges were reading properly. There had been an issue in May when the Star of New Mexico was lightly damaged in a landing accident at Reykjavik, but that had been properly attended to.

To Bainbridge's surprise, Blackie had very little to do with the actual flight plan calculations which the relief pilot and first officer worked up as far as Gander. Then the flight's navigator plotted the course over the North Atlantic paying particular attention to the winds aloft and establishing waypoints on the journey that would minimize headwinds and maximize tailwinds. After landing at Shannon, the two other pilots would work up the leg from Ireland to Paris. Blackie quickly reviewed the plan and tersely commented, "Another milk run." With the crew aboard, Blackie taxied the Connie over to the gate on the other side of the field and took his pilots and navigator to the Department of Commerce weather office on the terminal's top floor. There they received another, comprehensive weather analysis prepared specifically for Flight 830. Leaving that office with a folder of weather maps, Blackie headed for the plane and after conducting a pre-flight inspection quickly entered the cockpit. Soon, one of the pursers came forward and advised that all passengers were aboard and that the door had been locked. Unlike nearly every pilot on the line, Blackie did not depress his microphone and announce, "This is your captain speaking." Rather, he went back to the cabin and introduced himself directly to the twenty-two passengers. He firmly believed the majority was, at heart, afraid of flying or at least of the unknown that would be their companion until touching down along the River Shannon. After the greeting Blackie introduced the three members of the cabin crew, and mixed in a little humor while telling the passengers about the nearly eleven-hour flight and the weather they would encounter along the way.

After going through the forty-four items on the "Before Starting Engines Checklist" and the twenty items on the "Pre-Takeoff Checklist" with his relief pilot and flight engineer, Captain Blackburn was cleared to taxi into position on Runway 19 and hold. Within a few seconds came the takeoff clearance whereupon Blackie advanced all four throttle levers forward. With all engines bellowing, Blackie lifted the fifty-three ton Connie off the runway after hearing his first officer call out "VR" (velocity rotation speed). By the time they were over Nantucket, Blackie had called for two power reductions, set up the automatic pilot for Gander, and was cruising at 239 knots. He enjoyed a cup of coffee before going back to the passenger cabin to build some more rapport with the passengers who could not help but be receptive considering the filet mignon and champagne they were enjoying for dinner. Once back in the cockpit, Blackie enjoyed the same dinner less the bubbly. Blackie and Bainbridge got to talking about passengers in general which prompted Blackie to recall his most memorable passenger experiences. The most unusual were a pair of kids, mischievous brothers actually, who lit a fire in the wastepaper bin in the men's lavatory. Then there was the drunk who barged into the cockpit just as Blackie was making a night instrument approach into Zurich. The drunkard took a flash photograph that momentarily blinded both Blackie and his first officer as they turned around to see what the commotion was. Even in those days, the door to the cockpit was bolted, but most crews rarely used it. The lockable door had been required by the CAA after the crash analysis of one commercial flight revealed that both pilots had been shot in the back of the head. Blackie only occasionally used the bolt and then only when he suspected that a drunken passenger might try to come forward. Blackburn's most unusual, untoward incident involved an Italian tailor who went completely berserk shortly after takeoff. He was wrestled to the floor, tied up, and Blackie made an unscheduled stop at Gander before continuing on to Shannon.

Bainbridge was surprised to learn just how much paperwork attends a TWA flight. Blackie longed for the old days of Air Corps flying which did have its paperwork, but only to a minimal extent compared to the airlines. Blackie was required to keep a crew duty log, noting who was on duty and off for exactly how long. The flight engineers kept detailed logs: on powerplant performance, fuel consumption, and, on the workings of the plane other systems. To first officers fell the duty of maintaining the flight log itself and making position reports while over land. Once over the sea, such reports and all radio activity became the responsibility of the radio operator working in conjunction with the navigator. Flight 830 actually reached the sea off Gander thirteen minutes early, nicely compensating for a five minutes late takeoff. Shortly after overflying Gander, the navigator handed Blackie a new course whereupon Blackie set it up on the automatic pilot. Because Bainbridge was writing the profile of a professional pilot, he used what appeared to be a lull in cockpit activity to draw Blackie out. It may have appeared to be a lull, but Blackie and his relief pilot were constantly scanning the instrument panel to be certain all was as it should have been. Still, Bainbridge learned about the enterprising young man from the Great Plains, the Hal Blackburn Orchestra, the Kelly Field days, Blackie and Martha in the wilderness as small business owners, and inevitably the war.

Just as Blackie was starting to talk about the Blackburn Shell Station and the Red and White Store, the navigator handed him a form which was initialed and handed to the radio operator. It was the position and weather report which would be relayed back to Gander along with the estimate of exactly where the flight would be in one more hour's time. Rather than continue the dialogue with Bainbridge, Blackie asked the navigator to work up the equal time point, i.e., that point at which returning to Gander would be equal in distance to continuing onto Shannon. Blackie then explained the term to Bainbridge along with the point of no return which is worked out on the Howgozit form which records the flight's progress, although he noted that point of no return had lost some importance under normal circumstances due to the Constellation's enormous fuel capacity. Returning to his conversation with Bainbridge, Blackie talked about the war years, and since the war had been over for more than ten years, he felt comfortable in talking about what had once been covered up – his flights with the Seventh Photographic Reconnaissance Group in 1944. The conversation drifted to Blackie's more recent history with TWA and his aspirations for the future. Bainbridge learned that the two things Blackie treasured most, his family and flying, had suffered significantly when Blackie was in management positions and that although he had been offered even higher management jobs, he had turned them down in order to spend more time with Martha and his children. Flying the line involved being gone thirteen days a month which satisfied his need to fly and still give him plenty of time to be a homebody.

Around eleven, Blackie's rest period had arrived, and he swung himself up into the upper bunk for a three-hour nap. Bainbridge headed to the passenger cabin for a little shut-eye himself. While the two men were snoozing, the navigator discovered the shortcoming in the weather forecast. Rather than enjoying a sixty-knot tailwind, the wind had shifted and was blowing harder out of the north. They were also encountering cloud and turbulence at their assigned altitude of 19,000 where they were supposed to be cruising in the moonlight. Blackie was awakened by the turbulence, gazed nonchalantly at the Howgozit, and decided it was time for a snack which he proceeded to assemble in the crew galley. After finishing a cup of coffee, he settled into the left seat and discovered that the Connie was flying through snow. At about the same time the navigator gave him a new course to compensate for the northerly crosswind. At around seven-thirty Shannon time, the Connie broke out of the clouds into clear air, but faced thunderheads directly on the horizon. Blackie disconnected the autopilot when they got close to the towering cumulus and as Bainbridge observed, flew through the gaps in the clouds like a football player deftly dodging beefy tacklers. The navigator was not amused by all of this since it would certainly mess up his estimated time of arrival calculations. Finally the Connie broke out of the fleecy formations, although more of them could be seen to the north. Like many pilots, Blackie started comparing individual clouds to other shapes – an old man with a beard, a hunting dog at point, and a bulky church with a chubby steeple. Soon Blackie was in touch with Shannon approach and was cleared to descend to nine thousand which proved to be a bumpy ride down indeed. Shortly the Star of New Mexico was bolting through sleet with its shotgun pellet staccato pounding on the fuselage, and then it was freezing rain for which the flight engineer had a standard solution: alcohol which quickly restored visibility through the windshield. At nine-thirteen local time, the coast ahead could be seen while Blackie was descending to five thousand. The first officer brought out the sixteen-item Field Approach Checklist which was shortly followed by the Before Landing Checklist when they were much closer in. Flight 830 arrived twenty-eight minutes late due to the faulty weather forecast. For the passengers, Shannon meant a chance to stretch legs, buy some Irish items, and maybe have a hearty Irish breakfast for those not impressed with the TWA coffee and donuts that had been served on the plane. For Blackie and the cockpit crew, the landing was followed by another weather briefing and update to the flight plan which had to allow for more icing on the way to Paris and fog when they arrived. Blackburn, the dry humorist, made a point of commenting on the radio antenna wires extending up to the plane's tail. Within easy earshot of several passengers, he commented, "They sure don't build planes the way they used to. Hope those wires hold us together as far as Paris."

After a forty-seven minute stop in Shannon, the Star of New Mexico was airborne for Paris. Upon reaching his cruising altitude, Blackie put on his cap, turned the ship over to the relief pilot, and went back for a chat with passengers. Returning to the left seat, he explained to Bainbridge what the real danger of flying was: static fatigue which robs the senses and is based on a number of causes including boredom (repetitious scanning the same old instruments), noise and vibration, and low humidity resulting in dehydration. Blackie said that some people believed reading was a good antidote but that he did not believe in it, thinking it simply encouraged drowsiness. Blackie's approach to static fatigue was to get up, move around, and drink juices and coffee. By the time he finished explaining the perils of static fatigue, the flight was over Dieppe at one in the afternoon when Blackie checked in with Paris. Twenty-five minutes later the Eiffel Tower and Versailles were found not to be shrouded in fog as had been suggested by the weather briefers. The landing was a smooth one after a total flying time of fourteen hours and three minutes. Bainbridge was struck by Blackie's appearance which was lively and alert even though he had had only two hours of sleep on the flight over. After going through customs and a stop at TWA's operations office, the crew and Bainbridge caught a company bus to their hotel which on this trip was the Windsor, not the Celtic or the other "TWA hotel," the Reynolds, all of which were located close together and near the Arc de Triomphe. After years of flying to London, Paris, Athens, Cairo, and other European capitals, Blackie explained that he had pretty well "done" each city and no longer headed for the great tourist sights. Layovers were for relaxation, very long walks, and enjoying a good meal at the hotel or a sidewalk café. For relaxation, Blackie always had some reading material: some novel, the latest issue of *Time Magazine*, and a newsletter from his broker, Merrill Lynch. Blackburn and Bainbridge walked over to the nearby Reynolds Hotel for dinner (a small glass of wine, filet mignon, asparagus, a green salad, and coffee, but no dessert). The two went for an evening walk, but by eleven, Blackie was in bed after a couple hours of reading. His phone rang at eight the next morning which was annoying since Blackie had more sleep in mind because his return flight was not scheduled for takeoff until ten that night. After lunch, he went walking and following an early meal at the Reynolds again was on the crew bus back to Orly where he learned that he would be flying Flight 927, a tourist class flight. By the time he arrived for the briefing, he had had thirty hours off duty, much more than complying with CAA regulations for a minimum of twelve hours off duty between flights. Blackie left Paris with his crew and sixty passengers for what turned out to be a much smoother flight than the eastbound crossing, although it was an instrument landing back at Gander. Blackie explained to Bainbridge that the toughest part of a European roundtrip was the Gander to New York leg. After takeoff, the length is over 1,000 miles, the passengers are always restless by then, and there is the prospect of an hour of more paperwork and customs once back in New York.

Bainbridge had left his car at the Blackburns' and accompanied Blackie into the residence via the backdoor. Bebe, busy with a school poster in the kitchen said, "Hi Daddy," and Bonny, who was playing with friends also said "Hi," but quickly resumed her play with the other girls. Blackie greeted Martha and the two neighbor ladies she was entertaining, but soon quickly ascended the stairs to change out of his uniform. Bainbridge wrapped up the article with, "The intrepid birdman had returned to his nest." The article did not appear for several months, but when it did, Blackie had to tolerate some more good-natured humor about now becoming a print media star as if the big silver screen and the little screen were not enough for him.

In mid-December, one of the pilots who were loaned to Lufthansa made news by establishing a new speed record for an east to west crossing from Shannon to New York in spite of headwinds occasionally reaching 35 mph. The pilot was Clement Morehead, a neighbor of Blackie's in East Williston. At that point, there were only ten pilots on loan to Lufthansa and all but one were domiciled in New York. Nearly all had been flying the North Atlantic, and most, including Dick Sherwood, had flown for Blackie in ICD. In fact, Blackie had given check rides to all of them including the man who was selected to be the chief liaison with Lufthansa. That was Ernest Pretsch who was domiciled in Hamburg. He was not the most experienced pilot in that first group, although he had given Blackie instrument and line checks after Blackie returned to flying the line. Blackie had actually given Pretsch line checks on the DC-3 while based in Newark. Pretsch was selected because he had one quality to a greater extent than any of the other ten pilots. While the first pilots on the Lufthansa assignment could speak German, Pretsch was ultra-fluent, even idiomatic in the language. Even his English was spoken with a slight German accent. He was not popular with other pilots due to his gruff manner which could occasionally run to the intimidating. TWA management thought his Teutonic bearing would be perfect for dealing with the new Lufthansa pilots of whom nearly all were *Luftwaffe* veterans. The new Lufthansa pilots had been given an introduction to the Connie in Burbank and also were given a brief refresher course on twin-engined aircraft. In the early days of the TWA-Lufthansa relationship, they rode in the cockpit more as observers than anything else while the Americans demonstrated flying the routes. The loan of ten TWA pilots was found to be inadequate after a few months, and eventually the program was expanded although Blackie had no inkling of his eventual involvement in the early days of 1956.

The year 1956 got underway with stunning news about the death of Ralph Damon who had died of pneumonia on 4 January while Blackie was en route from Paris to Idlewild. Damon was run down and never noted for taking especially good care of himself. After dark, on 15 December, a night so raw and cold that prudent men dressed in layers, Damon introduced the massive, TWA combination neon sign/billboard at Times Square. The memorable sign, seventy five feet high, 100 feet long, and featuring a two-fifths scale Connie on top, was certainly worthy of a presidential introduction. It proved so striking that a second such sign was put up in downtown Chicago almost a year to the day later. Failing to wear a hat and overcoat, Damon was thoroughly chilled and after several days of what he thought was a bad chest cold rapidly developed pneumonia which his doctors were unable to reverse. The funeral was held at Garden City

Community Church in Minneola, Long Island, and interment was back in Damon's home state, Massachusetts. The funeral was attended by the greatest names in aviation and the business world, including many of Blackie's acquaintances such as Jimmy Doolittle and Eddie Rickenbacker. The immediate effect was to create a rudderless TWA while Howard Hughes searched for a successor. Early betting was that Jack Frye might return, and Hughes did contact him. The first of Frye's two terms was do-able, a $100,000 yearly salary, but the second was impossible: absolute control of TWA with no Noah Dietrich and Howard Hughes involvement. While Hughes pondered the succession issue, John Collings, the executive vice president, was appointed acting president.

Finding a new president was not yet on Howard Hughes' front burner. Although he had model 1649 Connies on order, on 7 February he announced the purchase of seven Boeing 707s which were to be the first of an eventual order totaling thirty 707s. Each had a price tag of $4,500,000 and would be bought by Hughes Tool Company for re-sale to TWA. The planes appealed to Hughes who was always enamored of speed, and the new 707s were to be capable 600 mph yet cruised at 575. The payload was pegged at 31,500 pounds and that was pushed along by something new for TWA – engines by Pratt and Whitney. Almost immediately after the order was announced, two of TWA's technical and flight managers, Marcy Farnon and Marv Horstman were dispatched to Wichita where Boeing put them through an orientation on the B-47 Stratojet. They were not alone – representatives from seven other airlines were taking the same plunge.

Although the CEO position was still vacant through the first quarter of 1956, those managing the company, John Collings, A.V. Leslie, and O.E. Cocke, were all elevated to senior vice president. The same announcement of those promotions also resulted in one of Blackie's closest friends making vice president. Blackie sent a warm note of congratulations to TWA's new vice president of operations, Frank Busch, who wrote a thank you note back to Blackie describing the promotion as a case of "the fat boy finally made good." Blackie was genuine in his comments, but hardly jealous. His good friend was not flying and had put on a few pounds since relocating to the executive offices. News coming out of those offices at the same time disappointed financial analysts who learned that TWA's first quarter profits were only $5,407,000 on sales of nearly $220,000,000. Costs were the culprit and most of those were associated with buying Super Connies and development of the massive, new repair and overhaul base at Mid-Continent International Airport north of Kansas City.

On 1 July 1956, shortly after Bob completed his junior year at Oberlin, the latest in the "love'em and leave'em" saga in TWA's relationship with its advertising agencies was announced. Batten, Barton, Durstine & Osborn (better known simply as "BBDO") was fired by TWA in spite of the agency's reputation as a creative powerhouse. The new agency was Foote, Cone, & Belding (FCB), a much-respected agency with headquarters in Chicago, but also with a major presence in New York and San Francisco among its other domestic locations. FCB immediately latched onto Captain Harold F. Blackburn like a black bear to honey once it learned that Blackie had just been portrayed as one of America's most eminent airline pilots among the readers of *The New Yorker* magazine. Full page ads with Blackie appearing in front of a Super G Connie promptly appeared in New York City's leading newspapers under a headline reading: "Fly With The Pilots Who Fly The World." The subhead continued, "TWA's Finest Service West." FCB was somewhat loose with the truth in the copy, not to mention the fact that when Blackie flew out of New York, he always flew east, not west, but in those days a modest amount of puffery was already accepted as the norm. Blackie, however, undoubtedly flinched in the face of the following caption copy:

TWA Captain Harold F. Blackburn (*New Yorker* profile May 5, 1956) has flown World War I Jennies, today's jets, and everything in between. For TWA he pioneered the first scheduled Atlantic flights and flew the first planeload of wounded back from D-Day. On Long Island where he lives with his lovely wife Martha and four active teenagers, his neighbors refer with envy to his lawn as Blackburn's wall-to-wall. Captain Blackburn with 18,000 flying hours to his credit (more than two years off the ground) is TWA's kind of pilot, the kind of man you like to have in command.

Martha was lovely to be sure, but there were only two teenagers at a home where the neighbors were, in fact, jealous of the radiantly green and well-manicured lawn at the Blackburn residence. The balance of the ad was filled with one-way, sky-tourist westward fares, including L.A. or San Francisco at $80, Chicago at $33, and Denver for $74 among others. The call to action gave the phone number at the main sales office at 120 Broadway as being just one of eleven convenient locations in the greater New York area. For the traveler on a budget, a ticket could be had for just ten percent of the full price with the balance paid in twenty months. The advertisement proved so successful that FCB turned it into a formula six months later and ran the ad in all of America's major metro markets, focusing on other International Division stars, such as Bob Buck, as each advertisement's role model. FCB had plenty of the right stuff for TWA, so much so that the romance lasted for the next twenty-one years or what amounts to three times the usual agent-client relationship.

TWA public relations also got on the Blackburn bandwagon before his fame flamed out by generating some good ink in Chicago papers concerning TWA's tenth year of offering flights from Chicagoland to Paris. Blackie was technically domiciled in New York, but a well-placed phone call to Flight Scheduling soon had Blackie taking a Connie from Chicago's main airport at the time, Midway, to Paris. The takeoff was preceded by a photo session with Blackie kneeling beside the left main landing gear, holding a small globe and explaining international flight to pretty little Louise Pineader, an eight-year old passenger.

Several weeks later, Chairman Pierson announced that the contract with Lufthansa would be extended another year. Before all was said and done, he would be repeating the announcement annually for several more years. TWA was, unfortunately, in the national news twice in the spring and early summer of 1956 for two tragic flying accidents of which the first, the crack-up of a Martin 404 at Pittsburgh barely made a ripple in the nation's news media. On 1 April, the Martin departed Pittsburgh, but very shortly following takeoff the fire warning alarm in the left engine went off just as the ship yawed sharply to the left. The pilot flying was actually a co-pilot on a line check, and he acted instinctively by quickly pulling the throttle back, feeding in right rudder, and reaching to manually feather the prop. The captain prevented the first officer from cycling the feathering button because he expected the prop would feather automatically. It did not, the plane, now with considerably drag from an unfeathered prop on a dead engine, plowed into the ground 1,690 feet past the runway, killing twenty-two people. The pilots, among the fourteen survivors, were blamed for an uncoordinated response to the emergency. Because the pilots were from TWA's Domestic Division, Blackie was not well acquainted with them although he was with Jack Gandy, the pilot on the next TWA disaster.

On 30 June 1956, the worst American aviation disaster to date occurred over the Grand Canyon when TWA Flight 2 collided with United Flight 718, a DC-6. Both aircraft had taken off within minutes of each other from Los Angeles bound for Kansas City and Chicago, respectively. United was on an instrument flight plan at 21,000 feet, but TWA had been flying at 19,000 and was in cloud when it filed for visual flight on top which was approved by Air Traffic Control. With that change, both aircraft were flying at 21,000 feet and were approaching the Winslow omni station from slightly different headings: fifty nine degrees magnetic for TWA and forty-six degrees for United. The planes were not under radar control simply because the CAA had no radar service in that remote area, although any quick thinking air traffic controller should have realized that there were two airliners on a converging course. From what could be determined from the wreckage, it appeared as though the overtaking aircraft, the United DC-6, saw the Constellation at the last second and tried to turn away from it, but at that point it was too late. The Connie's fuselage was severed just in front of the tail unit and the DC-6 lost most of its left wing in the collision. Both planes fell out of control to the Grand Canyon below. Blackie had gotten back from Paris on the twenty-sixth and was not flying again until July fourth, thus he heard about the accident as did most people: from television news. With 140 people killed, there was some blame for everybody, although the CAA did not identify either airline as the guilty party. Although Blackie had no great love for the CAA, it and ultimately the American flying public were the ultimate beneficiaries of the tragedy which eventually provided the impetus to create the

Federal aviation Administration. That agency was in time provided with enough funding to significantly expand radar coverage across the country. What made the accident particularly difficult to take was the composition of the passenger list; nearly a third was TWA Kansas City employees returning from an excursion to California.

With the company still without a CEO, Chairman Pierson was in the TWA news more frequently, and in late September it was announced that he had just received a high award from the German government. It was the German Commander's Cross of the Federal Republic of Germany which had been awarded for Pierson's role in supporting the launch of the new Lufthansa in addition to all of the good work he had done in rehabilitating Europe as the chairman of the American Council of the International Chamber of Commerce. Four days later it was announced that Blackie was headed to Hamburg as supervisor of flying at Lufthansa.

Chapter Eleven
The "Pore Farm"

In late August 1956, Ernie Pretsch was back in New York briefly and had asked to meet with Blackie about the possibility of Blackie's joining the team of TWA pilots on loan to Lufthansa. From Pretsch, Blackie learned that Lufthansa had already been preparing its first aircrews beginning with an orientation program at Taloa, the training subsidiary of Transocean Airlines of Oakland, California, in late 1954. That was well before TWA announced its involvement with Lufthansa in 1955, although the initial group was destined for short-haul flying in Germany using twin-engine Convair 340s, not on the international routes for which TWA had been employed. That first group of Lufthansa air crews, four pilots and four flight engineers, were all in their mid- to late forties and had on average about 8,000 hours in their logbooks. They had flown for the *Luftwaffe* during the war, but had been employees of Lufthansa before that. The Germans were eager students and, of course, very grateful for the new opportunity with Lufthansa, considering how many of them had spent the immediate post-war years as prisoners-of-war, truck drivers, factory workers, farm hands, etc. With pessimistic, dark humor, it was not unusual for one of them to have said to the other in early 1945, "Enjoy the war, the peace will be terrible." Nothing could have been worse than the war of course, but for nearly all Germans the first few years of peace were times of grinding hunger and unemployment.

Taloa's curriculum can best be described as a six-week "refresher course," consisting of ninety hours of classroom work, twenty-six hours of twin-engine flying, and a graduation flight to Hawaii. The German pilots with whom Prestch was working received similar training and the addition of a ground orientation on the Constellation in Burbank. Blackie had a lot to think about. Pretsch was accused of trying to recruit just his friends, although he did not have that many to begin with. Blackie had no problem with Prestch's recruitment policies and said he would do the exact same thing. There were several distinct advantages to working with Lufthansa and one major disadvantage. Prestch was not talking about Blackie's being just a combination instructor/line pilot. Blackie was offered the job of Lufthansa's supervisor of flying because the number of aircrews was expanding so rapidly. Blackie had once sworn off management jobs in 1948 and had said in public many times that he wanted absolutely no part in the management of his company or his union – that is until he became supervisor of flying in the International Division in 1955. There were several very appealing upsides to the job of which salary was an important consideration. Daughter Betty's education was basically tuition free since she went to school on a state scholarship, but Bob was still in college at Oberlin. Bonnie would start college in a few years, and there was a real possibility that her education might prove just as expensive as Bob's. At least Blackie would not have two daughters in school at the same time, but there was still Be's education to consider after Bonnie's. Although Blackie knew he could command a good price for his home in East Williston, he was also looking for another property, possibly for his retirement years, away from the immediate New York metro area. That could involve significant renovation costs, hence the need for more income.

From a professional viewpoint, the Lufthansa job offered the appealing dimension of new routes and new lands just as he had enjoyed when ICD had spread its wings in 1942. It appeared that Blackie's time with his family would at the worst be about the same as it had been. It occasionally resulted in more time at home since he would be gone for roughly two weeks a month and then home for two straight weeks. That was in contrast to how he had been flying – two or three trips a month which with layovers and delays meant he was gone at least seventeen or eighteen days a month. Pretsch was very anxious to see Blackie make the Lufthansa connection because he knew the German military psyche and he knew how Blackie would come across to the former *Luftwaffe* personnel. Blackie was a very firm, intense, and forthright person, and Pretsch knew that would play perfectly considering the German "*Befehl ist befehl*" (orders are orders) mentality. Although Blackie knew he would have to eat a little more crow on the "never again join management" pronouncements, he decided it was a very worthwhile offer and thus accepted with the appointment effective on 1 October 1956. It was, after all, in reality a lateral move, albeit one that offered more income and time at home.

Lufthansa was buying new model 1049 Constellations from Lockheed and had aggressive plans to renew its old pre-war routes to Africa, South America, the Middle East, and Asia in addition to European capitals and a new route to America. This time, the American State and War Departments had no objections to Germans flying South American routes. The story had been very different from the late 1930s through 1941 when the Germans, through their subsidiary company *Syndicato Condor,* had extensive routes throughout South America. Indeed, one of Lufthansa's Focke-Wulf Condors had shadowed the British cruisers HMS *Ajax* and HMS *Exeter* while they were blocking the German pocket battleship *Graf Spee* from escaping Montevideo harbor in late 1939.

Trans World Airlines

TWA was the trend-setter in commercial, piston-engined transport aircraft, progressing from DC-2s in the 1930s to sleek Constellations in the late 1940s.

Blackie's last regularly scheduled trip with TWA, Flight 963 with a model 749 Connie, was flown on 22 September into a stiff headwind from Germany back to Idlewild with stops in Shannon and Gander; it took eighteen hours. His next flight was with Lufthansa, out of Idlewild to Germany on 18 October with a fellow American check pilot, Dick Sherwood, and a German, Rolf Rossberg, to whom Blackie was giving a line check. During the war, Rosssberg had been a Junkers JU-88 light bomber pilot with KG 51, a unit that had flown on both the eastern and western fronts. Blackie learned that the "KG" stood for *Kampfgeschwader* [(bomber wing as opposed to a *Jagdgeschwader* ("JG") or fighter wing]. A *geschwader* consisted of several groups, each of which had several squadrons operating roughly a dozen airplanes apiece at full strength which was rarely the case in the field. For his first three days in Germany, Blackie flew instrument checks with his new German pilots and in doing so familiarized himself with the airports in Bonn and Dusseldorf.

When Blackie first met his German charges, he encountered a good cross section of the wartime *Luftwaffe* with one exception. He met very few former day fighter pilots. Plans to revive the old *Luftwaffe* go back to 1955 when West

Germany joined NATO because the Cold War showed no signs of going away. Unlike the old Nazi *Luftwaffe*, the new *Luftwaffe* was re-established as strictly a defensive force to bolster the NATO presence in Europe. Therefore, it recruited surviving day fighter and jet fighter pilots from the old wartime *Luftwaffe*. Lufthansa was much more interested in finding pilots who had flown multi-engine aircraft during the war, and if they had a lot of instrument or night flying in their logbooks, so much the better. Thus most of the pilots and flight engineers Blackie worked with had been members of transport or bomber crews. There were, however, quite a few night fighter pilots signing on with the new Lufthansa as well as a few long range, twin-engine fighter pilots. Those men at one time had been extremely dangerous enemies of the RAF and USAAF – any one of them would have tried to kill Blackie if given the chance. Many of the Germans had won the Iron Cross Second Class and subsequently the Iron Cross First Class (very roughly equivalent to an American Air Medal and Distinguished Flying Cross). Some of them had won the Knight's Cross of the Iron Cross and a few had been awarded higher degrees of that award including oak leaves and swords, placing their collective valor very roughly on par with that required for the American Distinguished Service Cross or British Distinguished Service Order. Several of those pilots flew in air lanes very close to Blackie's flights from Greenland to Scotland and from England to Casablanca. Those former *Luftwaffe* pilots, Bernhard Jope, Erich Stocker, Alfred Arzinger, Joachim Ohm, and Reinhold Wrede among them had, with the exception of the latter, all been squadron or group commanders within *Kampgeschwader* 40, the Brittany-based, long range interception unit. ICD navigators plotted courses as far west as possible to avoid contact on flights down to Africa from bases in Britain. As might be expected, many of the Germans with whom Blackie rode as a check pilot had been instructors during the war which is the principal reason why they survived the war. Unlike the RAF or the USAAF which rested its aircrews after a tour on operations or even assigned them to permanent non-combat duty, the *Luftwaffe* flew its frontline fighter, bomber, and attack pilots until they were killed, badly wounded, or were captured. One of the former instructors was the scholarly Hermann Terjung who, like Blackie, was a blind-flying expert. At the other end of the instructor spectrum was Wilhelm Kaiser who spent much of the war instructing on the Stuka dive bomber. Blackie's impression of the German pilots' airmanship was very positive, and over the course of the next three years, he flew with nearly all of the new German Constellation aircrews. Not all of the TWA pilots shared Blackie's view; some thought the Germans were regimented, unimaginative flyers and that others were not particularly skilled. That impression may have been realistic when speaking of some of the younger pilots whose training and experience on operations at the end of the war was limited. The older *Luftwaffe* pilots were very well trained; early in the war their training was more thorough than that of the RAF or USAAF. The Germans were slower to condense their training than the Allied air forces as the war progressed in 1940-

1942 simply because *Luftwaffe* leaders were told that the war was going to be a short one and not to skimp on training. By 1943, their error of not compressing training into a shorter period as had the RAF and USAAF became obvious, particularly to Allied fighter pilots who noticed the lack of skill in their *Luftwaffe* opponents by late 1944. Indeed, some of Blackie's new German acquaintances were still young men not yet age thirty-five. Another advantage enjoyed by the older *Luftwaffe* pilots was having initially flown sailplanes before moving to powered aircraft. Flying gliders provided them with as good a piloting foundation as can be had. Nearly all of the Germans were rusty as well they might have been considering their post-war prison terms and the near total lack of opportunity to fly anything in post-war Germany. Those new Lufthansa pilots with recent flying experience had usually kept up their skills by flying for airlines somewhere in the Third World after the war. Others became flight instructors in countries such as Syria and Argentina which were building up their air forces in the aftermath of World War II. A number of other former *Luftwaffe* pilots had only been released from Russian prisons in the early 1950s and had to take what work they could find when they returned.

Before Blackie came over to Hamburg, Prestch had explained exactly what Lufthansa's plans were: recreate all of their old-pre-war routes, including the one to and throughout South America via the South Atlantic although that would be flown with the Constellation, not seaplanes as in the 1930s. Lufthansa's plans were aggressive – its domestic network included Hamburg, Frankfurt, Cologne, Dusseldorf, and Munich. The one place it was neither flying to nor from was Berlin, and it would be many years before that would happen. When Blackie went to Lufthansa, it was already flying to London, Paris, and Madrid. In addition to New York, Lufthansa would soon be serving Chicago and Montreal. Flying via Dakar in Senegal, Lufthansa would initially reach Buenos Aires, Rio de Janeiro, and Sao Paulo. By the end of 1959, Lufthansa would extend from Baghdad and Tehran into India and as Far East as Bangkok.

When first notified of his assignment to Lufthansa, Ernie Pretsch tried to recruit as many German-speaking TWA pilots as possible, but it was found that such a capability was really not essential for several reasons. Nearly all of the Germans spoke English to begin with, and if they did not already have a good command of English, they were required to learn it practically overnight because English had become the language of international air traffic control. Two of the first German pilots Blackie flew with, Erwin Herms and Hans-Joachim Zantopp had learned their English during the war – as prisoners of war. Zantopp, unlike Herms, had become a prisoner of war by very conventional means; he was shot down. Zantopp, flying a Dornier 217 medium bomber, was downed down by an RAF Spitfire over Sicily in late July 1943 and was promptly captured by Allied ground forces which had just invaded Sicily. Although Blackie flew with quite a few former POWs, none had such an ironic road to a prisoner of war camp as did Erwin Herms, who came to grief in the dark of night on 26 November 1941.

He actually landed on an RAF base only to discover too late that he had not landed back in France. He was flying the fast Junkers JU-88 and had, to use the term of the day, been "meaconed" into believing that he had flown from his objective, shipping in the Irish Sea, back to the French coast. "Meacon" was the code name for the electronic countermeasures program which duped German pilots into thinking they were on course when, in reality, they were not. The system had first been used during nights with poor visibility to trick very low flying German night raiders into thinking the British coast was still many miles ahead. Most savvy German pilots quickly recognized the trickery, but those who did not flew directly into coastal bluffs or hit rising terrain immediately inland. At the time, the *Luftwaffe* was sending many of its bombers out and back by radio beam, frequently employing a second, intersecting beam from which the bomber crew could deduce that they were on or near the target. This was not lost upon RAF Signals Command, which promptly started sending stronger transmissions on the same frequencies, tricking some Germans into thinking they were nearing their French bases. In the process, the Royal Air Force fooled enough *Luftwaffe* crews into landing on British fields to create a nice, small collection of front-line German bombers. That cat and mouse game continued with each side getting the upper hand for a while before the *Luftwaffe* eventually scaled down its night bomber attacks on Britain.

Klaus Eidner, also an early student of Blackie's, was another former prisoner of war who was held much longer than either Zantopp or Herms although he had no opportunity to learn English while in prison for nearly six years. His POW camp was in Russia. Eidner had been a Heinkel 111 pilot with one of the *Luftwaffe*'s oldest bomber units, *Kampfgeschwader 55*.

Two of the Germans Blackie was grooming as captains went on to careers with Lufthansa that ended in tragedy. Horst Quednau, a highly decorated Heinkel HE-111 bomber pilot who had flown over 400 missions on both the Eastern and Western Fronts, was another one of the first Germans Blackie met. Thirteen years later Quednau was on short final approach into Mexico City when he suffered a massive heart attack. He rode it in, landed the Connie, and then expired. He was only fifty-two years old. Captain Christian Krack probably wished he could have died young as well because he had to live with the shame of having been the first Boeing 747 captain involved in a crash in which fatalities occurred. The date was 20 November 1974 on a flight from Frankfurt to Johannesburg with an intermediate stop in Nairobi where the takeoff accident occurred. The 747 took off in a semi-stalled condition and hit a low hill little more than a half mile from the end of the runway. Krack, his co-pilot, and engineer were found at fault because it was determined that the wings' leading edge slats had not been deployed, i.e., a checklist failure. Boeing was also criticized for not having a warning system annunciator for such problems, and it subsequently developed and installed one in all 747s. There had been two previous cases of 747s taking off with the slats retracted, but their pilots caught the mistake in time. Such quick action may not have worked at Nairobi considering how that airport's high altitude and midday heat worked to Krack's disadvantage by making the air less dense.

The first two Germans that Blackie flew with on a westbound transatlantic flight, on 25 October, were Heinz Niemann and Adelbert Schaffranek; both had been transport pilots during the war with the former flying the three-engine Junkers JU-52 and the latter the troop/cargo version of the FW-200 Condor. Blackie never had a direct confrontation with any German aircraft during the war, but from among the new Lufthansa pilots, the one whose path he may have come closest to crossing probably was Adelbert Schaffranek who flew with KG 200, a long range transport group. From May of 1943 right up until the very end of the war Schaffranek regularly flew from Germany to Greece, Rhodes, and Crete, bringing in mail and supplies and then flying wounded personnel home. It is a largely forgotten fact that Crete (an objective on one of Blackie's recce missions in 1944), Rhodes, and several other islands in the Aegean and Dodecanese remained in German hands right up until VE-Day. The British had wanted to take all of those islands, but Eisenhower saw little strategic value in them. He refused to allocate any resources to the effort, hence the German presence in the far eastern Mediterranean until war's end in early May 1945.

Many of Blackie's flights for the balance of 1956 were indeed the exact same routes he had already been flying: between New York and Frankfurt on the model 1049 Connie, although flights from Hamburg also started appearing in his logbook at the end of 1956. As far as Martha was concerned, Blackie's comings and goings to New York were essentially the same as they had always been – gone for a few days at a time, sometimes a week or more, but he always returned for at least a week or more. In time Martha adjusted to something that was, however, radically different. The insignia on Blackie's cap and jacket were Lufthansa's, not what he had been wearing for nearly the last twenty-five years. While most of Blackie's time with Lufthansa was flying regularly scheduled routes, he, as supervisor of flying, was frequently involved with flying local instrument and line checks. The first full flight check, flown on 6 November, was with a unique pilot: Hans Munsterer who had been the co-pilot on Adolf Hitler's personal FW-200 Condor. The captain of that Condor, Hans Baur, was not available for service in the new Lufthansa, although he had flown with the old Lufthansa prior to the war. The one-legged Baur had been captured by the Soviets in 1945 and in 1955 was turned over to the French who would not release him until 1957.

On 31 October, Blackie took off on a twenty-four hour flight into his past – to South America. The destination was not Natal, but rather to a place considerably more appealing: Rio de Janeiro. Rio was Lufthansa's South American hub as it had been before the war, and as the months progressed Blackie would also fly to Buenos Aires and still later to South America's west coast. Blackie's first flight to Rio was from Hamburg via Dusseldorf, Paris, and another city from

Blackie's wartime past: Dakar in northwest Africa. His companions on the nearly twenty-five hour flight were Erich Goetz, one of the many former JU-52 three-engine transport pilots joining Lufthansa, Max Brandenberg, and Peter Spoden. Max Brandenberg was a very successful bomber pilot who had sunk Allied ships in the Mediterranean and the Black Sea. Peter Spoden had been a night fighter pilot, was shot down and wounded several times, and was an ace who had brought down twenty-two Allied bombers of which nearly all were RAF heavies. After getting out of prisoner of war camp in 1946, Spoden was one of the lucky few who quickly landed a job; family connections soon had him working at his pre-*Luftwaffe* job on the regional trains of what had been the *Reichsbahn* (national railroad system). Prior to meeting Blackie, Spoden had refreshed his multi-engine experience on DC-3s in 1955 ten years after having flown Me-110 and Ju 88 night fighters in the *Luftwaffe*.

There were quite a few other night fighter pilots in the new Lufthansa which is not surprising considering their night flying, twin-engine, and instrument experience. Although many of Blackie's ex-night fighter German acquaintances had been shot down and wounded on more than one occasion they did not suffer quite the heavy loss rates that *Luftwaffe* day fighter pilots did once American Mustangs started appearing over Germany in great numbers by mid-1944. Still, the RAF Mosquitos that accompanied the RAF's Halifaxes and Lancasters to Germany at night certainly minimized the likelihood of Spoden and his colleagues surviving the war.

In late October and early November 1956, Blackie was flying roundtrips between Hamburg and New York or Rio de Janeiro while TWA's International Division was not flying to one of its regular stations. That was Cairo which was essentially shut down as a result of the Suez Crisis. TWA employees, including Blackie's buddy Swede Golien, got out of Dodge as best they could by car, boat, or plane. They would return after several weeks when TWA restored service through Cairo, but the crisis did put a blip in the International Division's income for the fourth quarter.

Blackie was back home in East Williston when TWA announced its new president nearly a year after Ralph Damon died. The new CEO, Carter Burgess, had actually worked at TWA during the Jack Frye days and had followed Frye to GAF. After that he had served in a number of management capacities, most recently as Assistant Secretary of Defense in Washington where he had the whole-hearted approval of President Eisenhower. Burgess' leadership of TWA could be summarized in two words: short and troubled. In one sense, it may have been a good thing that Blackie was somewhat removed from the TWA scene at this time because Burgess, although he got along famously with TWA's pilots, proved to be a bad-tempered bull in a china shop for other departments. Because he felt that TWA's on-time reputation was suffering and that the company was overloaded with deadwood, Burgess promptly sacked the executive vice president, John Collings. Naturally, this put considerable pressure on Frank

Busch who then became the top operations man. While Blackie did not suffer fools gladly, his old friend Frank Busch did not suffer fools, period. If the fool was the company CEO or a clumsy baggage handler, it made no difference. In spite of several loud confrontations with Burgess, Busch managed to keep his job and outlasted Burgess, although his being forced to move to the New York headquarters did little to facilitate a better working relationship between the two. Adding to Burgess' dyspepsia was the $2,300,000 loss for 1956. Financially, performance under Burgess would not improve much because 1957 would be another loss year as well, albeit not quite as bad as 1956. On the other hand, Burgess' temper was much like that of Larry Fritz; after the initial explosion, the fires tended to self-extinguish rather quickly.

Blackie had just gotten back from a grueling, headwind-dominated flight from Germany via Shannon and Gander with Horst Quednau and Joachim Ohm on 17 December 1956 when an action on his Air Force Reserve status was taken at the Continental Air Command Center in Denver, Colorado. On that day he, who was certainly the most experienced first lieutenant in the Inactive Air Force Reserve, was moved to the Retired Reserve Section. That effectively ended even the remotest remote possibility of his ever seeing action in the defense of his country.

Blackie was not at home to celebrate the New Year because he took off from New York for Hamburg on the twenty-eighth and would not be back from that flight and several others until 5 January. His companions on the outbound flight to Hamburg included an old American friend, Bill Sherwood, a flight engineer, and the two Lufthansa pilots: Erich Stocker, who had been an FW-200 pilot flying out over the Bay of Biscay during the war, and Klaus Scheer, a former twenty-four victory night fighter pilot. The flight was made in a model 1049 Connie carrying the German registration D-ALAK, a plane whose days were numbered. Almost exactly two years later, the plane was destroyed in a crash at Rio de Janeiro during a heavy thunderstorm in which the pilots had flown below the landing minimums and crashed into the bay short of the runway. Only three out of the thirty-nine passengers and crew aboard survived, including the German pilot Karl Frank, a night fighter pilot during the war. Among the dead was a friend of Blackie's, the check pilot, Wren McMains, who had joined TWA in 1941. The pilots were blamed for the accident which was attributed to pilot fatigue. The pilots had exceeded the number of hours at the controls allowed by Brazilian authorities, although the crew was still within the limits established by Lufthansa. While it was a bad accident, it was the only Connie lost from the time Lufthansa had been reorganized until the day all of its Connies were finally retired nearly ten years later. Compared to the multiple Connie accidents of other airlines, including TWA, Lufthansa's record tends to confirm Blackie's opinion of Lufthansa pilots as being skilled airmen.

As far back as mid-1955, Blackie and Martha concluded that with airline pilots being forced by federal regulations to retire upon reaching age sixty, they would prefer to reside outside of the greater New York area. They were not interested in a house in a more distant suburb, but were looking perhaps to relocate to a small farm – an idea that had been at the back of Blackie's mind as early as mid-1953. He had never really outgrown his small town western Nebraska origins and it was not unusual for pilots to be at least part-time farmers. His great friend Frank Busch already had a small ranch outside of Kansas City. TWA's former chief pilot, Otis Bryan, had grown up on an Indiana farm and upon retirement started ranching in Kansas on 3,200 acres with over 300 head of cattle. Other TWA pilots who were in Blackie's age group were looking at retirement took other approaches: studying for a law degree or opening a marina, service station, motel, or other small businesses among other second career options.

Blackie was looking for something more modest than a big cattle spread and began his search for a farm somewhere between Maine and Virginia during his days off. It did not take long to whittle that territory down to some location not more than a three-hour drive from Idlewild. The smaller radius put the Garden Spot State, New Jersey, into focus along with southern New England and eastern Pennsylvania. Early in the search, New Jersey seemed most promising, particularly as a number of Blackie's pilot friends had homes in rural communities there. The Keystone State, with its Pennsylvania Dutch farming tradition and proximity to New York, emerged as the favorite as did a nearly 100-acre property near Mertztown which is just west of Allentown. The property, located in Berks County, included several large fields, woods, a creek running on the southern and eastern border of the property, and a spring-fed pond. The farm, dating back exactly 200 years, was available as an absolute bargain, i.e., a real "fixer-upper." It included a very typical Pennsylvania Dutch two-story stone farm house, a large barn also of stone, and several smaller out buildings. The previous owner, Paul Dreibelbis, came from a long line of Berks County dairy farmers, but in his old age had really let the property go downhill. The Blackburns bought it for twenty grand and after twenty years of Blackie's industriousness and another twenty thousand in improvements sold it for $250,000. There was a minor and somewhat expensive misunderstanding about what was included in the farm price. Blackie had assumed that it included tools and farm equipment, but it in fact did not. Thus Blackie had to make attractive offers to neighboring farmers who had bought the equipment before the Blackburns arrived. When he completed the equipment purchases, Blackie became the proud owner of three tractors of which two, a little Farmall and a larger Massey-Ferguson, were used in the fields while the smallest was rigged with a snow plow for use in the winter. Joining the tractors in the barn was the Massey-Ferguson's accessory equipment including a mower, tedder, side rake, plows, harrow, planter, and cultivator. Blackie had a small hay baler which was towed by the larger tractor, but never acquired a combine since he had so few acres in corn.

On the assumption that he would have little problem in selling his East Williston house and the fact that his broker, Merrill Lynch, had done very well by him, Blackie and Martha purchased the farm for cash in December 1956. Always frugal, Blackie would not be faced with any more college tuition costs for Bob and because his mother, whom he had been supporting all along, had died in 1950, his savings grew significantly. Although he had no intention of moving until Bonnie and Be were out of school in June, he took the family down to look at the farm in January 1957. Bonnie's and Be's spirits plunged when they saw the place because it was poorly maintained, overgrown, and the lane leading from the road to the house was of the dirt variety and plentifully pot-holed. The place did have running water and electricity (barely), although there were more electrical outlets in the barn than the house. Whereas his daughters were dismayed at this backwater where even in the mid-twentieth century outhouses and one-room schools could still be found, Blackie saw past the run-down appearance and envisioned the place as it would look in the not so distant future. Blackie, while not a farmer to begin with, had always wanted to have a farm and began educating himself about crops and beef production as much as possible. To that end, Blackie spent many hours learning the beef business from one of his East Williston neighbors who owned a farm in Virginia where he had a herd of Black Angus. Blackie could also expect a lot of local advice once he got to Berks County's local office of the Pennsylvania State Agricultural Extension Service. In nearly all farming criteria such as milk production, number of farms, sales of agricultural products, Berks typically ranks fourth out of Pennsylvania's sixty-seven counties. Although the state of Pennsylvania is highly ranked for agriculture nationally in most farming categories, Berks County, prime dairy country, is not known for beef production. Still, the land is very fertile which is really what counted to Blackie insofar as he could grow grain to use as cattle feed.

Blackie's reading, up until recently had consisted of novels, general interest/news magazines, and *Aviation Week*. To that reading he added *Farm Journal* and several other farming magazines. He planned, through a combination of his own labor, local contractors, and hiring laborers, to have the place ready to be occupied by late June 1957. He already had a name for the spread – "Hidden Brook Farm." The name had come to him in a moment of epiphany when Lufthansa gave him a welcoming gift: a pair of gold cufflinks with the initials HB. The name was quite apropos; the small creek was nearly hidden by the woods on the southern and eastern edges of the property. On local maps, the brook in the woods was called Little Sacony Creek, but for the Blackburn family and friends it became Hidden Brook forever. In the middle of the creek, there was a large rock that had split in two. Naturally called "Split Rock," it became a favorite camping and picnic venue for the Blackburn grandchildren.

In personal correspondence, of which he was a master, Blackie, with his dry self-deprecating wit, always called the place "the pore farm," implying the poor place was so marginal that existence on it was at best hard scrabble. The name revealed another side of Captain Blackburn's humor and command of the language. With friends, he delighted in mangling the English language with all manner of Blackburnisms. When referring to the government, he would call it the "gummermint." When he moved to Long Island after having grown up in Nebraska where there is little if any regional accent, he would occasionally add "Long Island" as a pre-fix to New York. He did not spell the island's name properly; it came out as "Lawn Guy-landt" at the expense of his native New Yorker friends. "Saturday" became "Sattidee." When he walked to the garden with grandkids and Martha to apply insecticide to the vegetables, he would say "Lettuce spray." There were enough transmogrifications of proper English to establish Blackie as a stand-up comedian, had he the inclination.

In spite of the German reputation for precision and quality, Blackie experienced the occasional technical issue with the first occurring on 6 May 1957 when he lost the number two engine just after takeoff. He had to return to Shannon with his German pilots *Herren* von Konschegg *und* Meissenreider before completing the fourteen-hour flight to New York. Fortunately, the Lockheed Constellation flew well on three engines, and Lufthansa was in truth hardly the only airline to have problems with the Wright 3350 engines which most carriers overhauled earlier than recommended by the manufacturer. Indeed, Lufthansa was among those airlines that occasionally referred to the Connie as the "Lockheed Tri-Motor." There were other aggravations common to any airline, including weather issues. Occasionally, Blackie would be bound for Idlewild but due to bad weather he would be forced to go to an alternate which frequently was Philadelphia or McGuire Air Force Base in central New Jersey.

By March of 1957, Blackie's monthly flying hours had crept up due to the distance involved within Lufthansa's expanding routes which reached Asia and South America. But even with the increase, he was spending at least as much time at home which included work on the farm, selling his home in East Williston, and preparing for the move. The departure from Long Island took place in early July because Blackie's flights were scheduled at the end of the month. When the Blackburns moved into Hidden Brook Farm, the place was still undergoing a great transformation, but it had become much more family-friendly. Although his daughters had originally been dismayed, even crushed, by the place, they were now pleased to see that there were four large bedrooms (two with cedar-lined, walk-in closets), new bathrooms, and modern wiring throughout. A four-bay garage and a new porch had also been built. On the first floor, the kitchen had been dramatically expanded, and below that the basement, originally an earthen basement, was extended. Blackie would eventually complete the basement as a large woodworking shop. The farm would not really go fully operational with the Black Angus until the spring of 1959 which gave Blackie more time to learn farm management and to complete even more improvements. Because he did not see the place as a dairy farm, there was no need for all of the large pasture which was separated from the grain fields by a large stone fence. That was removed with the use of local neighbors, particularly by teenage boys from the Topton Home, a nearby Lutheran orphanage which encouraged its older boys to accept part-time employment where they could find it. Blackie had also acquired a mechanical stone crusher which was used to break the field stones into the gravel with which he improved the lane leading to the house. Because he did not need the pastures for dairy cows, he graded portions of the property to expand crop production. He also significantly graded the land near the house because, as it was, runoff water flowed directly at the house. Also on the drawing board was converting the porch to a large, screened-in Spanish style veranda.

There was a great deal of adjustment to their new community for Martha and the Blackburn daughters, beginning with their Pennsylvania Dutch neighbors whom they discovered did not have Dutch ancestors. The term Pennsylvania "Dutch" is an historical error arising in the late eighteenth century when British-Americans heard new immigrants from the continent speaking what was assumed to be Dutch. Actually, it was German as spoken primarily in the Rhineland from which most of the Pennsylvania Germans had emigrated. More significantly, when asked what language they were speaking, the immigrant would say "Deutsch" which was immediately assumed to be Dutch. Most Pennsylvania Dutch no longer speak German, however members of the Old Order Amish do, but they, having come primarily from Switzerland, are technically not speaking Pennsylvania Dutch. The Pennsylvania Dutch accent comes across as being about two-thirds Lawrence Welk and one-third Arnold Schwarzenegger. Blackie hardly noticed the accent, because he was spending about two weeks every month living with Germans whose English was heavily accented. But Martha and the girls may have initially been puzzled to hear about a neighbor's antagonistic dog being described as "wish-us" (vicious). That man, a resident of Berks County might describe his home state as "Pencil-Wane-Ya." There were other patterns of speech that left the Blackburns puzzled as was the case when Be was asked by a friend in the summer if she was "going swimming this after?" Here, the word "noon" was assumed as was the word "gone" when Martha might hear a neighbor remarking to a pharmacist, "My medicine is all" in lieu of saying "all gone." The same young friend would not refer to her Uncle John as anything but "Uncle Chon" nor would she usually call a donut a donut; rather it was a cruller or occasionally a *fasnacht.*

Considering the fact that Blackie was already very familiar with German food, there was little adjustment required in his case when moving to Pennsylvania Dutch country. For Martha and the girls, the new cuisine was yet another in what was a household marked by eclectic tastes. Pennsylvania Dutch cuisine runs to the very heavy side with plenty of butter, cream, lard, cheese, and organ meats. The danger of saturated fat and cholesterol were not yet widely understood in the 1950s which explains why many of the Blackburn neighbors were fond of main courses such as hog maw (the Pennsylvania Dutch take on haggis using pork instead of lamb), scrapple (a flat form of sausage made from spices, lard, corn meal, and the less appealing pork organs), liver dumplings, and *zitterting* (cooked meat from pig's feet, spiced, and pressed into a form).

The Blackburns enjoyed Mexican food as might be expected considering Martha's being a Texan. Since the1930s, Blackie had enjoyed also Oriental food, a cuisine that was not available anywhere except large cities, and certainly not in rural Berks County. The first experience with eastern cuisine came from the Blackburn's Japanese customers and neighbors in Los Angeles when Blackie was a small business owner. Before buying the farm, a familiar face from the past had come back into Blackie's life; it was Ben Chan, the Chinese-American cook whom Blackie had befriended in Chicago. After the war, Chan had returned to Hong Kong after having worked with General Electric. In Hong Kong, Chan had become an importer, principally of GE X-ray equipment and air conditioners. Unfortunately, the American government started restricting the exportation of those types of equipment, effectively shutting down Chan's business. Ben returned to America and was looking for financing to buy a promising property in Woonsocket, Rhode Island. Blackie became an investor in Chan's Chinese Food and Oriental Dining which became an overnight success. Ben repaid the loan quickly and invited all Blackburns available for an annual visit to a Chinese feast in Woonsocket at which no expense was spared. Blackie reciprocated and had the Chans down to the farm for a full week every summer. Blackie called Ben's son John "Ringo" because the young man wore his hair long as did the Beatles drummer, Ringo Starr. Blackie became "Uncle Blackie," John's "favorite Scottish uncle." Over the course of the week, the Chans cooked many Chinese meals utilizing to the fullest extent all of the fresh corn and other vegetables from the farm. The wild mallard population living on the pond was occasionally thinned to create Peking duck for the Blackburn table. Today, Chan's Chinese Food and Fine Oriental Dining is still going strong in Woonsocket with John Chan in charge and is a favorite among Rhode Islanders. The interest in foods from faraway was just one more dimension to the Blackburn image which was already viewed by the neighbors as exotic. That was certainly consistent with such strange household items as a monkey skin rug, more carpeting from the Middle East and American Southwest, and wall decorations from South America and North Africa.

The most remarkable transition to the farm was Martha's. She had grown up in a well-to-do family back in San Antonio where she had been a debutante with no shortage of suitors. She had adapted well to the lean years when Blackie was stuggling to succeed as a small businessman waiting for an opportunity to open with the airlines. With Blackie's becoming a captain in 1939, the household income took off and climbed rapidly as he moved up the management ladder. Although he was not in TWA's corporate management when the family moved to Pennsylvania, his income was far above the national individual employment income. However, living in Mertztown R.F.D. meant there was neither grocery store nor shopping mall nearby as there had been back on Long Island. Yet the farm became largely self-sufficient which is not surprising considering what was growing on it. There were two vegetable gardens whose bounty Martha learned how to can and store in the basement. For fruit, there were apple, cherry, and peach trees along with walnut, filbert, and pecan trees. Potatoes, grown in a patch near the corn crib, were stored in the basement at first where Blackie usually had a small keg of dandelion wine fermenting. From dealing with her butcher on Long Island, Martha was able to give precise instructions to the butchers who converted the Black Angus and sheep into many different cuts of beef and lamb, some of which were shared with neighbors and relatives. In late summer and autumn, Martha presided over the canning of tomatoes, pickles, beans, corn, peaches, and V-5 juice (her take on V-8 Juice). Martha occasionally made V-10 juice when enough ripe vegetables were at hand, but there were too few to make what she really wanted – something twice as good as V-8: the ultimate: V-16. With the help of neighbor friends Emily and Hester, Martha also made jam from the various berry bushes on the farm. Martha never won any blue ribbons for the jellies and jams, but the tomatoes were another matter. They were huge beefsteaks, so large that they had to be trimmed to fit sandwich bread. The plants alone were impressive – most were staked and reached over six feet high. It never ceased to amaze neighbors that somebody from a New York City suburb could grow prizewinning tomatoes in a traditional agricultural community. They had never, however, seen how well landscaped the Blackburn home was in East Williston or known that it was one of few in that community to have a prosperous garden. When Blackie was away, many of the farm chores fell to Martha, including keeping the feed trough full for the cattle. For heavy work, Blackie had the services of local farmers, including a young neighbor boy, Russell Seidel, who was in the same grade as Be Blackburn. Russell, many years later, recalled that Blackie was an experimental farmer, one who was always evaluating new techniques and seeds, yet on the surface he looked like any other Berks County farmer, complete to having a rain gauge not far from the front door (rainfall totals were recorded on the big TWA calendar in the kitchen). When Blackie went out the lane to the mailbox, he was not wearing a TWA captain's uniform. The former apparel salesman's work clothes now came from the Army-navy store in nearby Kutztown and consisted of a non-descript baseball cap, khaki work pants, knee-high rubber boots, and in winter a flannel-lined jacket. Like all of their neighbors, the Blackburns looked forward to receiving one catalog in particular every winter – it was the Burpee seed catalog.

Within a year after moving Mertztown, the novelty of a farming internationally flying jet captain made its way to the local newspaper, the *Kutztown Patriot*. The editor, Alliene DeChant was a well-recognized authority on the many aspects of Pennsylvania Dutch culture, including food. She was an early guest at Hidden Brook Farm where she was not served boiled tongue, *schmier kase* (cottage cheese) and *apfelschnitz pei* (dried apple pie) or shoo-fly pie for dessert as might be expected in a Pennsylvania Dutch home. What followed in the Blackburn kitchen and subsequently in DeChant's "A Little of This and a Little of That" column in the paper was a right proper Tex-Mex meal of home-made enchiladas and black beans. The editor was quite taken with the Chinese tea served with the meal, but considering the lack of praise, the date and nut loaf prepared by Be clearly had been overshadowed by the main course.

When Bob graduated from Oberlin in 1957, he was offered an Air Force commission after having taken AFROTC. Naturally, Blackie had assumed that Bob would complete flight training and go on to several decades of flying in the Air Force, an opportunity that had not come Blackie's way due to the Great Depression. Or perhaps Bob would put the military flying experience to good use by becoming an airline pilot. Just as those possibilities were considered, the Air Force dramatically increased the active duty component required of its newly commissioned pilot candidates. Unlike his father, Bob was not especially aviation-oriented nor was he attracted to piloting planes. Because he was unwilling to stay in the Air Force for almost twice the original amount of time, Bob declined the commission and set out to launch a career in social service, civil rights, and education. Bob felt very strongly about civil rights and had spent his junior year, as he said, "abroad" as the only white person at what is now Hampton University, the traditionally all black university in Virginia. Blackie and Bob were poles apart politically; Blackie was a Goldwater conservative and Bob was not. Yet Blackie's conservatism was not blind; he did share an interest in civil rights with his son and was a strong proponent of minority hiring at TWA – for pilots, hostesses, and management too. Bob's first job out of college was with the American Field Service from which he moved to the Philadelphia branch office of the Conference of Christians and Jews (later to become the International Conference for Community and Justice). By then Bob had married a classmate, Barbara Wendell, and when they moved to Philadelphia, it was as a family of three including their baby boy, Chris.

Blackie was naturally disappointed about Bob's career choice, but being the old stoic he accepted it and moved on. Compensating for the disappointment was the fact that Bob and his family, having settled in Philadelphia, were only an hour away from the farm. A turning point also arrived when the younger Blackburn could beat the elder Blackburn at tennis. It was going to happen sooner or later to be sure, but the day was inevitable considering the fact that Bob had been an outstanding jock in football, basketball, and baseball at Texas Military Academy. On the other hand, Blackie was almost age fifty-four when beaten which says a lot about the kind of shape he was in.

The second half of 1957 for Blackie passed much as the first half had with his flights being almost evenly divided between New York-Hamburg and those to South America. The only exceptions were those cities he reached as the result of Lufthansa's gateway cities being expanded in Germany, such as Dusseldorf, and new foreign route additions, including Montevideo in Uruguay. Lufthansa had initially placed orders for four Model 1649 Jetstream Connies, the last in the series and quite possibly the grandest propeller-driven airliner ever. Blackie's 15 October checkout on the 1649, by TWA check pilot Paul Porterfield, was a rather casual affair typical of Blackie's acquaintanceship with the B-24, Boeing 307, C-54, and original Connie: about three hours of airwork and landings. With that completed, Blackie waited almost three months before he flew another 1649 which he took to Burbank on 15 January 1958 to start the acceptance tests on the second of Lufthansa's four new 1649s. While he was in town, he had dinner with Clarence Shoop who had just been promoted to major general in the California Air National Guard. In another three years, Blackie's fellow reconnaissance adventurer would be the California ANG's commander. After two days of tests, Blackie and his crew consisting of Hans Grossklaus, Hans Klug, Walter Jung, and Franz Preuschoff took off for New York and then Hamburg, arriving on 21 January. Grossklaus and Klug had both been light bomber pilots during the war while Jung and Preuschoff had originally been aircrewmen who later trained as flight engineers. Of this group, Blackie, who certainly had more than just a little wanderlust in his soul, found Preuschoff by far the most interesting in conversation. Preuschoff, the oldest man in the group, had been a radioman during the war, but was an old Lufthansa hand dating back to 1936 and actually qualified as a genuine explorer. He had been a member of the German expedition to Northern Greenland in 1938 and in 1939 was the radioman on Germany's Antarctic expedition. Little more than a week later, Blackie checked out the first former *Luftwaffe* pilot to fly a 1649; it was Kurt Matzak, formerly a nineteen-victory ace night fighter pilot who had flown Messerschmidt 110s with *Nachtjagdgeschwader* 1.

Between talking with TWA pilots and the employee newspaper, Blackie was up-to-date on TWA's progress which for 1957 was significant. The new developments, including new seating and compartmentalizing the cabins on Connies and an expanding domestic route structure had helped boost the load factor. However, there were some traditional problems that culminated in December. Carter Burgess had finally tired of Howard Hughes' micro-management and increasingly bizarre behavior. A major factor in Burgess' tossing in the towel was Hughes' appropriating one of TWA's brand new Connie Jetstreams for his own personal use before the plane could be put into revenue service. Hughes had the plane for months and used it for junkets here and there from Montreal to California to Nassau. It absolutely curled Burgess' toes to realize he could not get the plane back at a time when TWA dearly needed it. Hughes wanted Burgess to stay on, but would not accept Burgess' terms: freedom from Hughes control of the airline.

Blackie was on the farm during the first week of January when he received a memo from R.C. Flornoy of the ALPA Master Council at the union's headquarters in Chicago. The memo called for all ALPA members in TWA to voluntarily contribute $10 ($5 from co-pilots) to purchase a suitable "going away" present for Burgess whom the TWA pilots within ALPA considered a friend. Blackie ponied up the ten bucks and sent it to his old captain, Don Terry who was collecting the money from the New York-based pilots, in spite of neither man having any contact with Burgess.

Ironically, Burgess had gotten along well with the one man most TWA insiders had come to dislike: Noah Dietrich. Although Dietrich was still on the TWA board, he had had a falling out with Hughes in early 1957 over compensation and resigned. With the dawning of the New Year, Warren Pierson became the acting president yet again, and as usual, there would be a long delay until a new CEO could be recruited. While Pierson was trying to recruit a new president, Lufthansa was expanding its routes which took Blackie into a few familiar airports, Lisbon, Madrid, Rome, and Cairo along with a new one, Cologne. Otherwise he was almost equally dividing his time between flying the North Atlantic, South Atlantic, and eastern Mediterranean routes until he left Lufthansa in late 1959. There is no evidence that Blackie ever looked up former TWA employees during his Lufthansa layovers in South America, but had he had any free days in Buenos Aires, he could have found a former TWA CEO – Carter Burgess had become ambassador to Argentina not long after resigning from TWA.

On the morning of 10 July 1958, Blackie took off from Lisbon bound for Hamburg with intermediate stops at Madrid and Frankfurt. On that day TWA announced its new President, Charles Thomas who had recently retired as Secretary of the Navy. Unlike Burgess, he brought one hallmark back to the presidency of TWA; although not an active pilot in the Jack Frye sense of the word, he had been a Navy pilot during World War I. More importantly, he was known as a corporate turn-around expert which is something Blackie could relate to since he too had taken a failing small business at the height of the Great Depression and turned it into a profit-maker. Blackie, as a perfectionist, was appalled at TWA's recent financial performance – the loss for the first half of 1958 exceeded $12,000,000. Thomas immediately put his finger on one of the major problems which was not sagging sales. Quite to the contrary, TWA was adding new domestic routes and filling its seats; the problem was mismanagement and costs spiraling out of control. Almost immediately, Thomas brought in management consultants and launched a survey of lower level managers to get a better feel of what could be done to improve efficiency. Even before Thomas arrived, TWA had a substantial awards program for cost-saving suggestions, but Blackie never won a dime. On the other hand, most of Blackie's astute observations related more to flying technique and airport facilities, neither of which could make an immediate, substantial difference in the bottom line. In the personnel changes that followed, several of Blackie's friends came out ahead, or at least their heads did not roll. Frank Busch remained at the head of operations and two of Blackie's old mentors were promoted into positions reporting to Busch. Joe Bartles became the western regional director and Larry Trimble was placed in charge of

European-Asian operations. Neither of those moves had any impact on Blackie at Lufthansa which continued to expand beyond Cairo to the east and from the east coast of South America to the Pacific coast.

Thomas had enjoyed little interference from Howard Hughes in the latter half of 1958 when he managed to turn a $12,000,000 first half deficit into a loss of roughly $700,000 for the full year. That performance is remarkable considering the fact that America was still in a mild recession that year and that there was an interruption of revenue when the machinists' union went out on strike and shut the domestic division down for two weeks at the end of the year. As the year 1958 closed, Blackie was fifty-seven years old, and in many ways on the periphery of what was going on at TWA. Most of TWA's energies then were focused putting the Boeing 707 into service and having both ground and flight personnel prepared for that new era.

In early January 1959 Blackie started flying with one of Lufthansa's more notable pilots – another night fighter ace: Karl-Friederich Muller. Many of Blackie's Lufthansa pilots were around twenty years younger than he, but Muller was only about ten years younger and had several distinctions in addition to downing nearly thirty RAF heavy bombers during the war. He was that rarity, a single engine night fighter pilot, belonging to the *Luftwaffe*'s *Wilde Sau* (Wild Boar) unit which sought out the RAF's bombers at night without the use of airborne radar. Muller flew occasional round trips to New York from Hamburg with Blackie who learned that his new acquaintance was among the few post-war Lufthansa pilots who had also flown for the pre-war Lufthansa. Considering all of Muller's instrument experience with Lufthansa, the *Luftwaffe* promptly recruited him in 1940 to become a blind-flying instructor after his initial service as a transport pilot. That instrument experience made him attractive to the JG 300 *Wilde Sau* group which is where Muller finished the war, although that is not what he was best known for among his Lufthansa peers. Because he had an extraordinarily large nose, he was better known as "*Nasen*" (Nose) Muller.

After the first 707 arrived in Kansas City on 1 February 1959 for training purposes, TWA put the 707 into service on 20 and 21 March by launching roundtrips between San Francisco and New York. Blackie, engaged in spring plowing at that point, may very well have seen both flights, or at least their vapor trails, in the sky north of Hidden Brook Farm.

Alfred Arzinger was in the left seat of a Super G Connie on 3 February when he and Blackie landed in Hamburg after a flight from South America. Arzinger had been a squadron commander and navigation officer in the Brittany-based KG 40 before that unit was basically shot to pieces by D-Day. He was lucky to survive and later became a JU-88 captain for the balance of the war. On the following morning, the newspapers in Tucson, Arizona, carried the details of a fatal broadside car accident which killed a local business executive. It was Jack Frye, who had left GAF in 1955 to start Frye Aircraft for the purpose of developing short takeoff and landing passenger aircraft. He had only recently written to Blackie announcing plans to take his company public and suggested that Blackie might buy a few shares. Blackie did not; but he joined many TWA old-timers in mourning Frye's sudden passing.

On 24 April, Blackie was back in Kansas City for some training on the Jetstream Connie flight simulator, but it proved to be a bittersweet four and one half hours in the machine. The Jetstream was no longer the queen bee in TWA's fleet. Just doors away, was a brand new Boeing 707 simulator which had been installed by Link Aviation. If Blackie felt any satisfaction in seeing the new 707 simulator, it may have been in knowing that another very good friend, Otis Bryan, was back from his brief tour at Philippine Airlines and working in Kansas City as a brewery manager, but would soon be joining Link as the international sales manager.

On 27 April, Blackie suffered one of his few inflight mechanical problems while at Lufthansa. He was taking off from Dakar with Arthur Schroeder and Josef Kreutzman in a 1049 Connie when they lost number the two engine, had to dump fuel, and return to the airport where they remained for the next five days while repairs were made. His German companions were both ex-twin engine fighter pilots; Kreutzman had flown JU-88s at night and Schroeder was quite literally a fugitive from the law of averages. He had flown the Messerschmidt Me-410 twin-engine day fighter, a type which almost always suffered fatally at the hands of Allied fighter escorts.

On 30 April President Charles Thomas rolled out his new organization in which he reduced the number of VPs reporting to the CEO and also reshuffled the executive ranks. Only one former pilot came out a winner; the up and coming Floyd Hall became the executive vice president in charge of flight operations. Like Blackie, he was a Kelly Field grad, but had joined TWA as a co-pilot in 1940. To Blackie's dismay, his friend Frank Busch was moved laterally to the "very important" job of vice president in charge of equipment retirement, i.e., finding new homes for old Connies. Hall reported to the new system general manager, E.O. Cocke, who had been the sales chief previously.

On 11 June, while Blackie was flying from Cairo back to Germany with the ill-starred Christian Krack, an event occurred in New York that Blackie would come to appreciate beginning in early 1960. It was the groundbreaking for TWA's spectacular new terminal at Idlewild. Whereas New York's airport was becoming more grand and certainly had the runways to accommodate the new jets, Chicago did not. Blackie flew several flights into Chicago's Midway Airport with the Germans in July, but after that, his remaining days with Lufthansa were all spent on South American flights. This was a period when TWA was riding high, announcing new cities on its domestic jet schedules and a profitable first half of 1959.

With Hughes digging a personally deeper financial hole for himself, Blackie's days at Lufthansa came to an end in early January 1960, and TWA's involvement terminated later that year. He went for a three-hour instrument check in a Jetstream out of Idlewild on 6 January and then flew his last Lufthansa flight to Hamburg. He returned to New York on 9 January with a familiar face in the left seat, Alfred Arzinger, and a new one, Co-pilot Martin Klein, yet another former JU-88 pilot. On the twenty-first he captained TWA flight 900 to France while also undergoing a line check. Blackie always looked back on his Lufthansa days with satisfaction and enjoyed the company of the German airmen with only one exception. That was on those occasions, typically following long flights to the Middle East or South America, when Blackie would join the Germans after hours for a drink or two. Blackie was well-known for being able to nurse a drink practically forever, unlike the Germans. As the evenings in Cairo or Montevideo wore on, the Germans would break into song and out would come all of the old wartime *Luftwaffe* drinking songs. Some of the songs' lyrics would appeal to any pilot regardless of nationality because they were about the sky and the joy of flying. But as the alcohol loosened the tongues of those former *Luftwaffe* flyers, songs with a more political or military content, such as *Bomben aufen Engeland*, would emerge – songs that were not tolerated in Germany or elsewhere in Europe. That is when *Herr Kommandant* Blackburn would bid adieu to his pilots and retire. In the course of the three years, Blackie had coached nearly 180 German airmen. They would not forget it, and in the not so distant future, Blackie and the other TWA pilots on loan to Lufthansa would receive formal recognition of their efforts in the late 1950s before Lufthansa was able to fly solo again.

After a two-week vacation, Blackie returned to the International Division flying the same schedules he had before the Lufthansa assignment with one exception, now he was regularly flying the ultimate Connie, the 1649 Jetstream. When he left the International Division in 1956 and again when he left Lufthansa, his title had been supervisor of flying, but when returning to Idlewild in 1959, there was no vacancy for the same position. So Blackie returned to his New York domicile as an ordinary, able-bodied line pilot for the time being. On the other hand, in bidding for preferred schedules, Blackie's chances of getting what he wanted were excellent because there were so few names ahead of him on the seniority list.

After being back for six weeks, Blackie ferried a Jetstream Connie from New York to the maintenance base at Kansas City, and brought another Jetstream back to New York on 26 March. Upon returning to Idlewild, he ran into Marv Horstman who was one of TWA's first Boeing 707 pilots. Horstman, who had flown the first 707 revenue flight to Paris and Rome in early December 1959, thought Blackie ought to consider bidding for jets, although Blackie was rather well-resigned to finishing what was left of his career on Connies. Blackie had not been to London in a while, but knew that TWA's first international jet flights were from New York to London and had been launched in late November. Horstman took Blackie out onto the hangar floor for a thorough inspection of a 707-331 Intercontinental which, unexpectedly, completely bowled Blackie over. It was an absolutely bedazzling experience. He promptly asked Horstman when the next 707 pilot's class was starting in Kansas City and was told that it commenced on 3 April. Contrary to earlier rumors that only pilots under age fifty-five would be eligible for jet training, Blackie was immediately accepted (no surprise considering his seniority). Blackie was off, running, and would graduate number one in his 707 class of which he was the oldest member, finishing with a score of ninety-five. Unlike his casual checkouts on every other big step up, i.e., the B-24, the Boeing 307, the C-54s, and several models of Connies, Blackie was now headed for over a month of intensive schooling because the 707 was more than just a bigger airplane; it was new technology and a new era. In many ways, learning the 707 was much like medical school: endless memorization and frequent tests.

Blackie had read about some of these new phenomena associated with the jets, including the meaning of Mach and how jet aircraft perform near, at, and beyond the speed of sound. His chances of exceeding Mach One, the speed of sound, were limited, but still he had to learn about "Mach tuck" or the tendency of aircraft to behave differently around the magic number, especially the tuck or tendency to pitch nose down. Another new concept was "Dutch roll" as associated with swept back wings in which the aircraft tends to wallow, i.e, when the nose turns one way, the wing dips but only to a certain point at which the process repeats itself in the other direction. Boeing engineers had, however, largely eliminated the problem through the use of sensors and a yaw damper which became part of the rudder. The tail was different; the entire horizontal stabilizer moved!

Blackie and his fellow classmates had been used to flying Connies at altitudes lower than the book said they could be flown simply because that invited engine problems when the Wright 3350s were operated at altitude in high blower. Because the jets absolutely performed best at high altitude, Blackie had to learn what happened in the event of a sudden decompression as was the case when the hapless TWA navigator was sucked out of his bubble over the North Atlantic years before. Blackie had to learn how to walk the line between getting down to around 10,000 feet as fast as possible from as much as 40,000 feet without overstressing the 707. There were new engine instruments to learn with strange acronyms, new subsystems such as spoilers, leading edge slats, and thrust reversers. One of Blackie's classmates described the cockpit in terms of a music rock concert because the alarm system consisted of series of mutually unfriendly horns, buzzers, klaxons, and chimes not to mention the inevitable bells and whistles. There was not just one autopilot, there was also a backup as there was for the flight director. New 707 pilots had to stop thinking about manifold pressure and revolutions and start thinking in terms of engine pressure ratios, turbine shaft speeds, exhaust temperature, and measuring fuel in terms of pounds, not gallons for their J57 turbo-jet engines. Horsepower was out, and pounds of thrust were in. Although primitive by today's standards, Blackie logged many hours in the Boeing 707 simulator of which many of the scenarios emphasized emergency management.

Blackie got fifteen hours of Boeing 707 flight time in school and completed the course with an FAA examiner who gave him his Boeing 707-331 type certificate. That came only after nearly two and a half hours of being put through the hoops with landing on three engines, instrument approaches, recovery from Dutch roll and Mach tuck, handling emergency decompressions, and a landing with no flaps. If there was one aspect of the 707 he enjoyed most, he always said it was the power; he loved the way the plane would fly just about as well on three engines as it did on four. To the average airport onlooker, the 707-131 (the domestic version 707) and the 707-331 looked about the same, but there were significant differences. Although both had wingspans wider than the model 1649 Connie, the 331 had a wingspan twelve feet greater than the 131 and with good reason. The 131's maximum takeoff weight was 247,000 pounds compared to the 331's 302,000 pounds. The 331 could carry much more fuel which is why it could fly non-stop across the Atlantic in either direction. The 331 was powered with Pratt and Whitney's JT-4 turbo-jet engine – the civil version of its J-75 engine which powered the Air Force's F-105 fighter. Unfortunately for Blackie and the many other operators of the 707, the JT-4s were so very loud that they were causing problems for which there were no easy, across-the-board solutions.

Trans World Airlines

Blackie at the controls of a Boeing 707-331 Intercontinental.

Still, Blackie was not qualified to fly the 707 on revenue service until he had flown it on a route familiarization flight and completed both line and instrument checks with a TWA check pilot. Those flight checks put forty-three hours with check captains Charles Swartzell and Neal Lytle into Blackie's logbook. Swartzell had flown the first revenue flight to Frankfurt via London and Lytle had recently become a check pilot at Ildlewild. Lytle was also known within TWA as something of an adventurer due to his taking a sailboat all the way from the Mediterranean up to the source of the Nile, a 2,600 mile journey. He had used his previous domicile in Cairo to become a world class salt water fisherman and SCUBA diver as well. Blackie's line check flight, flown with Lytle out of Idlewild for Paris on Flight 800, produced some excitement; they lost an engine on takeoff, thus providing Blackie with some early training on emergency management. Attending the jet transition was another reality; to get in the required number of hours each month meant more flights since the typical flight from Idlewild to Frankfurt in a 707 required on average about two-thirds the time a Connie did. For Blackie, that meant less time at home with his family and farm projects than he had previously enjoyed at Lufthansa. Nevertheless, flying the 707 on flights to Paris-Rome or London-Frankfurt was very satisfying for Blackie – he was content with his flying and also with life on the farm.

The farm had plenty of space – which had become a necessity, considering the way the grandchildren were coming along. Betty was soon going back to college to complete her degree, and was the mother of two children, seven-year old Karl and two-year old Rebecca, who were frequent visitors. Bob, Barbara, and their son Chris were also visitors to the farm, and Bonnie was home for the summer after completing her freshman year at Antioch College in Ohio. Be, the only daughter at home year around, had just finished the eighth grade.

On Saturday, 2 July, Blackie attended a banquet in New York hosted by Lufthansa in honor of the thirty-five TWA pilots, flight engineers, and ground staff who had helped Germany bring back its once-proud airline. TWA's five-year contract with Lufthansa had expired in May, and Lufthansa was now self-sufficient. The record compiled by the TWA coaches was impressive: 3,000 North Atlantic crossings, 600 across the South Atlantic, 350 to the Middle and Far East, and over 2,200 hours logged in instrument and line checks. When the flying hours were totaled, Blackie learned that the TWA aircrews had flown a sum of over 77,000 hours in the cockpits of Lufthansa's Connies. In the process, forty-seven new Lufthansa international captains had been created. Heading the Lufthansa delegation was the company's vice president, Gerhard Hoeltje, a leader who had come up through the technical and engineering ranks. Hoeltje praised the efforts of the TWA contingent as did Lufthansa's chief pilot, Rudy Mayr, another KG 40 survivor who had gone from flying FW-200s over the Bay of Biscay to JU-88s as a night fighter pilot with NJG 1 on homeland defense. Mayr, who was originally a captain on Lufthansa seaplanes and a veteran of Germany's Greenland expeditions in the 1930s, had been the first pilot to whom Blackie had given an instrument check after arriving in Hamburg in late 1956. Lufthansa had come a long way in the three years Blackie had flown with the line; in those three years Lufthansa's fleet grew from fifteen to thirty-two aircraft, the total number of flights tripled to 17,714 at the end of 1959, and the passenger count went from 229,670 to 786,626. The Lufthansa years had been a significant period in Blackie's life as a pilot – they put over 3,000 hours in his logbook or almost fifteen percent of his total hours at the time he returned to TWA in early January 1960. That was just as the books were closing on 1959 which appeared to be a very good year for TWA.

But all was not actually that well, and the growing unpleasantness was related to Howard Hughes' meddling in aircraft purchases for TWA. This new scenario, reminiscent of the Constellation acquisition days of the 1940s, would play out during Blackie's first year back at TWA. There was one major difference in this new storm in which Hughes had become convinced that TWA needed short-haul jets for shorter domestic routes and more 707s for the longest routes. Noah Dietrich was no longer available as a safety net.

Knowing that TWA needed a speedy jet, or at least a turbo-prop aircraft, to replace its aging Martin 404s and supplement the Boeing 707 fleet, Hughes took it upon himself to look at the field which at the time included the Bristol Brittania, the Lockheed Electra, and even the French Caravelle. Within the Hughes mindset, speed was foremost and with that necessity he was looking for a sort of slenderized, scaled down, souped-up 707 although such a bird did not exist in 1958. Lockheed was doing well with its Electra, but that was not a true jet and Douglas with its DC-8 and Boeing with its 707 had not publicly announced plans for little brothers to their big four-engine jets. The closed doors in Los Angeles and Seattle drove Hughes into the waiting arms of Convair, the old Consolidated Aircraft that had become a part of the giant General Dynamics combine which was anxious to get into the commercial jet market. Its entry, the Convair 880, looked very good on paper, but knowledgeable TWA'ers doubted that it had the endurance to make it in the commercial market because of its General Electric J79 engines. The 880 had an eloquent spokesman at Convair – Blackie's friend Bob Loomis who was on his way to becoming the company's CEO. Nevertheless, the 880's engines were in military use on such speed demons as the B-58 Hustler and F-102 Delta Dagger, but would not stand up to airline use, i.e., flying all day every day, according to TWA's top operations and maintenance authorities. Even so, Hughes went ahead and ordered thirty of them while he was already begging Boeing for payment extensions on the thirty-three 707s.

Because Hughes could not make payments, Boeing withheld shipments of new 707s after the first two were delivered. So for a while, the 707 that made the inaugural flight between San Francisco and New York was the only operational 707 in the fleet (the other was used for training in Kansas City). Hughes tried to create a company called Hughes Aircraft Leasing which would buy the 707s and 880s for subsequent lease to TWA, but that entity never flew. Charles Thomas objected vehemently, but more importantly so did a consortium of potential investment bankers: Dillon-Read, Lehman Brothers, and Lazard Freres. They had learned that the usual source of Hughes funding, Hughes Toolco, was suffering a downturn in its oil patch business. Another plan, to raise money in a stock offering to existing stockholders, was not greeted with any enthusiasm and had to be scrubbed. One reason for Hughes' desperation was that TWA's original lender, Equitable Life, said that it would give Hughes an extension on an existing $12,000,000 loan only if he arranged satisfactory financing for the $405,000,000 needed to finance the purchase of all of the new 707s and 880s. Through a combination of personal loans and selling delivery rights for six 707s to Pan Am, Hughes was able to borrow enough to temporarily keep his creditors at bay. TWA was doing quite well in terms of reducing costs under the direction of Charles Thomas, and the 707 was making money hand over fist. Hughes had staved off disaster, but kept on personally borrowing money under his own name, particularly from the California-based Bank of America, in order to pay for TWA's new 707s. The amount he borrowed would not begin to pay for all of the new planes that TWA had ordered. It simply delayed the inevitable financial showdown with the investment bankers who could put together enough money to buy all of the 707s and 880s for which TWA was obligated to buy.

By early March, the financial impasse reached the crisis point so Thomas called a board meeting. He leveled with the directors who were told that unless TWA could open up a financial pipeline to $260 million, the company would go under – very soon, especially since Hughes' banks had cut him off from any additional personal funding. The potential lenders, the Dillon, Read consortium, threw in another requirement to their potential loan: a ten year trust that would vote the shares owned by Hughes (about seventy-eight percent of the company's ownership) in the event that a TWA president resigned and was not replaced with a new president of whom the lenders approved within ninety days. The investment bankers gave Hughes until the end of July to come to terms, and Thomas advised Hughes that TWA would need a new president if the terms were not accepted.

On the evening of 28 July, Blackie took off from Idlewild for London and Frankfurt on what proved to be an uneventful eight-hour flight with the only instrument time occurring on the approach to London. Just a few hours before his takeoff, TWA announced what had happened in a turbulence-filled board-room just the day before. President Charles Thomas had resigned; it was yet another case of a CEO not tolerating being undercut by Howard Hughes who by then knew that Thomas' departure would hurt not only TWA, but himself as well. Because lenders always look to management stability when evaluating loans, Thomas' departure was a black mark considering the longevity of CEOs at Pan Am, United, American, and Eastern. The consortium of investment bankers and insurance companies had reached the point where they were about to directly influence both the management structure of TWA and create the voting trust, effectively neutralizing Hughes' control of the company. They informed Hughes that unless he promptly surrendered his stock to the voting trust, all deals were off. Hughes managed to secure a grace period and kept the lenders off guard by submitting one new funding proposal after another. The tactics worked for a while until 25 October when TWA's and the Hughes Tool Company's accounts were seized by their existing banks for loan default. Now desperate, Hughes tried courting private venture capitalists and proposed a public offering of TWA debentures. The drama played out until the snow started falling and was under the radar screens of Blackie and the majority of TWA's 20,000 employees whose interests focused on pride in their work and the assurance of sustained paychecks.

Blackie rarely discussed his personal view of TWA's fortunes with anybody other than his family and a few close, pilot friends. While he still loved flying, in his personal notes from the summer of 1960, he reveals his pessimism concerning TWA's future, and to a lesser extent, that of ALPA. By his count, there had been eight presidents and acting presidents at TWA since the end of World War II, and the company, in spite of Gordon Gilmore's expertise, was regularly getting beaten up in the business media and even such broad-based news weeklies as *Time Magazine*. Blackie likened TWA's situation to that of Capital Airlines, the eastern regional airline with hubs in Pittsburgh and Washington. Capital was reeling financially and had suffered mightily from a maintenance union walkout and also from a lawsuit over minority hiring. Vickers Aviation was about to foreclose on Capital's fleet of turbo-prop Viscounts thus ensuring bankruptcy. United stepped in and acquired Capital which led Blackie to wonder if he might be wearing Pan Am wings before the year was out. ALPA, as far as Blackie was concerned, was also suffering poor press, was doing an equally poor job of communicating with its own membership and appeared to have hardening of its arteries by failure to put talented, younger pilots in leadership positions.

By contrast to what was then circulating in corporate offices, spring and summer were pleasant times of the year in Pennsylvania Dutch country, and for farmers always very critical seasons. By the summer of 1960, Blackie had become agriculturally capable, was absorbing even more from the neighbors, and was turning Hidden Brook into a model, small beef farm. There had been very few bumps in the road, but one, for example, was the chickens that had not worked out. The small herd of Black Angus certainly had proved successful as had the sheep that helped keep the grass short. Blackie had had the good sense to not keep labor-intensive cows. The steers occasionally drank from Hidden Brook but much more frequently used a smaller creek just north of the pond or drank from the pond itself, largely ignoring a water trough on the far side of the pasture. There were nearly twenty barn cats of which most were hardly tame. One was favored enough to have a name; that was Bonnie's cat named "Piewacket" after the feline star in the 1958 movie, *Bell, Book, and Candle* with Jimmy Stewart and Kim Novak. The Blackburns always had a dog, usually named "Hey Boy"" after the Chinese bellhop in the then popular TV series, *Have Gun, Will Travel*. Then there was "Lucky" the horse which was used by Blackie and his daughters for riding, although the horse's size suggested that his recent ancestors may have been pulling hay wagons or other farm equipment.

Bob's work in Philadelphia changed that summer as he took on a new job, one that would really put him into a distinguished career track at the junction of civil rights and education. Eventually, it would also nearly cost him his life, but that was unimagined in 1960 when he was hired as executive director of the Citizens' Committee on Public Education (CCPE) in Philadelphia.

During the last quarter of 1960 Blackie was averaging about seventy hours a month, taking Flight 700 to Frankfurt and returning with Flight 701 to Idlewild, with the only exception being the occasional return via Boston. There were apprehensive and disappointed passengers on the 6 November flight to

Frankfurt when about an hour and fifteen minutes into the flight Blackie got a fire warning in one of the engines. He promptly dumped fuel and returned to Idlewild to have the problem checked out. The malfunction was not a major one, and the flight took off a second time that day for Germany. Toward the end of December word had spread among TWA's employees that the financial crisis had deepened with Hughes' refusal to buckle under to the condition of placing his TWA shares in a voting trust. Failure to do so would quickly place the company in bankruptcy which scared every employee from pilots to secretaries alike. That was in spite of the company's topline revenue which was actually quite strong thanks to new domestic routes and the existing 707s which were proving to be real cash cows.

To add to the corporate anxiety, on 16 December there was another horrific mid-air collision involving TWA and United – this time with a greater loss of life than the 1956 encounter between TWA and United planes over the Grand Canyon. Blackie was at home on Long Island when a United DC-8 bound for Idlewild exceeded its clearance limits, was therefore technically off course, and hit a TWA Connie in-bound to La Guardia over Staten Island. The Connie broke into three sections and plummeted directly to earth while the United DC-8 remained airborne briefly before going out of control and crashing in Brooklyn. Dead were 128 from the two airplanes and six people on the ground. Blackie was only slightly acquainted with the TWA crew, but had never flown with them since they were from TWA's domestic division.

Dillon, Read and the other lending institutions had given Hughes a final deadline of Friday 29 December 1960 to comply with their terms and had already named the potential members of the three-man voting trust. One might have been considered an insider: Raymond Holliday from the Hughes Tool Company. The other two had impeccable credentials, beginning with Ernest R. Breech a former TWA director from the 1930s who had gone on to the chairmanship of Bendix before moving to Ford Motor Company where had recently retired as chairman. The final member was a lawyer, Irving S. Olds, who had been chairman of United States Steel. At the last possible moment in the evening of the twenty-ninth, several hours after Blackie had flown in from Frankfurt and Boston, Hughes acquiesced. He gave permission to set up the trust and went along with several other distasteful requirements of which one was that the trust would be able to nominate new company directors. Warren Pierson was still chairman and acting president, but his days were numbered. On 20 March 1961 a new board of directors was elected, and they promptly went to work naming Breech as the new chairman and a virtual unknown as the new president. The new CEO was Charles Tillinghast who, although he was best known as a shrewd lawyer and a star collegiate football player, was the vice president of international operations at Bendix. Blackie was on his way back from London to New York on the day of that board meeting and very much aware that he was almost exactly nine months away from forced retirement as a jet captain. In those remaining months his flight schedules, with one exception, became dominated by flights to Paris and Rome instead of London and Germany. Because there was a major last minute crew availability issue, Blackie did fly one Jetstream flight to Cairo and back. On 6 October, a memo was waiting in his crew mail box at Idlewild when he got back from Kansas City and eight hours in the Boeing 707-331 simulator. The memo was confirmation of a discussion he had nearly a week earlier with Charles Swartzell who had become superintendent of flying for International. Blackie was promoted again to check captain.

Blackie was frequently in touch with Swartzell during 1960 and 1961 on a number of issues which could best be described as routine aggravations that come with the territory for airline captains and supervisors of flying alike. Some of the exchanges with Swartzell related to the occasional passenger who got sick on a flight and blamed the filet mignon dinner for making him sick, conveniently ignoring the fact that he was the only one among the passengers and crew who had been sickened. Other issues were related to late departures, mistakes in seat assignments, and delays caused by commissary personnel blocking aisles through galleys. Whenever a passenger complained, it unleashed a flurry of paperwork and in the end the matter all boiled down to what had been foremost in TWA's training for decades: excellence in service from the top to the bottom.

Other exchanges between Blackie and management were related to improvements in the 707's passenger and crew communications. As a result of Blackie's suggestions, improvements were eventually made in cabin signage for seatbelt and smoking restriction instructions. Stemming from another of his suggestions, a newer means of contacting the rear cabin crew was devised. Previously, the bell system might easily bring a hostess from the forward section when the captain specifically wanted a hostess from the rear section. From the flight deck crew's perspectives, Blackie's lobbying for changes in TWA's policy on alternate airports and noise abatement procedures hit closer to home. Back in the early Constellation days, Blackie had argued for changes in the instrument approach procedures for specific runways at both London and Paris. Ultimately, responsibility for implementing the improvements was that of the respective British and French aviation authorities, but eventually Blackie's suggestions did become reality. In another case, he found a flaw in TWA's route policies for the Model 1649 Constellation. Alternate airports were listed for use in TWA's operations manuals for specific TWA aircraft types and on 11 February 1960, Blackie encountered a situation that was not in the books. He had departed Boston in a Jetstream bound for Shannon which he found well below instrument minimums. He was then cleared to the alternate, Glasgow, which also went below minimums while he was en route. Both London and Paris were down and expected to remain so for hours. After holding at Glasgow, he was cleared back to Shannon where the visibility and ceiling appeared to be lifting, but then they went down again. Exasperated, Blackie requested and got a company clearance to land at Dublin which was not in the list of alternates approved for the Jetstream although it was for the model 1049 Connie. Dublin was very shortly thereafter approved for the 1649.

The 707 ushered in numerous benefits for the airlines and their passengers, but many communities complained about the noise. The noise issues would eventually be minimized by hush kits for the 707's engines and later by newer technology engines. The early 1960s saw only the initial means for reducing the sound issues – noise abatement procedures where air traffic patterns over sensitive residential areas were modified. It was an imperfect world at both London Heathrow and New York's Idlewild where the Air Ministry and the New York Port Authority had established procedures which allowed full power for takeoffs, but required throttling back once safely aloft while over heavily populated areas. Once well out over water or over less populated areas, pilots added power to resume the climb to altitude. The restrictions were also applicable to landing aircraft, but complying with the new requirements was much simpler. There were fines for failure to comply, but that was just the sore point. There were numerous variables involved including gross takeoff weight and atmospherical conditions – all of which were not part of the data acquisition process. What did count was the measured decibel level taken from a location at a specified point beyond the departure end of the runway. Blackie had on several occasions exceeded the permitted decibel level at both Heathrow and JFK, but knew he had been in compliance with the power settings and altitude requirements. He also made sure Marv Horstman and Neal Lytle were aware of one instance when both of them were flying together and had exceeded the permitted decibel level at New York. And they were among those people establishing the sound abatement procedures in the first place. From a pilot's viewpoint, the sound abatement procedures bordered on eroding flying safety, i.e., not allowing a swift climb to altitude at a time when some more altitude would be a helpful asset in case of engine problems. Much of the issue was variable and based on weather conditions; it was known that high humidity could make the difference between being in compliance and being fined. It was also demonstrated that being above low cloud could reduce the decibel level. Knowing that the sound abatement procedures were coming up for review in late 1961, Blackie became more vocal in his suggestions to management that the restrictions be eased. He also urged management to look at high approaches on landing as a means of maintaining reduced power settings, but in the end all of lobbying came to naught. Those on the ground had the political power to maintain the status quo.

In early November while between flights, Blackie received a phone call from a perfect stranger, a man who introduced himself as Robert Drew. The caller quickly mentioned that he had heard of John Bainbridge's outlining a new profile of Blackie which would appear in *New Yorker* and wondered if Blackie and TWA would be interested in a documentary film about Blackie's closing out his career as a TWA international jet captain. While New York City was and still is the world's epicenter for publishing, advertising, public relations, and print/broadcast journalism, its personal networks usually mean that members of one medium are aware of current projects in another medium. Drew was delighted to learn that Blackie was still flying for TWA, albeit there was not much time left to film the story since Blackie's days as a TWA pilot were to end within weeks. Blackie's story was appealing two Drew on several levels, particularly because Drew was a leader in the cinema verité movement. That form of documentary film sought to convey important events as they unfold, and as far as Drew was concerned the core of his films required a protagonist and an antagonist, i.e., an impending crisis. Blackie, Drew explained, was the protagonist and the Federal Aviation Administration with its mandatory retirement rule, was the antagonist. Another element of cinema verité was its lack of intrusiveness. That was an area trailblazed by Drew and his development of much lighter video and sound recording equipment that permitted a more intimate film unencumbered by bulky cameras, tripods, booms, cables, and the usual impedimenta associated with most video production of that era. In many ways, Drew's films were the real life versions of a popular televisions series of the day, *You Are There*.

Blackie had never heard of Robert Drew, and of the millions of people who saw Drew documentaries, few could remember his name either. Remembering Drew's work was another matter. Many of his documentaries were award winners. Among his best known works by the time he contacted Blackie were two from 1960: *On the Pole* (about the 1960 Indianapolis 500 auto race) and *Primary* (the Wisconsin presidential primary race between John F. Kennedy and Hubert Humphrey). Oddly, for a man with so much aviation in his background, Drew's only foray into aviation documentaries had been 1961's *X-Pilot* which was about test pilot Scott Crossfield's flights with the X-15 rocket plane. That film was never considered one of Drew's finest due largely to Crossfield's being a man of few words. With Blackie as his subject, an inarticulate central character was hardly a concern.

As the two men talked, Drew's connection with aviation surfaced as did the fact that his father had been killed in a B-24 crash at Gander in 1944 which Blackie vaguely remembered. Following his 1944 passage through the Italian front lines after having been shot down over Fondi, the younger Drew returned to the United States and with his assignment to the First Fighter Group became one of America's early jet fighter pilots. He was asked to contribute to a *Life Magazine* article on what it was really like to fly the hot, new jet fighters which led to his becoming more interested in journalism. While refining the article with editors in New York, Drew so impressed the magazine's editorial management that he was offered a job as a *Life* reporter in Los Angeles. Fast forward fifteen years and Drew had moved to Chicago with *Life*, spent a year studying at Harvard University, moved to New York to join Time-Life Films, and then had become an independent producer. He had, however, retained a close relationship with Time-Life Broadcast which would be the co-producer of a documentary about Captain Blackburn. The project was cleared through TWA Public Relations V.P. Gordon Gilmore and the legal department. ALPA whole-heartedly supported the project considering the film's core issue, the injustice of the government's forced retirement of a man who by all rights could have flown for another five years at least if not ten. The irony of ALPA's support was not lost on Blackie since it was the union that made his returning to the line so difficult in 1948.

Chapter Twelve
Heading West

The second Bainbridge profile of Blackie was launched before the Drew Associates project was initiated and was based on Blackie's taking Flight 800 to Rome on 18 October 1961 and a return to Idlewild on the twenty-first. Structurally, the article, which did not appear until nearly a year later (10 November 1962), was identical to the 1956 version, i.e., a writer meets a pilot at home, motors to the airport, and then enjoys the modern marvel of a fast and safe flight to Rome. Since the last article on Blackie which appeared in 1956, Bainbridge had written what was his bestseller, *The Super Americans*, the analysis of the Texan way of life which was published earlier in 1961. The book followed what was a very successful formula for Bainbridge: taking several articles on the same subject and re-casting them as a book. It certainly had worked for the publisher, Houghton-Mifflin, and unknown to Bainbridge and Blackie at the time, the same formula was going to be revived someday after *The New Yorker* published its second profile of Captain Blackburn.

The motoring to Idlewild with a writer in tow in 1961 was much longer since Blackie was driving from the Pennsylvania Dutch Country, and it was in a "new car," Blackie's 1950 Ford Sedan with 98,000 miles on the odometer. Bainbridge is amazed at how the car runs like a top, but he should not have, considering Blackie's wizardry as a mechanic. In the opening, Bainbridge marvels at how his pilot friend has hardly aged – just a little more gray hair and a couple new wrinkles in the crows' feet which merely enhances Blackie's airliner skipper appearance. The writer is pleasantly surprised by Blackie's renewed enthusiasm for transatlantic flying based on the differences between the old Connie and the new Boeing 707-331 Intercontinental jet. As Blackie looks back on twenty-five years with TWA, the Army Air Corps, and his private flying in California, he shares the observation that he quite literally grew up with aviation. He had been born two years before the Wright Brothers first flew, had flown a Curtiss Jenny, the fighters and bombers of the inter-war period, Lockheed's Lightning and Constellation, and now the grandest ship of them all: the Boeing 707. Blackie's enthusiasm for jets becomes contagious as he talks about the 707's speed, its climbing ability, and the inherent simplicity of its jet engines which create a new era in reliability and safety.

The ride from the rolling hills of Berks County past Manhattan to Idlewild is measured in hours as opposed to minutes when Bainbridge had first written about Blackie. In this longer journey, Bainbridge learns more about his traveling companion who gave up the executive life because he "didn't want to be strapped to a briefcase any longer. I decided to do what I like most – flying an airplane." The new reality is that Blackie had again become a check pilot which he always regarded as a mixed blessing since half of the pilots he checked resented being evaluated and the other half certainly did not need checking. The conversation drifted to what Blackie had been doing since Bainbridge last flew with him which allowed Blackie to explain his motivation for flying with Lufthansa for over three years. Blackie explained that he took the Lufthansa assignment because it represented change, new challenges, and flying to new destinations. It also involved flying many, many check flights, but they were very different from those he had flown with TWA. The Germans were eager to learn and enthusiastic throughout his assignment with Lufthansa. But that assignment came to an end as Lufthansa became self-sufficient and Blackie returned to his New York base, assuming he would finish out his years on Connies. He continued by telling Bainbridge about the transition from Connies to jets, and that the only previous time he felt such similar magic had occurred in 1919 when he sat in Ira Sloniger's Jenny.

Bainbridge follows Blackie through the usual pre-flight meetings and briefings and quickly learns how jets have impacted the flight crew, particularly in the cockpit where the radio operator and second flight engineer were no longer seen on flight decks. Modern radio technology eliminated the radio man, and the jet engines required much less instrumentation and therefore only half the manpower to monitor the gauges and move the controls. Bainbridge quickly noticed another difference; some of the other pilots getting weather briefings were wearing glasses, balding, and several were no longer at their college fighting weights. Blackie explained that the average transatlantic jet captain's age was fifty, compared to thirty-eight of the average airline pilot. The cabin crew numbers had increased to one purser and six hostesses or an increase of four compared to the Connies. That night's flight did have a full load: 140 souls due to so many of military passengers on their way to NATO or other European assignments. Blackie grumbled about another increase the paperwork, noting that the day had been reached when a pilot could be cleared for takeoff only when the weight of his paperwork equals the gross takeoff weight of his 707.

The weather briefing revealed a positive picture with Paris looking good for the following morning and Rome looking even better. A modest tailwind was promised after overflying Gander, but there would be a significant crosswind most of the way to Gander. Blackie quickly absorbed all of the numbers in the flight log that the dispatcher gave him and decided to file for an altitude of 33,000 feet. He requested a fuel load of 29,000 pounds based on looking at the projected route and the primary and secondary alternate fields. Bainbridge then got a basic lesson in airline economics; jets burn kerosene which then cost about ten cents a gallon compared to twenty-one cents for aviation gasoline. The jets guzzled the stuff – even while taxiing they gulped down more than a Connie did while cruising at altitude. They did not burn so much at altitude where the thin air also meant flying on a leaner mixture, not to mention the fact that reduced friction of less dense air essentially meant the same thing as less drag.

For much of the article, Blackie becomes positively radiant about the advantages of jets, especially observing that the Connies never truly could fly above the weather, but the 707s can all the time. Blackie noted that he had flown a 707 as high as 41,000 feet and that the plane was certificated for flight as high as 42,000. Yet he admitted that he had seen the very tops of storm clouds that high, but they were so isolated that it was very easy to fly around them. Another factor favoring jets was how winds have a decreasing impact on planes as their speed increases. The rest of the pre-flight routine and full crew briefing was much as Bainbridge remembered from his previous flight with Blackie, although it was obvious that the hostesses were less than euphoric about the ninety-one military passengers. Once aboard, Bainbridge was surprised how streamlined the cockpit seemed to be which was hardly noteworthy considering the smaller crew. The Boeing 707-331 Intercontinental instrumentation contained well over 100 fewer instruments and controls than a Connie. The pilots had just completed entering frequencies into the radios when the flight engineer showed up with some bad news; his pre-flight inspection revealed a defective fire extinguisher bottle in the number three engine, and because of that Flight 800 was not departing until it was fixed. Forty-five minutes were required for the fix. That was followed by another glitch. Blackie had to talk with the dispatcher whose final check of the 707's weight revealed that it was over the maximum allowed gross weight by 700 pounds. Blackie quickly dismissed the matter by opting for the quick solution – just burn it off before takeoff which was certainly more expeditious than de-fueling with one of the tanker trucks or offloading a few passengers and/or cargo. Fixing the fire bottle was not going well so Blackie continued to expound on the merits of jets. Had Gordon Gilmore, TWA's PR guru, been in the cockpit and remembering the several hundred thousand upscale *New Yorker* readers who would be hanging on Bainbridge's every word, he would have been overjoyed at this process of market development. The fact that the magazine's primary readership had increased seven percent since the last Blackie profile made the ink to flow all the more rewarding. But for Blackie, it was all very real because he was genuinely thrilled to be flying jets. "On takeoff, these engines are putting out the equivalent of better than sixty-seven thousand pounds of thrust. That's more than five times the power as we ever had before," Blackie noted. He told Bainbridge that he figured the jets were at least twice as reliable as the piston-engined Connies particularly because of the comparatively fewer moving parts. Because he wanted to accentuate the positive, Blackie did not get into other details about some of the 707's other systems. Had he chosen to go there, Bainbridge would have learned about Blackie's losing an engine after takeoff for his 707 certification flight or the occasion nearly a year before when he was taking off for London when he got an engine fire warning light, had to dump fuel, and return to Idlewild. Nor did Blackie mention the incident just a month before when he aborted a flight because a flap became unhinged and fell to the runway.

Then Blackie noted the importance of attitude flying compared to the Connie and that the 707 was very much a "fly by the book" airplane. He was quickly interrupted by the flight engineer who said the fix would be finished in ten minutes whereupon Blackie gave the order to start bringing the first class passengers aboard. Bainbridge now became an observer, not a participant in the terse conversations between pilot, first officer, and flight engineer as they started going through their pre-start checklists. While the second officer was reading off the checklist for the flight engineer, Blackie learned from ground control that they would be using Runway 13 Right, the longest at Idlewild at 14,511 feet. He quickly asked the second officer to look up the noise abatement procedure for that runway; "fourteen seconds" came the reply. By now the Boeing 707 was loaded and Blackie was performing a ritual that was a holdover from his Air Corps bomber days – putting on his leather gloves. He really did not need them of course, but it was just one of those little mannerisms or rituals that many pilots since time immemorial have performed, trusting that they will bring a good flight. The only other item that could approximate a Blackburn ritual or talisman was the wristwatch. When flying, just any watch would not do. It had to be the wristwatch set in a silver wristband with embedded turquoise, the one that told time in two dimensions. There was Zulu time, that is Greenwich Mean Time which was necessary for flying, and the other was "stomach time" or what the time was back home. The ritual and talisman brought little immediate luck as Blackie observed TWA Flight 700 pulling out ahead of him, thus almost certainly guaranteeing that his request for 33,000 feet would be denied. There were other Blackburn quirks that Bainbridge missed, although background noise may have obscured them. Blackie's radio conversations with the various controllers were always precise and terse. His transmissions over the company frequency were more conversational and were invariably preceded by few whistled notes which immediately followed his pressing the transmission button. As might be expected of a former band leader, Blackie invariably hummed music during a flight's quieter moments.

At 8:34 P.M., Flight 800 was holding short of 13 Right waiting for an Alitalia jet to clear the runway and for a BOAC jet which was on final approach. The collective mood on the flight deck was understated tension well under control as far as Bainbridge could tell. After a brief exchange with the tower and completing the "When Cleared For Takeoff" checklist, TWA's Boeing 707, aircraft number 761, came to takeoff power with the brakes applied. Blackie said "Time" to the first officer who punched the button on the instrument panel's stop clock to validate the aircraft's acceleration performance. Blackie then released the brakes and steered the 707 with a little hand tiller to maintain directional control until he could attain aerodynamic steering from the very large vertical stabilizer that was a very recognizable feature of Boeing subsonic jets. After hearing "eighty knots" Blackie again quickly scanned the instrument panel to ensure that all indicators were normal and at the same time nudged the throttles forward ever so slightly but still enough to raise the tone of the whining engines. With "Time Okay," followed several seconds later by "V-One," Blackie was committed to the takeoff. When the 707 reached 157 knots and Blackie heard "Rotate" he pulled the yoke back to achieve approximately eleven degrees pitch up from the end of the runway. Seconds later the big jet's nose wheel lifted off the runway. When hearing "V-Two" Blackie knew he had enough speed to maneuver – even if he suddenly lost one of the 707's engines. Once the 707 was up and climbing Blackie called "Up Gear" and pushed the nose down while simultaneously cutting back the power on the four Pratt and Whitney JT-4 engines for fourteen seconds to comply with the noise abatement requirements. Departure control came up on the radio with heading and altitude instructions.

What tension had existed during the takeoff slackened as Blackie was cleared to five and then twenty-two thousand feet. With his workload reduced, Blackie continued his observations on jet flying by indicating that they had dramatically increased passenger expectations of performance. In the old days, if a Connie were a half hour late, nobody seemed to mind, but now arriving a few minutes behind schedule seemed to cause irritation. In spite of airline late arrival performance, Blackie predicted a general increase in the speed of Atlantic crossings and mentioned the speed record that he had set the year before. He had flown New York to London in five hours and twenty-five minutes, but satisfaction was fleeting because his record was broken when a competitive flight landed ten minutes later after shaving three minutes off his record. Bainbridge began to better understand Blackie's preoccupation with getting the 33,000 altitude assignment, i.e., in the troposphere, over a dinner of filet mignon, white asparagus, and green salad just as the southeastern tip of Nova Scotia was coming into view. The dinner produced no heartburn although Blackie was exhibiting signs of impatience with air traffic control's dishing out altitude clearances. He observed that if flights didn't all take off at the same time the rules for vertical and horizontal separation for flights would not be such a limiting factor.

At this point, the predicted northwest crosswind was becoming more apparent as was the conclusion that the velocity was much higher than forecast. Blackie was disappointed in what was going to be a less than accurate call by the weather office, although he readily admitted theirs was a difficult job. On many nights he had encountered the jet stream where it was not forecast and vice versa. Blackie admitted that there was much to learn about the jet stream which at that time was a term applied to any sustained wind of seventy-five knots or more. The 707 was close enough to Gander to call for the ocean crossing altitude clearance which turned out to be 31,000 feet. That prompted another tutorial for Bainbridge on the matter of speed which at that point was Mach .82 or, at that altitude, 582 mph. Blackie liked that figure because speeds higher than that caused the 707 to rumble and bump, very much like an automobile running over a rough road. Passenger comfort was always on Blackie's mind, but if he personally wanted a really nice ride, it would have been in the first class compartment. That was better-insulated than the cockpit, forward of the engines, and not subject to head-on wind friction constantly buffeting the cockpit.

Blackie announced that he was about to conduct a tour of the cabin for a "little care and feeding of the people who pay our salaries." That was good news for the other two pilots who would get to do some of the actual flying. Personable as always, Blackie moved easily among the passengers making small talk for about twenty minutes. Once back in the cockpit, Blackie recounted some of the stories about his most unique passengers over the years. He had seen them all: the first-time ocean crossers, the drunks, the tense, the ultra-relaxed who slept from boarding to breakfast, the famous, the politicians, the movie stars, and the not so famous. He had found several of the leading European actresses overbearing and demanding; he wrote the most top heavy off as simply being "bovine" but treated them respectfully as he did all passengers.

Once back in the cockpit, Blackie asked the navigator to figure the equal time point (ETP), a term with which Bainbridge was less familiar than the better known "point of no return." He learned that in the day of jets with their immense fuel tanks, the latter term had lost much of its value and that the equal time point, always calculated on the basis of three-engine operation was more relevant. As the 707 was nearing the ETP, the navigator fearlessly got out his octant in preparation of fixing their position by shooting the stars. There no longer was a bubble on the 707's dorsal surface through which some malfunction he might be pushed out of the plane to fall thousands of feet to a watery grave. The 707 had a little periscope with a mirror system for the star shoot. The navigator was busy because he had to fix the location for every ten degrees of longitude and record them on his air report as well as give Blackie new headings to fly if there was any variation between planned and actual position. Thus the navigator was the busiest man among the crew between Gander and Shannon. TWA's Magellans would be busy shooting stars for the next three years, but then they were retired as victims of technology just as the radio operators had been.

There was a Doppler radar navigation system on board strictly for orientation purposes at that time, but it was still relatively new technology and as such its accuracy over land was much better than over water. Doppler radar alone was not responsible for the navigator phase-out; the primary cause was the inertial guidance system which was adopted in 1964, but for the time being Blackie relied on his navigator for the ocean crossings. TWA had equipped a few of its 707-331s with Doppler at a cost of $150,000 per unit, but was not expecting full certification of the system until well into 1962 when it planned on equipping twelve of its 707-331s with Doppler. At eleven-twenty the equal point time was reached – just off the southernmost tip of Iceland whereupon the navigator gave Blackie some bad news. They were twenty miles south of their planned flight position and the expected tailwind had not materialized to the extent advertised.

When the second officer tried to raise Shannon radio, there was so much atmospherical disturbance that he gave up and called "Item," one of the weather ships stationed at key intervals below. The tailwind component finally joined Flight 800 as the eastern sky was beginning to brighten. Blackie had recovered enough lost ground from the tailwind so as to be only two minutes behind schedule when overflying Belfast, Northern Ireland. Blackie began the descent into Paris with four rituals: announcing the arrival time to passengers, sticking some chewing gum into his mouth (because his ears were sensitive to pressure changes), putting on his leather gloves, and pulling out the Orly airport instrument approach plate and setting it on his lap. The crew went through the preliminary and final landing check lists, and when Blackie had the aerodrome in sight called "Down gear!" Two more technical advancements over the Connie simplified the landing: thrust reversers to save the brakes along with spoilers to create plenty of drag. The second officer complimented Blackie on the feather light touchdown which hardly surprised Bainbridge who was well aware of Blackie's reputation for textbook landings. The flight engineer, who had seen many more landings than Bainbridge exclaimed, "I've heard about them talk about rolling one on, but I've never seen one like that before."

They were three minutes late arriving in Paris, and Blackie wished that was only the time it took to complete the paperwork associated with the six hour and thirty-seven minute flight. After securing the plane, Blackie and crew went directly to the TWA dispatcher's office for the briefing on the remaining leg down to Rome where the weather was painted as beautiful. The first and second officers filled out the flight plan on which Blackie asked for an altitude of 27,000 feet. He also said goodbye to the cabin crew because they were not continuing to Rome. A new, all-French crew would be flying that leg as would be the first officer who promptly expressed his gratitude to Blackie. Twenty minutes after takeoff from Orly at eight-nineteen local time, the 707 was at 27,000 feet, high above a vast carpet of clouds. Over Turin, Italy, the ride became moderately bumpy which, Blackie explained, would have been a lot bumpier in a Connie; the 707 simply rides better in turbulence due to its wing shape and the fact that there is more flex in the wings themselves. At nine-thirty, the 707 started its descent to 22,000 and then when reaching that altitude a second clearance to 6,000 was received. Harold Blackburn, while still aircraft commander, moved into instructor mode and was giving the first officer a few pointers on the way down. The first officer had kept the approach a bit tight and as a result hurried the short final, but still pulled off an acceptable landing. Blackie observed, "Landing is like a golf shot. After it's over you know what you did wrong."

Blackie turned in his paperwork and then went through customs before catching the company bus to their assigned hotel downtown. The landing had been Blackie's approximately one hundredth in Rome which was satisfying, but the hotel choice was not. It was the Massimo d'Azeglio which was not the favorite that the Hotel Mediterraneo was. Nonethless he was glad to get there shortly before noon, drank a can of Metrecal after getting to his room, arranged for a wake-up call at six-thirty, and promptly hit the sack after unwinding with a crossword puzzle. Blackie and Bainbridge met in the lobby shortly after seven and went over to the Mediterraneo for dinner which was preceded by a bourbon and water. Dinner conversation focused on the safety of flying in the jets which because of their engines, speed, and fuel reserves were so much more reliable than the piston-engined airliners. After dinner, the two men walked for about a half-hour and then retired. "Leisure" was the byword for the following day, initiated by sleeping in for about an hour later than usual. Most of the day was spent just strolling around Rome, although the conversation invariably returned to flying and Blackie's gloom over the mandatory age sixty retirement rule.

At ten-fifteen in the morning, Blackie received the inevitable call from TWA Operations reminding him that he was scheduled for a one-thirty departure of Flight 801 that afternoon. Bainbridge recorded the return flight as completely uneventful, but noted that the Paris-New York segment was with a full load thanks to Air France whose ground workers who were out on strike. Blackie retarded the throttles over Boston while explaining that he thought the last half hour of the return flight was the most difficult simply due to the high density nature of New York air traffic. Bainbridge learned that the statistics showed forty-seven percent of airliner accidents occur during the landing phase as opposed to eighteen percent during takeoff, and he was not surprised to learn that fatigue was a major factor. Idlewild was handling 625 planes a day late in 1961 while Newark and La Guardia were each handling about 500. A pilot, fatigued after eight hours of ocean flying, still had to look out for two of Blackie's pet peeves: hot shot military jet pilots and private puddle jumpers in the same airspace. Nevertheless, Blackie eased the 707 back onto Mother Earth at seven minutes past eight – exactly as flight planned. Thirty minutes later, Blackie was in Hangar Twelve, turning in crew passports and the other paperwork. He would not arrive back at the farm until nearly midnight and had only six more roundtrips to Rome until being forced to retire. When he said "good bye" to Bainbridge, it might better have been "until next time" or possibly "see ya round," because there would be an encore of the Blackburn-Bainbridge partnership in little more than two years. The next collaboration would result in more than just pleasant, brief reading in *The New Yorker*.

The Drew production, simply called *Blackie*, was shot primarily on 22 and 23 December during a roundtrip flight to Rome and back, although the film, for impact, implies that the arrival back in New York was on Christmas Eve. Whereas the Bainbridge article could be considered a good blend of character study, travelogue, occupational analysis, and travel development the Drew production covered each of those themes in a collateral sense but compressed them to concentrate on the core issue, the injustice of depriving a man of his life's work and to a large extent, his self-identity. Structurally, the roughly forty-five minute film's production follows the Bainbridge storyboard, i.e., pilot leaves distant farm home, drives to Idlewild, and leads the flight to Rome and back, and then returns to home and hearth. The sequences, showing the sky, cockpit activity, thunderstorms over the Continent and near blizzard attending the landing, are among the most compelling visual scenes in any Drew film. However, *Blackie* presented some technical challenges of which not the least was how to record what goes on in the cramped space of a 707 cockpit. That was especially so during tense moments such as an instrument approach in bad weather, even with the most modern portable video equipment. It is important to note that the use of a storyboard was foreign to Drew because his goal was to capture unscripted events and, especially, a crisis, as it unfolded.

In addition to Robert Drew himself, cinematographers William Ray and Donn Pennebaker were also behind cameras for the shooting which ranged from Hidden Brook Farm to Idlewild, Paris, Rome, and back. Although the film implies that the flight was Blackie's last as an airline pilot, the actual shooting took place on Blackie's last *two* flights, those departing New York on 16 and 22 December 1961. When the shooting was completed, Drew realized that the film had failed to live up to his standards, i.e., there was no dramatic outcome to the crisis. All of the other hallmarks of Drew productions were there, however: a compelling central character in the person of Blackie, moving aerial scenes, tense moments during takeoffs, landings, and instrument conditions, and the romance of foreign lands. What Drew had was, in effect, a production that did not fly according to his own flight plan. Thus Drew Associates moved to Plan B in order to make the film come to life, and that was the extensive use of voice-overs by Blackie to convey his flying background, his beliefs, aspirations, and grinding disappointment that one day he was a hero and the next a zero just because he had reached his sixtieth birthday.

The film opens with scenes of Blackie on a tractor, plowing the north forty. Perhaps this was a subliminal technique to convey the message that Blackie had been put out to pasture by the antagonist. Nevertheless, after family and flying, the farm was nearest and dearest to Blackie. The action quickly moves into the cozy farmhouse as Blackie talks about the many pictures on the wall which convey his days in open cockpits, flying the line in DC-3s, the war years, friends made along the way, and finally his flying TWA's jets to foreign lands. The inflight shooting conveys the tension of takeoffs and landings, magnificent sea and skyscapes, and interaction within the crew on flight 900 to Lisbon, Madrid, and Rome. The other pilots on board, George Toop and Bill Shaw, had been flying with Blackie for years – all the way back to ICD days when they were very young co-pilots. Blackie covers some of the same subjects in his narration that Bainbridge did, such as the camaraderie of the air, but the theme of his career ending in several days is pervasive with the exception of a festive atmosphere in Rome. For dinner, Blackie's friends take him to a Blackburn favorite – the De Meo Patacca Restaurant, located on the Piazza dei Mercante (in center city Rome's Trastevere district, just south of the Tiber Island in the Tiber River). In the film, enthusiastic waiters sing "Arrviederci Roma" and do to this day at that establishment which is still going strong.

The flight home on TWA's Flight 801 is a repeat of that described in the Bainbridge article with four exceptions. First, Blackie's vivacious, twenty-year old daughter Bonnie joins the flight in Paris after her semester studying abroad in France. Secondly there is the tension of trying to land in New York before a major snowstorm descends. Thirdly, the roundtrip on the Drew project took one full hour longer due to adverse wind conditions. Finally, the hopeless situation Blackie faces, mandatory retirement, is emphasized not only with his own dialogue but also that with his crew. Even the passengers comment, including one who remarks that Blackie, although he was the oldest pilot flying for TWA at that time, does not look anywhere nearly old enough for retirement to which Blackie heartily replies, "I agree." Idlewild tower controllers had not forgotten that his landing back in New York would be his last as an airline pilot, and they gave him a special salute by momentarily cycling the approach strobe lights and runway lights up to full intensity. In his earphones Blackie heard, "Happy retirement from the control tower, Blackie."

The film continues into the following day which is Blackie's birthday complete with a splendid birthday party at home which draws attention to the bitter fact that Blackie's flying for TWA is finished for good. The film winds down at a festive birthday party, teenager daughter Be sledding down a snow-covered slope toward the pond, and Blackie admitting to that tugging feeling that comes over him every time he looks skyward at contrails or hears the sound of a jet. Considering the fact that several Jet and Victor Airways quite literally bracket the farm, one can imagine Blackie's looking skyward nearly every time he walked out of his house.

Was the film a success? Critics considered *Blackie* to be one of Drew's lesser films, and in the absence of audience viewership data now, a half-century later, it is hard to say what the reaction was among those watching the small screen. For the Blackburn family and Blackie's many friends, it was certainly an Academy Award winner. Considering the fact that Blackie always detested Hollywood's treatment of pilots and cockpits, the Drew project at least gave him the opportunity of seeing to it that audiences were given the ultimately accurate portrayal of what really happens on jet transport flight decks. ALPA, an enthusiastic, early backer of the project, promoted showing the film far and wide in its lobbying to extend pilot retirement age beyond sixty.

The *New Yorker* article and Drew Associates production did not play to universally enthusiastic reviews in other quarters – most notably in TWA's executive offices in New York. Floyd Hall, then a senior V.P. and systems general manager, asked Blackie in January 1962 to clarify a few points made in the article and film, particularly those in which TWA did not come off in a perfect light. That is, of course, a hazard dealing with the media which naturally wants to portray matters as they are. Blackie pointed out that a few items never made it into the article, i.e., the inflight movie which flickered so badly that the cabin attendant simply packed up the projector and called it a day, complaining of insufficient training. Nor did Bainbridge expound on the grumbling among some teetotalling passengers who did not want to wait for others to finish cocktails and snacks before having dinner. Overheard were other passenger observations that it might be better to use strictly European cabin staff on all flights since they seemed much more anxious to please than American crews. Blackie reminded Hall that service was a big issue and that the statistics told the story. Of the 3.3 million roundtrip passengers between the U.S.A. and Europe, only slightly more than 1.8 million flew on United States air carriers. Most of Blackie's criticism was reserved for baggage handling on the several flights in which some of it never made it onto the plane or was sent to the wrong destination, including much of it belonging to Drew's cameramen. Blackie's frustration boiled over when he wondered if "Lindbergh had lost his baggage on our first flight." Concluding in his usual no-holds-barred approach, Blackie observed that the problem was "insufficient supervision right down the line."

Blackie steadfastly resisted the inevitable retirement dinner and had been surprised by the coverage of his retirement in the *Skyliner* newspaper which put the stepping down on the front page with a large picture – something that had rarely, if ever, happened before. Forced by family and friends to relent, the well-attended retirement banquet was held in late January at Idlewild's TWA terminal. Most of Blackie's close friends were there – going back to the DC-2 days, Eagle Nest, ICD, and the International Division. Blackie's few detractors were not there that evening – not that he had that many, but over a quarter century of flying, much of it as a check pilot, his reputation for running the flight deck as a very tight ship had turned off a few lesser aviators. Company management was represented by Senior Vice President Floyd Hall and Vice President Gordon Gilmore among others. Hall used the old cliché, "The pilot's pilot," to describe Blackie but also went on to say that "Perhaps no greater tribute to Captain Blackburn can be made than to say that he epitomizes every standard of the piloting profession of which he is so justly proud." Hall had flown with ICD during the war years and had risen so rapidly through TWA's ranks that he was about to pop out of the top a year later and land as Eastern Airlines' CEO. Walter Gunn, who joined TWA in 1943 as a co-pilot on DC-3s and later retired as an international 747 skipper may have had the best and certainly most succinct summary of Blackie. He described Blackie as a "class act." Through the toasting and roasting, Blackie gave as good as he got; and, of all the accolades and gifts received, two stood out: The International Division hostesses presented him with a plaque in honor of his work with the ladies – dating all the way back to the days when the first hostess was hired. Blackie was a favorite of the hostesses because he was a true gentleman who never tried to take advantage of them. The other gift was a painting of an aged aviator in a leather flying jacket, ancient leather flying helmet with goggles, holding a very sooty torch which burned brightly with a fiery, golden flame, i.e., the original "Black Burn."

Fortunately, Blackie did not have a lot of time to brood about his career as a pilot having just made its last landing. He launched a second career and although it was shorter, it was every bit as successful. In the early 1960s, the market for intercontinental travel was developing but certainly had a long way to go. In the streamlining of TWA's departments under the Charles Thomas administration, Dr. John Furbay, the company's travel expert, became a part of PR vice president Gordon Gilmore's organization. Blackie joined that unit as director of TWA air travel development, although that was not the title on his business card. Rather it was, "Harold F. Blackburn, Captain Emeritus, Goodwill Ambassador." Since his starring role in 1955's Disney film, Blackie's appearance had become even more commanding to the extent that it would have been hard for central casting to round up a more experienced captain-like captain. It was all there – the crow's feet from years of flying, the gray-blue eyes (which still did not need glasses to pass an FAA physical), the military bearing, and the steady, confident voice. There were other senior captains retiring at about the same time, but many, although they may have had nerves of steel when making an approach right down to the bare minimums, suffered from varying degrees of glossophobia (fear of public speaking). That certainly had never been a Blackburn disability.

Although Blackie was quick on his feet and knew how to loosen up and work an audience with a few well-placed jokes, he took the time to study the styles of some of the twentieth century's leading public speakers such as Winston Churchill and Franklin Roosevelt. Because at the core of many of his speeches, Blackie was asking audiences to take a fresh look at the world, he frequently used a witty little opening joke that went along the lines of, "Minds are like parachutes – they only function when they are open." Occasionally, he would poke some fun at airline food by noting the story that went around about a couple of passengers who were disappointed in their filet mignon and trying to decide if it had been broiled, roasted, boiled, baked, or fried, but eventually concluded that it had been laundered. Blackie's humor, over the years, was not always so easily comprehended. Sometimes, his dry wit as applied to fine dining sailed right past the occasional co-pilot as Rusty DeoGuardia recounted years later. On 9 November 1940, DeoGuardia was a very young co-pilot much in awe of the captain and check pilot, Hal Blackburn, on Flight 19. That flight departed from where else but LaGuardia at 6:00 P.M. for the nearly six-hour journey to Chicago. Well east of Pittsburgh on that clear night, the hostess brought two dinner trays up to the cockpit where Blackie insisted that DeoGuardia eat his dinner first while Blackie flew the plane. About ten minutes after DeoGuardia finished, Blackie inquired of Deo Guardia's health which Rusty reported as just fine. Then Blackie consumed his fried chicken dinner. On a subsequent flight, DeoGuardia insisted out of politeness that Blackie have dinner first which Blackie instantly declined. DeoGuardia respectfully inquired just why his captain declined the offer to which Blackie quickly replied with an absolutely straight face, “To make sure the captain survives, of course. It’s company policy anyway.” There was no such policy at the time, but DeoGuardia accepted it as gospel truth.

For the next three years, Blackie was a guest speaker at a bewildering variety of venues; his speeches all carried a subliminal message of see the world with TWA but traveled under the cloak of broader subjects such as the future of air travel. Blackie had five standard talks which could be modified based on the audience. Among the speeches were, “Worldwide Wings,” “Only a Day Away,” and “Our World Neighborhood in the Jet Age.” Ironically, some of the material Blackie received from Gilmour and Furbay was from speeches he had re-written and given to them going back to the late 1940s. Gilmour had always requested a copy of Blackie’s talks when he became aware of them and never discarded a single one. In his first few months, Blackie made his own cue cards which were letter-sized sheets of paper with key subjects and secondary subjects printed in different ink colors. They quickly gave way to little index cards, and within a few months he needed no prompts at all. He soon established his credentials and unfailingly mentioned his origins that went back to the Jenny and DC-2, his experience during the war flying the world’s oceans along with the short tour as a combat pilot in Lightnings, and continued up to the present with intercontinental jets.

Blackie had the right to accept or decline speaking engagements based on his own preferences and personal schedule. As it developed, he accepted most of the requests through 1965 but became highly selective for the rest of his speaking days which effectively ceased in 1967. Blackie was compensated according to a formula that was essentially his hourly rate when he was flying 707s, and was given travel, lodging, meals, and incidental expenses. Occasionally, the expenses would also include those for Martha when she traveled with Blackie. The engagements were national in nature and heavily skewed toward being the speaker at banquets during trade shows or conventions. He also spoke to numerous smaller groups on a *pro bono* basis, i.e., banquets for Eagle Scout courts of honor, Quiet Birdman gatherings, etc. By the mid-60s, TWA marketing was becoming more sophisticated, particularly at identifying and going after specific demographic groups, working women for example. Gilmour tried to place Blackie at conventions and conferences of those target markets. Thus Blackie found himself speaking to associations of women accountants or women travel agents among other market segments. In that era, most large cities had one or more aviation clubs beyond what was probably the best known of them, the Wings Club in New York (of which Blackie was a member). Blackie spoke at many of their annual banquets and because aviation editors at newspapers were common in those days, TWA invariably enjoyed some good ink in the country’s leading dailies. TWA promoted Blackie nationally by placing a small color brochure in many locations – everywhere from ticket counters to seatback pockets. As might be expected, the state, regional, and national meetings of men’s service organizations such as Lions, Kiwanis, Rotarians, etc. frequently became Blackburn audiences. The remainder of the bread and butter speaking business came from such diverse sources as meetings of automotive engineers, state dairymen's associations, insurance groups, and motor carrier organizations among many, many others.

Most of Blackie's talks to aviation groups were based in the mid-Atlantic and frequently were not TWA assignments – he simply enjoyed talking to peers and aviation enthusiasts. One such meeting stands out because it was far away and represented a tie to Blackburn's early days with TWA. It was at the dedication of the Cavalcade of Wings exhibit at the Albuquerque Airport in 1965. The exhibit's founder, Harry Davidson, had been a Navy pilot who settled in Albuquerque as a civil engineer with a strong interest in local aviation history. The impressive exhibit is still in the airport as is a section of the display devoted to TWA's Eagle Nest in 1941. As with many of his speaking engagements, Blackie appeared on local radio and/or television, and in the Albuquerque visit, there was lengthy interview on the local PBS station. By 1968 Blackie completely gave up going on the road for TWA's travel development operation even though Gilmour pleaded with him to accept a limited number of engagements. By then, the Blackburn grandchildren were getting old enough to spend more time on the farm which was where Blackie preferred to spend most of his time.

Very shortly after retiring, Blacking and Martha traveled to Kansas City for a special event: the dedication of the Jack Frye International Training Center on 26 April 1962. For many attendees and certainly Blackie, the recognition was long overdue for Frye. TWA's president spoke, but the commemorative address was given by an old, close friend of Jack Frye's: former president Harry Truman. Dedication of the building and memorial plaques was by Jack's nine-year old daughter NevaJac Frye.

Almost as soon as Blackie launched his career as a public speaker, he was recruited into another venture: serving as an expert witness for law firms suing airlines for fatalities involved in accidents. That sideline activity was a brief one because Blackie was simply not comfortable in testifying against fellow airline pilots because he really believed in the brotherhood of flying. That was the case even though he was a perfectionist who could very quickly detect any pilot errors resulting in accidents. Blackie's first case as an expert witness was for a plaintiff who was suing Air France for loss of loved ones in the crash of a charter flight departing Paris for New York on 3 June 1962. The pilot had not confirmed that the elevator trim setting was correct for takeoff and then later aborted the takeoff when he should have known that there was insufficient runway ahead to land safely. In the ensuing runway excursion, 130 people out of the 132 aboard were killed. The year 1962, however, was a good one for expert witnesses testifying on behalf of plaintiffs suing airlines; in that year, the world's principal airlines recorded forty-one major accidents. Blackie's last case occurred in 1966 and involved the runway overshooting by a Pan Am 707 at John F. Kennedy International Airport (the renamed Idlewild Airport). The captain deviated from the glideslope during an instrument approach, landed long and fast, and came to rest in the shallow waters of Thurston Basin, an arm of Jamaica Bay on the southeast side of the airport. The 707 was a write-off, but at least there were no fatalities.

Collecting a fee based on damages paid by the airline with the world's worst safety record might have been difficult; that was Aeroflot, the Russian state airline. Blackie charged the law firms the exact same rate that he did as a public speaker, essentially what his hourly pay rate would have been as an airline captain. So, his second career as an expert witness was of very brief duration, especially so considering his feeling more comfortable in delivering an uplifting message to audiences which were substantially more friendly.

His third post-retirement career brought in next to no money, but that was hardly anticipated. With his beat-up portable Smith-Corona typewriter, he became an occasional aviation writer whose articles, typically on safety and training, were published in a number of aviation magazines, including *Flying* and *Business & Commercial Aviation.*

Blackie's speaking engagements and farming activities kept him very busy, but flying was something that he could not walk away from forever. In late 1962, he "captained" a Lufthansa flight, in the emeritus sense as a flight deck guest to Frankfurt for a week's European vacation with Be and Martha. He reprised that flight in May of the following year on those familiar TWA flights 800 and 801, to Paris, Rome, and back. Three weeks after returning from Rome, Blackie drove over to the Kutztown Airport and got checked out in one of the lightest planes he had ever flown: the Cessna 150 two-place trainer. The airport was among the smallest he had ever flown out of although it did have two runways of which one was paved. It was a busy little airport then, although it became a victim of the recession in 2008 and closed. But when Blackie flew from Kutztown, where he became a sort of local hero, the local fixed base operator also rented Cessna 172s and Piper Cherokees. Among the roughly forty aircraft based there were several gliders although Blackie never flew one. For the next ten years, he rented aircraft at Kutztown and gave neighbors, friends, and grandchildren pleasure flights in addition to frequently flying up to New Paltz and returning with Betty and her children. Other short cross countries over the years were flown to Allentown, Reading, Scranton, Lancaster, and over to New Jersey. Eventually, he checked out in the Cessna 172's big brother, the 182, which was a more comfortable cross country platform. At about the same time he checked out in the 182, in April 1965, Blackie got together with an old TWA buddy, Charles Kratovil, who flew into Kutztown in his Luscombe Silvaire, a delightful little high-wing monoplane from the late 1930s. Blackie flew the little ship for forty-five minutes, doubtless wishing he could swap places with Kratovil who still had two years of flying 707s left with TWA before reaching age 60. In September, Blackie checked out in the Citabria, a small taildragger that was aerobatic after a fashion; flights in the Citabria were strictly solo since he was looping and rolling just as he had done nearly four decades before near March and Kelly Fields as an aviation cadet.

1964 started with the first of two weddings that would be conducted at Hidden Brook Farm – on 11 January Bonnie married Jean-George Poucel, a medical student she had met at the beginning of her post-grad year at the University of Montpellier in France. Bonnie, who graduated from Antioch with a bachelor's degree in sociology, did not remain in Pennsylvania, but returned to France at about the same time Bob Blackburn was preparing to leave the country too. After Bob got his master's degree in intergroup relations from the University of Pennsylvania's Wharton School in 1964, he joined the Peace Corps and moved to faraway Somalia as the regional administrator. Daughter Betty, with two children in elementary school, completed her education that year with a bachelor's degree in English and would commence her career as an English teacher the following year. Of the four Blackburn children, Betty was the one to whom the mantle of musical leadership was passed. She excelled at the piano and had started playing the organ at her church. While in college she played piano in a jazz band led by a young man named Roger Thorpe, who, in his youth had been befriended by Sammy Kaye, the big band leader. Years later, it was Roger Thorpe who took over the "Swing and Sway with Sammy Kaye" band.

1964 was a relatively active year for Blackie's private flying, particularly considering the fact that he scarcely flew at all in 1963. It was a landmark month for TWA which had played a major role in the development of the Boeing 727 which went into service on TWA's short to medium haul routes that month. The plane would for all practical purposes become the DC-3 of its generation. Later in 1964, Blackie's pilot comrades were again pleased to find their friend back in print in something that had a John Bainbridge signature but substantially more shelf life than a fleeting article in *The New Yorker*. When Blackie and Bainbridge had flown together nearly three years before, Bainbridge was already thinking well into the future and the possibility of taking his articles on Blackie and re-casting them as a book. For that reason, he made sure his flight to Europe with Blackie was captured on film, although the only graphic in the magazine article was the caricature of Blackie which preceded the text. To record the flight, Bainbridge engaged Rudolf Janu, a talented photographer from Chicago who was making a name for himself as a magazine and architectural photographer. Janu is remembered primarily for his street scenes with people because he had a knack for capturing and bringing out personalities through candid shots while the subjects were hardly aware of their being photographed. The resulting book, *Like A Homesick Angel*, ran to 132 pages and included nearly thirty photographs. The publisher was Houghton-Mifflin with which Bainbridge collaborated on previous books, although *Like A Homesick Angel* was something of a departure because unlike Bainbridge's books about Toots Shore and the great Garbo, *Homesick Angel* was not a biography *per se* because so much of it was about flying. The book was essentially an expansion of his 1962 article on Blackie. Its page count and format, six by eight inches, put it into the small book category – one that could very easily be read on a jet from Kennedy to Heathrow. The book never made a mint for Bainbridge or Houghton-Mifflin but it was a hit with the Blackburns and the TWA community. It never received many reviews, and the few that it did receive played out largely along specific audience lines with the aviation press being the most critical. Bainbridge was savaged in *Flying Magazine,* the nation's leading magazine for the general aviation community. Its review praised the Janu photography but likened the book to the smooth surface of a deep lake upon which were the reflections of John Bainbridge's tape recorder. One could make the case for Houghton Mifflin not having sent a review copy to *Flying*, because its readers were not the target audience of the book which was intended for those whose familiarity with modern air travel was either non-existent or limited at best. Blackie was especially proud of the fact that the book was available in a Braille edition and a German-language edition which proved to have more shelf life than imagined at the time it was printed.

By the time the book was published in August 1964, Blackie and Martha had become a couple of empty-nesters. Be had left home and was attending Millersville University in nearby Lancaster County, but after two years would transfer to Lesley University in Cambridge, Massachusetts, where she studied for a degree in education. The Blackburns and Bainbridges enjoyed frequent reunions at Hidden Brook for years to come. Bainbridge, always very partial to Martha's V-5 juice, brought his brother-in-law and sister on one visit. The brother-in-law could not get enough of Martha's V-5 juice, particularly considering his periodic bouts with alcoholism for which he considered the V-5 as a tasty way to stay on the wagon. He was Richard Widmark, the Hollywood star.

On 19 June 1965, Blackie logged an hour and a half of local pleasure flying in the Citabria he occasionally rented at the Kutztown Airport. Almost exactly 4,000 miles to the east-northeast, the biennial Paris Airshow was starting to wind down, having opened a week before. The 1965 *Salon Internationale de L'Aeronautque et de L'Espace* was not an especially memorable show by previous standards because rather few new aircraft were displayed. The United States did exhibit its current missiles and the Russians did not. Marring the American participation was the crash of the supersonic jet bomber, the B-58 Hustler, in a landing accident which killed the pilot and seriously injured the other two crewmen. On the other hand, the USAF Thunderbirds put on another of their thrilling displays, and from a personality standpoint, the Americans gained considerable publicity when Vice President Hubert Humphrey arrived in Air Force One with two American astronauts, Ed White and Jim McDivitt. Russian Cosmonaut Yuri Gagarin also attended the show, but did not surface as a media favorite. What the media did focus on was the giant Antonov AN-22 transport which was capable of carrying 700 passengers.

The immense AN-22 completely overshadowed its smaller brother, the brand new Antonov AN-24B, a regional aircraft similar to the Fairchild F-27. The prestigious weekly news magazine, *Aviation Week & Space Technology* gave the AN-24B two pages in its follow-up coverage of the Paris Show and noted that the external appearance, with a few exceptions, was similar to the earlier AN-24A which had been introduced in 1960. Little else of the new aircraft was revealed at Paris, so any potentially new Soviet technology in regional turbo-prop aircraft remained a mystery. There were a very few people in the West who were intimately familiar with every aspect of the new Antonov AN-24B long before it arrived in Paris, and Harold F. Blackburn, retired TWA captain, was one of them.

Blackie's involvement with the Antonov AN-24B had begun several months before when he received an intriguing phone call from a very well placed airline friend. The caller was none other than the legendary William Patterson who was then in his last year as president of United Airlines. Blackie had gotten acquainted with Patterson during the war years when they both attended Air Transport Command contractor conferences. Patterson was a bit vague in his conversation with Blackie, but the gist of it was whether Blackie would be interested in a special assignment that could best be accomplished by a very experienced, recently retired international airline pilot. Patterson also hinted at the project being of value to the American government, free world airlines, and all jet transport manufacturers. Patterson continued by saying that if Blackie was interested, he would subsequently be contacted by another party which would have more details. Within several days Blackie took a phone call from a Mister "Smith" and a Mister "Jones" who said they had gotten Blackie's name from a leading airline official. They had asked for the name of a recently retired jet captain who was experienced in Atlantic crossings and familiar with flying in the Middle East. The two, whom Blackie always believed were CIA operatives, inquired if Blackie would be interested in ferrying a foreign-built, regional turbo-prop from the Caribbean to Cairo. Blackie was indeed interested but said he would not give a final answer until he had logged several hours flying it. That was agreeable to the CIA which then revealed the story behind the ferry flight.

The CIA, through intermediaries, had contracted with Lebanese Air Transport (Charter) Company (LATCO) to lease an Antonov AN-24B, registration code OD-AEN. LATCO had just become one of the very first AN-24B customers of whom most were located in Soviet client countries. The brand new plane had been spotted at Prestwick in January where *Flight International* reported that it was on its way to Nassau via Iceland. The plane, flown then by another retired jet transport pilot, was actually on its way to Puerto Rico. The CIA wanted the ferrying kept as low key as possible to avoid any unfavorable publicity, but did comply with all relevant regulations. There were two lawyers involved in setting up the lease contract: Pierre Sciclounoff in Geneva and George N. Lindsay in New York. Sciclounoff was then an up and coming lawyer who eventually became a world class collector of art and antiquities while representing many of Europe's power elite and serving on the boards of numerous prominent corporations. Lindsay, a former naval officer on destroyers in the Pacific during World War II, was a partner with Debevoise, Plimpton, Lyons and Gates. His brother was John V. Lindsay who was elected mayor of New York later in 1965.

Messrs. Smith and Jones asked Blackie to recruit a co-pilot and navigator which Blackie did: Vincent Olsen and Earl Korf. Olsen had flown with another airline, but Korf and Blackie went back to ICD days when Korf had started as a radio operator and later qualified as a navigator. Korf, at age sixty-one, was still climbing mountains and was well on his way to climbing all of America's highest peaks at that point. Blackie arrived in San Juan on 17 May and spent several days checking out on the Antonov. LATCO had made arrangements for its Latin American affiliate, Compania Silva de Fomento, Ltda. to send its chief pilot, Esteban D. Bosch, to check Blackie out on the Antonov. Blackie needed a type certificate to fly the plane, and that was acquired through Jaime D. Serra, the FAA's regional manager in San Juan.

After the plane's arrival in San Juan, the CIA and its contract overhaul company, Mark Callahan Associates from Miami, completely disassembled the Antonov looking for new technology and made its final report on 16 April. There were a few minor problems which the Callahan firm and Compania Silva corrected. The original landing in San Juan resulted in two blown tires associated with a brake system issue. There were hydraulic and weather radar problems as well, but they were properly repaired, which was important because LATCO intended to sell the plane when it was returned to the Middle East. The plane's troublesome environmental system was not repaired which meant that it was essentially an unpressurized aircraft, and that created a problem because Blackie had no intention of a low altitude North Atlantic crossing.

Blackie and crew were ready to take off on 20 May. Although they were satisfied with the plane's airworthiness, there was one mildly distressing item left in the cargo hold: a galvanized, a two-gallon pail full of parts. The mechanics could not figure out where the components were to be re-installed, so they just put them in a bucket and hoped for the best. The CIA had provided Blackie with a wad of cash to pay for gas and services along the way. They also provided some emergency contact information in case the money ran out. The route was one very familiar to Blackie after stops in West Palm Beach and Baltimore – the old North Atlantic route which he had first flown twenty-three years before. One day was lost due to bad weather in Newfoundland where Blackie had a layover day at St. Johns. The problems had already started to mount, beginning with the cost of portable oxygen due to the plane's pressurization system being down. That unforeseen problem depleted some of their cash early in the game. Another aggravation was the constant questioning from air route traffic controllers and airport towers when Blackie identified himself as "Antonov Alpha Echo

November." The full registration was OD-AEN (October Delta Alpha, Echo November). Most controllers were used to hearing alphabetic registration call signs because of occasional Canadian air traffic, which also used letters instead of numbers, but the "Antonov" raised eyebrows and prompted questions.

A more constant headache was the Antonov's instrumentation which was based on the metric system. Blackie had to create little notes and tape them next to key instruments because conversions from metric to English measurements were substantial. The difference impacted all of the air data instruments, many of the engine instruments, and each one of the instruments that measured temperature, pressure, and quantity. About the only system not impacted was navigation.

The crew flew from St. Johns to Iceland and then to the Azores where a cable was waiting for them concerning the status of their cash. Blackie replied that it was rapidly dwindling; their funds were, after an irritating delay, later replenished via a cable to American Express in Lisbon. While in Lisbon, they were instructed to halt in Rome since LATCO was re-evaluating crewing and alternate airports for final delivery of the plane. Blackie was advised, however, to make sure that he and his crew had visas for Egypt. Blackie headed for a familiar haunt after landing in Rome – the Hotel Mediterraneo. On 27 May, he was advised that the next stage of the flight would not be to Cairo, but Athens instead. In Athens a LATCO crew, including the company's chief pilot, would not only fly the plane from there to Cairo, but would bring a new, six-man crew on the flight from Rome to Athens as part of that crew's familiarization with the new aircraft. As it turned out, the weather was exceptionally good once leaving St. Johns; Blackie never so much as logged a minute of time on instruments during the twenty hours he flew the plane after departing Newfoundland.

Thus Blackie's CIA caper came to a close on 28 May when he and his crew started their journey home from Athens. Although Blackie did not think highly of the plane due to its inferior performance and rough construction, he had enjoyed a short vacation along some of the routes of his past. He was also richer by nearly $2,000, although he did not get to see old friends in Cairo. Blackie never learned what the CIA paid in total for ferrying the plane to Puerto Rico, having it torn down, reassembled, and returned, but his portion of the total cost came to $8,500 which included his fee, crew pay, and expenses along the way. The CIA learned everything it wanted to know about the AN-24B long before *Aviation Week* reported on the "new" Soviet aircraft, but, of course, remained utterly silent about the entire matter, including the fact that the AN-24B was bereft of any meaningful new technology at all. The only differences between the AN-24A and B models were the addition of two more passenger windows and a modified wing center section to accommodate newly-designed flaps and greater fuel capacity.

There was, however, a lingering mystery associated with incident. Not long after returning to the farm, a large black car would frequently be parked within easy viewing distance on the road beyond Blackie's lane. Blackie sensed that he was being watched and finally called Messrs. "Smith" and "Jones" at their New York offices. "Don't worry Captain – those men who are watching you. We're watching them as well." Messrs. Smith and Jones never revealed who the original watchers were, but the mysterious car out on road quickly went away and never appeared again.

In 1966, son Bob Blackburn and his family came home from serving two years as an administrator in the Peace Corps in Somalia. Daughter Betty's son Karl was twelve and her daughter Rebecca was eight – and they all enjoyed spending time on the farm. The farm had become an even more fascinating place, especially considering the two pets that Bob had brought back from Somalia. There was a fascinating little hedgehog that amused the children by rolling up into a little ball, but it was the other pet which created all of the excitement. That was a young male cheetah Bob had adopted. The cat, named "Ranjit" (a typical boy's name in Somalia) had been orphaned at a very early age and adopted by the Blackburns. When the family returned home, all of the paperwork and customs forms were in order, but Ranjit somehow got loose in the Rome airport. This created a major stir, particularly in the media which credited the tall American tourist with single-handedly subduing the wild and vicious predator. Ranjit was a very frequent visitor to Hidden Brook Farm, but his first appearance was the most memorable based on what happened when he encountered the Black Angus steers in the meadow. Ranjit spotted the prime beef, crept towards them, then broke into the speeding charge for which cheetahs are well known, and came to a screeching halt as soon as he reached the steers. The beefiest steer just took a nonchalant look at Ranjit and then returned to grazing. Ranjit had a cheetah's basic prey-chasing instincts down all right, but because he had been orphaned so young had never been taught how to kill his food.

Blackburn Family Collection

Blackie and "Ranjit" at Hidden Brook Farm.

Ranjit's favorite activity was rough-housing and wrestling with Bob and Blackie on the farm's front lawn. He never scratched or bit anybody, or for that matter, any other animal on the farm; he simply enjoyed a good free for all frolic, although some of the neighbors were incredulous at first. Sadly, the cheetah eventually developed an intestinal condition for which the local veterinarian had no cure. The Blackburns then involved the University of Pennsylvania, College of Veterinary Medicine, but even its specialists could not turn the disease around.

By the mid-60s, Blackie and Martha were attending more than just a few retirement parties as some of Blackie's friends who were in management retired at age sixty-five and other pilots hit the wall at age sixty and had to retire. One of the most pleasant retirement parties in 1966 was that for Swede Golien. He was feted particularly in celebration of his eight years of work as general manager of Ethiopian Airlines. Twenty years after its first flight in 1946, the airline was flying to twenty locations on three continents. Most of the men honoring Golien had become TWA vice presidents after originally serving in ground or pilot positions back in the 1940s. Blackie probably had the best line of the night when he said of Swede that, "In eight years of management, he helped advance that country eight centuries."

When he came back from Somalia that same year, Bob located in Philadelphia and went to work for the Philadelphia School District as head of the Office of Integration and Intergroup Education. At Bob Blackburn's suggestion, the school superintendent, Mark Shedd, combined the Office of Integration and Intergroup Education with several other related departments at the district office all of which reported to the associate superintendent of community affairs. The man Bob suggested to lead the new group, Marcus Foster, took the position. Foster was a black educator who had come up through the Philadelphia public school system and had been a successful principal at several schools in the district. He was working on his Ph.D. at the University of Pennsylvania at the same time Bob was pursuing a Ph.D. in educational leadership at Union College Graduate School. Foster was also emerging as a national leader in charting a course through what was an educational minefield – between those who demanded too little of public schools in poor, crime-infested urban areas and those who unrealistically expected too much from schools as the means by which minority students could achieve their potential.

Interestingly, all of Blackie and Martha's children followed careers in somewhere in education. Be had graduated from Lesley College in 1968 with a degree in education and was married to David Hubbard at Hidden Brook Farm that year. She eventually joined the administrative staff at the Massachusetts Institute of Technology where she worked part time while her daughter Amanda was young, and eventually became a full-time employee of MIT. Bonnie got her M.A. in education from the University of Denver in 1969 and became a French teacher in Colorado Springs. Betty continued to teach English at Solebury after receiving an M.A. in English literature. Blackie's life revolved around the farm and family, although only his two oldest children lived nearby (Betty and Bob).

Many of Blackie's flights in the late 1960s consisted of his taking a Cessna 172 up to New Paltz and returning with Betty and the grandkids. One such flight took place on 7 April 1967, although Blackie made no special note of the date other than a routine logbook entry. On that date, TWA's DC-9s went into service following the last flight of a TWA Connie the day before. The transition made TWA the first operator of an all-jet domestic fleet.

While Blackie was still accepting the occasional speaking engagement to a limited number of audiences in late 1966, there were big changes in the wind at TWA's New York office as to just who would be addressing the company's mass audiences. Foote, Cone & Belding (FCB) was TWA's advertising agency and had been for many years. Indeed, FCB was then at the top of its game and was one of the oldest agencies in the country, enjoyed considerable respect in its industry, and was among the ten largest agencies in America. Still it came as no earthshaking news when TWA announced that it was reviewing its advertising program and would accept presentations from qualified agencies which could support its far-flung operations. FCB was told that it too could participate in the competition. Such an insult is usually the kiss of death in the advertising game, but the agency dug in its heels and defying conventional wisdom came up with a campaign that saved the account: "Up, Up and Away TWA!"

The renewed relationship continued for more than a decade, but TWA eventually did cut Foote, Cone & Belding loose. The agency's grief was brief: it soon had the British Airways account. There was a droll touch to the happier days when TWA and FCB were flying high together. FCB's Manhattan's offices were on the thirty-sixth floor of, what else, the Pan Am building at 200 Park Avenue.

It is now the Met Life Building, but in 1977 the skyscraper was better known for something else: a gruesomely tragic accident on the helipad atop the building in which an airport shuttle helicopter's landing gear collapsed, lowering the rotor blades to the extent that several people near the chopper were sliced apart. Another blade flew over the edge of the building and slashed into a Foote, Cone & Belding account exec's office on the thirty-sixth floor. Luckily for him, he was down the hall when the accident happened. After crashing into the office window and releasing a shower of metal and glass, half of the blade broke off, fell another thirty-six stories, and killed an unfortunate woman walking on Madison Avenue below. To top it off, Pan Am was currently struggling financially and was already in severe belt-tightening mode. Its demise would come within fourteen years. And TWA would follow Pan Am a decade later although Blackie would not live to see the passage of those two titans of the skies.

Blackie did live long enough to see many of his old comrades fold their wings. Among the first to fly west was his close wartime buddy Clarence Shoop who died from stomach cancer on 27 January 1968 in Beverly Hills at age sixty. At the time of his death Shoop was no longer involved with the test flying or helicopter operations at Hughes Aircraft, but had become a vice president responsible for international marketing of Hughes' aerospace components and microelectronics. It was a loss deeply felt by Blackie whose friend had made it possible for him to accomplish one of his most important goals – combat flying over Nazi Europe in World War II.

Blackie's flying had become sporadic by the late 1960s; he would fly several times each month for a couple of months in a row and then not go near Kutztown Airport for six or seven months. In April of 1969, he treated himself to a checkout in the Cessna 177 Cardinal which was essential a 172 that had been fitted with a fuel pump, had been structurally strengthened, had its wing struts removed, and been aerodynamically restyled. His first passengers were Martha, Be, and her husband on a two and a half hour pleasure flight in early 1969. His next flight, just some local flying in the Cardinal around Kutztown on 29 August, ended uneventfully after an hour and a half. That was more than could be said for a TWA Flight 840 out of Rome which did not land in Athens that day as planned. It was hijacked and diverted to Syria by Palestinian terrorists who released all but two Israeli passengers. The hijackers then attempted to blow up the 707, but succeeded only in destroying the nose section. That hijacking was not the first time an airliner had been hijacked by terrorists considering El Al's having suffered the theft of one of its planes on a London to Rome flight the year before. And it would not be the last Palestinian hijacking of a TWA jet either. Clearly, Blackie had retired during an age when the skies were definitely friendlier.

In the spring of 1970, Marcus Foster, Bob Blackburn's boss in the Philadelphia School District office, accepted the position of school superintendent in Oakland, California. It was an opportunity where he felt he could make a real difference in a setting that had many of the same problems as Philadelphia, including urban crime and lack of community involvement in education. Foster was anxious to be in a position of real control which he had not been in while in Philadelphia. The fact that the Oakland School System was only about a quarter of the size of Philadelphia suggested that reforms might be more easily implemented. At first, Bob was disappointed in Foster's resignation because he thought there was so much left to do in Philadelphia, but was eventually persuaded to move to Oakland as the deputy school superintendent. Blackie and Martha were pleased at their son's new role, but disappointed since Bob, Barbara, Samantha, their baby daughter born in 1968, and son Chris would be so far away from the farm.

In 1971, ten years had passed since Blackie had flown a TWA 707 and his involvement with the company's air travel development had ended four years previously. But Captain Blackburn had not been forgotten by TWA. The company had historically done a very good job of reminding current employees of the company's great achievements of which the first scheduled flight to Paris was still regarded as a major milestone. In 1966, the company newspaper reprised the twentieth anniversary of Blackie's Paris flight in 1946 with a front page article. In 1971 the editor of the new company newspaper, *TWA Today,* requested a current photograph of the skipper who had led the epic flight. In the 1 February issue's full back cover story, the editor ran the photo which the former captain supplied, although it was not exactly what the editor was expecting. Blackie did not send his "official" TWA photograph showing the very senior airline captain, but instead supplied a send up of the Grant Wood painting "American Gothic." The painting features the gaunt, farmer with the 1,000 yard stare and a farm house in the background. The photograph sent from Hidden Brook farm shows a barn with customized hex sign on one end in the background. In the foreground there is a straight-faced Blackie holding a pitchfork and wearing a 1920s leather flying jacket, great flowing silk scarf around his neck, and helmet with goggles pushed up on his forehead. The only other photo on the page was of a Connie with the Egyptian pyramids below – the framing of the photo was identical to the one Blackie had photographed twenty-seven years before of Clarence Shoop's F-5 Lightning over the same pyramids. That issue also carried a few other noteworthy items which vividly portrayed TWA's situation ten years after Blackie flew his last flight as a TWA captain. Blackie's generation of TWA pilots were fading in more ways than one; the death of a TWA pioneer, John Collings at age sixty-seven, was noted in the paper which also had a brief story about the great jet airliner that Blackie never got to fly: the Boeing

747. Captain Ernest Pretsch was shown presenting the key to New York to German officials in Frankfurt who welcomed the first TWA 747 to Germany. The front page article was a summary of the speech given by President Charles Tillinghast who had announced that the nation's airlines were standing at a strategic crossroads after the calamitous, recession year of 1970 when the nation's eleven largest airlines lost a pre-tax total of $180 million due to over-capacity and rapidly rising costs. TWA was a big loser in that 1970 bloodbath, losing $73 million after net earnings in 1969 were $12 million. Tillinghast called for finding a compromise between stifling regulation and excessive competition that were causing so much red ink which in turn forced Tillinghast to furlough TWA employees instead of across the board pay cuts.

The company newspaper was Blackie's link to TWA which had undergone major changes since Blackie's Christmas Eve return from Paris in 1961. Charles Tillinghast enjoyed the advantage of guiding a TWA completely independent of Howard Hughes. Although he owned nearly 80 percent of the company's stock, Hughes could not vote it; only the independent voting trust could. As soon as he could, Tillinghast conducted a stem to stern review of all of TWA's operations, including an image study done among the flying public. On the outside, TWA enjoyed a very positive perception due in no small part to its extensive advertising and public relations which created the image of an airline that was larger than it really was. On the inside, there was considerable turmoil about the Convair 880s being too expensive to operate and about TWA's lack of sophistication in buying equipment and services. The study showed Tillinghast that TWA's greatest asset, by far, was its route structure to Europe (no surprise in that finding). There was also the growing realization that the piston-engined aircraft in the fleet were a distinct disadvantage. That reality drove TWA to look for a replacement for the Connies and old Convairs, but at the time, the Boeing 727 was still only on the drawing boards. When Tillinghast and Breech were installed at TWA's helm, the board had contemplated suing Howard Hughes for breaking anti-trust laws by forcing TWA to buy its planes from Hughes Tool Company. Tillinghast was seriously considering following through on the suit when Hughes threatened to sue TWA in the summer of 1961 on the basis that the voting trust had acquired Hughes' TWA shares by illegal means. That TWA was ordering new Boeing 707s instead of the Convair jets Hughes particularly favored simply added more fuel to the fire from which Hughes hoped to regain control of "his" airline.

With that threat, the TWA board authorized Tillinghast to follow through and sue Hughes which it did in the summer of 1961. The attempts to settle out of court, in-court proceedings, counter suits, appeals, and abortive merger talks with Pan Am went on for a decade after Blackie retired from flying. The original judgment was in TWA's favor, and the damages hearings opened in 1966 with TWA being awarded $138 million. That was appealed and the award was sustained, but Hughes' lawyer managed to get it to the Supreme Court which ruled in Hughes' favor. In the end, TWA had spent a bundle on legal fees and gained nothing in damages. The great gain was that enjoyed by Howard Hughes who finally sold all of his TWA stock and netted nearly $550 million, much of which he used to acquire the regional airline, Air West in 1970.

President Tillinghast and Chairman Breech, free of Howard Hughes, made the decision to back out of merger talks with Pan Am which allowed them to focus entirely on TWA's future. 1962 was a loss year, but the second half had been stellar and pointed to a return to profitability which did happen in following years. That was important to Blackie who, although he was not receiving paychecks anymore, was receiving TWA pension checks. He was actually regarded by his Merrill-Lynch broker practically as a peer because Blackie and Martha studied their investments and potential purchases very carefully. In their investment deliberations, they were also assisted by Martha's brother, Dr. William Bondurant, a shrewd investor himself.

CEO Charles Tillinghast was still running the company as the new decade of the 1970s dawned. Blackie's flying was infrequent and he would be the first to say it was too infrequent to maintain the skills of an average pilot. His head had not really been turned by TWA's acquisition of the Douglas DC-9 or Boeing 727, but on 25 February 1970 TWA put a new airplane into service – one that had to turn Blackie's head because some of his close friends, many with a year or so of retirement, became the first captains. It was the Boeing 747 and according to Blackie's acquaintance, Bob Buck, it flew just like a mega-707. It would not be TWA's last widebody because in another two years the tri-jet Lockheed L-1011 went into service on domestic long haul routes. The L-1011 incorporated much of the technology that Lockheed had developed in the competition to build an American SST. The desire to not be left out of the race caused Tillinghast to hedge his bets; he placed orders for the Anglo-French Concorde but also bought options for the Boeing SST. Cleverly, Tillinghast's contract for the Concorde had a money-back cancellation clause if the plane's cost exceeded a certain level. That mark was exceeded, TWA got its money back, and Boeing never built an SST.

Blackie's TWA was changing in ways that he could not relate to as a pilot, although its diversification beginning in the late 1960s was something he could understand as an investor. Tillinghast could see that although the company had been profitable in the early 1960s, the realities of the airline business included potentially devastating strikes, sky-high costs of new airplanes, periodic overcapacity, and business recessions. And airline deregulation was not even on the horizon yet; that would follow in 1978. Therefore TWA moved into non-traditional areas by forming a holding company, Trans World Corporation, and then acquiring the Hilton hotel business and Century 21 real estate among other purchases.

Hidden Brook Farm was a magnet for Blackie's many friends, even going back to his band years. Mike Derby from the Hal Blackburn Orchestra was an occasional visitor as were many former aviation cadet classmates and airline pilots. Sadly, one planned visit resulted in a searing tragedy for Blackie who had lost only a few acquaintances in flying accidents since the ICD days. Blackie had gotten together with Bob Loomis who had flown into Kutztown on 11 September 1970 in his new, twin-engined Beech Baron and let Blackie take it up for some airwork. In May of 1972 during Blackie and Martha's trip to California, Blackie flew the same Baron with Loomis on a roundtrip between Los Angeles and San Diego. In mid-September, Loomis called Blackie and was invited to spend a couple of days at the farm while Loomis and his wife Louise were on their trip to Plymouth, Massachusetts, where the two planned to celebrate their thirty-fifth wedding anniversary. Loomis, then in his mid- to late fifties, called from St. Louis on 29 September to inquire about local weather and decided that landing at Kutztown probably would not be advisable. After landing in Pittsburgh he called Blackie again and said he was filing an instrument flight plan for Reading and that he would call as soon as they landed. Shortly after six P.M. Loomis was cleared for the approach into Reading, but after that radio exchange all contact was lost. Alarmed, Blackie called the airport later in the evening and learned that Loomis was missing. Late that night, the remains of the Baron were found; it had flown into a ridge in the Blue Mountains about twelve miles north of Reading, killing all aboard including the Loomis' daughter, Laurie, and her two children. Blackie was incredulous since Loomis was an accomplished instrument pilot; witnesses on the ground all agreed that the plane sounded as though it was having engine problems. Reading weather at the time of the crash was a 500 foot ceiling and two and one-half miles of visibility near the minimums but certainly not challenging to a pilot of Loomis' experience.

On the evening of 7 November 1973, Blackie was watching the television news when he heard the unbelievable. The broadcast included a report from Oakland where the school superintendent and the deputy superintendent had been shot during the late evening of the day before. The news clip clearly indicated that Marcus Foster was dead, but left Blackie and Martha completely in the dark about Bob. Rather than try to call Bob's residence, Blackie called a pilot friend in Alameda, Captain Bill Heller, who called the Oakland police to get the latest update because the first reports had both Foster and Blackburn being killed. Heller quickly got back to Blackie that his son was at Highland Hospital, had undergone emergency surgery, and was expected to live. Blackie was able to talk to Barbara later that day, but decided to let the dust settle before coming out to Oakland. Bob was in the hospital for a month and then home for another month; then he went back to work as the interim school superintendent.

When the full story of Foster's assassination came out, it was as shocking as the shooting itself. After three years on the job, Foster still faced discipline and security issues in the Oakland schools, and had to address critics who were split over whom to blame for the problems. Was crime in the schools inherently the students' fault or was it because the community and school system failed the students? Foster and Blackburn had to find solutions to school violence issues that would not only work but also satisfy demands from political groups ranging from citizens' committees to the Black Panther Party – all on top of the normal work of a school system central office. On the evening of 6 November, Foster and Bob were walking to the parking lot following the conclusion of a monthly school board meeting when they were ambushed by members of the Symbionese Liberation Army (SLA). Foster was killed instantly by a hail of revolver bullets while Bob was hit by a shotgun loaded with buckshot shells. Bob managed to stagger back into the building and was soon on his way to the hospital and several hours of touch and go surgery in which he suffered cardiac arrest three times. The young surgeon, Coyness Ennix, repaired the damage done to Bob's liver, kidneys, and spleen that night and went on to a brilliant career as a cardiovascular surgeon.

Three days later, the SLA took credit for the killing, which they thought would be the opening salvo in its revolution that would purge the country of people they considered a threat to society. They had considered other highly visible leaders as their first victim, but had focused on Foster, mistakenly, in the belief that he was about to install police in every school and require every student in the district to carry a photo I.D., i.e., he was a threat to society as they perceived it should be. The SLA promptly realized what a blunder it had committed. The kidnapping and brainwashing of heiress Patty Hearst soon followed the Foster assassination, but justice eventually caught up with the SLA; its leader, Donald DeFreeze, was killed in a confrontation with police and others were imprisoned. Bob Blackburn was asked to become the permanent superintendent, but he declined to do so, and in 1977 resigned from the Oakland school system to go into consulting and university education which he still pursues with vigor at this writing. He did serve as an interim superintendent for several other school districts, leading to his humorously describing himself as the place to go to "Rent-a-Superintendent."

The year 1974 proved to be a bittersweet one for Blackie; after 12 July of that year, he never flew an airplane again. But go out in style he did, beginning in early May when he was attending a convention at the Greenbrier Resort in West Virginia. Through his "old buddy" connections he flew six roundtrips in a classic, twin-engine Beech model 18 between Greenbrier and Roanoke, bringing other convention goers to the resort. Blackie and Martha headed to California to see Bob and family in July with a stop in Mitchell on the way to see old friends and relatives. He took his brother Charles out to the Western Nebraska Regional Airport at Scottsbluff, rented a Piper Cherokee, and flew around the haunts of his youth for the better part of two hours, then completing the last page in his tenth and final logbook with a lifetime total of 22,795 flying hours. His last flight was over the same locale where, almost exactly fifty-five years to the day, he had been so thrilled with his very first flight.

Hidden Brook Farm was still the fun-filled venue for the grandchildren who were delighted to play in the yard, run in the fields, or listen to their grandfather's tales of flights over icebergs, deserts, mountain ranges, oceans, meandering rivers, and jungles in faraway lands. Of the seven grandchildren only Betty and Be's kids were living in the East. Bonnie had started her family, but was living with her elementary age kids, Jean-Jacques and Genevieve, in Colorado. Still the family tried to have one big reunion back at the farm every year. Of course Blackie had a larger family too – his wide circle of TWA friends. Blackie had no interest in becoming an officer of the leading TWA alumni organization of the day, the TWA Seniors Club, or TARPA (The Active and Retired Pilots of TWA) when it was founded in 1979. However, he was a very visible participant at their meetings and conventions as he was with other aviation organizations such as the Quiet Birdmen. In his mid-70s, Blackie was still a very fit man which was due in part to the nature of running a successful farm with its many chores and occasional hard, physical labor. Blackie swam when he could and had taken to riding a bike – occasionally up to ten miles a day.

The farm was visited in the summer of 1974 by a family that was complete strangers to Blackie. The guests were a link to the book *Like A Homesick Angel* which had been published ten years before. In the late spring, Blackie had gotten a letter that was simply addressed to Captain Harold Blackburn, Mertztown, Pennsylvania, USA. Blackie, by this point well known at the Mertztown post office, got the letter in spite of the incomplete address. The letter had been sent from Vienna, Austria, a city in which Blackie had no acquaintances. The letter had been written by Angelika Krischke, a young school girl who would be visiting America with her parents and brother in the summer. She had read the German edition of *Like A Homesick Angel* and wanted to meet the man who was the subject of the book. The Krischkes did visit the farm, although communications were somewhat difficult because Angelika's English had not progressed to the point where she was comfortable with it – and her parents and younger brother spoke no English at all. They enjoyed their visit, but it was obvious to the Blackburns that their guests had not allowed enough time in America to accomplish their goals of seeing the entire East Coast in less than two weeks. Young Angelika stayed in touch with Blackie, and when the Krischkes visited the West Coast several years later, they did allow more time for their journey.

Although he had been an avid hunter in his youth, Blackie never seriously took advantage of the small game potential on the farm. Pheasants were still very plentiful in Berks County in the 1970s, but the only birds Blackie ever shot on the farm were the occasional mallards which alighted on the pond; they wound up on the dinner table. Blackie thoroughly taught his grandsons, Karl and Chris, gun safety, but there was an ulterior motive in that. Like most farmers, Blackie held woodchucks, "groundhogs" to the locals, in very low regard since their holes with mounds at the entrances frequently fouled mower blades when cutting clover. Blackie rarely shot woodchucks himself; but he paid Karl and

Chris a fifty-cent bounty for each one they shot. In their teenage years, the grandsons proved invaluable performing other chores; especially pruning the ivy on the barn gables and repairing the barn roof because Blackie's terrible fear of heights had not improved with age. One visitor, the elderly wife of one of Blackie's friends, was on the back porch talking with Blackie and Martha when she noticed that the .22 slide action rifle had been returned to its rack after grandson Karl had finished with his target practice. That practice occasionally consisted of tossing a tin can out into the backyard which sloped up, thus serving as a good backstop. The lady suggested that she would like to see if she could hit a tin can as well as Karl had, so Blackie loaded the tubular magazine and gave it to the lady who followed Blackie's every instruction on how to slowly squeeze the trigger while keeping the bead front sight squarely on the target and in the little notch on the rear sight. It was Blackie's Model 1906 Winchester, one he had since his childhood in Mitchell. Blackie's guest hit the tin can squarely on her first try and in rapid succession hit the can another seven times as it was being blasted to the back of the yard. Blackie had been taken in! The guest as a young teenager had been befriended and taught trick shooting by none other than America's "Little Miss Sure Shot," Annie Oakley.

There were no Grange Halls in the immediate area, although the nearby communities of Topton, Fleetwood, and Kutztown all had them. Hidden Book Farm became, for all practical purposes, a small Grange Hall by itself considering how many farming neighbors would swing by on practically a daily basis. The conversation was predictable: weather, crop yields per acre, or the latest catastrophe such as a hay wagon turning over. With the Blackburn children and grandchildren coming back frequently, all of the home's bedrooms would frequently be occupied, and when they were not in residence, friends would be there. Blackie's annual Christmas card list was nudging right up against a total of 200, thus creating a large base of invitees. With the marriages of the Blackburn kids, the in-laws were very frequent guests as well. Then there were Blackie's brothers and their families along with former neighbors from previous residences going back to the early 1930s, including Blackie's Japanese-American friends from Los Angeles.

In mid-April 1975, Blackie and Martha began their annual trip to California to visit Bob and his family in Oakland. This journey was always via San Antonio where they stopped to see Martha's folks and get their annual physical exams, courtesy of William Bondurant, M.D., who was also the chief medical officer at Nix Hospital in downtown San Antonio. Martha had been coughing over the last month or two, but it had not been deep or frequent enough to be especially alarming. Doctor Bondurant was suspicious and ordered X-rays which showed lung cancer. Martha was promptly operated on and had one lung removed, but it was too late. The cancer had already spread to her brain and she died at the hospital six weeks later on 9 May, leaving a completely devastated family in the aftermath. Her ashes were scattered back at Hidden Brook Farm.

TWA was also in a bad way at the time and there were doubts about its survival. At one point, it had to sell off some 747s just to meet its payroll, and the flight attendant strike of 1973 had been devastating. Unprofitable routes had to be abandoned, including those through the Pacific region that had in 1968 made TWA truly a global carrier. The recession years of 1970-71 made matters worse. By 1975 TWA was starting to climb out of the depths, thanks largely to profits coming from Blackie's old routes to Europe.

While Blackie was in Texas, the farm work was in left the capable hands of Russell Seidel, but even with the distraction of his many daily chores upon returning, Hidden Brook Farm had become a very quiet place for the once independent, self-reliant aviator. Blackie was lonely. Most of his children and grandchildren were far, far away with the exception of Betty who had become a teacher at the Solebury School in New Hope, Pennsylvania, (just north of Philadelphia). Blackie had always been a prolific correspondent, and during this period he was in touch with old flying buddies and friends from his childhood days in Mitchell. The Nebraska acquaintances had always corresponded regularly over the years and had occasional reunions in Mitchell. Blackie's best friend from the old days in Mitchell, Dave Hilyard, did graduate from Colorado A&M, came back to the family farm outside of Mitchell, and farmed the place with his family for the rest of his life until dying at age ninety. From his last contact with Dave and other Mitchell friends, Blackie knew that his childhood sweetheart, the former Helen Jones, had very recently been widowed. In the years since she and Blackie were dating, she had married, was widowed in 1943, remarried in 1946, and then became a widow again in 1975. After exchanging several letters, Blackie asked her to come for a visit to Hidden Brook Farm to which Helen agreed – under the condition that she, as a very proper lady, bring one of her long-time friends, Elizabeth Heiss, to act as a chaperone. So Helen and Elizabeth did come to the Pennsylvania Dutch Country and became house guests at the farm.

After having been there for several days, Helen came into the kitchen one morning and found Blackie busy cooking breakfast. He turned around and said, "Where's my morning kiss." Blackie got the kiss and Helen soon got a proposal of marriage. Helen was concerned that a marriage so soon after both spouses had died only months before might be inappropriate; so she inquired of her family how they felt about it. The response was almost immediate and certainly positive from both families and ran along the lines of "You're not getting any younger. Tomorrow may never come. So why not enjoy as much happiness as possible now. Forget what others may think." That was exactly what Blackie and Helen hoped for, but neither one of them wanted a splashy, large wedding with many family members and friends attending a service and elaborate reception in Mitchell. Blackie and Helen were married on 3 November 1975 in Torrington, Wyoming (just across the border, about twenty-five miles west of

Mitchell). Blackie's brother Charles and his wife were the only witnesses and as soon as the ceremony was over, Blackie and Helen drove back to Pennsylvania.

Helen's adjustment to life down on the farm was rapid. Although she had lived in Mitchell and Scottsbluff most of her life, she had spent a brief period on her father's farm. Like Dr. Charles Blackburn, Archie Jones had acquired some land under the Homestead Act, but had a rough time making a go of it as a farmer and later settled in Mitchell. Helen immensely enjoyed travel and accompanied Blackie on several European trips with the TWA Seniors Club and, later, going to TARPA gatherings. Blackie rarely experienced any leisure moments at the farm; he always had something going which ranged from sawing up firewood for the home's three fireplaces to whitewashing the fences. After his farm chores were completed and his shower had been taken, Blackie always joined Helen for a pre-dinner, bourbon and water cocktail. Helen had always enjoyed cats, and at Hidden brook she was never short of feline company. At that point, the barn cats had lost most of their wildness and would even sleep with the dog, Hey Boy. There was one exception, a cat that never became truly tame, although it would allow Helen to approach with food. In matters of the household, Helen's approach was much like Blackie's in the cockpit. Even though Blackie insisted on having a cleaning lady come to Hidden Brook Farm, Helen gave the place a pre-cleaning before the lady got there, not wanting anybody to think that she was a slack housekeeper. Music was still an important part of the Blackburn home, and Helen was just as accomplished with the piano as Blackie was, although she did not play the Hammond organ as frequently as he did.

If the year 1978 was tranquil at Hidden Brook Farm, it was anything but that at TWA's headquarters in New York. Charles Tillinghast had by then become chairman of TWA and had overseen the installation of Carl Meyer, Jr., as president on 6 January 1976, just thirty-seven days before the death of Howard Hughes. Meyer had been on the job for two years when the Airline Deregulation Act of 1978 was passed. The impact was felt most heavily by the established majors such as TWA, American, Pan Am, and United, but of those companies TWA was better positioned to withstand the onslaught of new, low-cost competition. Nonetheless, there was a substantial downsizing of the roughly 36,000 employment and a reduction in the aircraft inventory. TWA's saving graces were its utilization of the hub and spoke concept, with New York and St. Louis as the hubs, and a relatively young fleet of efficient aircraft.

Eventually the realities of the "golden years" caught up with Blackie and Helen. In 1979, Blackie turned seventy-eight and found that it took much more effort to maintain the farm. He and Helen had attended enough TWA Seniors and TARPA gatherings in Arizona to seriously consider selling the farm and possibly moving to the Phoenix area where Helen's son Stan was living. That would certainly satisfy one requirement: having close access to TWA flights. And the winter weather was certainly a lot more conducive to pleasant living than in Berks County which has all of the cold, dreary dampness of a typical mid-Atlantic winter. Blackie had considered what might become of Helen if he were the first to go onto his final reward and thought that it would make a lot of sense if Helen were widowed again that she ought to be somewhere near family. Arizona would take Blackie away from daughters Betty and Be, but on the other hand Arizona was much closer to Bob and Bonnie. When Blackie and Helen attended the 1980 TWA Seniors gathering in Wickenburg, Arizona, they stopped in Phoenix and bought a house in Sun City, which surprised everybody at first, but knowing Blackie they realized that the purchase had been in the works for several months. Blackie, with occasional black humor, called their new home, "God's holding pen." One of the more laborious tasks attending the move to Arizona in 1980 was deciding what the movers would and would not take. One item certainly would not be left behind or given to his kids: the glass-topped coffee table with a Hal Blackburn Orchestra bass drum for a base. Blackie's old aviation cadet buddy and United pilot, Pinky Stainback, had a good suggestion for the immense collection of Army Air Corps and TWA memorabilia: donate it to what was becoming a nationally significant repository of aviation memorabilia, particularly airline material. Soon, Blackie had boxed up all of his logbooks (except the last one), charts, reams of correspondence, orders, yearbooks, memos, and all of the material from his youth and band days for shipment to the American Heritage Center at the University of Wyoming at Laramie. The collection was put into nearly fifty archive boxes while his uniforms and the RKO, Disney, and Time-Life films were stored in separate, specialty boxes. The farm was sold to one of the country's more prominent metallurgists whose company was located in Reading; he proved to be a superb steward of Blackie's hard work and now, decades later, the farm is still in wonderful shape.

TARPA and the TWA Seniors Club were not the only organizations to remember the old days. TWA itself flew Blackie and Helen to Paris in early February 1981 for a special reception on 6 February to celebrate the thirty-fifth anniversary of TWA's first regular flight to Paris. The reception honored four men who were prominently involved in the flight and the subsequent roll out of the International Division. The group also included Larry Trimble whose TWA career began in 1929, Bob Berle, and Jerry Thornton. Of the group, French television was most interested in the pilot of the flight and interviewed Blackie the same day. French papers ran the article and the photo of the four men cutting the four-tiered cake at the reception. There was only one barely noticed downside in the celebration; the cake was topped with a Lockheed L-1011 model, not the Connie named "Star of Paris."

On Sunday night, 2 January 1984, Helen was in the family room reading the paper while Blackie headed to the utility room to take the garbage out. He was reaching up to a cabinet to get a garbage bag when he lost control of his right arm and immediately thereafter found himself on the floor, the victim of a minor stroke. He managed to gather himself up, put the garbage container out by the curb, came back into the house, advised Helen of his problem, and said he was still dizzy. In the subsequent examination, the doctor also found what Blackie called "a slight touch of dementia" although his health otherwise was still very robust for his age. At that point, the dementia, in its very early stage, was more apparent to Blackie than other people, but he was frustrated by occasionally having to hunt for the right word. He was still very active, went swimming frequently, had a regular exercise program, and rode his bicycle for several miles nearly every day. He also maintained a very active correspondence with friends – and occasionally opened letters with an original salutation – "Greetings from Wrinklesville."

By this point, Blackie was very much one of TWA's silver eagles and was quite aware of how many of his old friends had folded their wings in recent years. The Active and Retired Pilots of TWA ran an article on the ten oldest living pilots in its magazine, *TARPA TOPICS*. Blackie, at eighty-four was only the eighth oldest, but the names of the older seven were composed of the pilots Blackie had served as a co-pilot. The names read like a list of TWA's pioneers: Lew Goss, Dutch Holloway, Tommy Tomlinson, Walt Smiley, Si Morehouse, Ted Weaver, and Sonny Boy Hall. There were two other "Eagles" which is what TARPA called its member over the age of seventy-five and both were younger than Blackie and also had lower seniority numbers: Cliff Abbott and Joe Bartles.

Blackie and Helen had been enjoying life in Sun City for over five years when Helen, who rarely saw a doctor, was diagnosed with advanced breast cancer. Blackie immediately cancelled a long-planned trip across the Pacific that would have taken them as far as Bali. After spending her last months in a nursing home, Helen died on 20 January 1986 and was cremated according to her wishes which also included instructions to scatter her ashes from the top of Scotts Bluff south of Scottsbluff, Nebraska. Blackie did not feel up to making the trip and asked Stan Delahoyde, Helen's son who had earlier lived in Scottsbluff, to honor her request. Stan, accompanied by his wife and young son, did so on Thanksgiving Day 1986.

Blackie soldiered on for a year in Sun City, but his mental condition progressed to the point where living alone had become inadvisable. During that year, Stan Delahoyde and his wife Tudy kept an eye on the old aviator and treated him as though he had been their own father instead of a step-father. Stan Smith, formerly a TWA sales executive and past president of the Arizona chapter of the TWA Seniors Club, had made arrangements with both the TWA newspaper and *Phoenix Gazette* to interview Blackie in conjunction with the fortieth anniversary of Blackie's groundbreaking flights to Paris, but the interview was delayed due to Helen's death. The TWA paper did not participate in the subsequent interview but instead reprinted an earlier article on the flight. The *Phoenix Gazette* did run the article, but it was the last time Blackie appeared in the news media prior to his death.

The Sun City house was sold and Blackie moved to Bob Blackburn's home in Oakland in March of 1987. At that point, Bob and Barbara were basically empty-nesters; son Chris had graduated from college and was well into a career in real estate fifty miles north in Sonoma County while his sister Samantha was back East at Swarthmore College. Bob's house is located on a high ridge along the very southwest edge of Oakland and has a commanding view of the San Francisco skyline and the Bay. Very quickly, a bathroom and broad picture window were added to the former family room to create a spacious bedroom-den for Blackie. One aspect of the Blackburn home was very familiar to Blackie, almost like being back at the farm; there were many pets including another dog named "Hey Boy," "Kitty" the cat, "George" a friendly ferret, and "El Birdo," the parrot which enjoyed an occasional ride on Blackie's shoulder.

By this point, Blackie's stroke, Alzheimer's disease, and poor hearing had largely robbed him of the ability to communicate effectively by letter or telephone. The growing sense of being muzzled was a source of burning frustration for Blackie who had always been so articulate with the written or spoken word, but he immensely enjoyed receiving letters and visits from his many friends. In most ways, the old trailblazing pilot coped with his setbacks very well; physical exams showed that he was otherwise remarkably fit for a man in his mid-80s. He took long, two-hour power walks, on the roads with their spectacular views of the Bay area and helped with the gardening and such daily tasks as doing the dishes. The mementos of a very exciting life overflowed from the walls of Blackie's room out into the rest of the home, although Bob humorously advised Blackie's friends in the letter announcing the move that the shrunken heads of Pan Am pilots were restricted solely to Blackie's room. With all of the other children being involved in education one way or another, the daughters were able to spend long periods in the summers of 1987-1989 at Bob's home, and each daughter created what was a periodic newsletter for Blackie's many friends.

Ultimately, Blackie withdrew increasingly unto himself and became very frail. Still, he would sit on the deck or look through the great picture window in his room at what were constant reminders of his past. Far below what amounted to the eagle's perch that was Bob's home, there was San Francisco International Airport, Alameda Naval Air Station, and Oakland International Airport. Although hidden behind San Francisco's downtown skyline and by then a county park, Crissy Field was still out there in Blackie's memory, and he could easily visualize his flights to and from what had been that important Army Air Corps base. In his mind's eye were TWA's flights from SFO to LAX in the DC-3s of another era although what Blackie could see in the skies over the bay were TWA's progressions of Boeings – the 727s, 737s, and giant 747s taking off for many of Blackie's old destinations ranging from Kansas City to London.

Blackie was not a formally religious person, and he positively revealed that to John Bainbridge. Blackie only very infrequently accompanied Martha and

Bebe to the Methodist Church near the farm; and indeed, daughters Bonny and Be had been married at the farm, not in a church. Yet the pages of *Like a Homesick Angel* reveal a very spiritual man, one who certainly does not have all of the answers, but one who believes there is a master plan out there somewhere. The wonders of the universe were uniquely revealed to him, his fellow pilots, and sailors who intimately experience the earth, seas, and skies as opposed to that felt by what Blackie and many airmen referred to as "groundlings." While Blackie's religion could probably best be described as equal parts of agnosticism and pantheism with a modest infusion of Protestantism, there is some indication that as the day for his final takeoff approached he may have had a change of heart. As best he could communicate, he said that he was looking forward to the day when he would be united with the three women who had meant the most to him: his mother Marion, Martha, and Helen.

Blackie slipped away and flew west on 4 August 1989. There was no question about where his ashes would go – back East with Martha's at Hidden Brook Farm.

Epilogue

The goose that laid the golden egg, as Blackie considered Trans World Airlines, was in a serious decline during Blackie's last days. Incapable of dealing with the huge fixed cost burden and low-cost competition, Trans World Corporation found itself between the inevitable rock and hard place in 1985, i.e., between corporate raiders Carl Ichan and Frank Lorenzo. In the end, Trans World Corporation's airline business was acquired by Ichan who had a reputation of dismembering his acquisitions and then selling off the parts. Unexpectedly, Ichan took over with the promise of returning TWA to profitability. One of his first moves was buying Ozark Airlines in 1986 which had the effect of strengthening its existing hub at St. Louis. Veteran TWA employees still did not trust Ichan and feared that he would simply use TWA as a means of dramatically increasing his own personal wealth. Those fears were confirmed when Ichan took TWA private in 1988. In that process, Ichan's personal wealth was increased by nearly $470 million while TWA's debt soared to over $500 million. The debt effectively made it impossible for TWA to modernize what was becoming an aging fleet. And there was still the matter of high labor and fuel costs which Ichan addressed in 1991 by selling one of TWA's greatest assets, its route to London, to American Airlines for nearly $450 million. Twice in the 1990s, TWA went into and out of bankruptcy, emerging as only a shadow of what it had been at the peak of Blackie's career.

In 1996, TWA Flight 800 exploded over Long Island Sound which further injured the line. TWA fought back with new vigor that was boosted by having outlasted Carl Ichan who had left the chairmanship in 1993. Before leaving, Ichan obligated TWA to a ten-year deal which sold discounted tickets to a subsidiary he controlled with the net effect that the airline lost an estimated $150 million annually to that Ichan entity. During the late 1990s, TWA shrank to the point that it became primarily a domestic carrier, vexed by the traditional problems of high operating costs, labor agreements that proved to be strategic cripplers, and interest rates heading upward. There was still some vitality and pride in the line as some of the statistics proved. In 1996, it had the worst record in the industry for on time performance, but in 1997 it climbed up to second, and its load factor had become enviable again. In January 2000, however, the final crossroad was reached when President William F. Compton, a one-time TWA pilot, called a press conference to announce what the very short term future held for the Lindbergh Line: enter a third, more difficult bankruptcy for which the climb out could be doubtful or sell out to AMR Corporation, American Airlines' parent. For TWA employees, American had always come across as a lean and mean, slash and burn type corporation; thus there was not much support for selling out among most TWA employees. There was a lot of Monday morning quarterbacking after the sale to American; many TWA'ers thought the company could have toughed it out and survived. Some airline analysts believed that Compton had no other choice but to sell to American and try to save as many jobs as possible while his detractors believed Compton sold TWA employees down river.

After 9/11, American's days also became numbered; the descent was hastened by the severe recession which began in 2008. Steps such as downsizing the St. Louis hub, closing the Kansas City maintenance base, furloughing aircrews, and layoffs kept American aloft for a while, but the company filed for bankruptcy in November 2011. US Airways acquired American in February 2013, and decided to move forward under the American name. Ironically, US Airways' ancestor company, All American Airways, had been created in Pittsburgh in 1939 during the days when Blackie was flying through that city in the employment of an airline that saw nothing but blue skies in its future forever.

Bibliography/Sources

Books

Ashcroft Bruce. *We Wanted Wings*. San Antonio, Texas: USAF Air Education Training Command. 2005.

Bainbridge, John. *Like a Homesick Angel*. Cambridge, Massachusetts: Houghton Mifflin Company. 1964.

Buck, Robert. *North Star Over My Shoulder*. New York: Simon and Schuster. 2002.

Cameron, Rebecca. *Training to Fly, Military Flight Training 1907-1945*. San Antonio, Texas: USAF Air Education Training Command, Air Force History and Museums Program. 1999.

Cate, Curtis. *Antoine de Saint-Exupéry, His Life and Times*. New York: G.P. Putnam's Sons. 1970.

Cave, High B. *Wings Across the World, The Story of the Air Transport Command*. New York: Dodd, Mead and Company. 1945.

Cearly, George W. Jr. *TWA, A Pictorial and Illustrated History of Trans World Airlines 1925-1987*. Dallas: 1988.

Christie, Carl A. with Hatch, Fred *Ocean Bridge: The History of RAF Ferry Command*. Toronto: University of Toronto Press. 1995.

Craven, W.F. and Cate, J.L. *The Army Air Forces in World War II* (seven vols.). Chicago: University of Chicago Press. 1951.

Daley, Robert. *An American Saga, Juan Trippe and his Pan Am Empire*. New York: Random House.1980.

Dietrich, Noah. *Howard, The Amazing Mr. Hughes*. Greenwich, Connecticut: Fawcett. 1976.

Dimitri, Ivan (West, Levon). *Flight to Everywhere, A Picture Journey over 32,000 Miles of Air Transport Command Routes through Jungle, Desert, and Arctic*. New York: McGraw-Hill. 1944.

Downing, Taylor. *Spies in the Sky*. London: Little Brown. 2011.

Dugan, James and Stewart, Carroll. *Ploesti, The Great Ground-Air Battles of 1 August 1943*. New York: Random House.1962.

Follis, Thomas K. *He Wore a Pair of Silver Wings, The World War II Memoir of a Lightning Recon Pilot*. Bennington, Vermont: Merriam Press. 2012.

Freeman, Roger A. *The Mighty Eighth War Manual*. London: Janes Publishing Company. 1984.

Gann, Ernest K. *A Hostage to Fortune*. New York: Alfred A. Knopf. 1978.

—. *Ernest K. Gann's Flying Circus*. New York: Macmillan. 1974.

—. *Fate is the Hunter*. New York: Simon and Schuster. 1960.

Goss, Christopher H. *Bloody Biscay, The History of V Gruppe/Kampfgeschwader 40*. Manchester, England: Crécy Publishing Limited. 1997.

Green, William. *Famous Fighters of the Second World War*. New York: Hanover House. 1957.

Gunn, Walter. *A Life Aloft, from DC-3 to 747*. Mission, Kansas: Wings Publishing. 1998.

Guy, John E. *One Guy's Journey*. (privately published), 2001.

Hansen, Chris. *The Times and Schemes of Enfant Terrible, General Elliott Roosevelt*. Tucson, Arizona: Able Baker Press. 2012.

Headquarters Army Air Forces. *Pilot Training Manual for the P-38 Lightning* (AAF Manual 51-127-1). Washington, D.C.: 1945.

Ivie, Thomas. *Aerial Reconnaissance, The 10th Photo Recon Group in World War II*. Fallbrook, California: Aero Publishers. 1981.

Karash, Julius A. and Montgomery, Rick. *TWA, Kansas City's Hometown Airline*. Marceline, Missouri: Kansas City Star Books. 2001.

Keen, Patricia F. *Eyes of the Eighth. A Story of the 7th Photographic Reconnaissance Group. 1942-1945*. Sun City, Arizona: CAVU Publishers. 1996.

Klotz, Alexis. *Three Years Off This Earth.* New York: Doubleday. 1960.

Kucera, Dennis C. *In a Now Forgotten Sky. The 31st Fighter Group in WW2.* Stratford, Connecticut: Flying Machine Press. 1997.

Lay, Beirne. *I Wanted Wings.* New York: Grossett and Dunlap. 1937.

Leonard, Royal. *I Flew for China.* Garden City, New York: Doubleday, Doran and Company. 1942.

Lindbergh, Charles A. *The Wartime Journals of Charles A. Lindbergh.* New York: Harcourt Brace Jovanovich. 1970.

—. *We.* (New York: G.P. Putnam's Sons.1927). Guilford, Connecticut: The Globe Pequot Press. 2002.

Ten Years Lufthansa. Cologne: Lufthansa German Airlines. 1965.

Maguire, John A. *Gear Up.* Atglen, Pennsylvania: Schiffer Publishing Company.1995.

Martin, Barry S. *Forgotten Aviator, The Adventures of Royal Leonard.* Indianapolis: Dog Ear Publishing. 2011.

Mauer, M. *Combat Squadrons of the Air Force in World War II.* Washington, D.C.: Zenger Publishing Co. 1969.

—. *The U.S. Air Service in World War One.* Washington, D.C.: Government Printing Office. 1978.

Murchie, Guy. *Song of the Sky.* Cambridge, Massachusetts: The Riverside Press.1954.

Obermaier, Ernst. *Die Ritter Kreuz Träger der Luftwaffe 1939-1945, Band 1, Jagdflieger.* Mainz, Germany: Verlag Dieter Hoffman. 1966.

O'Connell, P.J. *Robert Drew and the Development of Cinema Verite in America.* Carbondale, Illinois: Southern Illinois University Press. 1992.

Pearcy, Arthur. *Lend-Lease Aircraft in World War II.* Osceola, Wisconsin: Motorbooks International. 1996.

Pyle, Ernie. *Brave Men*. New York: Henry Holt and Company. 1944.

Roll, David L. *The Hopkins Touch*. Oxford, England: Oxford University Press. 2011.

Seagrave, Sterling. *The Soong Dynasty*. New York: Harper & Row, 1985.

Serling, Robert. *Howard Hughes' Airline, An Informal History of TWA*. New York: St. Martin's/Marek. 1982.

——. *When The Airlines Went To War*. New York: Kensington Books. 1997.

Severe, Errol. *The Last of a Breed*. Eureka Springs, Arkansas: Lighthouse Productions. 1997.

Sloniger, Jerrold. *One Pilot's Log, The Career of E.L. "Slonnie' Sloniger*. Charlottesville, Virginia: Howell Press. 1997.

Spencer, John P. *In The Crossfire, Marcus Foster and the Troubled History of American School Reform*. Philadelphia: University of Pennsylvania Press. 2012.

Spoden, Peter. *Enemy in the Dark*. Bristol, England: Cerebrus Publishing. 2013.

Spooner, Tony. *Warburton's War*. London: William Kimber. 1987.

STANAVO Pilot's Handbook. San Francisco: Standard Oil Company of California. 1938.

Stevens, Leonard. *New York to Rome. Jet Flight 808*. New York: Harper and Row. 1962.

Taylor, Michael J. H. (Editor). *Jane's Encyclopedia of Aviation*. New York: Portland House. 1989.

Transcontinental and Western Air, Inc. (TWA). *Outlines of Ground and Flight Instruction*. Albuquerque, New Mexico: 1942.

Tunis, John B. *Million Miler, The Story of An Air Pilot*. New York: Julian Meissner, Inc. 1942.

Wachtel, Joachim. *Lufthansa History*. Cologne, Germany: Lufthansa German Airlines. 1975

Watt, Sholto. *I'll Take the High Road, A History of the Beginning of the Atlantic Ferry in Wartime*. Fredericton, New Brunswick: Brunswick Press.1960.

Wells, Trent (Editor). *The History of Aviation Cadet Class 42-B.* Coulterville, California: Cadet Class 42-B Association. 1997.

Winchester, James. *Lockheed Constellation.* London: Airlife Publishing. 2011.

Articles

Bainbridge, John. "You Come, You Go." *The New Yorker.* May 5, 1956.

—. "Like a Homesick Angel." *The New Yorker*. November 10, 1962.

Burble Point. (Aviation Cadet Newspaper). March Field, Riverside California. 1931. Vols. One through Four.

Flying Magazine. February, 1965. New York: Ziff Davis Publishing Co.

Headquarters Army Air Forces. *Pilots Information Manual* 1944 (AAF Manual 24). Washington, D.C.: 1944.

Hotz, Robert. "North Atlantic," *Air Force*. May 1943.

New York Times. 17 December 1954, "Reich trainees in USA."

TARPA TOPICS (selected issues)

TWA Skyliner (employee newspaper), selected issues, mid-1935 through third quarter 1989.

United States Department of Commerce, Civil Aeronautics Administration. *Air Navigation Radio Aids.* Washington: July 1, 1940.

"Wings for the Trojan Horse." *Foreign Affairs*. Washington, DC. January 1941.

Unpublished Material

Blackburn, Harold. Correspondence and Pilot Logbooks Numbers One through Ten. Blackburn Papers. Laramie, Wyoming: University of Wyoming, American Heritage Center.

Bliss, Kermit. *Transcript of an Oral History Interview with Kermit E. Bliss, Eighth Air Force Photo-Reconnaissance Pilot, European Theater, WWII. 1995.*

Shoop, Clarence. *From the Diary of Clarence A. Shoop, Colonel, Air Corps.*

Documents

"Airplane Accidents in the Continental U.S., by Principal Model of Airplane." Publications Branch, United States Air Force Historical Office. Washington, D.C.

"Blackburn Testimony." Presidential Emergency Board Airline Negotiations Committee. 7 May 1946. New York. Pages 1890-1930.

"Statistical Handbook." Washington, D.C.: Transcontinental & Western Air, Inc. 1946.

"Technical Report of Aircraft Accident Classification Committee" (reports on forty-one flying accidents involving members of the Kelly Field pilot graduating class of February 1931 from 1930 through 1943), B-24 D Accident on Trail Peak, Cimarron, New Mexico, April 22, 1942, Col. Jack Zimmerman Catalina Accident November 2,1942. Air Force Historical Research Agency, Maxwell Field. Alabama.

Websites

Fifty-fifth Fighter Group
http://www.55th.org

Fifty-fifth Fighter Group
http://www.cieldegloire.com/fg_55.php

134th Infantry Regiment, Nebraska National Guard
http://coulthart.com/134

440th Troop Carrier Group in WW II
http://440thtroopcarriergroup.com

Air Corps newsletters from the 1920s into the 1940s
http://www.afhso.af.mil/aircorpsnewsletter

Aircraft, crash reports
http://aviation-safety.net

Airfields, abandoned and forgotten in Europe
http://www.forgottenfield.com

Airports, American, abandoned
http://www.airfields.freeman.com

Airports, Southern California
http://qnet.com

Airline timetables
http://www.timetableimages.com

Aviation, milestones calendar
http://www.thisdayinaviation.com

Aviators, prominent pilots at Davis-Monthan Field
http://dmairfield.org

Blackburn genealogy
http://www.houseofnames.com/blackburn

Boeing Company history
http//www.boeing.com/boeing/history

Eighth Air Force fighter and reconnaissance groups
http://www.littlefriends.co.uk

Eighth Air Force War Diary
http://www.scottyfire.com

Frye, Jack
http:jack-frye.blogspot.com

Frye, Jack and Helen
http://www.sedona/legendhelenfrye.com

Hudson, Lockheed, flying characteristics
http://anu.edu.qv

Ireland, Foreign aircraft landings in during WWII
http://www.csn.ul.ie/~dan/war/crashes/htm

Knight, Major Eric mini-bio
http://lassiecomehome.info

Lightning, Lockheed, development of photo-recon models
http://www.34prs.org

Lockheed Constellation
http://www.conniesurvivors.com

Luftwaffe aircrew archives
http://aircrewremembered.com/krackerdatabase

Luftwaffe airfields 1935-1945 (including occupied countries)
http//:www.ww2.dk/lwarfields

Luftwaffe officer career summaries
http://www.ww2.dk/lwoffz.html

Luftwaffe bases in Germany
http://www.wg.org

Mountain, Joseph D. collection
http://airandspace.si/edu

Prestwick and Ayr, Scotland, historic photos
http://www.mcintyre.demon.com.co.uk/local

Roosevelt, President's daily journal
http://fdrlibrarymarist.edu/daybyday

Spitfire aircraft production, serial #s and a/c history
http://www.airhistory.org.uk/spitfire

TWA, Active and Retired Pilots
http://www.Tarpa.com

TWA Seniors Club
http://www.twaseniorsclub.org/memories

World War II aviation, combat chronology by Jack McKillop
http://milhist.net/usaaf

World War II, daily calendar of important dates
http://ww2db.com

Personal Interviews

Family
Bob Blackburn (son)
Betty Tjossem (daughter)
Bonnie Poucel
Be Blackburn
Chris Blackburn (grandson)
Karl Beard

Trans World Airline Pilots
Ed Betts
Joseph Carr
Walter Gunn
Howard Kincheloe
Harold Thurston

Seventh Photographic Reconnaissance Group
John Blyth
Pamela Shoop Sweeney

Friends
John Chan
Harry Davidson
Cecil Folmar, M.D.
Russell Seidel

Philmont Ranch
Elliott "Chope" Phillips

Harold F. Blackburn's first flight was in a Curtiss Jenny one year after the Great War ended. His last flight as a Trans World Airlines Boeing 707 transatlantc captain was forty-two years later, after which he kept on flying light aircraft for another fourteen years. His passage in the skies over five continents truly mirrored aviation in the twentieth century. His logbook reflects flights in a dazzling array of aircraft: the Curtiss B-2 bomber, the DC-2, 3, and 4, along with the heavy bombers of World War II, the B-17 Flying Fortress and the B-24 Liberator. He carried America's top brass to and from Europe during World War II in Boeing Stratoliners and Douglas C-54s and also flew clandestine photo reconnaissance missions over Nazi-Occupied Europe in Lockheed Lightnings. His reputation was made on the most glamorous piston-engined airliner of all time: the Lockheed Constellation. His place as a pioneering twentieth century pilot was sealed when he became one of very few pilots to win both the prestigious Collier and Mackay Trophies.

Index

58568438R00265

Made in the USA
Charleston, SC
13 July 2016